Springer Monographs in Mathematics

Springer
*New York
Berlin
Heidelberg
Hong Kong
London
Milan
Paris
Tokyo*

James Murdock

Normal Forms and Unfoldings for Local Dynamical Systems

With 15 Illustrations

Springer

James Murdock
Mathematics Department
Iowa State University
Ames, IA 50011
USA
jmurdock@iastate.edu

Mathematics Subject Classification (2000): 37-xx, 58F36, 34Cxx

Library of Congress Cataloging-in-Publication Data
Murdock, James A.
    Normal forms and unfoldings for local dynamical systems/James Murdock.
        p. cm. — (Springer monographs in mathematics)
    Includes bibliographical references and index.

    1. Differentiable dynamical systems.   2. Normal forms (Mathematics)   I. Title.
    II. Series
    QA614.8 .M87 2002   2002024169

ISBN 978-1-4419-3013-2   e-ISBN 978-0-387-21785-7

Printed in the United States of America.

9 8 7 6 5 4 3 2 1

www.springer-ny.com

Springer-Verlag   New York Berlin Heidelberg
A member of BertelsmannSpringer Science+Business Media GmbH

Preface

The subject of local dynamical systems is concerned with the following two questions:

1. Given an $n \times n$ matrix A, describe the behavior, in a neighborhood of the origin, of the solutions of all systems of differential equations having a rest point at the origin with linear part Ax, that is, all systems of the form

$$\dot{x} = Ax + \cdots,$$

   where $x \in \mathbb{R}^n$ and the dots denote terms of quadratic and higher order.

2. Describe the behavior (near the origin) of all systems close to a system of the type just described.

To answer these questions, the following steps are employed:

1. A normal form is obtained for the general system with linear part Ax. The normal form is intended to be the simplest form into which any system of the intended type can be transformed by changing the coordinates in a prescribed manner.

2. An unfolding of the normal form is obtained. This is intended to be the simplest form into which all systems close to the original system can be transformed. It will contain parameters, called *unfolding parameters*, that are not present in the normal form found in step 1.

3. The normal form, or its unfolding, is truncated at some degree k, and the behavior of the truncated system is studied.

4. Finally, it is shown (if possible) that the full (untruncated) system has the same behavior established for the truncated system (or, in certain cases, the value of k for which this is the case is computed before determining the behavior of the truncated system). This is called *establishing k-determinacy* or *k-sufficiency*.

As the title of this book suggests, our focus will be primarily on the first two of these steps; a secondary focus will be the determinacy question. That is, we will study in detail the procedures available for obtaining normal forms and unfoldings (in the sense described later in this preface). To a smaller (but still substantial) degree, we will study procedures for deciding how many terms of a system are sufficient to establish its behavior. When it comes to *actually establishing* the behavior of systems, considerations of time and space prevent us from going beyond a limited number of examples, which it is hoped will serve to illustrate the techniques presented. The justification for this is that there are a number of books available (which will be cited at the appropriate time) that treat more complicated examples, often with the aim of obtaining conclusions in the quickest way possible. These books tend to treat each problem in an ad hoc manner, without laying a careful foundation of general principles useful for all problems. It is that need that is addressed here. Most of these general principles exist either in the journal literature (mostly of a "pure" mathematical character) or in the "folklore" of the subject, but have never been written down in book form. Armed with the ideas presented in this book, one should be able to understand much more clearly the books and papers that present specific conclusions for complicated examples not presented here.

As in my earlier book on perturbation theory, my goal in this book is to say the things that I found I needed to understand, but that other books did not say. Unlike that earlier book, this one is not an "introduction" to its subject. The book is self-contained, and does not require any specific knowledge other than the basic theories of advanced calculus, differential equations, and linear algebra. But it is unlikely that a student will find this book to be understandable without some previous exposure to the subject of dynamical systems, at least enough to motivate the kinds of questions that are asked and the kinds of solutions that are sought.

An Outline of the Chapters

Chapter 1 examines two easy two-dimensional examples of normal forms, one semisimple and the other nilpotent. For each example, the normal form and the unfolding are computed, leading to the Hopf and Takens–Bogdanov bifurcations (the study of which is continued in Sections 6.5 and

6.6). The treatment of these examples in Chapter 1 begins with the elementary methods used in such texts as [52] and [111] and goes on to suggest the advantages of more advanced methods to be developed in Chapter 4.

Chapter 2 is an interlude devoted to certain topics in linear algebra. It will have become clear in Chapter 1 that the study of normal forms is concerned with the complement to the image of a certain linear operator, the so-called *homological operator*, which we denote by the British monetary symbol £ (to be read as "pounds"). Chapter 2 collects four methods of finding a complement to the image of a linear operator. The development includes self-contained treatments of the Fredholm alternative (for finite-dimensional spaces), the Jordan canonical form, and the representation theory of the Lie algebra sl(2). (This representation theory is simplified considerably by adopting an unnecessary assumption, true in the general case but difficult to prove. It happens that this assumption is self-evident in our applications, so its adoption is harmless.) Readers who are already familiar with these topics may skip the chapter, but are advised that we occasionally introduce unfamiliar terminology (such as *entry point, triad, chain-weight basis*, and *pressure*) when no convenient term seems to exist. It will be useful to skim the relevant sections of Chapter 2 when these terms are encountered later in the book. Readers for whom parts of Chapter 2 are new may omit the starred sections (dealing with sl(2) representation theory) on a first reading; mostly, these are needed only for the starred sections in Chapters 3 and 4.

Chapter 3 is concerned with matrix perturbation theory, presented as a linear special case of normal form theory. This chapter should be taken seriously, even by readers mainly interested in the nonlinear case (Chapter 4). One reason is that every idea and notation developed in Chapter 4 is introduced first in Chapter 3. Many of these ideas, in particular those connected with Lie theory, are easier to understand in the linear case; the Lie groups and Lie algebras that arise in the linear case are finite-dimensional, and the matrix exponential suffices to explain them. (In Chapter 4, the one-parameter groups defined by matrix exponentials must be replaced by local flows.) A second reason for paying attention to linear normal forms is that they play an important role in the unfolding theory of Chapter 6. In addition to normal forms, Chapter 3 introduces the ideas of *hypernormal forms* (in Section 3.3) and *metanormal forms* (Section 3.7). Hypernormal forms exploit the arbitrariness remaining after ordinary normalization to achieve additional simplifications, ultimately leading to what are sometimes called *unique normal forms*. Metanormal forms introduce fractional powers of parameters (shearing or scaling, related to Newton diagrams) to obtain yet further simplifications.

Chapter 4 is the central chapter of the book, where the project outlined in Chapter 1 is carried out. This project is to explain the structure of normal forms using the language of a module of equivariants (of some one-parameter group) over a ring of invariants, and (as a secondary goal) to give algorithms suitable for use in symbolic computation systems. The treatment includes both semisimple and nonsemisimple cases, and includes both the inner product (or "Elphick–Iooss") and sl(2) (or "Cushman–Sanders") normal form styles, as well as my own "simplified" normal form style. The work of Richard Cushman and Jan Sanders is already at least 15 years old, but has not (until now) received a systematic treatment in monograph form. Section 4.7 contains very recent work extending the Cushman–Sanders theory to the inner product and simplified normal form styles, and providing an algorithmic approach to the determination of Stanley decompositions for normal form modules. Section 4.10 introduces the hypernormal form theory of Alberto Baider and others. (Metanormal forms are not considered in Chapter 4, but reappear in Section 6.6. See also the open problems list at the end of this preface.)

With regard to the algorithmic portions of Chapters 2, 3, and 4, it is unfortunate that no specific implementations of these algorithms are available. They have been successfully implemented in Maple by Jan Sanders (who originated most of them), but he informs me that his programs no longer work, because of changes in the Maple software since they were written.

Until this point in the book, no attention has been given to the possible application of normal forms (except briefly in Chapter 1). Chapters 5 and 6 are an attempt to remedy this omission, as far as time and space allow. Although Chapters 2, 3, and 4 are intended to be exhaustive treatments of their subjects, Chapters 5 and 6 (as already mentioned at the beginning of this preface) are not. Instead, I have presented selected topics in which I have a personal interest and feel that I have something to say. Some of these topics end with open problems and suggestions for further research.

Chapter 5 concerns the geometrical structures (invariant manifolds and preserved fibrations and foliations) that exist in truncated systems in normal form, and the estimates that relate these structures, and the solutions lying in them, to the corresponding structures and solutions in the full (not truncated) system (which is in normal form only up to the truncation point).

Chapter 6 is a collection of results related to bifurcation theory. Sections 6.1 and 6.2 contain two developments of the basic theory of bifurcation from a single zero eigenvalue, one using Newton diagrams and fractional power series, and the other using singularity theory; the developments are parallel, so that one can see the relative strengths and weaknesses of the two methods. In both sections, the theme is sufficiency of jets. Sections 6.3 and 6.4 contain a treatment of my theory of "asymptotic unfoldings," somewhat improved from the original paper (because of results from Sections 3.4 and 4.6). Sections 6.5 through 6.7 compute the unfoldings of certain specific bifurcation problems by the methods of Section 6.4, and either analyze the bifurcation in detail (Section 6.5) or give annotated references to the literature for this analysis (Sections 6.6 and 6.7).

Appendices A and B contain background material on rings and modules useful in Chapters 4 and 6. Topics include smooth germs, flat functions, formal power series, relations and syzygies, an introduction to Gröbner bases, and the formal and Malgrange preparation theorems. Appendices C and D contain additional sections that originally belonged to Chapters 3 and 4, but were moved to the appendices because of their optional character. (References to these appendices are given at the appropriate points in Chapters 3 and 4.)

Formats, Styles, Description, and Computation

Two distinctions are maintained throughout this book, one between the "description problem" and the "computation problem" for normal forms, and the other between "formats" and "styles."

1. If a normal form is thought of as the "simplest" form into which a given system can be placed, there might be disagreement as to what form is considered simplest. A systematic policy for deciding what counts as simplest is called a *normal form style*. The important normal form styles are the semisimple, inner product, simplified, and sl(2), or triad, styles.

2. In order to put a system into normal form, one or more near-identity transformations are employed. A *format* is a scheme for handling these transformations, and includes the decisions as to whether a single transformation (with coefficients determined recursively) or a sequence of transformations (applied iteratively) is used, and whether the transformations are handled directly or via generators (in the sense of Lie theory).

3. The *description problem* for a given normal form style is to describe what that normal form style "looks like" in a given situation (usually given the leading term or linear part, depending on the context). For linear problems (Chapter 3), what a style "looks like" is usually answered by giving a diagonal structure, or block structure, or stripe structure. For nonlinear problems (Chapter 4), it is best answered by describing the vector fields in normal form as a module over a ring of scalar functions, using such algebraic devices as Stanley decompositions and Gröbner bases.

4. The *computation problem* is the problem of placing a specific system (having either numerical or symbolic coefficients) into normal form of a particular style, using a particular format. This entails determining the coefficients of the normalized system, either numerically or as functions of the symbolic coefficients in the original system.

It is very helpful to keep styles separate from formats and the description problem separate from the computation problem. Normal forms of any style can be worked out using any format. (There is one exception to this: A generated, or Lie-theoretic, format must be used if the original system and its normalization are both required to belong to a specific Lie subalgebra.) It follows that one can think about styles without worrying which format is to be used, and about formats without regard to the style. The description and computation problems are not quite as distinct as this, but are more so than might be expected. They share a good deal of machinery, such as the homological equation. But it is possible to write a computer program (for a symbolic processing system) that puts a system into normal form (that is, solves the computation problem) without having a theoretical description in advance of what the normal form will look like. It is equally possible to describe the normal form without giving algorithms that generate the coefficients. In fact, the mathematical techniques required for the two problems are rather different, at least superficially: The computation problem involves algorithms and questions of algorithmic efficiency, while the description problem brings in abstract algebra (theory of rings and modules). At a deeper level, these subjects do interact with one another, as the recent burst of activity in the application of Gröbner bases to computational algorithms attests.

Unfoldings

This book adopts and advocates what might be called a "pragmatic" approach to unfoldings, rather than an "ideological" approach. The "ideological" approach is to choose an equivalence relation, and then to define a *universal unfolding* of a system to be (roughly) a system with parameters that (a) reduces to the given system when the parameters are zero, and (b) exhibits all possible behavior, up to the given equivalence relation, that can occur in systems close (in some topology) to the given system. The equivalence relations in common use fall into two classes that might be called *static* and *dynamic* equivalence relations; the former means that the sets of equilibrium states for two equivalent systems are topologically similar, while the latter requires that the full dynamics of the two systems be topologically similar. Neither equivalence relation takes into account any asymptotic properties of the systems, such as rates of approach to a stable rest point; only the stability itself matters. In addition, there may not exist a universal unfolding having finitely many parameters (finite codimension); for the dynamic equivalence relations, there almost never is. This approach is "ideological" in the sense that it begins from an a priori decision as to what behavior is interesting and what is expected from an unfolding.

The "pragmatic" approach, in contrast, begins from the fact that normal form calculations must always be terminated at some degree k (that is, with a normalized jet). Once k is chosen, it is always possible to consider an arbitrary perturbation of the jet (within its jet space). This perturbation already contains only finitely many parameters, and the number can be reduced through normal form (and hypernormal form) calculations. The resulting simplified perturbed jet is what we take as the unfolding. The next step in this approach is to ask what properties of this jet (or its associated flow) are k-determined; these are the properties that are so solidly established by the k-jet that they cannot be changed by the addition of higher-order terms. The unfolding that has been calculated is then a true (universal) unfolding *with regard to these properties*, although not necessarily with regard to all properties. (It can be shown that the collection of all properties that are k-determined does define an equivalence relation, and the unfolding is universal with respect to this relation. See the Notes and References to Section 6.4.) Some of the properties that are k-determined by this jet will be topological in character, and others will be asymptotic. Frequently, the unfolding computed by this method coincides with the universal unfolding in one of the original senses, when such an unfolding exists, or can be reduced to that unfolding by applying one or two additional coordinate changes, such as time scalings, that do not fit within the framework of normal forms. It can be the case that some interesting properties are fully unfolded at one value of k and others require a higher value. The case of "infinite codimension" is exactly the case in which no choice of k is sufficient to unfold the system fully with respect to one of the classical equivalence relations.

As already pointed out at the beginning of this preface, the treatment of bifurcation problems in this book is not intended to be at all complete. Especially for the more complicated bifurcations, such as mode interactions, the dynamical behavior is quite complicated; often it becomes understandable only under the assumption of various symmetry conditions. Symmetry is a subject that is quite important in itself, but is not addressed at all in this book (except with regard to the symmetries introduced by normalization, manifested as equivariance of vector fields). It does not fall within the scope of this book to discuss complicated bifurcation diagrams. In those complicated problems that we treat, we compute the normal form and unfolding, discuss to some extent the question of jet sufficiency, and refer to the literature for details.

Is the Nilpotent Normal Form Useful?

It is sometimes said that only the semisimple normal form is useful. (This includes what we call the "extended semisimple normal form," that is, the result of normalizing a nonsemisimple system with respect to the semisimple part of its linear term.) Since a fair portion of this book is concerned with the nilpotent (or, more generally, nonsemisimple) normal form, it is worthwhile to address the question of its usefulness. The answer has two parts: a list of existing results that use the nonsemisimple normal form, and a list of promising open problems. First is the list of the existing results (which may not be complete):

1. The nonsemisimple normal form for linear problems (matrix perturbations) is the starting point for the transplanting, or shearing, method described in Section 3.7.

2. The "simplified normal form," one version of the nonsemisimple normal form, is the starting point for the computation of unfoldings described in Section 6.4.

3. The nilpotent normal form in two dimensions is the basis for the various shearing, scaling, and blowup methods used to study the Takens–Bogdanov problem. These methods are touched upon in Sections 5.4 and 6.6. It is to be emphasized that in these approaches one does not study the system directly in its nilpotent normal form. Instead, one first creates the nilpotent normal form and then performs a transplanting operation leading to a new organizing center that is not nilpotent (and does not even have a rest point at the origin). The resulting problem is studied by an entirely different set of methods, which are global in character. So the difficulty of working with the nilpotent normal form itself does not even arise. Instead, the simplifications achieved by the nilpotent normal form are passed on to the new system in a different guise.

4. The nonsemisimple normal form for Hamiltonian systems in two degrees of freedom is the starting point for the analysis of the Hamiltonian Hopf bifurcation. See Notes and References to Section 4.9.

Next is the list of open problems. Solution of any or all of these problems would increase the usefulness of the nonsemisimple normal form.

1. What is the correct general nonlinear transplanting (shearing) theory? Such a theory should generalize the linear shearing theory of Section 3.7 and include the scaling for the Takens–Bogdanov problem given in Section 6.6.

2. What is the complete theory of the following methods, best understood in two dimensions (see Notes and References to Sections 5.4 and 6.6):

   a. the blowup method (probably best explained in the writings of Freddy Dumortier)?

      b. the power transformation method (introduced in the writings of Alexander Bruno)?

3. How do these methods (shearing, blowup, and power transform) relate to one another?

4. What is the correct general error estimate for nilpotent problems, and how do its time scales relate to the correct shearing scales? (In Chapter 5, error estimates are derived only for cases in which there is no nilpotent part on the center manifold.)

5. Is there a way to extract useful geometrical information from Lemma 5.4.2? This lemma describes the near-preservation of a foliation in the sl(2) normal form for nilpotent problems.

6. How can the unfolding theory of Section 6.4 be carried out using the sl(2) normal form? This question is significant in view of the previous one, which suggests that the sl(2) normal form contains significant dynamical information. The problem comes down to the following: to find a construction such that the analogue of Lemma 6.4.3 holds for the sl(2) normal form and preferably to state this construction in a coordinate-free manner. (Lemma 6.4.3 itself holds only for the simplified normal form and cannot be stated in a coordinate-free manner, since the definition of that normal form is itself not coordinate-free.)

What Is New?

The expert will want to know which results in this book, if any, are actually new (apart from matters of exposition and terminology). Here is a list of the items that appear to me to be new or possibly new:

1. The development of the simplified normal form given here is more complete than the other available presentations. This discussion occurs in Sections 3.4, 4.6, and 6.4.

2. The treatment of linear shearing (or transplanting) given in Section 3.7 is new. It extends results obtained by Bogaevsky and Povsner so that they hold for the inner product normal form style as well as the sl(2) style; the original arguments were valid for the sl(2) style only.

3. Section 3.7 gives an algorithmic procedure for deciding k-determined hyperbolicity. My previous work on this topic (some of it joint with Clark Robinson) was not algorithmic in character.

4. Section 4.7 is entirely new, although the paper (which contains additional material) may appear shortly before the book. This provides an algorithm to produce a Stanley decomposition for the normal form module in the nilpotent case, either in the inner product, simplified,

or sl(2) normal form styles, given the Stanley decomposition for the corresponding ring of invariants.

5. Chapter 5 (especially Sections 5.1 and 5.3) contains a number of general results that have not (to my knowledge) appeared in the literature except in particular examples. Most of the results that cannot be referenced should probably be classed as folklore rather than as new results, although it is not clear how widely the folklore is known. I have shown these results to several researchers and received responses ranging from surprise to "yes, I thought that was true, but I don't know any references." On the other hand, some ideas leading to open problems stated in this chapter may be new.

6. Although it is a rather minor point, I don't think the role of seminormality (Definition 3.4.10) in the theory of the inner product normal form has been noticed; this is because the condition is automatically satisfied when the leading term is in Jordan form, as is usually assumed. The sl(2) normal form never requires an assumption of seminormality, either in the linear or nonlinear setting.

7. The unfolding theory developed in Sections 6.3 and 6.4 is a few years old and has appeared previously in only one paper.

Acknowledgments and Apologies

This book would never have come into existence in its present form without extensive discussions with Jan Sanders and Richard Cushman, and input from the following mathematicians over a number of years (listed in alphabetical order): Alberto Baider, Alexander Bruno, Kenneth Driessel, Karin Gatermann, Marty Golubitsky, George Haller, Irwin Hentzel, Alexander Kopanskii, Bill Langford, Victor Leblanc, A.H.M. Levelt, John Little, Ian Melbourne, Sri Namachchivaya, Peter Olver, Charlie Pugh, and Jonathan Smith. Special thanks are due to my graduate student David Malonza for his careful proofreading of much of this book, and for his many questions, which forced me to improve the exposition in several places. Thanks are also due to David Kramer, who copyedited the manuscript and also laboriously transformed my files from the dialect of TEX in which I had written them to another preferred by the publisher, in the process making a number of suggestions that improved the clarity of the exposition as well as the layout of the pages.

All of the shortcomings of this book, and any outright errors, are to be attributed to my own vast ignorance. Any errata that come to my attention after the book appears in print will be posted to my Web site,

http:\\www.math.iastate.edu \jmurdock\.

Any contributions to the open problems mentioned in this book will also be referenced there (with the authors' permission). If you discover errors or publish contributions that should be cited, please notify me at jmurdock@iastate.edu.

A Note About Notations

The following remarks are given to aid the reader who is already familiar with the ideas to recognize quickly certain notations used in this book, without having to find the passages in which these notations are introduced.

There are many places in normal form theory where a decision must be made that affects the signs that appear in many equations. Decisions that are convenient in one part of the theory are often inconvenient elsewhere. The following three principles have been adopted, because of their simplicity and because, together, they settle all of these decisions:

1. The Lie bracket of matrices (and of linear operators in general) should agree with the usual commutator bracket, so that

$$[A, B] = AB - BA$$

rather than the negative of this.

2. The Lie bracket of vector fields should reduce to the Lie bracket of matrices when the vector fields are linear. Thus,

$$[u, v] = u'v - v'u$$

rather than the negative of this. (Here, u and v are column vectors, and u' and v' are matrices of partial derivatives.)

3. The Lie operator L_u should agree with the Lie derivative as it is usually defined, so that $\mathsf{L}_u v$ is the rate of change (in the Lie sense) of v in the forward direction along the flow of u.

Unfortunately, the last two of these principles, taken together, imply that

$$\mathsf{L}_u v = [v, u] \quad \text{rather than} \quad [u, v].$$

Taking the first principle into account as well (to fix the meaning of the bracket of Lie operators), it follows that

$$\mathsf{L}_{[u,v]} = [\mathsf{L}_v, \mathsf{L}_u].$$

That is, L turns out to be a Lie algebra antihomomorphism, which is occasionally awkward.

We regard vector fields as column vectors and write these as

$$v = (v_1, v_2, \ldots, v_n) = \begin{bmatrix} v_1 \\ v_2 \\ \vdots \\ v_n \end{bmatrix}.$$

(Thus, any vector written with parentheses and commas is automatically a column vector; if we want a row vector, it must be written $\begin{bmatrix} v_1 & v_2 & \ldots & v_n \end{bmatrix}$). We do not identify vectors with differential operators; instead, the differential operator associated with a vector field v is written

$$\mathcal{D}_v = v_1 \frac{\partial}{\partial x_1} + v_2 \frac{\partial}{\partial x_2} + \cdots + v_n \frac{\partial}{\partial x_n}.$$

By the first principle above, the bracket of two such operators is their commutator. This implies that

$$\mathcal{D}_{[u,v]} = [\mathcal{D}_v, \mathcal{D}_u].$$

Therefore, $\mathcal{D}$ is also an antihomomorphism.

The usual way to avoid these problems (in advanced work) is to identify v and $\mathcal{D}_v$ and to eliminate the bracket that we call $[u, v]$; the Lie operator then reduces to the "Lie algebra adjoint" operator, defined by $\mathrm{ad}_u v = [u, v]$; this is a Lie algebra homomorphism. This works well for abstract considerations, but clashes with matrix notation (our principle 1 above) as soon as systems of differential equations are written in their familiar form.

The transpose of a matrix is denoted by $A^\dagger$, and the conjugate transpose by A^*. More generally, L^* denotes the adjoint of a linear operator with respect to some specified inner product.

Ames, Iowa   James Murdock

Contents

Preface   v

1 **Two Examples**   1
  1.1   The (Single) Nonlinear Center   1
  1.2   The Nonsemisimple Double-Zero Eigenvalue   21

2 **The Splitting Problem for Linear Operators**   27
  2.1   The Splitting Problem in the Semisimple Case   28
  2.2   Splitting by Means of an Inner Product   32
  2.3   Nilpotent Operators   34
  2.4   Canonical Forms   39
  2.5   * An Introduction to sl(2) Representation Theory   47
  2.6   * Algorithms for the sl(2) Splittings   57
  2.7   * Obtaining sl(2) Triads   63

3 **Linear Normal Forms**   69
  3.1   Perturbations of Matrices   69
  3.2   An Introduction to the Five Formats   74
  3.3   Normal and Hypernormal Forms When A_0 Is Semisimple   87
  3.4   Inner Product and Simplified Normal Forms   99
  3.5   * The sl(2) Normal Form   118
  3.6   Lie Theory and the Generated Formats   129
  3.7   Metanormal Forms and k-Determined Hyperbolicity . . .   142

4 Nonlinear Normal Forms   **157**
   4.1   Preliminaries   157
   4.2   Settings for Nonlinear Normal Forms   160
   4.3   The Direct Formats (1a and 1b)   164
   4.4   Lie Theory and the Generated Formats   174
   4.5   The Semisimple Normal Form   190
   4.6   The Inner Product and Simplified Normal Forms   221
   4.7   The Module Structure of Inner Product and
        Simplified Normal Forms   242
   4.8   * The sl(2) Normal Form   265
   4.9   The Hamiltonian Case   271
   4.10 Hypernormal Forms for Vector Fields   283

5 Geometrical Structures in Normal Forms   **295**
   5.1   Preserved Structures in Truncated Normal Forms   297
   5.2   Geometrical Structures in the Full System   316
   5.3   Error Estimates   323
   5.4   The Nilpotent Center Manifold Case   335

6 Selected Topics in Local Bifurcation Theory   **339**
   6.1   Bifurcations from a Single-Zero Eigenvalue:
        A "Neoclassical" Approach   341
   6.2   Bifurcations from a Single-Zero Eigenvalue:
        A "Modern" Approach   356
   6.3   Unfolding the Single-Zero Eigenvalue   366
   6.4   Unfolding in the Presence of
        Generic Quadratic Terms   371
   6.5   Bifurcations from a Single Center (Hopf and
        Degenerate Hopf Bifurcations)   382
   6.6   Bifurcations from the Nonsemisimple Double-Zero
        Eigenvalue (Takens–Bogdanov Bifurcations)   389
   6.7   Unfoldings of Mode Interactions   396

A Rings   **405**
   A.1   Rings, Ideals, and Division   405
   A.2   Monomials and Monomial Ideals   412
   A.3   Flat Functions and Formal Power Series   421
   A.4   Orderings of Monomials   424
   A.5   Division in Polynomial Rings; Gröbner Bases   427
   A.6   Division in Power Series Rings; Standard Bases   438
   A.7   Division in the Ring of Germs   444

B Modules   **447**
   B.1   Submodules of $\mathbb{Z}^n$ 447
   B.2   Modules of Vector Fields   449

C Format 2b: Generated Recursive (Hori)   **451**
   C.1   Format 2b, Linear Case (for Chapter 3)   451
   C.2   Format 2b, Nonlinear Case (for Chapter 4)   457

D Format 2c: Generated Recursive (Deprit)   **463**
   D.1   Format 2c, Linear Case (for Chapter 3)   463
   D.2   Format 2c, Nonlinear Case (for Chapter 4)   471

E On Some Algorithms in Linear Algebra   **477**

References   **481**

Index   **489**

1

Two Examples

In this chapter, two examples, one semisimple and the other not, will be treated from an elementary point of view. The purpose of the treatment is to motivate the concerns and themes of the remainder of the book. The semisimple example is the nonlinear center; when unfolded, this becomes the Hopf bifurcation. The nonsemisimple example is the generic double-zero eigenvalue, which unfolds to the Takens–Bogdanov bifurcation.

Although the discussion in this chapter is generally elementary, some topics are meant as previews of subjects to be discussed in detail later. In these cases the treatment here may appear somewhat sketchy, but it is not necessary to understand every point completely at this stage.

1.1   The (Single) Nonlinear Center

Consider the system of differential equations

$$\dot{u} = Au + Q(u) + C(u), \qquad (1.1.1)$$

where $u = (x, y)$ (regarded as a column vector),

$$A = \begin{bmatrix} 0 & -1 \\ 1 & 0 \end{bmatrix}, \qquad (1.1.2)$$

and Q and C are quadratic and cubic terms, respectively. (We understand the words quadratic and cubic to mean *homogeneous* quadratic and cubic. That is, Q contains only quadratic terms, and C only cubic terms. It is understood here, and below, that any vector written with parentheses and commas is actually a column vector; if we need a row vector it will be written as a row matrix, with square brackets and no commas.) The system (1.1.1) has a rest point at the origin, and the linear terms have eigenvalues $\pm i$. If these eigenvalues were in the left half-plane, the origin would be asymptotically stable; the rest point would be what is called a *sink*, or an *attracting rest point*, meaning that all orbits beginning near the origin approach the origin as $t \to \infty$. If one or both were in the right half-plane, the rest point would be unstable, either a *saddle* or a *source*. But none of these is the case, because the eigenvalues are on the imaginary axis. If the system were linear (that is, if the nonlinear terms were zero), the pure imaginary eigenvalues would imply that the origin was neutrally stable and surrounded by periodic solutions, a configuration called a *linear center*. But this also is not the case, because we do not assume that the nonlinear terms vanish. In fact, the stability or instability of the rest point at the origin for the system (1.1.1) cannot be determined from the linear terms alone, but will depend on the higher-order terms. This rest point is called a *nonlinear center*, or more precisely the *two-dimensional cubic nonlinear center* (since $u \in \mathbb{R}^2$ and (1.1.1) contains no terms of higher order than cubic). We propose the following problem: to introduce new coordinates $v = (\xi, \eta)$ into (1.1.1) so as to bring the quadratic and cubic terms into the simplest possible form, in the hope that this will facilitate determining the stability of the origin. This strategy is based on the fact that changing coordinates cannot affect the stability of the origin: If orbits of (1.1.1) beginning near the rest point approach the rest point as $t \to \infty$, this will be remain true no matter what coordinates are used. (We will, in fact, use only coordinate changes that map the origin to itself, so the words "rest point" can be replaced by "origin" in the last sentence.)

1.1.1. Remark. In elementary differential equations courses one studies the *linear spring*, or *harmonic oscillator*, governed by the equation

$$\ddot{x} + x = 0.$$

Introducing $y = \dot{x}$, this can be written as the system of differential equations

$$\begin{bmatrix} \dot{x} \\ \dot{y} \end{bmatrix} = \begin{bmatrix} 0 & 1 \\ -1 & 0 \end{bmatrix} \begin{bmatrix} x \\ y \end{bmatrix}.$$

In polar coordinates ($x = r\cos\theta$, $y = r\sin\theta$) this becomes

$$\dot{r} = 0,$$

$$\dot{\theta} = -1.$$

The negative sign in the last equation is somewhat awkward, and can be avoided by interchanging the roles of x and y, so that we begin with

$$\ddot{y} + y = 0$$

and set $x = \dot{y}$ to obtain

$$\begin{bmatrix} \dot{x} \\ \dot{y} \end{bmatrix} = \begin{bmatrix} 0 & -1 \\ 1 & 0 \end{bmatrix} \begin{bmatrix} x \\ y \end{bmatrix},$$

or

$$\dot{r} = 0,$$

$$\dot{\theta} = 1.$$

This fits the pattern of (1.1.1) with vanishing nonlinear terms. The full system (1.1.1) can arise as a model of a *nonlinear spring* in which the restoring forces are not exactly proportional to the extension.

The Normal Form for the Nonlinear Center

The simplification will be carried out in two stages, using an approach called *format 1a* in later chapters. The first stage is to perform a change of variables of the form

$$u = v + q(v), \tag{1.1.3}$$

where $q : \mathbb{R}^2 \to \mathbb{R}^2$ is homogeneous quadratic. Since this transformation has no constant term, it does not move the origin; the origin is still the rest point. Since "$u = v$ up to first order" (since q is strictly quadratic), the linear term Au of the differential equation (1.1.1) should be unchanged in the new coordinates (that is, it should become Av). Therefore, the new equations should have the form

$$\dot{v} = Av + \widehat{Q}(v) + \widehat{C}(v) + \cdots, \tag{1.1.4}$$

where the dots denote higher-order terms introduced by the transformation. This can be proved, and an expression for $\widehat{Q}(v)$ can be obtained, by differentiating (1.1.3) with respect to time and using (1.1.4) to obtain, through quadratic terms,

$$\dot{u} = \dot{v} + q'(v)\dot{v} = Av + \left(\widehat{Q}(v) + q'(v)Av \right) + \cdots,$$

and comparing this with

$$\dot{u} = A(v + q(v)) + Q(v + q(v)) + \cdots = Av + (Aq(v) + Q(v)) + \cdots,$$

which results from substituting (1.1.3) into (1.1.1). The result is

$$\widehat{Q}(v) = Q(v) + Aq(v) - q'(v)Av. \tag{1.1.5}$$

Here q' denotes the matrix of partial derivatives of q. (If this derivation is not convincing, a more rigorous approach will be given in Lemma 4.3.1.) If we define the *homological operator* $\mathcal{L}$ acting on mappings $q : \mathbb{R}^2 \to \mathbb{R}^2$ by

$$(\mathcal{L}q)(u) = q'(u)Au - Aq(u), \tag{1.1.6}$$

then (1.1.5) can be written

$$\pounds q = Q - \widehat{Q}. \tag{1.1.7}$$

Equation (1.1.7) is called a *homological equation*. We have not indicated the variables in (1.1.7); it is irrelevant whether calculations using this equation are performed using the letter u or v, as long as the correct variables are used in (1.1.4). Written out in greater detail, with $u = (x, y)$, $q(u) = (f(x,y), g(x,y))$, and A as in (1.1.2), the $\pounds$ operator (1.1.6) takes the form

$$\pounds \begin{bmatrix} f(x,y) \\ g(x,y) \end{bmatrix} = \begin{bmatrix} -yf_x + xf_y + g \\ -yg_x + xg_y - f \end{bmatrix}. \tag{1.1.8}$$

1.1.2. Remark. The omission of the terms indicated by $\cdots$ in (1.1.4) has an important consequence: Conclusions drawn from this equation (or any later modifications of this equation) are valid only in a neighborhood of the origin. Near the origin, the terms of low degree are the dominant terms, and it is reasonable to hope that these terms determine the behavior of the system. When x and y are large, on the other hand, terms of high degree dominate terms of lower degree, and it would not be reasonable to expect the truncated system to determine the behavior. This is the reason for the name *local* dynamical systems: We deal only with behavior local to the origin. Examples of local behavior include the stability or instability of the rest point at the origin, and the bifurcation of a limit cycle from the origin, both of which will be studied in this section.

1.1.3. Remark. Any mapping of $\mathbb{R}^2 \to \mathbb{R}^2$, such as q or Q, can be regarded in several ways:

1. as a *transformation* from one "copy" of $\mathbb{R}^2$ to another;
2. as a *deformation* within the same "copy" of $\mathbb{R}^2$ (for instance, a stretching of a rubber sheet); or
3. as a *vector field* on $\mathbb{R}^2$.

In our situation, Q is naturally regarded as a vector field and q as a transformation. Notice that $\pounds$ simply acts on a quadratic mapping $\mathbb{R}^2 \to \mathbb{R}^2$ and produces another such mapping. The fact that q in (1.1.7) is thought of as a transformation, while Q and $\widehat{Q}$ are viewed as vector fields, is of no consequence. There is a vector space $\mathcal{V}_1^2$ whose elements are quadratic maps of $\mathbb{R}^2 \to \mathbb{R}^2$, and $\pounds$ is an operator on this vector space; that is, $\pounds : \mathcal{V}_1^2 \to \mathcal{V}_1^2$. We will refer to $\mathcal{V}_1^2$ as the "space of quadratic vector fields on $\mathbb{R}^2$." It is six-dimensional, and a basis for it is given in (1.1.9) below. The notation is that $\mathcal{V}_j^n$ is the space of vector fields on $\mathbb{R}^n$ of degree $j + 1$; see Definition 4.1.1.

1.1.4. Remark. Throughout this book we will use the British monetary symbol $\pounds$ (which can be read "pounds") for the homological operator. The symbol $\pounds$ is chosen because it resembles the letter L (which stands for *Lie operator*, to be introduced later, of which $\pounds$ is a special case). For those who have seen a little homological algebra, the *homological equation* (1.1.7) can be pictured as follows. Let $\mathcal{V}_j^n$

be the space of homogeneous vector fields of degree $j + 1$ on $\mathbb{R}^n$, let $\mathcal{L}$ be the homological operator defined by an $n \times n$ matrix A, and consider the (very short) chain complex

$$0 \to \mathcal{V}_j^n \to \mathcal{V}_j^n \to 0,$$

where the only nontrivial map is $\mathcal{L} : \mathcal{V}_j^n \to \mathcal{V}_j^n$. The homology space $\mathcal{V}_j^n / \operatorname{im} \mathcal{L}$ of this chain complex is an abstract version of the normal form space for degree $j + 1$, in the sense that it is isomorphic to any normal form style (or complement of $\operatorname{im} \mathcal{L}$) but does not, by itself, select a style. This remark is entirely trivial and does not contribute to the study of normal forms, but it does justify the terminology. The connection between normal forms and homological algebra becomes more serious in connection with hypernormalization; see the Notes and References to Section 4.10.

We now ask whether it is possible to choose q in (1.1.7) so as to eliminate the quadratic terms entirely, that is, to make $\widehat{Q} = 0$. According to (1.1.7), this requires finding q such that $\mathcal{L}q = Q$. This is possible, for arbitrary Q, if and only if $\mathcal{L} : \mathcal{V}_1^2 \to \mathcal{V}_1^2$ is onto, where $\mathcal{V}_1^2$ is the space of quadratic vector fields mentioned in Remark 1.1.3. Since $\mathcal{L}$ is linear, and the space of quadratic vector fields on $\mathbb{R}^2$ is 6-dimensional, $\mathcal{L}$ can be represented as a 6×6 matrix. With respect to the basis

$$\begin{bmatrix} x^2 \\ 0 \end{bmatrix}, \begin{bmatrix} xy \\ 0 \end{bmatrix}, \begin{bmatrix} y^2 \\ 0 \end{bmatrix}, \begin{bmatrix} 0 \\ x^2 \end{bmatrix}, \begin{bmatrix} 0 \\ xy \end{bmatrix}, \begin{bmatrix} 0 \\ y^2 \end{bmatrix} \tag{1.1.9}$$

for the space of (homogeneous) quadratic vector fields, the matrix of $\mathcal{L}$ is easily (but tediously) calculated to be

$$\mathcal{L} = \begin{bmatrix} 0 & 1 & 0 & 1 & 0 & 0 \\ -2 & 0 & 2 & 0 & 1 & 0 \\ 0 & -1 & 0 & 0 & 0 & 1 \\ -1 & 0 & 0 & 0 & 1 & 0 \\ 0 & -1 & 0 & -2 & 0 & 2 \\ 0 & 0 & -1 & 0 & -1 & 0 \end{bmatrix}. \tag{1.1.10}$$

Further tedious processes of linear algebra, such as row reduction to echelon form, can be used to verify that this matrix is indeed invertible, showing that $\mathcal{L}$ is onto and the quadratic term can be eliminated. (A much easier way of seeing this will be explained later in this section.) At this stage our system has been simplified to the form (1.1.4) with $\widehat{Q}(v) = 0$. Notice that we have not calculated the expression for $\widehat{C}$. This expression will be quite complicated, and will depend on Q and q in addition to C. (In fact, we have not even calculated q, only shown that it exists.) At this point a crucial decision must be made: Do we wish to carry out the *computations* necessary to reduce any actual *specific* system of the form (1.1.1) to simplest form, or do we wish only to *describe* the simplest form to which any *arbitrary* system of the form (1.1.1) can be reduced? We call the latter problem the *description problem* for the normal form of (1.1.1), and the former the *computation problem*. Only the description problem will be addressed in this introductory chapter. For this purpose we need not calculate q or $\widehat{C}$, but merely treat $\widehat{C}$ as an arbitrary cubic vector field. In fact, we do not even need to keep track of the coordinate changes that we use (which is of course essential for the computation problem). So we can replace v by u, drop the hat on $\widehat{C}$, and express our results so far in the form

$$\dot{u} = Au + C(u) + \cdots . \tag{1.1.11}$$

That is, the general equation of form (1.1.1) can be put into form (1.1.11) by changing variables and changing the cubic terms C.

The change in C that has been made so far was uncontrolled, and does not constitute a simplification of the cubic terms. The next step (the second of the two stages mentioned above) is to make a change of variables in (1.1.11) targeted at simplifying C. To that end we introduce new variables, again temporarily called $v = (\xi, \eta)$, by

$$u = v + c(v),$$

where c is cubic. This time the linear and (absent) quadratic terms in (1.1.11) are unchanged, while the cubic term C is replaced by $\widehat{C}$ with

$$\mathcal{L}c = C - \widehat{C}, \tag{1.1.12}$$

and the higher-order terms (represented by dots) are modified in an un-controllable way. The operator $\mathcal{L}$ is the same as in (1.1.6), except that now it is taken as mapping the vector space $\mathcal{V}_2^3$ of (homogeneous) cubic vector fields on $\mathbb{R}^2$ into itself. With respect to the basis

$$\begin{bmatrix} x^3 \\ 0 \end{bmatrix}, \begin{bmatrix} x^2y \\ 0 \end{bmatrix}, \begin{bmatrix} xy^2 \\ 0 \end{bmatrix}, \begin{bmatrix} y^3 \\ 0 \end{bmatrix}, \begin{bmatrix} 0 \\ x^3 \end{bmatrix}, \begin{bmatrix} 0 \\ x^2y \end{bmatrix}, \begin{bmatrix} 0 \\ xy^2 \end{bmatrix}, \begin{bmatrix} 0 \\ y^3 \end{bmatrix} \tag{1.1.13}$$

the matrix of $\mathcal{L}$ is

$$\mathcal{L} = \begin{bmatrix} 0 & 1 & 0 & 0 & 1 & 0 & 0 & 0 \\ -3 & 0 & 2 & 0 & 0 & 1 & 0 & 0 \\ 0 & -2 & 0 & 3 & 0 & 0 & 1 & 0 \\ 0 & 0 & -1 & 0 & 0 & 0 & 0 & 1 \\ -1 & 0 & 0 & 0 & 0 & 1 & 0 & 0 \\ 0 & -1 & 0 & 0 & -3 & 0 & 2 & 0 \\ 0 & 0 & -1 & 0 & 0 & -2 & 0 & 3 \\ 0 & 0 & 0 & -1 & 0 & 0 & -1 & 0 \end{bmatrix} . \tag{1.1.14}$$

The reduced row echelon form of this matrix is

$$\begin{bmatrix} 1 & 0 & 0 & 0 & 0 & 0 & 0 & -1 \\ 0 & 1 & 0 & 0 & 0 & 0 & 1 & 0 \\ 0 & 0 & 1 & 0 & 0 & 0 & 0 & -1 \\ 0 & 0 & 0 & 1 & 0 & 0 & 1 & 0 \\ 0 & 0 & 0 & 0 & 1 & 0 & -1 & 0 \\ 0 & 0 & 0 & 0 & 0 & 1 & 0 & -1 \\ 0 & 0 & 0 & 0 & 0 & 0 & 0 & 0 \\ 0 & 0 & 0 & 0 & 0 & 0 & 0 & 0 \end{bmatrix}. \qquad (1.1.15)$$

The two rows of zeros at the bottom shows that this matrix has a two-dimensional kernel (or null space), so $\mathcal{L}$ acting on cubic vector fields is not invertible, and the cubic terms cannot be entirely eliminated.

We have now arrived at an issue so crucial to the subject of normal forms that it will occupy all of Chapter 2 as well as large parts of Chapters 3 and 4. This issue is, how do we handle the homological equation (1.1.12) when $\mathcal{L}$ is not invertible? Remember that our goal is to simplify C. That means we want to choose $\widehat{C}$ to be as simple as possible, subject to the condition that (1.1.12) must be solvable for c. In other words, $C - \widehat{C}$ must lie in the image (or range) of $\mathcal{L}$. Briefly (and this will be developed at length later in the book), the image im $\mathcal{L}$ is a subspace of the space $\mathcal{V}_2^2$ of cubic vector fields; if we select a complement to this image and let $\widehat{C}$ be the projection of C into this complement, then $C - \widehat{C}$ will belong to the image, and (1.1.12) will be solvable. The problem, then, comes down to selecting a complement to im $\mathcal{L}$. We call this the choice of a normal form *style* , and it is here (as the idea of "style" suggests) that one's notion of simplicity comes into play.

In the present example, these issues can be handled fairly easily, since there is only one normal form style (called the *semisimple normal form*) that is accepted by everyone as the best. Recall that according to (1.1.15), $\dim \ker \mathcal{L} = 2$. Since $\dim \operatorname{im} \mathcal{L} + \dim \ker \mathcal{L} = \dim \mathcal{V}_2^2 = 8$, it follows that $\dim \operatorname{im} \mathcal{L} = 6$. It is therefore plausible, based on dimension count alone, that ker $\mathcal{L}$ is a subspace of $\mathcal{V}_2^2$ complementary to im $\mathcal{L}$:

$$\mathcal{V}_2^2 = \operatorname{im} \mathcal{L} \oplus \ker \mathcal{L}. \qquad (1.1.16)$$

It will be shown (in Theorem 2.1.3 and Lemma 4.5.2) that this conjecture is true when the matrix A is semisimple, that is, diagonalizable over the complex numbers; this is the case for the matrix A given by (1.1.2). The validity of (1.1.16) can be verified (by tedious but elementary methods) in the current instance, by the following calculations: First, from the row reduced matrix (1.1.15), determine that ker $\mathcal{L}$ is spanned by $(1, 0, 1, 0, 0, 1, 0, 1)$ and $(0, 1, 0, 1, -1, 0, -1, 0)$. Next, check that these are linearly independent of the columns of (1.1.14), which span im $\mathcal{L}$. This can be done, for instance, by appending these vectors as additional columns in (1.1.14) and repeating the row reduction.

Having established that (1.1.16) is true, it follows that $\widehat{C}$ may be taken to lie in ker $\mathcal{L}$. The vectors in $\mathbb{R}^8$ that we have just calculated as a basis for ker $\mathcal{L}$ are, of course, coordinate vectors with respect to the basis (1.1.13) for $\mathcal{V}_2^2$, and correspond to the vector fields

$$\left(x^2 + y^2\right) \begin{bmatrix} x \\ y \end{bmatrix} \quad \text{and} \quad \left(x^2 + y^2\right) \begin{bmatrix} -y \\ x \end{bmatrix}. \tag{1.1.17}$$

Thus, we may take

$$\widehat{C}(\xi, \eta) = \alpha \left(\xi^2 + \eta^2\right) \begin{bmatrix} \xi \\ \eta \end{bmatrix} + \beta \left(\xi^2 + \eta^2\right) \begin{bmatrix} -\eta \\ \xi \end{bmatrix}.$$

Since we are solving only the description problem, we may rename the variables as x and y, and state our result as follows: The normal form for (1.1.1), calculated to degree 3, is

$$\begin{bmatrix} \dot{x} \\ \dot{y} \end{bmatrix} = \begin{bmatrix} 0 & -1 \\ 1 & 0 \end{bmatrix} \begin{bmatrix} x \\ y \end{bmatrix} + \alpha \left(x^2 + y^2\right) \begin{bmatrix} x \\ y \end{bmatrix} + \beta \left(x^2 + y^2\right) \begin{bmatrix} -y \\ x \end{bmatrix} + \cdots.$$
$$\tag{1.1.18}$$

Stability and Sufficiency of Jets

Now let us see whether this normal form calculation enables us to solve the stability problem for (1.1.1). For a preliminary investigation it is convenient to drop the terms of (1.1.18) that are represented by dots; this is called passing to the *truncated system*, or, more precisely, taking the *3-jet* of (1.1.18). (A *k-jet* is simply a Taylor polynomial of degree k, or the truncation of a power series after the kth term.) Introducing polar coordinates ($x = r\cos\theta$, $y = r\sin\theta$) brings the truncated system into the form

$$\dot{r} = \alpha r^3, \tag{1.1.19}$$
$$\dot{\theta} = 1 + \beta r^2.$$

It is easy to understand the behavior of (1.1.19): If $\alpha < 0$, the origin is asymptotically stable (because r is decreasing); if $\alpha > 0$, the origin is unstable; and if $\alpha = 0$, the origin is neutrally stable, surrounded by periodic orbits. The period of these orbits is not usually independent of the amplitude, as it would be in a linear system, but increases with amplitude (a *soft center*) if $\beta < 0$ and decreases (*hard center*) if $\beta > 0$. The terms "hard and soft center" are borrowed from the more familiar "hard and soft spring," and have meaning even when $\alpha \neq 0$, provided that the "period" is understood as the time for one complete rotation of the angle.

This discussion of (1.1.19) is not conclusive for (1.1.18), but instead raises the important question of *sufficiency of jets*, which in this case takes the following form: Is the 3-jet of (1.1.18) sufficient to determine the stability of the system? The answer in this case is fairly easy to guess (details will be given below): If $\alpha < 0$ the origin of (1.1.18) is asymptotically stable; if

$\alpha > 0$, it is unstable; and if $\alpha = 0$, the stability of the origin cannot be determined, because it depends on the higher-order (dotted) terms. Thus, we say that *the 3-jet is sufficient to determine stability* if $\alpha \neq 0$, insufficient if $\alpha = 0$. In addition, if $\alpha < 0$, we will say that *the stability of the origin is 3-determined*. It is important to understand that the sufficiency of a jet is never absolute, but is relative to the question being asked. For example, if $\beta > 0$, the origin of (1.1.18) is hard; if $\beta < 0$, it is soft; and if $\beta = 0$, the 3-jet is insufficient to determine hardness or softness; therefore the 3-jet may be sufficient to determine stability but not sufficient to determine hardness (if $\alpha \neq 0$ but $\beta = 0$), or the reverse.

Here are the necessary details to show that if $\alpha < 0$, the origin of (1.1.18) is asymptotically stable. By the monomial division theorem (see Section A.2, especially Theorem A.2.16 and Remark A.2.17), the first equation of (1.1.18) may be written in full as $\dot{x} = -y + (\alpha x - \beta y)(x^2 + y^2) + f_0(x,y)x^4 + f_1(x,y)x^3y + f_2(x,y)x^2y^2 + f_3(x,y)xy^3 + f_4(x,y)y^4$, where $f_0, \ldots, f_4$ are unknown but smooth. Consider the term $f_1(x,y)x^3y$, and let M_1 be the maximum of $|f_1(x,y)|$ on the unit disk $r \leq 1$; then $|f_1(x,y)x^3y| \leq M_1 r^4 |\cos^3 \theta \sin \theta| \leq M_1 r^4$. The other unknown terms, and those of $\dot{y}$, may be treated similarly. Since $\dot{r} = \dot{x}\cos\theta + \dot{y}\sin\theta$, we conclude after a little calculation that $\dot{r} = \alpha r^3 + g(r,\theta)$ with $|g(r,\theta)| \leq M r^4$ on the unit disk, for some $M > 0$. Since $\alpha < 0$, it follows that $\dot{r} \leq (\alpha + Mr)r^3$ is negative for $0 < r < |\alpha|/M$, and the origin is asymptotically stable. (The final step requires a reference to the technical definition of asymptotic stability and an argument from the theory of Lyapunov functions, or simply a reliance on the intuitive meaning of asymptotic stability. Strictly speaking it is $r^2 = x^2 + y^2$, rather than r, that is the Lyapunov function, because r^2 is smooth at the origin.)

Finally, it should again be mentioned that if it is required to find the stability or instability of the origin in a *specific* system of the form (1.1.1), then it is necessary to reduce *that system* to normal form so that the sign of α can be determined. For this it is not sufficient to have only the *description* (1.1.18) of the normal form, but also the full *computation* of the normal form. This requires working out the details of several steps that we were able to omit in solving the description problem. These steps are best handled by recursive algorithms to be given in Appendices C and D.

A Complete Solution of the Description Problem

Up to this point, our discussion of the nonlinear center has followed the pattern of most elementary presentations of the subject. But there are many shortcomings to this approach. For example, consider our discussion of the homological equation (1.1.7) for the quadratic terms. In order to show that $\hat{Q}$ could be taken to be zero, we showed that $\mathcal{L}$ is invertible by writing down its matrix and performing row operations. We promised

at the time that there was an easier way. It will be shown in Chapter 4
(Lemma 4.5.2) that if A is semisimple (that is, diagonalizable when the use
of complex numbers is allowed), then the operator $\mathcal{L}$ defined by (1.1.6) is
also semisimple, and its eigenvalues, when acting on homogeneous vector
fields of a given degree, can be computed from those of A. In the present
problem, the eigenvalues of $\mathcal{L}$ acting on vector fields of degree m turn out
to be the numbers $(m_1 - m_2 \pm 1)i$, where m_1 and m_2 are nonnegative
integers with $m_1 + m_2 = m$. Thus, we can compute the eigenvalues in the
quadratic terms by writing down the table

m_1	m_2	$(m_1 - m_2 - 1)i$	$(m_1 - m_2 + 1)i$
2	0	i	$3i$
1	1	$-i$	i
0	2	$-3i$	$-i$

Since none of the eigenvalues are zero, $\mathcal{L}$ is invertible on the quadratic
terms (that is, on the space of homogeneous quadratic vector fields), and
we can take $\widehat{Q} = 0$.

The table of eigenvalues for the cubic terms is as follows:

m_1	m_2	$(m_1 - m_2 - 1)i$	$(m_1 - m_2 + 1)i$
3	0	$2i$	$4i$
2	1	0	$2i$
1	2	$-2i$	0
0	3	$-4i$	$-2i$

Since $\mathcal{L}$ is diagonalizable and the eigenvalue zero occurs twice, there are
two linearly independent eigenvectors of $\mathcal{L}$ associated with the eigenvalue
zero. That is, the dimension of the kernel of $\mathcal{L}$ is 2; it follows that there will
be two parameters (α and β) in the cubic terms of the normal form. The
eigenvalue table, by itself, is not sufficient to complete the computation
of the normal form; it is still necessary to calculate the kernel in order to
determine the explicit form of the terms in which α and β appear. But there
is a way to find the kernel without first finding the large matrix (1.1.14).

The computation of this kernel can be viewed as a problem in partial
differential equations, instead of one in linear algebra. According to (1.1.8),
a vector field (f, g) lies in $\ker \mathcal{L}$ precisely when

$$-yf_x + xf_y = -g, \tag{1.1.20}$$
$$-yg_x + xg_y = f.$$

In polar coordinates,

$$-y\frac{\partial}{\partial x} + x\frac{\partial}{\partial y} = \frac{\partial}{\partial \theta},$$

so (1.1.20) can be written as $f_\theta = -g$, $g_\theta = f$. Together these imply
$f_{\theta\theta} + f = 0$, which is the familiar equation of a spring, with solution
$f(r, \theta) = A(r)\cos\theta + B(r)\sin\theta$; since $g = f_\theta$ it follows that $g(r, \theta) =$

$A(r)\sin\theta - B(r)\cos\theta$. Returning to rectangular coordinates,

$$f(x,y) = \frac{A(\sqrt{x^2+y^2})}{\sqrt{x^2+y^2}}x + \frac{B(\sqrt{x^2+y^2})}{\sqrt{x^2+y^2}}y, \qquad (1.1.21)$$

$$g(x,y) = \frac{A(\sqrt{x^2+y^2})}{\sqrt{x^2+y^2}}y - \frac{B(\sqrt{x^2+y^2})}{\sqrt{x^2+y^2}}x.$$

To obtain solutions that are homogeneous cubic polynomials, it is necessary to take $A = a(x^2+y^2)^{3/2}$ and $B = b(x^2+y^2)^{3/2}$, for constants a and b. This gives two linearly independent cubic vector fields (f,g) in ker $\pounds$, which is the correct number according to our eigenvalue table, and once again yields the normal form (1.1.18).

One of the advantages of (1.1.21) is that we can deduce from it not only the cubic terms of the normal form, but the terms of any order. But before doing this, it is worthwhile to reconsider the original system (1.1.1) that we have been studying, which contains no terms beyond cubic. Notice that terms of higher order are introduced at once in (1.1.4), and these remain present (with modifications) throughout all subsequent steps. When a particular feature that we wish to study (such as the stability of the origin) is not 3-determined, it is natural to normalize to a higher order. But it is pointless to do this if the original system (1.1.1) was obtained by dropping the higher-order terms of a more complete system. Many equations studied in applications arise in exactly this way, and then the dotted terms in (1.1.4) cannot be considered to be correct. In order to make sure that normalizing the higher-order terms is meaningful, let us replace (1.1.1) with a system of the form

$$\dot{x} = -y + f(x,y), \qquad (1.1.22)$$
$$\dot{y} = x + g(x,y),$$

where f and g are smooth (infinitely differentiable) functions whose power series begin with quadratic terms:

$$f(x,y) \sim f_2(x,y) + f_3(x,y) + f_4(x,y) + \cdots, \qquad (1.1.23)$$
$$g(x,y) \sim g_2(x,y) + g_3(x,y) + g_4(x,y) + \cdots,$$

where f_j and g_j are homogeneous of degree j. (Warning: In Chapter 4, for technical reasons, the indices will denote *one less than the degree*, but for now we will use the degree.)

1.1.5. Remark. The power series (or Taylor series) of a smooth function f need not converge (hence the symbol $\sim$, rather than $=$), and even when it does converge, it need not converge to f. This comes about because of the existence of *flat functions*, also called *exponentially small* or *transcendentally small* functions, which vanish at the origin together with all of their derivatives, although the function itself is not identically zero. (The basic example in one variable is

$f(x) = e^{-(1/x)^2}$ for $x \neq 0$, $f(0) = 0$.) Any two smooth functions having the same Taylor series differ by a flat function. In particular, if the Taylor series converges, the function itself may differ from the sum of its Taylor series by a flat function. Since the full Taylor series does not adequately represent a smooth function, the most important fact is that the Taylor series is asymptotic. That is, f differs from its k-jet $j^k f$ (the truncation of its Taylor series at degree k) by a smooth function of the order of the first omitted term:

$$|f(x) - (j^k f)(x)| \leq c|x|^{k+1}$$

in the case of functions of a single variable, with analogous formulas (with $|\cdot|$ replaced by $\|\cdot\|$) if x or f is a vector. It is sometimes of interest to know that every formal power series is the Taylor series of some smooth function, that is, that the mapping from smooth functions to power series is onto. This is called the Borel–Ritt theorem, proved in Appendix A (Theorem A.3.2).

Replacing (1.1.1) by (1.1.22) and (1.1.23) is a natural generalization that does not affect any of our previous calculations. It is not hard to extend these calculations to show that for any given positive integer $k \geq 2$, the vector fields (f_j, g_j) for $j = 2, 3, \ldots, k$ can be brought into a normal form characterized by the condition that $\mathcal{L}(f_j, g_j) = 0$. (Several ways of doing this will be developed in Chapter 4.) It is not, in general, possible to bring all terms into normal form, only a finite number of them.

1.1.6. Remark. Since we will soon find it useful to speak of the "normal form to all orders," in spite of its purported impossibility, it is worthwhile to point out the subtleties involved here. First, it is standard practice in mathematics to speak of mathematical objects as existing if they are uniquely determined, even if they are not calculated. Thus, we regard the decimal digits of π as existing, and think of π as an infinitely long decimal. (Constructivists are entitled to their own opinions.) Now, there is clearly no obstacle in principle to the computation of the terms of the normal form to any order, and therefore all of the terms exist in the mathematical sense. Therefore, it is possible to speak of the formal power series that would be obtained if the system were put into normal form to all orders. Of course, this series may not converge, and is subject to the limitations discussed in Remark 1.1.5. Even if the normal form to all orders converges, the sequence of transformations that normalize the successive terms need not converge. These difficulties can be overcome (in principle) by careful use of the Borel–Ritt theorem to show the existence of smooth functions filling the required roles, but since these smooth functions are not computable, invocation of the Borel–Ritt theorem is not actually helpful here. So when we do speak of the normal form to all orders, it should be kept in mind that in the end this must be truncated at some degree k, and then the original system can be transformed into a system that agrees with the normal form up to that degree.

Now we can turn to (1.1.21) to solve the description problem for the normal form of (1.1.22) to order k. It is not possible to choose A and B in (1.1.21) so as to obtain a homogeneous polynomial of even order; therefore all even terms in the normal form must vanish. Terms of odd order arise by taking A and B to be constants multiplied by odd powers of $\sqrt{x^2 + y^2}$, and they have the same form as the cubic terms except that they contain a higher power of $(x^2 + y^2)$. That is, (1.1.18) becomes

$$\begin{bmatrix} \dot{x} \\ \dot{y} \end{bmatrix} = \begin{bmatrix} 0 & -1 \\ 1 & 0 \end{bmatrix} \begin{bmatrix} x \\ y \end{bmatrix} + \alpha_1 \left(x^2 + y^2 \right) \begin{bmatrix} x \\ y \end{bmatrix} + \beta_1 \left(x^2 + y^2 \right) \begin{bmatrix} -y \\ x \end{bmatrix} \qquad (1.1.24)$$
$$+ \alpha_2 \left(x^2 + y^2 \right)^2 \begin{bmatrix} x \\ y \end{bmatrix} + \beta_2 \left(x^2 + y^2 \right)^2 \begin{bmatrix} -y \\ x \end{bmatrix} + \cdots .$$

When normalized to degree $2k+1$, truncated (taking the $(2k+1)$-jet), and expressed in polar coordinates, the result is

$$\dot{r} = \alpha_1 r^3 + \alpha_2 r^5 + \cdots + \alpha_k r^{2k+1}, \qquad (1.1.25)$$
$$\dot{\theta} = 1 + \beta_1 r^2 + \beta_2 r^4 + \cdots + \beta_k r^{2k}.$$

Now it is easy to extend the previously given stability criterion: If $\alpha_1 < 0$, the origin is stable and the stability is 3-determined; if $\alpha_1 = 0$ and $\alpha_2 < 0$, it is stable and the stability is 5-determined, and so forth. Of course, if the first nonzero α_j is positive, the origin is unstable. If all of the α_j through $j = k$ are zero, the stability is not $(2k + 1)$-determined, and the normalization must be continued to higher k. There remains the possibility that absolutely all α_j are zero (in the sense of Remark 1.1.6); in that case the stability of the origin is said to be *not finitely determined*.

1.1.7. Remark. In order to show that in this case the origin may be either asymptotically stable or unstable (and need not be neutrally stable), one may take f and g in (1.1.22) to be flat functions. Then all terms of (1.1.23) are zero, the system is already in normal form to all orders, and in polar coordinates $\dot{r} = h(r, \theta)$ where h is flat in r. Since h may be either positive or negative, the origin may be asymptotically stable or unstable. The existence of problems that are not finitely determined places an ultimate limitation upon all methods that will be presented in this book.

The Normal Form Module; Invariants and Equivariants

Equation (1.1.25) gives a complete solution of the description problem for the normal form of a nonlinear center to any order k, but does not yet give a complete description of the mathematical structure of the normal form. To do this, it is necessary to pass to the *normal form to all orders*, bearing in mind that (according to Remark 1.1.6) this is only a convenient manner of speaking, and that all "actual" normal forms are truncated at some order k. It turns out that the set of differential equations that have a given linear term and are in normal form to all orders possesses the structure of a module over a ring.

> **1.1.8. Remark.** Most of what we do will be clear without a technical knowledge of ring and module theory. Very briefly, a *field* is a set of objects that can be added, subtracted, multiplied, and divided, with the ordinary rules of algebra holding. The rational numbers, the real numbers, and the complex numbers form fields. A *ring* (by which we always mean a commutative ring with identity) is a similar structure that permits addition, subtraction, and multiplication, but not necessarily division. Examples are the ring of integers, the ring of polynomials in a specified number of variables, and the ring of formal power series ("infinitely long polynomials") in a specified number of variables. See Appendix A for more information about rings. Just as a vector space is a set that admits addition of vectors and multiplication of vectors by scalars taken from some field, a *module* is a set that admits addition of its elements, and multiplication of its elements by elements of a specified ring. The most important difference between a vector space and a module is that a vector space spanned by a finite number of elements always has a basis (linearly dependent elements can be eliminated from the spanning set). This is not always true for finitely generated modules (because the elimination process would require division in the ring), and in general there will be relations called *syzygies* among the generators of a module. Examples of this phenomenon will appear in Sections 4.5 and 4.7; see also Appendix B.

If the normal form (1.1.24) is extended to all orders and rearranged, it may be written (formally, that is, without regard to convergence) as

$$\begin{bmatrix} \dot{x} \\ \dot{y} \end{bmatrix} = \begin{bmatrix} 0 & -1 \\ 1 & 0 \end{bmatrix} \begin{bmatrix} x \\ y \end{bmatrix} + \varphi \left(x^2 + y^2 \right) \begin{bmatrix} x \\ y \end{bmatrix} + \psi \left(x^2 + y^2 \right) \begin{bmatrix} -y \\ x \end{bmatrix},$$

where φ and ψ are formal power series in $s = x^2 + y^2$ beginning with the linear term (in s):

$$\varphi(s) = \alpha_1 s + \alpha_2 s^2 + \alpha_3 s^3 + \cdots,$$
$$\psi(s) = \beta_1 s + \beta_2 s^2 + \beta_3 s^3 + \cdots.$$

It is convenient to allow constant terms in these series even though they do not appear in the normal form (they will appear later in the unfolding), so that

$$\varphi(s) = \alpha_0 + \alpha_1 s + \alpha_2 s^2 + \cdots,$$
$$\psi(s) = \beta_0 + \beta_1 s + \beta_2 s^2 + \cdots.$$

The collection of formal power series in $s = x^2 + y^2$ forms a ring. (Such formal power series can be added, subtracted, and multiplied, but not always divided.) The vector fields

$$\varphi(s) \begin{bmatrix} x \\ y \end{bmatrix} + \psi(s) \begin{bmatrix} -y \\ x \end{bmatrix} \tag{1.1.26}$$

appearing in the normal form are linear combinations of the basic vector fields (x, y) and $(-y, x)$ with coefficients taken from the ring (in the same way that vectors in the plane are linear combinations of two unit vectors with coefficients taken from the field of real numbers). Thus, the set of vector fields forms a module over the ring, with generators (x, y) and $(-y, x)$. (This is a particularly simple module, in that there are no syzygies among the generators.)

This example illustrates what is, in principle, the best possible solution of the description problem for a normal form. In the simplest cases such a description has the following ingredients:

1. A collection of functions $I_1, \ldots, I_p$ of the coordinates. These functions are called the *basic invariants*, for a reason to be explained shortly. In the present example $p = 1$ and $I_1 = x^2 + y^2$.

2. The ring of formal power series in $I_1, \ldots, I_p$. This is called the *ring of (formal) invariants*.

3. A collection of vector fields $v_1, \ldots, v_q$ called the *basic equivariants*. In our example $q = 2$, $v_1 = (x, y)$, and $v_2 = (-y, x)$.

4. The *normal form module*, or *module of equivariants*, which is the set of linear combinations of the basic equivariants, with coefficients taken from the ring of invariants.

In more complicated cases, the same pattern holds, except that there may be relations among the basic invariants (so that the ring of invariants is a quotient ring of a power series ring by an ideal of relations) and also relations among the basic equivariants. These issues will be dealt with as they arise.

It remains to explain the terms *invariant* and *equivariant*. The function $I = x^2 + y^2$ is invariant under rotation, in the sense that if the vector (x_1, y_1), thought of as an arrow from the origin, is rotated around the origin to give (x_2, y_2), then $I(x_1, y_1) = I(x_2, y_2)$. The vector fields $v_1 = (x, y)$ and $v_2 = (-y, x)$ also appear to be invariant under rotation, *when thought of as whole vector fields*. That is, if the vectors of the vector field are drawn on a sheet of paper and the paper is rotated, the vector field will appear the same. But this "invariance" is different from the invariance of I, because if (x_1, y_1) rotates to (x_2, y_2), then $v_i(x_1, y_1) \neq v_i(x_2, y_2)$. Instead, the vector $v_i(x_1, y_1)$ *rotates into* $v_i(x_2, y_2)$. This type of invariance is technically called equivariance.

But why are rotations involved at all? This comes from the fact that the solutions of the *linear part* of our original equation (1.1.1) are rotations.

Remember that normalizing a system of differential equations means normalizing it *relative to its linear part*. Our entire discussion has been based on the operator $\mathcal{L}$, defined in (1.1.6) in terms of the matrix A that gives the linear part of (1.1.1). It will turn out that for each A there is a group of transformations e^{At} under which the associated normal form is equivariant (in the present case, the group of rotations), and the vector fields equivariant under this group form a module over the scalar invariants for the same group. Details will have to await the development of sufficient machinery (in Chapters 3 and 4).

Unfolding the Nonlinear Center

There are two reasons for introducing what are called *unfoldings* of a dynamical system such as (1.1.1):

1. *Imperfection in modeling.* The given differential equation may be an imperfect mathematical model of a real system, that is, the model may have been derived using certain simplifying assumptions (such as an absence of friction). If the predictions of the model do not exactly match the behavior of the real system, but it is not feasible to improve the model (perhaps because the neglected quantities are too small to measure or too difficult to model), a natural step is to consider the behavior of *all* differential equations that are close to the original model, to see whether one of these will predict the observed behavior. This leads directly to the notion of unfolding.

2. *Bifurcation theory.* The given differential equation may have been obtained by fixing the value of one or more parameters in a more general equation. In this case one may be interested in studying how the behavior of the solutions changes as the parameters are varied. Varying the parameters produces systems close to the original one, which again belong to the unfolding of the given system. This is most important when the original values of the parameters have been taken to be *critical* values (or a *bifurcation point* in parameter space) at which some important transition in behavior takes place.

It is important to understand that there are many ways in which one system of differential equations can be considered "close" to another, and not all of these are covered by the idea of an unfolding. In particular, if a model contains fewer state variables than the actual system, this imperfection cannot be corrected by passing to the unfolding.

In order to illustrate these ideas, which are developed more completely in Chapter 6, we will briefly and sketchily treat the unfolding of the nonlinear center and the resultant bifurcation, known as the *Hopf bifurcation*. We begin by adding certain arbitrary terms, multiplied by a small parameter

ε, to (1.1.1) to obtain

$$\begin{bmatrix} \dot{x} \\ \dot{y} \end{bmatrix} \equiv \begin{bmatrix} 0 & -1 \\ 1 & 0 \end{bmatrix} \begin{bmatrix} x \\ y \end{bmatrix} + Q(x,y) + C(x,y) + \varepsilon \left\{ \begin{bmatrix} p \\ q \end{bmatrix} + \begin{bmatrix} a & b \\ c & d \end{bmatrix} \begin{bmatrix} x \\ y \end{bmatrix} \right\}.$$

(1.1.27)

The added terms represent the constant and linear terms of an arbitrary smooth function of (x,y), and the symbol $\equiv$ has been used in place of $=$ to indicate that we are omitting (or "calculating modulo") terms of degree 4 or higher in x and y, terms of degree 2 or higher in ε, and also products of ε times terms of degree 2 or higher in x and y. This represents a (somewhat arbitrary) choice of a (somewhat asymmetrical in x, y, and ε) jet, which we hope will be a sufficient jet to determine the behavior of the system. (Of course, the choice is not actually arbitrary, but is based on the knowledge that it gives useful results. But for exploratory purposes such a choice can be tried arbitrarily.)

Thus, *every* system that is close to (1.1.1) can be written in the form (1.1.27), up to the jet that we have just described. Our task is to simplify this system as much as possible, determine its behavior (that is, determine the behavior of the truncated system), and then study whether this behavior is the same as that of the full system (with the deleted terms restored). There are six arbitrary quantities in (1.1.27), namely, p, q, a, b, c, and d. In the course of simplifying the system these will be reduced to two arbitrary quantities, the *unfolding parameters*, which in this instance have the physical meaning of damping and frequency modulation.

> **1.1.9. Remark.** Until Chapter 6, where a more precise definition of *asymptotic unfolding* will be given, we will use the word *unfolding* in a somewhat loose sense, to mean a system obtained by adding an arbitrary perturbation, taking a specified jet, and simplifying as much as possible. The arbitrary parameters that remain will be called the unfolding parameters, and the number of unfolding parameters is called the *codimension*. For those who are familiar with the usual technical definitions, we point out that our unfoldings are not necessarily miniversal unfoldings with respect to topological equivalence, and our codimensions are not necessarily the true codimensions in this sense. In fact, most dynamical systems do not have true miniversal unfoldings, because their codimension is infinite. In our approach this will be reflected in the facts that the number of unfolding parameters usually increases when the order of the jet is increased, and that in most cases no jet is sufficient to determine the complete behavior of the full system (up to topological equivalence). Nevertheless, the unfoldings that we calculate using jets of specified order can be sufficient to determine certain aspects of the behavior of the full system, such as (in the present example) the existence and stability of a Hopf bifurcation.

The first step in simplifying (1.1.27) is to perform the changes of variables needed to place the unperturbed ($\varepsilon = 0$) terms into normal form (1.1.18). These changes of variables may modify the perturbation terms, but since these are arbitrary anyway at this point, we need not be concerned. The result is

$$\begin{bmatrix} \dot{x} \\ \dot{y} \end{bmatrix} \equiv \begin{bmatrix} 0 & -1 \\ 1 & 0 \end{bmatrix} \begin{bmatrix} x \\ y \end{bmatrix} + \alpha \left(x^2 + y^2 \right) \begin{bmatrix} x \\ y \end{bmatrix} + \beta \left(x^2 + y^2 \right) \begin{bmatrix} -y \\ x \end{bmatrix} \qquad (1.1.28)$$
$$+ \varepsilon \left\{ \begin{bmatrix} p \\ q \end{bmatrix} + \begin{bmatrix} a & b \\ c & d \end{bmatrix} \begin{bmatrix} x \\ y \end{bmatrix} \right\}.$$

The presence of the small constant term $\varepsilon(p, q)$ means that there is no longer a rest point at the origin, but in fact there is still a rest point close to the origin, and it may be moved back to the origin by the small shift of coordinates

$$x = \xi - \varepsilon q,$$
$$y = \eta + \varepsilon p.$$

In fact, this transformation, followed by renaming (ξ, η) as (x, y), eliminates the constant term and reduces our equation to

$$\begin{bmatrix} \dot{x} \\ \dot{y} \end{bmatrix} \equiv \begin{bmatrix} 0 & -1 \\ 1 & 0 \end{bmatrix} \begin{bmatrix} x \\ y \end{bmatrix} + \alpha \left(x^2 + y^2 \right) \begin{bmatrix} x \\ y \end{bmatrix} + \beta \left(x^2 + y^2 \right) \begin{bmatrix} -y \\ x \end{bmatrix} \qquad (1.1.29)$$
$$+ \varepsilon \begin{bmatrix} a & b \\ c & d \end{bmatrix} \begin{bmatrix} x \\ y \end{bmatrix}.$$

(The quantities a, b, c, d are again modified by this change of variables, as are higher-order terms that are deleted. But, since a, b, c, d are still arbitrary, we do not care. The general rule is that once the terms of a given order have been simplified, we must be careful to preserve them, but until then it does not matter. Of course, to solve the *computation problem* it would be necessary to record all the changes so that they can be implemented for a system with specific numerical coefficients, but again we are only solving the *description problem* at the moment.)

The next step is to simplify the linear terms in the perturbation. Putting $u = (x, y)$, (1.1.29) may be written as

$$\dot{u} \equiv Au + f(u) + \varepsilon Bu,$$

with

$$B = \begin{bmatrix} a & b \\ c & d \end{bmatrix}.$$

To this equation, we will apply a transformation of the form

$$u = (I + \varepsilon T)v,$$

with T initially unspecified; at the end we will choose T to make the transformed equation as simple as possible. Using the fact that the inverse transformation can be expressed as $v = (I - \varepsilon T + \cdots)u$ (where the omitted terms are of higher order in ε), the transformed equation is seen to be

$$\dot{v} \equiv Av + f(v) + \varepsilon \widehat{B} v,$$

where $\widehat{B} = B + AT - TA$. (Near-identity linear transformations of this type will be studied in great detail in Chapter 3.) If we define a new homological operator $\mathcal{L}'$, acting on 2×2 matrices, by

$$\mathcal{L}'Q = [Q, A] = QA - AQ,$$

then the relation between T and $\widehat{B}$ may be expressed as the homological equation

$$\mathcal{L}'T = B - \widehat{B}, \tag{1.1.30}$$

which is similar in structure to (1.1.7). In fact, the similarity is more than superficial: $\mathcal{L}'$ is the same as the operator $\mathcal{L}$ defined in (1.1.6), applied to linear vector fields represented by their matrices. More precisely, if Q is a matrix and Qu is the linear vector field that it defines, then $\mathcal{L}(Qu) = (\mathcal{L}'Q)u$.

Because the structure of this problem is the same as that of the normal form problem, the work necessary to describe the simplest form for $\widehat{B}$ has already been done: $\widehat{B}$ must belong to $\ker \mathcal{L}'$, or equivalently, $\widehat{B}v$ must belong to $\ker \mathcal{L}$, which means that $\widehat{B}v$ must be equivariant under the group of rotations. Thus, returning (as we always do after changing variables) to the original notation (replacing v by $u = (x, y)$ and $\widehat{B}$ by B), the normalized linear vector field Bu must be formed from the basic equivariant vector fields (x, y) and $(-y, x)$ by linear combinations using invariants as coefficients (see (1.1.26)). But the only invariants that will give linear vector fields are constants. So (with the conventional choice of signs, so that δ may be called the *damping*) we must have $Bu = -\delta(x, y) + \nu(-y, x)$, that is,

$$B = \begin{bmatrix} a & b \\ c & d \end{bmatrix} = \begin{bmatrix} -\delta & -\nu \\ \nu & -\delta \end{bmatrix}.$$

Thus, (1.1.29) finally simplifies to

$$\begin{bmatrix} \dot{x} \\ \dot{y} \end{bmatrix} \equiv \begin{bmatrix} 0 & -1 \\ 1 & 0 \end{bmatrix} \begin{bmatrix} x \\ y \end{bmatrix} + \alpha \left(x^2 + y^2 \right) \begin{bmatrix} x \\ y \end{bmatrix} + \beta \left(x^2 + y^2 \right) \begin{bmatrix} -y \\ x \end{bmatrix} \tag{1.1.31}$$
$$+ \varepsilon \begin{bmatrix} -\delta & -\nu \\ \nu & -\delta \end{bmatrix} \begin{bmatrix} x \\ y \end{bmatrix},$$

containing (as promised) only the two unfolding parameters δ and ν. To determine the physical meaning of these parameters, we put (1.1.31) into

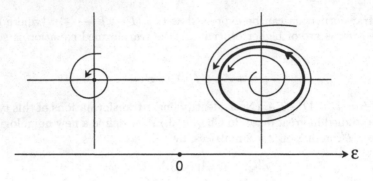

Figure 1.1. Hopf bifurcation with $\alpha < 0$ and $\delta < 0$.

polar coordinates to obtain (compare (1.1.19))

$$\dot{r} = -\varepsilon\delta r + \alpha r^3, \qquad (1.1.32)$$
$$\dot{\theta} = 1 + \varepsilon\nu + \beta r^2.$$

Thus, δ represents linear damping ($\delta > 0$ stabilizes the origin, regardless of the sign of α that determines nonlinear damping or excitation), while ν represents frequency modulation.

> **1.1.10. Remark.** By scaling time (introducing a new independent variable τ by $\tau = (1+\nu)t$), it is possible to obtain equations having the same form as (1.1.32) with ν absent. Then δ is the only remaining unfolding parameter. Because of this, the Hopf bifurcation is usually classified as a codimension-one bifurcation. In fact, as the following discussion shows, δ controls the topological features of the bifurcation and ν affects only the period of the periodic orbit. Scaling time does not fit within the framework of normal form transformations, which involve only the dependent variables.

Observe that, if $\varepsilon\delta/\alpha$ is positive, there is a circle of radius

$$r^* = \sqrt{\frac{\varepsilon\delta}{\alpha}}$$

on which $\dot{r} = 0$; see Figure 1.1. This represents an invariant circle, or *limit cycle*, for (1.1.32). In particular, suppose $\alpha < 0$, so that the origin of the unperturbed ($\varepsilon = 0$) system is stable. Suppose $\delta < 0$ is fixed, and let ε be gradually increased from zero. Then the radius r^* of the limit cycle gradually increases from zero, as if the origin were blowing a smoke ring. The origin becomes unstable (because $\delta < 0$), but outside the limit cycle r is still decreasing (because $\alpha < 0$). Therefore, the limit cycle itself is asymptotically orbitally stable, all solutions of (1.1.32) approaching the limit cycle either from the inside or the outside. This process of "blowing a smoke ring" is the famous *Hopf bifurcation*.

1.1.11. Remark. When we say "let ε be gradually increased from zero," we do not mean that ε increases gradually *with time*. Instead we mean to consider a number of distinct systems of the form (1.1.32), in each of which ε is constant, but ε increases from zero as we pass from one system to the next. The question of what happens in a single system in which ε is varied with time is a much more difficult question, referred to in the literature as the problem of a "slowly varying bifurcation parameter" (or sometimes as "dynamic bifurcation," although this phrase usually has a different meaning). It will not be addressed here.

Of course, the discussion we have given so far does not prove the existence of a Hopf bifurcation in the original system. Equation (1.1.32) is a truncated system (because of the $\equiv$ in (1.1.31)). In order to prove the existence of the Hopf bifurcation, it is necessary to show that the (asymmetrical) jet used to define $\equiv$ is sufficient to determine the behavior that we have described. This will be done in Section 6.5. Because our analysis is local (Remark 1.1.2) we can expect the limit cycle to exist only for small ε, with r^* small, and we can expect the stability of the limit cycle to be only local. Solutions beginning at a fair distance from the origin may tend to other attractors that do not appear in (1.1.32).

Notes and References

Introductory treatments of normal form theory using matrices such as (1.1.10) are contained in Guckenheimer and Holmes [52], Wiggins [111], and Arrowsmith and Place [7]. Another useful introductory book is Kuznetsov [67], although the normal form work in this book is carried out on an ad hoc basis for each example and is not separated from the rest of the analysis. An extensive treatment of normal forms for (mostly) two-dimensional systems is given in Chow, Li, and Wang [29]. The partial differential equations approach, as in (1.1.20), is included in Elphick et al. [41]. Further references for normal forms will be found in Chapter 4 below, and for unfoldings in Chapter 6.

1.2   The Nonsemisimple Double-Zero Eigenvalue

The system to be studied in this section is

$$\begin{bmatrix} \dot{x} \\ \dot{y} \end{bmatrix} = \begin{bmatrix} 0 & 1 \\ 0 & 0 \end{bmatrix} \begin{bmatrix} x \\ y \end{bmatrix} + Q(x,y) + \cdots, \qquad (1.2.1)$$

where Q is the homogeneous quadratic term.

Normal Form Styles for the Nonsemisimple Double Zero

To simplify the quadratic term, the discussion in the previous section beginning with (1.1.3) down to (1.1.7) applies word for word, except that now

$$A = \begin{bmatrix} 0 & 1 \\ 0 & 0 \end{bmatrix}, \tag{1.2.2}$$

which entails, by way of (1.1.6), that

$$\pounds \begin{bmatrix} f(x, y) \\ g(x, y) \end{bmatrix} = \begin{bmatrix} y f_x - g \\ y g_x \end{bmatrix}. \tag{1.2.3}$$

But the solution of the homological equation

$$\pounds q = Q - \widehat{Q} \tag{1.2.4}$$

goes rather differently, because A is not semisimple (diagonalizable). The normal form cannot be taken to belong to ker $\pounds$, because this does not form a complement to im $\pounds$.

With respect to the basis (1.1.9), the matrix of $\pounds$ on the quadratic terms is

$$\pounds = \begin{bmatrix} 0 & 0 & 0 & -1 & 0 & 0 \\ 2 & 0 & 0 & 0 & -1 & 0 \\ 0 & 1 & 0 & 0 & 0 & -1 \\ 0 & 0 & 0 & 0 & 0 & 0 \\ 0 & 0 & 0 & 2 & 0 & 0 \\ 0 & 0 & 0 & 0 & 1 & 0 \end{bmatrix}. \tag{1.2.5}$$

It is easy to see by inspection that the third and sixth columns of this matrix are linear combinations of the other four, which are linearly independent and therefore form a basis for the image of $\pounds$. Expressing these four columns as quadratic vector fields (after dividing the first column by two) yields

$$\begin{bmatrix} x^2 \\ -2xy \end{bmatrix}, \quad \begin{bmatrix} xy \\ 0 \end{bmatrix}, \quad \begin{bmatrix} y^2 \\ 0 \end{bmatrix}, \quad \begin{bmatrix} -xy \\ y^2 \end{bmatrix}.$$

Any two quadratic vector fields linearly independent of these may be chosen as a basis for a complement to im $\pounds$, and therefore as the basis for a possible *style* for the quadratic normal form. Two simple choices are

$$\begin{bmatrix} 0 \\ x^2 \end{bmatrix}, \quad \begin{bmatrix} 0 \\ xy \end{bmatrix} \tag{1.2.6}$$

and

$$\begin{bmatrix} x^2 \\ 0 \end{bmatrix}, \quad \begin{bmatrix} 0 \\ x^2 \end{bmatrix}. \tag{1.2.7}$$

With the first choice (1.2.6), the normal form becomes

$$\begin{bmatrix} \dot{x} \\ \dot{y} \end{bmatrix} = \begin{bmatrix} 0 & 1 \\ 0 & 0 \end{bmatrix} \begin{bmatrix} x \\ y \end{bmatrix} + \alpha \begin{bmatrix} 0 \\ x^2 \end{bmatrix} + \beta \begin{bmatrix} 0 \\ xy \end{bmatrix} + \cdots . \qquad (1.2.8)$$

This normal form style will be called a *simplified normal form*. One of its distinguishing features is that it contains nonzero entries only in rows corresponding to the bottom rows of Jordan blocks of A (assumed to be in Jordan normal form).

One drawback to this ad hoc manner of selecting a normal form style (or a complement to im $\mathcal{L}$) is that it has to be done again from scratch in each new degree. Thus, if we wanted to find the normal form for cubic terms, we would write down an 8×8 matrix, similar to (1.1.14), calculate its image, and select a complement. There would be no relationship between this complement and the one found for the quadratic terms, and hence little likelihood that the resulting normal form would have a nice mathematical description, such as a module structure involving invariants and equivariants. There has to be a better way.

In Section 4.6, it will be shown that a complement to im $\mathcal{L}$ is always given by ker $\mathcal{L}^*$, where

$$(\mathcal{L}^* q)(u) = q'(u)A^*(u) - A^* q(u), \qquad (1.2.9)$$

A^* being the adjoint (or conjugate transpose) of A. (When A is real, as here, this is simply the transpose.) The normal form style resulting from this choice of complement will be called the *inner product normal form* (because it is related to a special inner product on vector fields, explained in Section 4.6). In our case

$$\mathcal{L}^* \begin{bmatrix} f(x,y) \\ g(x,y) \end{bmatrix} = \begin{bmatrix} x f_y \\ x g_y - f \end{bmatrix}, \qquad (1.2.10)$$

so the inner product normal form (to any degree) consists of polynomial vector fields (of the required degree) satisfying the partial differential equations

$$x f_y(x,y) = 0, \qquad (1.2.11)$$
$$x g_y = f.$$

Calling such vector fields *equivariants*, and allowing formal power series (rather than only polynomials), it is not hard to see that the equivariants form a module over the *invariants*, or (scalar) formal power series $\varphi(x,y)$ satisfying $x\varphi_x(x,y) = 0$. (We can, of course, drop the x factor and write simply $\varphi_x(x,y) = 0$, and similarly in the first, but not the second, line of (1.2.11). We have written the longer form to emphasize the recurrence of the operator $x\partial/\partial y$, which is differentiation along the flow of the linear vector field $A^*(x,y)$. This flow defines the group with respect to which the invariants and equivariants are, respectively, invariant and equivariant.) In

Section 6.6, it will be seen that the only basic invariant is $I_1 = x$, and the basic equivariants are $v_1 = (0, 1)$ and $v_2 = (x, y)$; the normal form is

$$\begin{bmatrix} \dot{x} \\ \dot{y} \end{bmatrix} = \begin{bmatrix} 0 & 1 \\ 0 & 0 \end{bmatrix} \begin{bmatrix} x \\ y \end{bmatrix} + \sum_{i=1}^{\infty} \left(\alpha_i x^i \begin{bmatrix} x \\ y \end{bmatrix} + \beta_i x^{i+1} \begin{bmatrix} 0 \\ 1 \end{bmatrix} \right). \tag{1.2.12}$$

The inner product normal form is more complicated than the simplified normal form (1.2.8), having nonzero entries outside the bottom row, but it has stronger mathematical properties. In particular, the simplified normal form is not equivariant under the flow of A^*.

A third normal form style, to be developed in Section 4.8, is what we call the sl(2) normal form (because it involves the "special linear algebra" of 2×2 matrices with zero trace). In the present problem it coincides with the inner product normal form, but in general it is different. (Once again, in appearance it is usually more complicated even than the inner product normal form, but has additional advantageous structural features.) The development of the sl(2) normal form depends upon a substantial amount of mathematical theory. This theory, and the sl(2) normal form itself, is presented in the starred sections, which may be omitted without loss of continuity.

Unfolding the Nonsemisimple Double Zero

We begin with the simplified normal form (1.2.8), and consider a system to be "close" to (1.2.8) if it is a perturbation of (1.2.8), obtained by adding terms multiplied by a perturbation parameter ε. Calculating modulo cubic terms in x and y, quadratic terms in ε, and products that are linear in ε and quadratic in x and y, the "arbitrary perturbation" of (1.2.8) is

$$\begin{bmatrix} \dot{x} \\ \dot{y} \end{bmatrix} \equiv \begin{bmatrix} 0 & 1 \\ 0 & 0 \end{bmatrix} \begin{bmatrix} x \\ y \end{bmatrix} + \begin{bmatrix} 0 \\ \alpha x^2 + \beta xy \end{bmatrix} + \varepsilon \left\{ \begin{bmatrix} p \\ q \end{bmatrix} + \begin{bmatrix} a & b \\ c & d \end{bmatrix} \begin{bmatrix} x \\ y \end{bmatrix} \right\}. \tag{1.2.13}$$

Our goal is to reduce the number of arbitrary parameters p, q, a, b, c, d from six to two (in the generic case) or three (in all cases) by performing a series of coordinate changes. As usual, each coordinate change will be written as a change from (x, y) to (ξ, η), after which the notation reverts to (x, y). The first such change is

$$y = \eta + \varepsilon k,$$

with x unchanged; we call this transformation a *primary shift*, and it carries (1.2.13) into the following form:

$$\begin{bmatrix} \dot{x} \\ \dot{y} \end{bmatrix} \equiv \begin{bmatrix} 0 & 1 \\ 0 & 0 \end{bmatrix} \begin{bmatrix} x \\ y \end{bmatrix} + \begin{bmatrix} 0 \\ \alpha x^2 + \beta xy \end{bmatrix} + \varepsilon \left\{ \begin{bmatrix} p + k \\ q \end{bmatrix} + \begin{bmatrix} a & b \\ c & d + k\beta \end{bmatrix} \begin{bmatrix} x \\ y \end{bmatrix} \right\}.$$

Choosing $k = -p$ removes p from the constant term. Renaming $d + k\beta$ as d yields the 5-parameter expression

$$\begin{bmatrix} \dot{x} \\ \dot{y} \end{bmatrix} \equiv \begin{bmatrix} 0 & 1 \\ 0 & 0 \end{bmatrix} \begin{bmatrix} x \\ y \end{bmatrix} + \begin{bmatrix} 0 \\ \alpha x^2 + \beta xy \end{bmatrix} + \varepsilon \left\{ \begin{bmatrix} 0 \\ q \end{bmatrix} + \begin{bmatrix} a & b \\ c & d \end{bmatrix} \begin{bmatrix} x \\ y \end{bmatrix} \right\}.$$

Next, by the same methods used to obtain (1.1.31), a linear transformation

$$\begin{bmatrix} x \\ y \end{bmatrix} = (I + \varepsilon T) \begin{bmatrix} \xi \\ \eta \end{bmatrix}$$

can be found that reduces the system further to the three-parameter form

$$\begin{bmatrix} \dot{x} \\ \dot{y} \end{bmatrix} \equiv \begin{bmatrix} 0 & 1 \\ 0 & 0 \end{bmatrix} \begin{bmatrix} x \\ y \end{bmatrix} + \begin{bmatrix} 0 \\ \alpha x^2 + \beta xy \end{bmatrix} + \varepsilon \left\{ \begin{bmatrix} 0 \\ q \end{bmatrix} + \begin{bmatrix} 0 & 0 \\ c & d \end{bmatrix} \begin{bmatrix} x \\ y \end{bmatrix} \right\}, \qquad (1.2.14)$$

where c and d are again allowed to differ from their original values. The next, and final, transformation does not fit into the usual normal-form framework, but instead is an example of "hypernormalization" or "normalization beyond the normal form." We call it a *secondary shift*:

$$x = \xi + \varepsilon h$$

carries (1.2.14) into

$$\begin{bmatrix} \dot{x} \\ \dot{y} \end{bmatrix} \equiv \begin{bmatrix} 0 & 1 \\ 0 & 0 \end{bmatrix} \begin{bmatrix} x \\ y \end{bmatrix} + \begin{bmatrix} 0 \\ \alpha x^2 + \beta xy \end{bmatrix} + \varepsilon \left\{ \begin{bmatrix} 0 \\ q \end{bmatrix} + \begin{bmatrix} 0 & 0 \\ c + 2\alpha h & d + \beta h \end{bmatrix} \begin{bmatrix} x \\ y \end{bmatrix} \right\}.$$

$$(1.2.15)$$

Observe that unlike the primary shift, the secondary shift has no effect on the constant term of the perturbation. It would have been useless to incorporate this shift into the first step, where our goal was to simplify the constant term. But now the secondary shift is useful. It has the effect of "injecting" the quantities $2\alpha h$ and βh into the bottom row of the matrix for the perturbed linear terms, *without disrupting the simplification of this matrix that has already been achieved* (that is, the a and b entries are still zero). It remains to choose h, which we do in one of the following ways: If $\alpha \neq 0$, choosing $h = -c/2\alpha$ eliminates the parameter c; if $\beta \neq 0$, choosing $h = -a/\beta$ eliminates the parameter d; if both α and β are nonzero, we can choose between eliminating c and d; if both α and β are zero, the secondary shift has no effect and we must retain both c and d.

Let us now assume that $\alpha \neq 0$; this is a generic condition on the unperturbed quadratic term. In this case, we can eliminate c (and once again modify d), and the resulting system can be rearranged as

$$\begin{bmatrix} \dot{x} \\ \dot{y} \end{bmatrix} \equiv \begin{bmatrix} 0 \\ \varepsilon q \end{bmatrix} + \begin{bmatrix} 0 & 1 \\ 0 & \varepsilon d \end{bmatrix} \begin{bmatrix} x \\ y \end{bmatrix} + \begin{bmatrix} 0 \\ \alpha x^2 + \beta y^2 \end{bmatrix}. \qquad (1.2.16)$$

It is convenient to introduce new independent small unfolding parameters $\mu_1 = \varepsilon q$, $\mu_2 = \varepsilon d$, and write this unfolded system as

$$\begin{bmatrix} \dot{x} \\ \dot{y} \end{bmatrix} \equiv \begin{bmatrix} 0 \\ \mu_1 \end{bmatrix} + \begin{bmatrix} 0 & 1 \\ 0 & \mu_2 \end{bmatrix} \begin{bmatrix} x \\ y \end{bmatrix} + \begin{bmatrix} 0 \\ \alpha x^2 + \beta y^2 \end{bmatrix}. \qquad (1.2.17)$$

On the other hand, if $\beta \neq 0$, we can achieve instead the unfolding

$$\begin{bmatrix} \dot{x} \\ \dot{y} \end{bmatrix} \equiv \begin{bmatrix} 0 \\ \mu_1 \end{bmatrix} + \begin{bmatrix} 0 & 1 \\ \mu_2 & 0 \end{bmatrix} \begin{bmatrix} x \\ y \end{bmatrix} + \begin{bmatrix} 0 \\ \alpha x^2 + \beta xy \end{bmatrix}, \qquad (1.2.18)$$

or (if we wish to permit arbitrary α and β with no condition imposed) we must accept the codimension-three unfolding given by

$$\begin{bmatrix} \dot{x} \\ \dot{y} \end{bmatrix} \equiv \begin{bmatrix} 0 \\ \mu_1 \end{bmatrix} + \begin{bmatrix} 0 & 1 \\ \mu_2 & \mu_3 \end{bmatrix} \begin{bmatrix} x \\ y \end{bmatrix} + \begin{bmatrix} 0 \\ \alpha x^2 + \beta xy \end{bmatrix}. \qquad (1.2.19)$$

The first of these unfoldings, (1.2.17), is the most commonly studied; as the two parameters μ_1 and μ_2 are varied (in a neighborhood of $(0,0)$), saddle-node, Hopf, and homoclinic bifurcations are discovered. This is called the *Takens–Bogdanov bifurcation* and will be discussed in greater detail in Section 6.6, along with the question of the sufficiency of the jet indicated by $\equiv$. This is perhaps the best understood of all local dynamical systems, in the sense that there exists a complete proof that this jet is sufficient for all topological features of the motion.

Notes and References

The example studied in this section is treated in all of the references listed at the end of Section 1.1. Further treatment of this example, and further references, will be found in Sections 5.4 and 6.6.

2
The Splitting Problem for Linear Operators

The examples in Chapter 1 have made it clear that it is important to be able to find a complement $\mathcal{C}$ to the image of a linear map $L : V \to V$, that is, a space $\mathcal{C}$ such that

$$V = (\operatorname{im} L) \oplus \mathcal{C}. \qquad (2.0.1)$$

This is purely a problem in linear algebra, and so can be discussed without any explicit mention of normal forms. This chapter contains four solutions to this problem:

1. If L is semisimple, the easiest solution is to take $\mathcal{C} = \ker L$ (Section 2.1).

2. If V has an inner product, we can take $\mathcal{C} = \ker L^*$, where L^* is the adjoint of L with respect to the inner product (Section 2.2).

3. If a Jordan chain structure is known for L, $\mathcal{C}$ can be taken to be the span of the tops of the Jordan chains. This is worked out in Section 2.3 for the nilpotent case and in Section 2.4 in general. These sections include a proof of the Jordan canonical form theorem.

4. The case frequently arises later in this book that L belongs to an "sl(2) triad" consisting of three operators on V having specific properties (stated in Section 2.5). These properties lead to algorithms enabling easy calculation of a Jordan chain structure for L as well as the associated complement $\mathcal{C}$ and the projection maps into that complement. This theory, which is a part of the representation theory of the Lie algebra sl(2), is presented in the starred Sections 2.5–2.7,

and may be omitted on a first reading. This material is (mostly) used only in the starred sections of later chapters.

In addition to finding the complement $\mathcal{C}$, it is important to be able to compute the projections

$$P : V \to \mathcal{C} \quad \text{and} \quad (I - P) : V \to \operatorname{im} L \qquad (2.0.2)$$

associated with the splitting $V = \operatorname{im} L \oplus \mathcal{C}$. (Actually, P and $I - P$ are best thought of as maps $V \to V$; writing $P : V \to \mathcal{C}$ is meant only to emphasize that $\operatorname{im} P = \mathcal{C}$. When viewed as a matrix, P is square.) Each of the approaches to the splitting problem (listed above) is associated with a particular approach to calculating the projections.

2.1   The Splitting Problem in the Semisimple Case

The following theorem is well known from elementary linear algebra:

2.1.1. Theorem. Let $L : V \to W$ be a linear mapping of finite-dimensional vector spaces. Then

$$\dim V = \dim \operatorname{im} L + \dim \ker L.$$

Proof. Choose a basis $v_1, \ldots, v_k$ for $\ker L$ and extend it with vectors $v_{k+1}, \ldots, v_n$ to form a basis for V. Since $Lv_1, \ldots, Lv_k$ vanish, the vectors $Lv_{k+1}, \ldots, Lv_n$ span $\operatorname{im} L$. If these vectors were not linearly independent, there would be a linear relation $c_{k+1} Lv_{k+1} + \cdots + c_n Lv_n = 0$, which would imply that $c_{k+1} v_{k+1} + \cdots + c_n v_n$ belonged to the kernel of L, contradicting the construction of the v_j. So $\dim \operatorname{im} L = n - k$. $\square$

In general, the image and kernel in this theorem belong to different spaces, but if L is an *operator* (or *endomorphism*) on V (i.e., if $W = V$), it is natural to wonder whether

$$V = \operatorname{im} L \oplus \ker L. \qquad (2.1.1)$$

This is not always true: The example $L : \mathbb{R}^2 \to \mathbb{R}^2$ with

$$Lx = \begin{bmatrix} 0 & 1 \\ 0 & 0 \end{bmatrix} x$$

shows that the image and kernel of L may even coincide. But there is one circumstance under which (2.1.1) is true, namely, when L is *semisimple*. Roughly speaking, semisimple means "diagonalizable over $\mathbb{C}$." In order to spell this out in greater detail, we quickly review the notions of eigenvalue and eigenvector, paying special attention to the meaning of a complex eigenvalue for a real operator.

Semisimplicity

Taking the (easier) complex case first, suppose that V is a vector space of dimension n over $\mathbb{C}$. An eigenvalue/eigenvector pair for L is a complex number λ and a nonzero vector $v \in V$ such that $Lv = \lambda v$. If we choose any basis for V, and represent L with respect to this basis as a (complex) $n \times n$ matrix A, the eigenvalues of L are the roots of the characteristic equation $\det(A - \lambda I) = 0$, which, as an nth-degree polynomial, has n (complex) roots counting multiplicity. These eigenvalues will be written $\lambda_1, \ldots, \lambda_n$ with repetitions, or $\lambda_{(1)}, \ldots, \lambda_{(r)}$ without repetitions (with $\lambda_{(i)}$ having multiplicity m_i and $m_1 + \cdots + m_r = n$). If L has n linearly independent eigenvectors (which is guaranteed if the n eigenvalues are distinct, but not otherwise) and S is the matrix having these eigenvectors as columns (expressed as coordinates with respect to the chosen basis), then $S^{-1}AS = \Lambda$ will be diagonal, with the eigenvalues appearing as the diagonal entries. In this case L is called *diagonalizable* or *semisimple*.

If V is a vector space of dimension n over $\mathbb{R}$, the story is a little more complicated. An eigenvalue/eigenvector pair (in the strict sense) is now a *real* number λ and a nonzero vector $v \in V$ with $Lv = \lambda v$. If a basis is chosen and L is represented by a real $n \times n$ matrix A, the eigenvalues of L will be among the roots of $\det(A - \lambda I) = 0$, but this polynomial may have strictly complex roots, which cannot be counted as true eigenvalues of L, because there are no corresponding eigenvectors v in V. (Nevertheless, it is common to refer to these as "complex eigenvalues of L.") We now ask what is the significance, for the real operator L, of the existence of these complex eigenvalues.

The matrix A can be viewed as defining a linear operator $A : \mathbb{C}^n \to \mathbb{C}^n$ that is an extension of $A : \mathbb{R}^n \to \mathbb{R}^n$. The complex roots of the characteristic equation are true eigenvalues of this extended map. Complex eigenvalues and eigenvectors of real matrices come in complex conjugate pairs, such that if $\alpha \pm i\beta$ is a pair of eigenvalues (with $\alpha, \beta \in \mathbb{R}$) and if $u + iv$ (with $u, v \in \mathbb{R}^n$) is an eigenvector for $\alpha + i\beta$, then $u - iv$ will be an eigenvector for $\alpha - i\beta$. Given such a pair of eigenvalues and eigenvectors, the complex equation $A(u + iv) = (\alpha + i\beta)(u + iv)$ implies the two real equations $Au = \alpha u - \beta v$ and $Av = \beta u + \alpha v$. Then, corresponding to the vectors u and v in $\mathbb{R}^n$ there are vectors $\hat{u}$ and $\hat{v}$ in V with $L\hat{u} = \alpha\hat{u} - \beta\hat{v}$ and $L\hat{v} = \beta\hat{u} + \alpha\hat{v}$. The subspace U of V spanned by $\hat{u}$ and $\hat{v}$ is invariant under L. For our purposes (compare Remark 1.1.1) it is convenient to take the basis for U in the reverse order, $\{\hat{v}, \hat{u}\}$; the restriction $L|_U$ of L to this subspace, expressed as a matrix with respect to the basis $\{\hat{v}, \hat{u}\}$, is

$$\begin{bmatrix} \alpha & -\beta \\ \beta & \alpha \end{bmatrix}. \tag{2.1.2}$$

This will be our "standard real 2×2 block" corresponding to the conjugate pair of complex eigenvalues $\alpha \pm i\beta$. Thus, if the real matrix A is "diagonalizable over $\mathbb{C}$" (i.e., diagonalizable when regarded as $A : \mathbb{C}^n \to \mathbb{C}^n$), then L is expressible by a real matrix having block diagonal form, with a 1×1 block (a single real number) for each real eigenvalue and a 2×2 block of the form (2.1.2) for each pair of conjugate eigenvalues. This block diagonal matrix will equal $S^{-1}AS$, where S is an invertible real matrix whose columns are the "new" basis vectors (expressed in the "old" basis). Under these circumstances, L is again called *semisimple*.

This discussion may be summarized by the following definitions:

2.1.2. Definition. A linear transformation $L : V \to V$ is called *semisimple* if one of the following two cases holds:

1. V is a complex vector space and L is diagonalizable; or

2. V is a real vector space and L is block-diagonalizable in the form

$$\begin{bmatrix} \lambda_1 & & & & & & & \\ & \ddots & & & & & & \\ & & \lambda_s & & & & & \\ & & & \alpha_1 & -\beta_1 & & & \\ & & & \beta_1 & \alpha_1 & & & \\ & & & & & \ddots & & \\ & & & & & & \alpha_t & -\beta_t \\ & & & & & & \beta_t & \alpha_t \end{bmatrix}.$$

A matrix of this form is said to be in *real semisimple canonical form*.

Semisimple Splitting

Having defined "semisimple," we can state the main result of this section:

2.1.3. Theorem. If $L : V \to V$ is semisimple, then

$$V = \operatorname{im} L \oplus \ker L.$$

Proof. Choose a basis in which L is either diagonal or block diagonal (of the form described above). Then $\ker L$ is the span of those elements of the basis that are eigenvectors with eigenvalue zero, and $\operatorname{im} L$ is the span of the remaining elements of the basis. $\square$

The projections (2.0.2) associated with Theorem 2.1.3 could be calculated by finding a basis of eigenvectors for V (or, in the case of a real matrix with complex eigenvalues, a basis that block-diagonalizes L) and resolving any vector v into its components using this basis. Then $Pv \in \ker L$ is the sum of the components along the eigenvectors with eigenvalue zero. But this procedure requires the solution of many systems of linear equations (first to find the eigenvectors, then to resolve each v into components). The following algorithm, suitable for use with a computer algebra system, computes P without solving any equations. Recall that if $f(t)$ is a polynomial in a scalar variable t, and L is an operator, then $f(L)$ is obtained by substituting L for t and including a factor I (the identity operator) in the constant term of f. Thus, if $f(t) = 3t^2 + 5t + 2$, then $f(L) = 3L^2 + 5L + 2I$. The theorem as stated below requires that the eigenvalues be known. In most of our applications this is the case, but it is not actually necessary: Dividing the characteristic polynomial $\chi(t)$ of L by the greatest common divisor of $\chi(t)$ and $\chi'(t)$ yields a polynomial that has each eigenvalue as a simple root, and the f of the theorem can then be found easily. The corollary following the theorem gives the projections into eigenspaces other than the kernel.

2.1.4. Theorem. Let $L : V \to V$ be a semisimple linear operator with nontrivial kernel, having eigenvalues $\lambda_{(1)}, \lambda_{(2)}, \ldots, \lambda_{(r)}$, where $\lambda_{(1)} = 0$ and multiple eigenvalues are listed only once. Let f be the polynomial

$$f(t) = \frac{(\lambda_{(2)} - t)(\lambda_{(3)} - t) \cdots (\lambda_{(r)} - t)}{\lambda_{(2)} \lambda_{(3)} \cdots \lambda_{(r)}} = \left(1 - \frac{t}{\lambda_{(2)}}\right) \cdots \left(1 - \frac{t}{\lambda_{(r)}}\right).$$

Then

$$P = f(L) : V \to \ker L$$

is the projection associated with Theorem 2.1.3.

Proof. First suppose V is a complex vector space. Choose any basis for V and represent L by its matrix A in this basis. Let S be a matrix such that $S^{-1}AS = \Lambda$ is diagonal. It is simple to compute $f(\Lambda)$: Just apply f to each diagonal entry. By construction, $f(\lambda_{(1)}) = f(0) = 1$ and $f(\lambda_{(i)}) = 0$ for $i \neq 1$. Therefore, $f(\Lambda)$ and $I - f(\Lambda)$ are diagonal matrices with only zeros and ones in the diagonal; the former has its ones where Λ has zero eigenvalues, the latter, where Λ has nonzero eigenvalues. Thus, $f(\Lambda)$ and $I - f(\Lambda)$ are projections onto the kernel and image of L, respectively (expressed as matrices in the diagonalizing basis). Since $f(\Lambda) = f(S^{-1}AS) = S^{-1}f(A)S$, $f(A)$ is the same projection onto the kernel, expressed in the original basis. Since the choice of basis is arbitrary, the result can be stated, independently of any basis, in the form $P = f(L)$.

If V is real and L has only real eigenvalues, the proof is the same (with $\mathbb{R}^n$ in place of $\mathbb{C}^n$). If V is real and L has complex eigenvalues, let A be a matrix for L and regard A as mapping $\mathbb{C}^n$ to $\mathbb{C}^n$. As above, $f(A)$ and $I - f(A)$ are the projections onto $\ker A$ and $\operatorname{im} A$. Since eigenvalues of the real matrix A occur in conjugate pairs, $f(t)$ is a real polynomial, so $f(A)$ is a real matrix that, regarded as a map of $\mathbb{R}^n \to \mathbb{R}^n$, projects onto the (real) $\ker A$, so $f(L)$ projects onto $\ker L$. □

2.1.5. Corollary. Let $L : V \to V$ be a semisimple operator with eigenvalues $\lambda_{(1)}, \ldots, \lambda_{(r)}$, where multiple eigenvalues are listed only once. (It is not required that zero be an eigenvalue.) Let

$$f_i(t) = \prod_{j \neq i} \left(1 - \frac{t}{\lambda_{(j)} - \lambda_{(i)}} \right).$$

Then

$$P_i = f_i(L) : V \to E_{(i)}$$

is the spectral projection of V into the eigenspace associated with $\lambda_{(i)}$.

Notes and References

There are many excellent linear algebra texts that contain the material in this section and the following three (Sections 2.1–2.4) in some form. Two of my favorites are Axler [8] and Hoffman and Kunze [60].

2.2   Splitting by Means of an Inner Product

Let V be a real or complex n-dimensional vector space, and let $\langle \, , \, \rangle$ be an inner product on V. If V is complex, we require the inner product to be Hermitian, meaning that

$$\langle v, w \rangle = \overline{\langle w, v \rangle}.$$

This implies that $\langle v, v \rangle$ is real for $v \in V$, which allows the norm to be defined:

$$\|v\| = \langle v, v \rangle^{1/2}.$$

We will show that associated with any such inner product on V there is a procedure for solving the splitting problem for any linear operator $L : V \to V$.

The first step is to define the *adjoint* of L with respect to $\langle \, , \, \rangle$, denoted by L^*. The word "adjoint" is, unfortunately, used in many different senses in mathematics, and even in linear algebra. The adjoint we have in mind is the unique linear operator $L^* : V \to V$ satisfying the condition that

$$\langle Lv, w \rangle = \langle v, L^*w \rangle$$

for all $v, w \in V$. The existence and uniqueness of such an adjoint operator (in a finite-dimensional setting) is easy to prove. First, recall that there must exist an orthonormal basis for V (because the Gram–Schmidt process can be applied to any basis to create one that is orthonormal). Using an orthonormal basis makes V isomorphic with $\mathbb{R}^n$ or $\mathbb{C}^n$ in such a way that the inner product $\langle \, , \, \rangle$ corresponds to the ordinary dot product (in the case of $\mathbb{R}^n$) or the standard Hermitian inner product (for $\mathbb{C}^n$); recall that this last is

$$z \cdot w = z_1 \overline{w}_1 + \cdots + z_n \overline{w}_n.$$

Representing L in the same basis by a matrix A, it is easy to see that the transpose $A^\dagger$ in the real case, or the conjugate transpose A^* in the complex case, serves as an adjoint, and no other matrix will.

The following theorem states that $\ker L^*$ serves as a complement to $\operatorname{im} L$. This theorem is sometimes referred to as (or included as part of) the "Fredholm alternative theorem," although that name is often reserved for related results in infinite-dimensional settings (Hilbert and Banach spaces) that are much deeper:

2.2.1. Theorem. Let $L : V \to V$ be a linear operator on a finite-dimensional real or complex vector space with an inner product (Hermitian if V is complex). Then

$$V = \operatorname{im} L \oplus_\perp \ker L^*.$$

The symbol $\oplus_\perp$ means that $V = \operatorname{im} L \oplus \ker L^*$ and that in addition, if $v \in \operatorname{im} L$ and $w \in \ker L^*$, then $v \perp w$, that is, $\langle v, w \rangle = 0$.

Proof. The existence of a subspace W such that $V = \operatorname{im} L \oplus_\perp W$ is elementary: It simply consists of all those vectors in V that are perpendicular to $\operatorname{im} L$. The proof proceeds by showing that $W = \ker L^*$.

Supposing first that $w \in \ker L^*$, we claim $w \in W$, that is, $w \perp u$ for all $u \in \operatorname{im} L$. But $u = Lv$ for some $v \in V$, and $\langle u, w \rangle = \langle Lv, w \rangle = \langle v, L^*w \rangle = \langle v, 0 \rangle = 0$.

Next suppose $w \in W$, so that $w \perp Lv$ for all v. In particular, this will be true for $v = L^*w$, so that $0 = \langle Lv, w \rangle = \langle LL^*w, w \rangle = \langle L^*w, L^*w \rangle = \|L^*w\|^2$, implying $L^*w = 0$, so that $w \in \ker L^*$.   □

2.2.2. Remark. Different inner products on the same space V will give rise to different adjoints L^* and therefore to different splittings. This raises the question of whether every possible splitting arises from some inner product. The answer is yes: Given a splitting, choose a basis for each of the summands, and define an inner product by declaring the combined basis to be orthonormal.

2.2.3. Remark. If $L^* = L$ (L is self-adjoint), Theorem 2.2.1 yields $V = \operatorname{im} L \oplus \ker L$, as already proved in Theorem 2.1.3. (If L is self-adjoint, then it is semisimple, by the spectral theorem.) The same is true if L is normal ($LL^* = L^*L$); see Lemma 3.4.12. On the other hand, if L is semisimple but not normal, there are in general two different splittings, $V = \operatorname{im} L \oplus \ker L = \operatorname{im} L \oplus \ker L^*$.

2.2.4. Remark. If $u_1, \ldots, u_k$ is an orthonormal basis for $\ker L^*$, then the projection P onto $\ker L^*$ associated with Theorem 2.2.1 is

$$Pv = \langle v, u_1 \rangle u_1 + \cdots + \langle v, u_k \rangle u_k.$$

It is necessary to solve equations in order to find $u_1, \ldots, u_k$, but once this is done, it is not necessary to solve further equations to project an individual vector v. There does not appear to be a simple way to construct projections without solving equations at all, analogously to Theorem 2.1.4. This is one reason why the methods of Sections 2.5–2.7 have advantages when L is not semisimple: The algorithm of Section 2.6 provides the projections without solving any equations.

2.3 Nilpotent Operators

In this section and the next, we will prove the Jordan canonical form theorem and see how it provides another answer to the splitting question. The first step is to understand the special case of nilpotent operators. Everything in this section applies equally to real and complex vector spaces.

Jordan Chains

If $L : V \to V$ is a linear operator, the *generalized kernel* of L is the set of v such that $L^j v = 0$ for some nonnegative integer j. (Note that this concept does not make sense for $L : V \to W$ with $W \neq V$.) The least such j is called the *height* of v.

A linear operator $N : V \to V$ is *nilpotent* if its generalized kernel is the whole space V. If N is nilpotent, every $v \in V$ will have a height, and if V is finite-dimensional (which we henceforth assume), there will exist a maximum height k, which can also be characterized as the least integer such that $N^k = 0$. To prove the existence of k, observe that the height of a linear combination of vectors cannot be greater than the maximum height of the vectors, so k equals the maximum height occurring in a basis for V. We call k the *index of nilpotence* of N.

Let $N : V \to V$ be nilpotent. (All of the following definitions will be extended to general linear transformations in the next section.) A *lower Jordan chain* for N of *length* ℓ is an ordered sequence of ℓ nonzero vectors of the form $\{v, Nv, N^2 v, \ldots, N^{\ell-1} v\}$ with $N^\ell v = 0$ and $v \notin \operatorname{im} N$. The vector v is called the *top* of the chain, $N^{\ell-1} v$ is called the *bottom*, the height of the top vector equals the chain length ℓ, and the notions of top, bottom, and height are represented by listing the vectors from top to bottom with the map N pointing downward, as shown in Figure 2.1.

The vectors of a lower Jordan chain for N form a basis for a subspace $W \subset V$ that is invariant under N, and the matrix of N restricted to W,

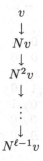

Figure 2.1. A Jordan chain.

expressed in this basis, takes the following form:

$$N|_W = \begin{bmatrix} 0 & & & & \\ 1 & 0 & & & \\ & 1 & 0 & & \\ & & & \ddots & \\ & & & 1 & 0 \end{bmatrix}. \qquad (2.3.1)$$

If the elements of the Jordan chain are listed in the opposite order, from the bottom to the top, the resulting ordered basis $\{N^{\ell-1}v, \ldots, Nv, v\}$ for W is called an *upper Jordan chain*, and the off-diagonal ones appear above the diagonal:

$$N|_W = \begin{bmatrix} 0 & 1 & & & \\ & 0 & 1 & & \\ & & 0 & 1 & \\ & & & \ddots & 1 \\ & & & & 0 \end{bmatrix}. \qquad (2.3.2)$$

It will be shown (Theorem 2.3.3) that given any nilpotent map $N : V \to V$, there exists a *chain basis* for V, that is, a basis consisting of one or more Jordan chains for N. Then V is the direct sum of invariant subspaces $W_1 \oplus \cdots \oplus W_r$ spanned by the Jordan chains, and the full matrix of N is block-diagonal with blocks of the form (2.3.1) or (2.3.2), depending on how the basis is ordered. Such a matrix is said to be in (lower or upper) *Jordan canonical form.* (Jordan canonical form for matrices that are not nilpotent is defined in the next section.)

Elimination of Entry Points

The following example illustrates a process that we call "elimination of entry points." This process is an important step in the proof of the existence of Jordan chains. Let $N : \mathbb{R}^7 \to \mathbb{R}^7$ be the operator given by

$$N = \begin{bmatrix} 0 & 0 & 0 & 0 & 0 & 0 & 0 \\ 1 & 0 & 0 & 0 & 0 & 0 & 0 \\ 0 & 1 & 0 & 0 & 0 & 0 & 0 \\ 0 & 0 & 1 & 0 & 0 & 0 & 1 \\ 0 & 0 & 0 & 1 & 0 & 0 & 0 \\ 0 & 0 & 0 & 0 & 0 & 0 & 0 \\ 0 & 0 & 0 & 0 & 0 & 1 & 0 \end{bmatrix}$$

with respect to the standard basis $\{e_1, \ldots, e_7\}$. This operator is not quite in Jordan canonical form, because of the 1 in the last column. N maps the standard basis elements as follows:

$$e_1 \mapsto e_2 \mapsto e_3 \mapsto e_4 \mapsto e_5 \mapsto 0,$$
$$e_6 \mapsto e_7 \mapsto e_4 \mapsto e_5 \mapsto 0,$$

giving two (lower) Jordan chains, $\{e_1, e_2, e_3, e_4, e_5\}$ and $\{e_6, e_7, e_4, e_5\}$. However, taken together, these chains give a linearly dependent set (because of the repetition of e_4 and e_5) and not a basis for $\mathbb{R}^7$. To remedy this, we keep the first of these chains (because it is longest), and notice that the second chain "enters" the first at the "entry point" e_4, as illustrated in the following diagram:

Noticing that the top vector e_6 of the second chain is at the same height in the diagram as e_2, we replace e_6 by $e_6 - e_2$, and observe that applying N to the modified element yields

$$(e_6 - e_2) \mapsto (e_7 - e_3) \mapsto 0.$$

Thus, we have created a new Jordan chain $\{e_6 - e_2, e_7 - e_3\}$ whose length is the same as the length of the old chain *up to its entry point*; the new

diagram is

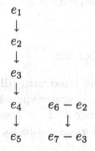

With respect to the new basis $\{f_1,\ldots,f_7\} = \{e_1, e_2, e_3, e_4, e_5, e_6 - e_2, e_7 - e_3\}$, consisting of the two Jordan chains combined, the matrix of N takes the Jordan canonical form

$$\begin{bmatrix} 0 & 0 & 0 & 0 & 0 & & \\ 1 & 0 & 0 & 0 & 0 & & \\ 0 & 1 & 0 & 0 & 0 & & \\ 0 & 0 & 1 & 0 & 0 & & \\ 0 & 0 & 0 & 1 & 0 & & \\ & & & & & 0 & 0 \\ & & & & & 1 & 0 \end{bmatrix}.$$

Existence of a Chain Basis

Two lemmas will be useful in the proof of the general theorem. If $U \subset V$ is an invariant subspace, and $v \in V$, the *height of v relative to U*, denoted by $\mathrm{ht}(v, U)$, is defined to be the smallest j such that $N^j v \in U$. (Since $0 \in U$, this includes the case that $N^j v = 0$. In this case there is no entry point.)

2.3.1. Lemma. If $N : V \to V$ is nilpotent, $U_1 \subset U_2 \subset V$ are invariant subspaces, and $v \in V$, then

$$\mathrm{ht}(v, U_2) \leq \mathrm{ht}(v, U_1).$$

Proof. Trivial. □

2.3.2. Lemma. If $N : V \to V$ is nilpotent, $U \subset V$ is an invariant subspace spanned by a single Jordan chain with top v and length ℓ, and j is an integer with $0 \leq j \leq \ell$, then an element $u \in U$ lies in the image of N^j, provided that the height of u is less than or equal to $\ell - j$.

Proof. Ordering the chain basis of U by height, we can write

$$u = a_1 N^{\ell-1} v + a_2 N^{\ell-2} v + \cdots + a_\ell v.$$

If the height of u is $\leq \ell - j$, then $a_{\ell-j+1} = \cdots = a_\ell = 0$, in which case

$$u = N^j \left(a_1 N^{\ell-j-1} v + \cdots + a_{\ell-j} v\right)$$

expresses u as an element of im N^j. In effect we have pushed the element u backwards up the Jordan chain until it has "hit the roof" and cannot be pushed farther. □

Now we turn to the main theorem:

2.3.3. Theorem (Chain basis theorem, nilpotent case). Let V be a finite-dimensional real or complex vector space and $N : V \to V$ a nilpotent operator, with index of nilpotence k. Then there exists a chain basis, that is, a basis for V consisting of Jordan chains for N.

Proof. By the definition of index of nilpotence, there must exist a vector $v \in V$ of height k. Then the sequence $v, Nv, \ldots, N^{k-1}v$ is a lower Jordan chain for N spanning an invariant subspace $W_1 \subset V$. If $W_1 = V$, we are done. If not, it is necessary to construct a second Jordan chain; if this still does not suffice, we proceed recursively. Since the second chain is slightly easier to construct than the third (and later) chains, we will describe it separately before giving the recursive step.

Choose an element $v \in V$ having the greatest possible height j relative to the subspace W_1, and recall that $j \leq k$. If $v, Nv, \ldots$ does not enter W_1 until it reaches zero, then $v, Nv, \ldots N^{j-1}v$ is our second Jordan chain; this will always be the case if $j = k$. Suppose, then, that $j < k$ and there is an "entry point" (as in the example above); the second chain is obtained by eliminating the entry point, that is, by subtracting from v the element of the first chain having the same (absolute) height and using this as the top of the new chain, which will have length j. Notice that when we have successfully found the top of the new chain, its absolute height equals its relative height.

It remains to describe the recursive step, together with the procedure for eliminating an entry point when there is more than one already-constructed chain. Suppose, then, that we have already constructed independent invariant subspaces $W_1, \ldots, W_r$ of V, and that each W_i has a basis consisting of a single Jordan chain with top w_i and length ℓ_i. Notice that ℓ_i is equal to the (absolute) height of w_i, and also to its height relative to the direct sum $W_1 \oplus \cdots \oplus W_{i-1}$ of the previously constructed subspaces. Assume that up to this point, the construction has been carried out in such a way that each ℓ_i is the maximum height of any vector in V relative to $W_1 \oplus \cdots \oplus W_{i-1}$. (We have seen that this is true at the second stage.) It follows from this assumption and Lemma 2.3.1 that $\ell_1 \geq \ell_2 \geq \cdots \geq \ell_i$; for instance, $\mathrm{ht}(w_3, W_2) \leq \mathrm{ht}(w_3, W_1) \leq \mathrm{ht}(w_2, W_1)$.

Let $W = W_1 \oplus \cdots \oplus W_r$. Choose an element v with the maximum possible height j relative to W, and notice that $j \leq \ell_r$; in fact, again using Lemma 2.3.1,

$$j = \mathrm{ht}(v, W) \leq \mathrm{ht}(v, W_1 \oplus \cdots \oplus W_{r-1}) \leq \mathrm{ht}(w_r, W_1 \oplus \cdots \oplus W_{r-1}) = \ell_r.$$

Now $N^j v = z \in W$, and we may uniquely decompose z as $z = z_1 + \cdots + z_r$ with $z_i \in W_i$. We claim that there exists $u \in W$ such that $N^j u = z$. Once this is shown, we are finished: The element $v - u$ has both absolute and relative height j, so we can take $w_{r+1} = v - u$ as the top of a new chain of length $\ell_{r+1} = j$ spanning an invariant subspace W_{r+1}. Since $j < \ell_r$, the requirements of the recursion are satisfied.

To show that u exists, it suffices to show that each z_i for $i = 1, \ldots, r$ belongs to im N^j. Lemma 2.3.2, applied to W_i, shows that this will be true if the height of z_i is $\le \ell_i - j$. If this is not true, then $N^{\ell_i - j} z_i \ne 0$. But $N^{\ell_i - j} z$ is the component of $N^{\ell_i - j} z = N^{\ell_i - j} N^j v = N^{\ell_i} v$ in W_i, and if $N^{\ell_i} v$ has a nonzero component in W_i, then it does not belong to $W_1 \oplus \cdots \oplus W_{i-1}$. This would mean that $\mathrm{ht}(v, W_1 \oplus \cdots \oplus W_{i-1})$ is greater than ℓ_i, contradicting the hypothesis of the recursion. $\qquad\square$

Nilpotent Splitting

The Jordan structure of a nilpotent map N provides another answer to the splitting problem for N. Suppose that we have obtained a chain basis for V. Then it is clear that:

1. The *bottom* vectors of the chains form a basis for ker N.

2. The chain vectors *other than the tops* form a basis for im N.

The next result follows at once:

2.3.4. Corollary. Let $N : V \to V$ be nilpotent. Then the tops of any complete set of Jordan chains for N form a basis for a complementary subspace to im N.

2.4 Canonical Forms

Let $L : V \to V$ be any linear operator on a real or complex vector space V of dimension n. We will generalize the results of the last section (for the nilpotent case) to establish the *Jordan canonical form* of L, and, in the real case, the *real canonical form*. Then we will see how these give another way to find a complement to im L.

Generalized Eigenspaces

Fix a basis for V, and let L be represented by a matrix A with respect to this basis. Regard A as a mapping $A : \mathbb{C}^n \to \mathbb{C}^n$, as in Section 2.1, even if V is a real vector space. Let the characteristic equation

$$\det(A - \lambda I) = 0$$

have distinct roots $\lambda_{(1)}, \ldots, \lambda_{(r)}$ with multiplicities $m_1, \ldots, m_r$ (so that $r \leq n$ and $m_1 + \cdots + m_r = n$). Choose any j, $1 \leq j \leq r$, and let $F_j \subset \mathbb{C}^n$ be the generalized kernel of $A - \lambda_{(j)}I$; F_j is called the *generalized eigenspace with eigenvalue* $\lambda_{(j)}$, and its elements are *generalized eigenvectors*. Note that F_j is an invariant subspace of $A : \mathbb{C}^n \to \mathbb{C}^n$, and the restriction $N_j = (A - \lambda_{(j)}I)|_{F_j}$ of $A - \lambda_{(j)}I$ to this subspace is nilpotent. By Theorem 2.1.4 there is a basis for F_j consisting of Jordan chains for N_j, with respect to which N is block diagonal with blocks as in (2.3.1) or (2.3.2). It follows that $A|_{F_j}$, expressed in this chain basis, becomes a block diagonal matrix B having $\lambda_{(j)}$ down the diagonal, together with certain off-diagonal ones coming from N_j. From what we have done so far, it is not clear what the size of B is, except that it is square and has side equal to $\dim F_j$. (Since we are in $\mathbb{C}^n$, dimension here always means complex dimension.)

2.4.1. Lemma. $\dim F_j = m_j$.

Proof. Extend the chain basis for F_j, arbitrarily, to obtain a basis for all of V. With respect to this basis,

$$A = \begin{bmatrix} B & C \\ & D \end{bmatrix}, \tag{2.4.1}$$

where B is as above. (Since the basis has been changed from the original arbitrary basis, strictly speaking we should write $S^{-1}AS$ instead of A here. Notice that the zero block below B arises from the fact that F_j is invariant.)

The characteristic polynomial of an operator is independent of the basis in which the operator is expressed, so

$$\det(A - \lambda I) = \det(B - \lambda I) \det(D - \lambda I).$$

(This follows from the definition of determinants as alternating sums of products formed using an element from each row and column: The presence of an element from C in such a product implies the presence of a factor from the zero block.) The first factor has $\lambda_{(j)}$ as its only root, with multiplicity $\dim F_j$. The lemma will be proved if we show that $\det(D - \lambda I)$ does not have $\lambda_{(j)}$ as a root. If it did, then $(D - \lambda_{(j)}I)q$ would equal zero for some $q \neq 0$, and we would have

$$\begin{bmatrix} B - \lambda_{(j)}I & C \\ & D - \lambda_{(j)}I \end{bmatrix} \begin{bmatrix} 0 \\ q \end{bmatrix} = \begin{bmatrix} Cq \\ 0 \end{bmatrix}.$$

Since

$$\begin{bmatrix} B - \lambda_{(j)}I & C \\ & D - \lambda_{(j)}I \end{bmatrix} \begin{bmatrix} Cq \\ 0 \end{bmatrix} = \begin{bmatrix} (B - \lambda_{(j)}I)Cq \\ 0 \end{bmatrix},$$

it would follow that for any k,

$$\begin{bmatrix} B - \lambda_{(j)}I & C \\ & D - \lambda_{(j)}I \end{bmatrix}^k \begin{bmatrix} 0 \\ q \end{bmatrix} = \begin{bmatrix} (B - \lambda_{(j)}I)^{k-1}Cq \\ 0 \end{bmatrix}.$$

Taking k large enough that $(B-\lambda_{(j)}I)^{k-1} = 0$ (since $B-\lambda_{(j)}I$ is nilpotent), we have that

$$\begin{bmatrix} 0 \\ q \end{bmatrix} \in \text{ generalized kernel of } \begin{bmatrix} B - \lambda_{(j)}I & C \\ & D - \lambda_{(j)}I \end{bmatrix} = A - \lambda_{(j)}I.$$

But this generalized kernel is just F_j, and elements of F_j have the form $(p, 0)$ (in the basis we are using). So $(0, q)$ can belong to F_j only if $q = 0$, which contradicts the definition of q.   □

The number $m_j = \dim F_j$ is called the *algebraic multiplicity* of the eigenvalue $\lambda_{(j)}$, in contrast to its *geometric multiplicity*, which is the (complex) dimension $n_j = \dim E_j$ of the (true) eigenspace E_j (in $\mathbb{C}^n$) of $\lambda_{(j)}$. Since $\dim E_j \geq 1$ and $E_j \subset F_j$, it is clear that

$$1 \leq n_j \leq m_j. \tag{2.4.2}$$

The next theorem shows that

$$m_1 + \cdots + m_r = n. \tag{2.4.3}$$

Define a *Jordan chain* for A in $\mathbb{C}^n$ to be any Jordan chain (in the sense of Section 2.3) for the (nilpotent) restriction of $A - \lambda_{(j)}I$ to $F_j \subset \mathbb{C}^n$, for any j.

2.4.2. Theorem (Chain basis theorem, general case). There exists a basis for $\mathbb{C}^n$ consisting of Jordan chains of A.

Proof. If we can show that

$$\mathbb{C}^n = F_1 \oplus \cdots \oplus F_r, \tag{2.4.4}$$

we will be finished: The bases of Jordan chains for each F_j, taken together, form the desired basis for $\mathbb{C}^n$. From Lemma 2.4.1 and the fundamental theorem of algebra, we have

$$n = \dim F_1 + \cdots + \dim F_r,$$

so the only way for (2.4.4) to fail would be for the sum of the subspaces not to be direct. In this case, let

$$U = F_1 + \cdots + F_r,$$

and note that $\dim U = k < n$, and U is invariant under A. Now the proof of Lemma 2.4.1 can be applied to $A|_U$. The eigenvalues of $A|_U$ are $\lambda_{(1)}, \ldots, \lambda_{(r)}$, since each eigenvalue of $A|_U$ must be an eigenvalue of A, and every eigenvector of A belongs to one of the F_j and hence to U. Let the multiplicity of each eigenvalue $\lambda_{(j)}$ (for $A|_U$) be k_j, and let $\widehat{F}_j$ be the generalized eigenspace (in U). On the one hand, $k_1 + \cdots k_r = k < n$, but on the other hand, $F_j \subset \widehat{F}_j$, so that $k_j \geq m_j$ and $k_1 + \cdots + k_r \geq n$. This contradiction establishes that (2.4.4) holds.   □

Jordan Canonical Form

Let S be a matrix whose columns form an (upper) chain basis for C^n for the matrix A. Then $S^{-1}AS$ takes a particularly simple form, called *(upper) Jordan canonical form*, which is defined as follows (but is best understood from the example following the definition).

2.4.3. Definition. A matrix A is in *upper Jordan canonical form* if it has two "nested" block-diagonal structures, as follows:

1. A is block-diagonal with blocks (on the diagonal) called *large blocks*. Each large block (considered as a square matrix in its own right) has only one eigenvalue, and distinct large blocks have distinct eigenvalues. All entries of A outside the large blocks are zero.

2. Each large block is itself block-diagonal, with one or more blocks (on the diagonal) called *small blocks* or simply *Jordan blocks*. Each diagonal entry of a small block is equal to the eigenvalue of its large block; each superdiagonal entry of a small block is equal to 1; and all other entries of a small block, and all entries of a large block outside of its small blocks, are zero. (In the case that the small block is 1×1, its only entry is the eigenvalue.)

If the off-diagonal ones appear in the subdiagonal rather than the superdiagonal, the matrix is in *lower Jordan canonical form*.

An example of a matrix in upper Jordan canonical form is

$$\begin{bmatrix} 5 & 1 & 0 & 0 & & & & & \\ 0 & 5 & 1 & 0 & & & & & \\ 0 & 0 & 5 & 1 & & & & & \\ 0 & 0 & 0 & 5 & & & & & \\ & & & & 5 & 1 & & & \\ & & & & 0 & 5 & & & \\ & & & & & & 2 & 1 & \\ & & & & & & 0 & 2 & \\ & & & & & & & & 2 \end{bmatrix}. \tag{2.4.5}$$

There are two large blocks, a 6×6 block with eigenvalue 5 and a 3×3 block with eigenvalue 2; there are four small blocks, of sizes 4×4, 2×2, 2×2, and 1×1.

It is clear from the proof of Theorem 2.4.2 that when a matrix A is put into Jordan canonical form, the size of the large block having a given eigenvalue equals the algebraic multiplicity of that eigenvalue, and the number of small blocks within the large block equals the geometric multiplicity of the eigenvalue.

It remains to transfer the result of Theorem 2.4.2 from C^n back to V. (Remember that we began this discussion with a linear transformation

$L : V \to V$.) If V is real but L has complex eigenvalues, we cannot expect that L will have a chain basis in V.

2.4.4. Theorem. If V is a complex vector space, or if V is a real vector space and L has only real eigenvalues, then there is a basis for V in terms of which the matrix of L takes Jordan canonical form.

Proof. If V is complex, then the initial choice of a basis expressing L as a matrix A provides an isomorphism between V and $\mathbb{C}^n$. Then the basis for $\mathbb{C}^n$ found in Theorem 2.4.2 corresponds to a basis for V. If V is real and L has only real eigenvalues, then the arguments leading to Theorem 2.4.2 can be carried out without ever introducing $\mathbb{C}^n$. Instead, choosing a basis for V leads to a real matrix A for L, and also an isomorphism between V and $\mathbb{R}^n$. The spaces F_j can be taken as real subspaces of $\mathbb{R}^n$, and everything proceeds as before, so Theorem 2.4.2 gives a basis for $\mathbb{R}^n$ that corresponds by isomorphism to a basis for V. $\qquad\square$

Real Canonical Form

If V is a real vector space but L has complex eigenvalues, $\mathbb{C}^n$ has a basis with respect to which A takes Jordan canonical form, but some vectors in that basis are not real and do not correspond to vectors in V. In this case we proceed as we did in Section 2.1 under similar circumstances. First, notice that if A is a real matrix, $\lambda \in \mathbb{C}$, $z \in \mathbb{C}^n$, and $(A - \lambda I)^j z = 0$, then $(A - \overline{\lambda} I)^j \overline{z} = 0$. It follows that if $\lambda = \alpha + i\beta$, $\beta \neq 0$, is an eigenvalue of A and $z_1, \ldots, z_m$ is a chain basis for the generalized eigenspace associated with λ, then $\overline{z}_1, \ldots, \overline{z}_m$ is a chain basis for the generalized eigenspace for the eigenvalue $\overline{\lambda} = \alpha - i\beta$. (Remember that the vectors $z_1, \ldots, z_m$ may comprise several Jordan chains; this is the situation whenever a large block contains several small blocks.) Thus, any complex Jordan canonical form for a real matrix A has a special structure: Each large block associated with a (strictly) complex eigenvalue λ corresponds exactly, both in size and in the structure of its decomposition into small blocks, with the large block for the conjugate eigenvalue $\overline{\lambda}$. We now focus attention on a single Jordan chain among the vectors $z_1, \ldots, z_m$, and for convenience assume that this chain is $z_1, \ldots, z_k$. Let $z_j = u_j + iv_j$, with u_j and v_j in $\mathbb{R}^n$. The subspace of $\mathbb{C}^n$ spanned by $z_1, \ldots, z_k, \overline{z}_1, \ldots, \overline{z}_k$ is invariant under A and contains the real vectors $u_1, v_1, \ldots, u_k, v_k$. We claim that the span of these vectors over $\mathbb{R}$ (that is, linear combinations of these vectors with real coefficients) is an invariant subspace of $\mathbb{R}^n$ under A. In verifying this, we will also determine the matrix of A, restricted to this subspace and expressed in this basis.

First, since $z_1, \ldots, z_k$ is an upper Jordan chain, z_1 and $\overline{z}_1$ are eigenvectors. Therefore, the results of Section 2.1 apply, and the (real) space spanned by u_1 and v_1 is invariant, with A (on this subspace) taking the form (2.1.2). Next, $Az_2 = \lambda z_2 + z_1$. (This can be read off from the Jordan canonical form of A.) Therefore, $A(u_2 + iv_2) = (\alpha + i\beta)(u_2 + iv_2) + (u_1 + iv_1)$.

Separating real and imaginary parts gives

$$Au_2 = \alpha u_2 - \beta v_2 + u_1,$$
$$Av_2 = \beta u_2 + \alpha v_2 + v_1.$$

Therefore, the (real) space U spanned by u_1, v_1, u_2, v_2 is invariant. The restriction of A to U has the following matrix with respect to the ordered basis $\{v_1, u_1, v_2, u_2\}$:

$$\begin{bmatrix} \alpha & -\beta & 1 & 0 \\ \beta & \alpha & 0 & 1 \\ 0 & 0 & \alpha & -\beta \\ 0 & 0 & \beta & \alpha \end{bmatrix}. \qquad (2.4.6)$$

Continuing in this way, we add two vectors at a time until we have arrived at the full basis $\{v_1, u_1, \ldots, v_k, u_k\}$ determined by the conjugate Jordan chains $z_1, \ldots, z_k, \overline{z}_1, \ldots, \overline{z}_k$, and then proceed in the same manner with other conjugate pairs of Jordan chains. Remember that there may also be real eigenvalues of A, and that for these the generalized eigenvectors may be taken to be real without more ado. Thus, a full basis for $\mathbb{R}^n$ is constructed, which corresponds under isomorphism to a basis for V. The matrix of L with respect to this basis will be in *real canonical form*, defined as follows:

2.4.5. Definition. A real matrix A is in *(upper) real canonical form* if it has two nested block-diagonal structures, as follows:

1. It is block-diagonal with *large blocks*, each large block (considered as a matrix in its own right) having either one real eigenvalue or a pair of complex conjugate eigenvalues. Distinct large blocks have different eigenvalues. All entries of A outside the large blocks are zero.

2. Each large block is block-diagonal with *small blocks* having one of the following types. All entries of a large block outside of its small blocks are zero.

   a. For a large block having a single real eigenvalue, its small blocks are ordinary Jordan blocks, having the eigenvalue on the diagonal, 1 in the superdiagonal, and zero elsewhere.

   b. For a large block having a conjugate pair $\alpha \pm i\beta$ of eigenvalues, its small blocks are even-dimensional. Divide the entire small block into 2×2 cells, and define the "diagonal" and "superdiagonal" as usual except that they consist of cells rather than single entries. Then each small block has matrices of the form (2.1.2) in the diagonal cells, identity matrices in the superdiagonal cells, and zero elsewhere.

In the case of a semisimple matrix, *real canonical form* coincides with *real semisimple canonical form* (Definition 2.1.2).

Solution of the Splitting Problem

The canonical form (either real or Jordan) of a matrix can be viewed as a block-diagonal matrix with only two blocks, one nilpotent and one nonsingular:

$$A = \begin{bmatrix} B & 0 \\ 0 & C \end{bmatrix}. \tag{2.4.7}$$

Here B is simply the large block in the canonical form associated with the eigenvalue zero, and C comprises the remaining large blocks. If $L : V \to V$ and a basis for V has been found according to which the matrix of V takes this form, then V is decomposed into a direct sum of subspaces, $V = U \oplus W$, with L acting on U by B and on W by C. Since C is invertible (because it does not have zero as an eigenvalue), W is contained in im L. By Section 2.3, U is the direct sum of im B and a complement spanned by the tops of the Jordan chains for B. But U, being the generalized eigenspace of L with eigenvalue zero, is simply the generalized kernel of L. Thus, we have proved the following splitting theorem:

2.4.6. Corollary. If $L : V \to V$ is a linear operator on a real or complex vector space, the tops of the Jordan chains with eigenvalue zero span a complement of im L.

The Semisimple and Nilpotent Parts of an Operator

It might seem natural to call B in (2.4.7) the "nilpotent part" of A, and C the "invertible part." But the phrase "nilpotent part" is usually used in a different sense. Suppose that A is any matrix in Jordan canonical form. We can write

$$A = A_s + A_n, \tag{2.4.8}$$

where A_s (the "semisimple part" of A) is the diagonal matrix having the same diagonal as A, and A_n (the "nilpotent part") is the matrix whose only nonzero elements are the off-diagonal ones occurring in A. If A is in real canonical form, a similar decomposition is possible, with A_s containing the 1×1 and 2×2 blocks in the diagonal of A, and A_n the off-diagonal ones and 2×2 identity matrices. In either case, it is clear that A_s is semisimple and A_n is nilpotent, and it can be checked that A_s and A_n commute. By changing basis to Jordan form, splitting into $A_s + A_n$, and changing back to the original bases, any matrix can be expressed as the sum of a semisimple and a nilpotent matrix that commute, and therefore the same is true of any linear operator. What is not so obvious (but is occasionally useful) is that the decomposition is unique. Since a detailed proof takes us away from the subject at hand, the proof below is a bit sketchy. Additional details can be found in any advanced linear algebra book, usually in connection with the "primary decomposition theorem." (Many books treat this theorem *before* the Jordan form theorem, and therefore have to work rather hard. Since we derive it from the Jordan form theorem, our proof is easier, but still surprisingly nontrivial.)

2.4.7. Theorem. If $L : V \to V$ is any linear operator on a finite-dimensional real or complex vector space, there is a unique decomposition

$$L = L_s + L_n$$

such that L_s is semisimple, L_n is nilpotent, and $L_sL_n = L_nL_s$. The operators L_s and L_n are called the *semisimple part* and *nilpotent part* of L, respectively.

Proof. Represent L by a matrix A in Jordan canonical form and decompose A as in (2.4.8); this establishes existence. Then suppose there is another decomposition $A = A'_s + A'_n$ into commuting semisimple and nilpotent matrices. Then

$$A_s - A'_s = A'_n - A_n,$$

and we must show that both sides are zero. This will follow if we show that the left-hand side is semisimple and the right-hand side is nilpotent, because the only semisimple matrix that is also nilpotent is the zero matrix.

The essential point is that A_s, the semisimple part obtained from the Jordan form, can be expressed as a polynomial in A:

$$A_s = P(A).$$

(There is no "universal" polynomial P that accomplishes this, but there is a different P for each A.) Assuming this for the moment, it follows from the fact that A'_s and A'_n commute with each other that they also commute with A, hence with $P(A) = A_s$, and hence also with A_n. Two commuting semisimple matrices are *simultaneously diagonalizable*, and therefore $A_s - A'_s$ is semisimple. A suitably high power of $A'_n - A_n$ will vanish because each term in its binomial expansion will contain a high power of either A'_n or A_n (after the factors are commuted to bring like factors together); therefore $A'_n - A_n$ is nilpotent, and we are finished.

It remains to see that there is a polynomial such that $A_s = P(A)$. Remember that A is in Jordan form. If $\lambda_{(1)}, \ldots, \lambda_{(r)}$ are the eigenvalues of A, each $A - \lambda_{(j)}I$ will have a nilpotent block (which may be a "large" block, i.e., one composed of several small blocks), and some power $P_j(A) = (A - \lambda_{(j)}I)^{k_j}$, which is a polynomial in A, will have a zero block in that position. There exist polynomials $Q_j(t)$ such that

$$P_1(t)Q_1(t) + \cdots + P_r(t)Q_r(t) = 1.$$

Then $P_1(A)Q_1(A) + \cdots + P_r(A)Q_r(A) = I$, and each $P_j(A)Q_j(A)$ is the projection onto the jth generalized eigenspace of A. (This is a generalization of Theorem 2.1.4, but unfortunately not one that is useful in constructing the projections needed for Remark 2.2.4.) It follows that

$$A_s = \lambda_{(1)}P_1(A)Q_1(A) + \cdots + \lambda_{(r)}P_r(A)Q_r(A) = P(A).$$

□

The proof of Theorem 2.4.7 provides one method for achieving the decomposition of L into its semisimple and nilpotent parts: Put L into Jordan form and separate the diagonal and off-diagonal portions. But from an algorithmic point of view, computation of the Jordan form of a matrix is difficult, and there are ways to obtain the splitting into semisimple and nilpotent part without first finding the Jordan form. This topic is explored briefly in Appendix E.

2.5   * An Introduction to sl(2) Representation Theory

The matrices

$$x = \begin{bmatrix} 0 & 1 \\ 0 & 0 \end{bmatrix}, \qquad y = \begin{bmatrix} 0 & 0 \\ 1 & 0 \end{bmatrix}, \qquad z = \begin{bmatrix} 1 & 0 \\ 0 & -1 \end{bmatrix} \qquad (2.5.1)$$

play a special role in the starred sections of this book. The vector space spanned by these matrices (over $\mathbb{R}$ or over $\mathbb{C}$) is the space of 2×2 (real or complex) matrices with trace zero, and is called the 2×2 *special linear Lie algebra*, denoted by sl(2). (When it is necessary to distinguish between the real and complex cases, we write sl(2,$\mathbb{R}$) or sl(2,$\mathbb{C}$).) The name *Lie algebra* expresses the fact that this space of matrices is closed under the operation of *commutator bracket*, defined as

$$[P, Q] = PQ - QP. \qquad (2.5.2)$$

In fact, x, y, and z satisfy the relations

$$[x, y] = z, \qquad [z, x] = 2x, \qquad [z, y] = -2y, \qquad (2.5.3)$$

which imply the closure of sl(2) under the bracket. (Lie algebras in general will be defined in Section 3.6.)

It frequently happens that a finite-dimensional vector space V is equipped with three operators, X, Y, and Z, satisfying the following properties:

1. $[X, Y] = Z, \quad [Z, X] = 2X, \quad [Z, Y] = -2Y.$

2. Z is diagonalizable (over $\mathbb{R}$ if V is a real vector space, over $\mathbb{C}$ if V is complex).

In such a case, we will say that $\{X, Y, Z\}$ form an *sl(2) triad*, or simply a *triad*. The second hypothesis here (that Z is diagonalizable) is actually redundant, and can be proved from the bracket relations, but the proof is difficult and (for our purposes) unnecessary since in our applications it will be clear that Z is diagonalizable. If $\{X, Y, Z\}$ is an sl(2) triad, the vector space spanned by these operators is closed under commutator bracket and is isomorphic to sl(2). (There is one exceptional case, $X = Y = Z = 0$.) An sl(2) triad is equivalent to what is usually called an *sl(2) representation*, although the technical definitions are different.

> **2.5.1. Remark.** A *representation* is any homomorphism of a group or algebra into a group or algebra of linear operators on a vector space; the vector space is called the *representation space*, and the group or algebra is said to *act on* the space. As a homomorphism, the mapping must preserve whatever group or algebra operations are present. Thus, an sl(2) representation is a mapping of sl(2), the Lie algebra of 2×2 matrices with zero trace, into a space of operators (or matrices) acting on some other vector space; the mapping must preserve the operations of addition, multiplication by scalars, and commutator bracket. In particular, the mapping will send x, y, and z to certain operators (or matrices) X, Y, and Z. Rather than focus on the homomorphism, we prefer to deal with X, Y, and Z themselves, and call them an *sl(2) triad*.

Special terminology is used for the eigenvalues and eigenvectors of Z: Eigenvalues are called *weights*, eigenvectors are called *weight vectors*, and eigenspaces, *weight spaces*; if v is a weight vector, its weight will be denoted by $\mathrm{wt}(v)$.

Given a vector space V and a triad $\{X, Y, Z\}$, there exist two splittings that provide complements to the images of X and Y, namely,

$$V = \operatorname{im} X \oplus \ker Y = \operatorname{im} Y \oplus \ker X. \tag{2.5.4}$$

The existence of these splittings is established in this section, along with much additional information about the structure of the operators X, Y, and Z. Section 2.6 provides algorithms for the computation of the projections associated with (2.5.4). Section 2.7 discusses various ways in which sl(2) triads arise or can be constructed.

The fundamental fact about sl(2) triads is that there exists a special basis (called a *chain-weight basis*) for V in which X and Y, which are nilpotent, are in modified Jordan form, and Z is diagonal. (A matrix is in *modified Jordan canonical form* if it is in Jordan form except that the nonzero off-diagonal elements are not necessarily ones.) The chain-weight basis is not unique, and there are several possible forms for X and Y. The version given in the following theorem places Y in "true" Jordan form, so that the only "modifications" occur in X.

2.5.2. Theorem. Let $\{X, Y, Z\}$ be an sl(2) triad of operators on a real or complex n-dimensional vector space V. Then:

1. X and Y are nilpotent.

2. The restriction of Z to $\ker X$ is diagonalizable, with eigenvalues (weights) that are nonnegative integers.

3. Any basis $v_1, \ldots, v_k$ of $\ker X$ consisting of weight vectors (eigenvectors of Z) can be used as a set of Jordan chain tops for Y. More precisely, setting

$$\ell_j = 1 + \text{wt}(v_j), \qquad (2.5.5)$$

the vectors

$$Y^i v_j \qquad \text{for} \qquad i = 0, \ldots, \ell_j - 1 \qquad (2.5.6)$$

form a basis for V. Such a basis will be called a *chain-weight basis*. The number $\text{wt}(v_j)$ is called the *top weight* of the jth chain, and ℓ_j is the *length* of the chain. The matrix of Y with respect to the basis (2.5.6) is in Jordan form (for nilpotent matrices).

4. Each of the chain-weight basis vectors (2.5.6) is itself a weight vector, with

$$\text{wt}(Y^i v_j) = \text{wt}(v_j) - 2i = \ell_j - 2i - 1. \qquad (2.5.7)$$

As i ranges over $0, \ldots, \ell_j - 1$, these weights form a sequence, called a *string of weights*, of even or odd integers, symmetrically situated around zero. The matrix of Z with respect to the basis $Y^i v_j$ is diagonal, with these strings of weights occurring in the diagonal in blocks corresponding to the Jordan blocks of Y.

5. For $i = 1, \ldots, \ell_j - 1$, define the *pressure* on $Y^i v_j$ to be

$$p_{ij} = \sum_{m=0}^{i-1} \text{wt}(Y^m v_j) = i(\ell_j - i). \qquad (2.5.8)$$

Then

$$X(Y^i v_j) = p_{ij} Y^{i-1} v_j \qquad (2.5.9)$$

for $i = 1, \ldots, \ell_j - 1$. (Recall that $X v_j = 0$ by definition. If we define $p_{0j} = 0$, (2.5.9) can be said to hold when $i = 0$, even though $Y^{-1} v_j$ does not exist. Another way to state (2.5.9) is that XY is semisimple and the p_{ij} are its eigenvalues.) The matrix of X with respect to the basis $Y^i v_j$ is in modified Jordan form with the pressures in the off-diagonal positions (see examples below).

6. The kernel of X is the span of the chain tops, and its image is the span of the chain vectors other than the tops. The kernel of Y is the span of the chain bottoms, and its image is the span of the chain vectors other than the bottoms. The splittings (2.5.4) follow.

Before proving the theorem it is helpful to look at some examples. The following tables give the height, weight, and pressure for a chain of length 5 or 6 with chain top v:

	Height	Weight	Pressure
v	5	4	0
Yv	4	2	4
Y^2v	3	0	6
Y^3v	2	-2	6
Y^4v	1	-4	4

	Height	Weight	Pressure
v	6	5	0
Yv	5	3	5
Y^2v	4	1	8
Y^3v	3	-1	9
Y^4v	2	-3	8
Y^5v	1	-5	5

It is easy to fill in the numbers in such a table, knowing the length of the chain. The height column can be filled in from the bottom up; next, the top entry in the weight column is one less than the height to its left, and from there the column descends by 2. The pressure column begins with 0, and each succeeding pressure entry is the sum of the pressure above it and the weight to the left of that, or equivalently, the sum of the weights above it. (The word "pressure" is a play on words suggested by picturing the chain vectors as a stack of bricks, the pressure on each brick being the sum of the weights of the bricks above it. There are, however, bricks with negative weight.) From such a table, it is easy to write down the matrices of X, Y, and Z with respect to a chain-weight basis. The simplest examples are those in which V contains a single chain. (In the language of representation theory of sl(2), these are called *irreducible representations*.) For instance, if there is a single chain of length $\ell = 5$, and if the chain-weight basis is ordered v, Yv, Y^2v, Y^3v, Y^4v, then (from the first table above) the matrices are

$$X = \begin{bmatrix} 0 & 4 & & & \\ & 0 & 6 & & \\ & & 0 & 6 & \\ & & & 0 & 4 \\ & & & & 0 \end{bmatrix}, \quad Y = \begin{bmatrix} 0 & & & & \\ 1 & 0 & & & \\ & 1 & 0 & & \\ & & 1 & 0 & \\ & & & 1 & 0 \end{bmatrix}, \quad (2.5.10)$$

$$Z = \begin{bmatrix} 4 & & & & \\ & 2 & & & \\ & & 0 & & \\ & & & -2 & \\ & & & & -4 \end{bmatrix}.$$

It is also possible to order the same chain-weight basis as Y^4v, Y^3v, Y^2v, Yv, v. The matrices will then be given by

$$
X = \begin{bmatrix} 0 & & & & \\ 4 & 0 & & & \\ & 6 & 0 & & \\ & & 6 & 0 & \\ & & & 4 & 0 \end{bmatrix}, \quad
Y = \begin{bmatrix} 0 & 1 & & & \\ & 0 & 1 & & \\ & & 0 & 1 & \\ & & & 0 & 1 \\ & & & & 0 \end{bmatrix}, \qquad (2.5.11)
$$

$$
Z = \begin{bmatrix} -4 & & & & \\ & -2 & & & \\ & & 0 & & \\ & & & 2 & \\ & & & & 4 \end{bmatrix}.
$$

Notice that the ordering of the weights in Z depends on whether X or Y is in "upper" (modified) Jordan form.

When V contains more than one chain, the matrices will be in block-diagonal form, with the blocks having either the form (2.5.10) or (2.5.11). For instance, suppose V has two chains, of length 3 and 2. Then, with the ordering $v_1, Yv_1, Y^2v_1, v_2, Yv_2$, the matrices will be

$$
X = \begin{bmatrix} 0 & 0 & 0 & & \\ 2 & 0 & 0 & & \\ 0 & 2 & 0 & & \\ & & & 0 & 0 \\ & & & 1 & 0 \end{bmatrix}, \quad
Y = \begin{bmatrix} 0 & 1 & 0 & & \\ 0 & 0 & 1 & & \\ 0 & 0 & 0 & & \\ & & & 0 & 1 \\ & & & 0 & 0 \end{bmatrix}, \qquad (2.5.12)
$$

$$
Z = \begin{bmatrix} -2 & & & & \\ & 0 & & & \\ & & 2 & & \\ & & & -1 & \\ & & & & 1 \end{bmatrix}.
$$

There are three natural ways to display the chains in this example, by weight, by height, or by "depth" (the height with respect to X). Each of these arrangements will be useful later in this book. By weight they appear as

Weight		
2	v_1	
1		v_2
0	Yv_1	
-1		Yv_2
-2	Y^2v_1	

(2.5.13)

By height or depth, they are

Height			Depth		
3	v_1		1	v_1	v_2
2	Yv_1	v_2	2	Yv_1	Yv_2
1	Y^2v_1	Yv_2	3	Y^2v_1	

$$(2.5.14)$$

There are two standard variations on Theorem 2.5.2. The roles of X and Y may be interchanged, so that a chain-weight basis is generated by applying X repeatedly to weight vectors spanning $\ker Y$; in this case the matrix of X will be in "true" Jordan form, and Y will contain the "pressures." Or it is possible to adopt a symmetrical arrangement, in which neither X nor Y is simply iterated to generate the basis, but "fudge factors" are included in both directions. This method leads to matrices such as the following:

$$X = \begin{bmatrix} 0 & 3 & 0 & 0 \\ 0 & 0 & 2 & 0 \\ 0 & 0 & 0 & 1 \\ 0 & 0 & 0 & 0 \end{bmatrix}, \quad Y = \begin{bmatrix} 0 & 0 & 0 & 0 \\ 1 & 0 & 0 & 0 \\ 0 & 2 & 0 & 0 \\ 0 & 0 & 3 & 0 \end{bmatrix}, \quad Z = \begin{bmatrix} 3 & & & \\ & 1 & & \\ & & -1 & \\ & & & -3 \end{bmatrix}.$$

It could be argued that the symmetrical form is preferable, because the mathematical roles of X and Y are completely symmetrical. We choose to use the form of Theorem 2.5.2, leading to matrices like (2.5.10) or (2.5.11). Most applications in this book have been arranged so that Y is of greater applied significance than X, and it is then convenient to have Y in true Jordan form. Notice that the three variant forms of Theorem 2.5.2 lead to six possible forms for the matrices of $\{X, Y, Z\}$, because for each form of the theorem there are two natural orderings of the chain-weight basis vectors.

It is time to turn to the proof of the theorem:

Proof. Let v be any weight vector, with $\mathrm{wt}(v) = w$, so that $Zv = wv$. Then $2Xv = [Z, X]v = Z(Xv) - XZv = Z(Xv) - X(wv) = Z(Xv) - w(Xv)$, so $Z(Xv) = 2(Xv) + w(Xv) = (w + 2)(Xv)$. Therefore, Xv is either zero, or is a weight vector with

$$\mathrm{wt}(Xv) = \mathrm{wt}(v) + 2. \tag{2.5.15}$$

Then the sequence $v, Xv, X^2v, \ldots$ must eventually reach zero. (If not, there would exist an infinite set of weight vectors with different weights. But eigenvectors with different eigenvalues are linearly independent, and V is finite-dimensional.) Since Z is diagonalizable, every vector in V can be written as a linear combination of weight vectors, and is therefore annihilated by a high enough power of X. Therefore, X is nilpotent. The argument for Y is similar, with

$$\mathrm{wt}(Yv) = \mathrm{wt}(v) - 2. \tag{2.5.16}$$

Now suppose that v is any vector in $\ker X$. Then $0 = 2Xv = [Z, X]v = Z(Xv) - X(Zv) = 0 - X(Zv)$, so $Zv \in \ker X$. In other words, Z maps ker X into itself, so ker X is an invariant subspace of V under Z. But Z is diagonalizable on V, so its invariant subspaces are spanned by eigenvectors (weight vectors). Therefore, ker X is spanned by weight vectors. That is, Z restricted to ker X is diagonalizable. The fact that the weights for ker X are nonnegative integers will appear shortly.

Since Z is diagonalizable on ker X, there is a (nonunique) basis $v_1, \ldots, v_k$ for ker X consisting of weight vectors. The strategy for the remainder of the proof is as follows. Let u denote any weight vector in ker X, and let ℓ be its height under Y. Let U be the subspace of V spanned by the linearly independent vectors $u, Yu, \ldots, Y^{\ell-1}u$. We claim that U is invariant under X, Y, and Z, and compute the action of these operators on U in terms of this chain basis. This will establish the basic relationships between weights, heights, and pressures in a single chain, and prove that the weights of a chain are integers forming a string centered around zero. Then we show that the chains headed by $v_1, \ldots, v_k$ are independent. The remaining claims of the theorem are then obvious.

Clearly, U is invariant for Y, and the matrix of $Y : U \to U$ with respect to the basis $Y^i u$ is in Jordan form (upper or lower depending on the chosen ordering). Moreover, U is invariant for Z because by (2.5.16) each vector $Y^i u$ is a weight vector; the matrix of $Z : U \to U$ is diagonal, with these weights as the diagonal entries. To show that U is invariant for X we calculate X times each basis vector in turn, using $[X, Y] = Z$, or $XY = YX + Z$. Let the weight of u (the top weight of the chain) be w. Then, recursively,

$$X(u) = 0,$$
$$X(Yu) = Y(Xu) + Zu = Y(0) + wu = wu,$$
$$X\left(Y^2u\right) = YX(Yu) + Z(Yu)$$
$$= Y(wu) + (w-2)(Yu) = \{w + (w-2)\}(Yu),$$
$$X\left(Y^3u\right) = YX\left(Y^2u\right) + Z\left(Y^2u\right)$$
$$= Y\{w + (w-2)\}(Yu) + (w-4)\left(Y^2u\right)$$
$$= \{w + (w-2) + (w-4)\}\left(Y^2u\right),$$

and so on. Thus, X maps each chain vector $Y^i u$ to the vector above it, $Y^{i-1}u$, multiplied by the "pressure on $Y^i u$," that is, the sum of the weights of the vectors above it. Carrying the calculation one step beyond the bottom of the chain gives

$$X\left(Y^\ell u\right) = \{w + (w-2) + (w-4) + \cdots + (w - 2\ell + 2)\}\left(Y^{\ell-1}u\right).$$

Since $Y^\ell u = 0$ and $Y^{\ell-1}u \neq 0$, it follows that

$$0 = w + (w-2) + \cdots + (w - 2\ell + 2) = \frac{\ell}{2}\{w + (w - 2\ell + 2)\} = \ell(w - \ell + 1),$$

using the formula for the sum of a finite arithmetic series (half the number of terms times the sum of the first and last terms). Therefore, $w = \ell - 1$. This implies that the top weight is a nonnegative integer, and gives the relation between chain length and top weight. It follows from (2.5.16) that all the weights are integers, and that they are symmetrically arranged around zero.

It only remains to show that the chains with tops $v_1, \ldots, v_k$ are independent. Assume that these are ordered so that $\mathrm{wt}(v_1) \geq \cdots \geq \mathrm{wt}(v_k)$, so that their heights satisfy $\ell_1 \geq \cdots \geq \ell_k$. Following the pattern of the proof of Theorem 2.3.3, let W be the span of the chains headed by $v_1, \ldots, v_r$ for any $r = 1, \ldots, k-1$; it is enough to prove that the chain headed by v_{r+1} has no entry point into W. Suppose that it did; that is, suppose that there were an i such that $Y^i v_{r+1} \notin W$ and $0 \neq Y^{i+1} v_{r+1} \in W$. Then $XY^{i+1} v_{r+1} \in W$ because W is invariant under X, but also $XY^{i+1} v_{r+1}$ would be a multiple of $Y^i v_{r+1}$ by a suitable pressure factor. This is a contradiction.   □

Theorem 2.5.2 suggests a method for constructing a chain-weight basis, given a vector space V and a triad $\{X, Y, Z\}$, namely, find $\ker X$, find the eigenvectors of Z in $\ker X$, and iterate Y on these vectors. This is not usually the most efficient way to proceed. In most applications the weights of Z are known in advance, and very often a basis of weight vectors (a *weight basis*) is known as well. When these are known, they are of significant help in simplifying the calculation of a chain-weight basis.

2.5.3. Theorem. For each nonnegative integer w, let m_w be the multiplicity of w as a weight of Z (with $m_w = 0$ if w is not a weight), and let E_w be the eigenspace of Z with eigenvalue w. Then the number of chain tops of weight w equals $m_w - m_{w+2}$, and these chain tops may be taken to be any basis for $\ker(X : E_w \to E_{w+2})$, that is, the kernel of the restriction of X to E_w (which maps into E_{w+2}). When w is the highest odd weight or the highest even weight, this kernel is all of E_w.

Proof. Let w be the highest odd or even weight, and let $v \in E_w$. Since $Z(Xv) = (w + 2)Xv$ and $w + 2$ is not a weight, it follows that $Xv = 0$. Thus, all of E_w belongs to $\ker X$, and any basis for E_w provides part of a set of chain tops $v_1, \ldots, v_k$ as described in Theorem 2.5.2. Now suppose that w is a nonnegative integer. If it is odd, assume that it is less than the highest odd weight; if it is even, less than the highest even weight. The multiplicity of w must be at least as great as that of $w + 2$, because Y maps E_{w+2} one-to-one into E_w; if the multiplicity is greater, then $\dim E_w > \dim E_{w+2}$ and there must be new chain tops of weight w. Since these chain tops must belong to $\ker X$, they can be found by examining $\ker(X : E_w \to E_{w+2})$.   □

The importance of this theorem in practice is that it reduces the size of the systems of equations that must be solved in order to find the chain tops; it is not necessary to solve $Xv = 0$ on the entire space V, only on the relevant E_w. It is convenient to begin by making a *weight table* listing the nonnegative weights and their multiplicities. From this a *top weight list* may be constructed, showing the weights of the chain tops that must be found, and, by implication, the dimension of ker X in each weight space. We will now illustrate this with an example typical of those that will occur later in this book.

Let $V = \mathcal{F}_2^5$ be the vector space of real homogeneous quadratic polynomials in the variables x, y, z, u, v; this is a space of 15 dimensions. Let

$$\mathcal{X} = y\frac{\partial}{\partial x} + z\frac{\partial}{\partial y} + v\frac{\partial}{\partial u}, \tag{2.5.17}$$

$$\mathcal{Y} = 2x\frac{\partial}{\partial y} + 2y\frac{\partial}{\partial z} + u\frac{\partial}{\partial v},$$

$$\mathcal{Z} = -2x\frac{\partial}{\partial x} + 2z\frac{\partial}{\partial z} - u\frac{\partial}{\partial u} + v\frac{\partial}{\partial v}.$$

In Section 2.7, it will become clear that operators such as (2.5.17) form triads, but for now it is easy to check this directly: Property 1 of a triad (the commutation relations) is a simple calculation, and property 2 (diagonalizability of Z) follows from the fact that the monomials $x^i y^j z^k u^l v^m$ are eigenvectors of $\mathcal{Z}$ with weight $w = -2i + 2k - l + m$:

$$\mathcal{Z}(x^i y^j z^k u^l v^m) = (-2i + 2k - l + m)(x^i y^j z^k u^l v^m). \tag{2.5.18}$$

Thus, the monomials with $i + j + k + l + m = 2$ form a weight basis for $\mathcal{F}_2^5$.

To find the chain structure of $\mathcal{F}_2^5$ under the triad (2.5.17), begin by using (2.5.18) to compute the weights of the 15 monomials $x^i y^j z^k u^l v^m$ in $\mathcal{F}_2^5$ (missing entries being zero):

i	j	k	l	m	w	
2					-4	
1	1				-2	
1		1			0	
1			1		-3	
1				1	-1	
	2				0	
	1	1			2	(2.5.19)
	1		1		-1	
	1			1	1	
		2			4	
		1	1		1	
		1		1	3	
			2		-2	
			1	1	0	
				2	2	

Next, construct the weight table, listing only the multiplicities of nonnegative weights, since negative weights are symmetrical:

$$\begin{array}{lccccc} \text{Weight} & 4 & 3 & 2 & 1 & 0 \\ \text{Multiplicity} & 1 & 1 & 2 & 2 & 3 \end{array} \qquad (2.5.20)$$

It follows from Theorem 2.5.3 that the top weight list is

$$4 \quad 3 \quad 2 \quad 1 \quad 0. \qquad (2.5.21)$$

In the highest even and odd weights, there is no need to calculate a kernel, and the chain tops that fill these positions in the top weight list will be z^2 and zv (from (2.5.19)). The chain of top weight 2 will have its top in the kernel of X restricted to E_2, which is spanned by yz and v^2, so it is necessary to find the kernel of X only on this restricted subspace, which we can do by computing

$$X(ayz + bv^2) = az^2.$$

This is zero if $a = 0$, so v^2 can be taken as the top of the chain. Next, the chain top of weight 1 must be found in the span of yv and zu, where

$$X(ayv + bzu) = (a + b)zv.$$

This vanishes if $b = -a$, so $yv - zu$ can be selected. Finally, the chain top (which is also the bottom) of weight zero lies in the span of xz, y^2, and uv. But

$$X(axz + by^2 + cuv) = (a + 2b)yz + cv^2,$$

and we can take $a = 2, b = -1, c = 0$ to give $2xz - y^2$. Thus, the "filled" top weight list becomes

$$\begin{array}{ccccc} 4 & 3 & 2 & 1 & 0 \\ z^2 & zv & v^2 & yv - zu & 2xz - y^2. \end{array} \qquad (2.5.22)$$

The chains under these tops can be generated by iteration of y to produce the following table, arranged by height:

$$\begin{array}{llllll} 5 & z^2 \\ 4 & 4yz & zv \\ 3 & 8xz + 8y^2 & 2yv + uz & v^2 & & & (2.5.23) \\ 2 & 48xy & 4xv + 2yu & 2uv & yv - zu \\ 1 & 96x^2 & 8xu & 2u^2 & 2xv - yu & 2xz - y^2. \end{array}$$

As a check, notice that y applied to each element of the bottom row gives zero, as expected.

Notes and References

Some of the terminology in this section, including the words "triad" and "pressure," is nonstandard. The theory of sl(2) representations may be found in Fulton and Harris [46], Chapter 11; this chapter is mostly self-contained, but uses a certain Theorem 9.20 (stated in Chapter 9 but proved in Appendix C) whose proof is rather long. We have avoided the need for Theorem 9.20 by adding the assumption that Z is diagonalizable.

2.6 * Algorithms for the sl(2) Splittings

The goal of this section is to develop efficient algorithms for the computation of the projections associated with the splittings (2.5.4), given a vector space V with an sl(2) triad $\{X, Y, Z\}$. In principle, once a chain-weight basis is known, it is possible to express any vector in terms of this basis by solving a system of linear equations. Then the projections of that vector can be found by collecting the terms that lie in the span of the chain tops (for the projection into $\ker X$) or chain bottoms (for $\ker Y$). But even in the small example at the end of the last section, it is not completely easy to resolve an arbitrary polynomial in $\mathcal{F}_2^5$ into its components with respect to the basis (2.5.23) in this way; it requires solving 15 linear equations in 15 unknowns (although in this case many of these are trivial). For an algorithm that will work efficiently on larger problems using a computer algebra system, it is best to find a method that avoids the solution of equations altogether.

The Subspaces $V_{\ell h}$

The first step is to define certain subspaces of V that exist because of the results in Section 2.5. Define $V_{\ell h} \subset V$ to be the subspace spanned by the basis vectors of height h in the chains of length ℓ:

$$V_{\ell h} = \text{span}\{Y^i v_j : \ell_j = \ell, \ell_j - i = h\}. \tag{2.6.1}$$

Also define

$$V_h = \bigoplus_\ell V_{\ell h} = \text{span}\{Y^i v_j : \ell_j - i = h\}. \tag{2.6.2}$$

Thus, V_h is the subspace spanned by all of the basis vectors of height h (regardless of the length of the chain to which they belong). Notice that the elements of V_h are not the only vectors of height h; a vector v has height h if and only if $v = v_1 + \cdots + v_h$, with $v_i \in V_i$ and $v_h \neq 0$.

At first sight, these spaces seem to depend on the particular choice of a chain-weight basis (which is not unique, particularly when there are several chains of the same length), but in fact they do not:

2.6.1. Lemma. The spaces $V_{\ell h}$ do not depend on the choice of chain-weight basis.

Proof. It is clear that $V_{\ell 1}$ does not depend on the choice of chain-weight basis, because vectors in $V_{\ell 1}$ are in $\ker Y$ and have weight $-(\ell - 1)$, so

$$V_{\ell 1} = E_{1-\ell} \cap \ker Y.$$

For $h > 1$, $V_{\ell h}$ can then be characterized by

$$V_{\ell h} = X^{h-1} V_{\ell 1}.$$

$\square$

A vector $v \in V$ will be called *completely split* if it has been decomposed into a sum of the form

$$v = \sum v_{\ell h} \quad \text{with} \quad v_{\ell h} \in V_{\ell h}. \tag{2.6.3}$$

The summation extends over all chain lengths ℓ that exist in V and, for each ℓ, $h = 1, \ldots, \ell$. If v has been completely split, then the projection into $\ker X$ associated with (2.5.4) is given by

$$Pv = \sum_{\ell} v_{\ell \ell}, \tag{2.6.4}$$

and the other projections are expressed similarly.

The remainder of this section is devoted to algorithms for the complete splitting. There is a simple way to do the splitting for vectors of height one. The general case follows by "pushing down" into height one by iterating Y, doing the splitting there, and "pulling up" to the original height by iterating X.

Splitting a Vector of Height 1

The first algorithm shows how to split a vector of height one when the vector is expressed in a known weight basis.

2.6.2. Algorithm (Height-one sorting algorithm). Suppose that u is a vector of height one and that

$$u = \sum_{i=1}^{n} a_i e_i,$$

where $n = \dim V$ and $e_1, \ldots, e_n$ is a basis of weight vectors of Z. Then for any ℓ, the projection of u into $V_{\ell 1}$ is the sum of the terms $a_i e_i$ for which $\text{wt}(e_i) = 1 - \ell$. In other words, the terms $a_i e_i$ may be "sorted" into the spaces $V_{\ell 1}$ according to the rule

$$\ell = 1 - \text{wt}(e_i).$$

Proof. V can be written as a direct sum of weight spaces,

$$V = \oplus E_w,$$

and the terms $a_i e_i$ may be sorted into these spaces (several terms, perhaps, going into the same weight space), so that $u = \sum u_w$, $u_w \in E_w$. Since u is of height one, $\sum Y u_w = Y u = 0$; since each $Y u_w$ belongs to a different E_{w-2}, there can be no cancellation among the terms, so each $Y u_w$ is 0. Introduce an arbitrary chain-weight basis in V. Then it is clear that each nonzero u_w must belong to the span of the chain bottoms of weight w; in particular, it follows that $u_w = 0$ for $w > 0$. The chains whose bottoms have weight w have top weight $-w$ and length $\ell = 1 - w$. Therefore, each u_w belongs to $V_{\ell 1}$ with $\ell = 1 - w$. If the terms $a_i e_i$ are sorted directly into these spaces according to the rule $\ell = 1 - \mathrm{wt}(e_i)$, the same sums u_w will result. □

If a weight basis $e_1, \ldots, e_n$ is not known (or if u is not presented in terms of such a basis, so that to express u in the basis would require the solution of equations), but the weights of Z on V are known, then Corollary 2.1.5 can be used to construct projections of V onto the eigenspaces of Z. That is, for each weight w a polynomial $f_w(t)$ may be constructed such that $f_w(Z)$ is the projection onto the weight space of weight w. Since $w = 1 - \ell$ for vectors of height one, we have the following algorithm, which does not require solving equations:

2.6.3. Algorithm. If $u \in V$ is of height one, its component in each $V_{\ell 1}$ will be $f_{1-\ell}(Z)u$.

> **2.6.4. Remark.** Although Algorithm 2.6.3 does not require solving equations, it is still not the most efficient algorithm for the purpose in most of our applications. For instance, in Section 3.5, we work with a triad $\{X, Y, Z\}$ of $n \times n$ matrices and another triad $\{\mathbb{X}, \mathbb{Y}, \mathbb{Z}\}$ of $n^2 \times n^2$ matrices. For the latter triad, Algorithm 2.6.3 would have us use polynomials in $\mathbb{Z}$ to find the projections, whereas it is possible to obtain these projections indirectly from Z without ever using $n^2 \times n^2$ matrices. See steps 7 and 8 in Section 3.5. Similar remarks hold in Chapter 4, where the large matrices, if it were necessary to use them, would become even larger than $n^2 \times n^2$.

The Operator $Y^\dagger$

It is convenient to introduce the operator

$$Y^\dagger : V \to V$$

defined by the equation

$$Y^\dagger v = \sum \frac{1}{h(\ell - h)} X v_{\ell h}, \qquad (2.6.5)$$

where $v = \sum v_{\ell h}$ as in (2.6.3). (It will be seen shortly that $Y^\dagger$ is actually the transpose of Y, as the notation suggests, provided that Y is written as a matrix with respect to a chain-weight basis.) Note that $Y^\dagger v$ is effectively defined only for those vectors v that have already been completely split; an important aspect of (2.6.5) is that when $Y^\dagger$ is applied to a completely split vector, the result appears in completely split form (with $Xv_{\ell h} \in V_{\ell, h+1}$). Another important fact is that $Y^\dagger$ is a partial inverse of Y:

$$YY^\dagger v = \begin{cases} v & \text{if } v \in \operatorname{im} Y, \\ 0 & \text{if } v \in \ker X, \end{cases} \qquad (2.6.6)$$

$$Y^\dagger Y v = \begin{cases} v & \text{if } v \in \operatorname{im} X, \\ 0 & \text{if } v \in \ker Y. \end{cases}$$

(In fact, $Y^\dagger Y$ and $YY^\dagger$ are the projections into $\operatorname{im} Y$ and $\operatorname{im} X$, respectively, associated with the splittings (2.5.4).) To see that (2.6.6) holds, notice that in terms of a chain-weight basis $Y^i v_j$, Y^{-1} is simply X with the pressure factor removed. That is, since $Y^i v_j$ has height $h = \ell_j - i$, the pressure on $Y^i v_j$ is $p_{ij} = \ell_j(h - \ell_j)$, and $Y^\dagger(Y^i v_j) = Y^{i-1}v_j$. Thus, the matrix of $Y^\dagger$ (with respect to a chain-weight basis) is the same as the matrix of X with the pressures replaced by ones, and this is just the transpose of the matrix of Y.

Splitting a Vector of Height 3

In order to split a vector v into its components $v_{\ell h}$ in the spaces $V_{\ell h}$, the first step is to determine its height by applying Y to v repeatedly until zero is obtained. It is convenient to motivate the general splitting algorithm (which we call the *pushdown algorithm*) by considering the case of a vector of height 3. If the height of v is 3, we claim that v can be split by successively calculating the following vectors $a, b, c, \ldots$:

$$
\begin{aligned}
a &= v, \\
b &= Ya = Yv, \\
c &= Yb = Y^2 v, \\
d &= Y^\dagger c, \\
e &= b - d, \\
f &= d + e, \\
g &= Y^\dagger f, \\
h &= a - g, \\
i &= g + h = v.
\end{aligned}
$$

This sequence begins and ends with the vector v. The essential point (which will become clear in a moment) is that from c on, each of these vectors is known in completely split form. Thus, when we reach $i = v$ at the end of the calculations, v will be known in completely split form.

To see what is going on "behind the scenes" in these calculations, notice that since v is of height 3, it must be the sum $v = v_1 + v_2 + v_3$ with $v_i \in V_i$ (see (2.6.2)). It follows that the vectors $a, b, c, \ldots$ have the following components of each height:

	a	b	c	d	e	f	g	h	i
V_3	v_3	0	0	0	0	0	v_3	0	v_3
V_2	v_2	Yv_3	0	Yv_3	0	Yv_3	v_2	0	v_2
V_1	v_1	Yv_2	Y^2v_3	0	Yv_2	Yv_2	0	v_1	v_1

Notice that c, e, and h have components in V_1 only, and therefore can be completely split by Algorithm 2.6.2 or 2.6.3. Once any vector has been split, $Y^\dagger$ can be applied to it (and produces a split vector). Therefore, from column c on, each vector in the table is completely split.

Splitting a Vector of Any Height

The case $k = 3$ should be sufficient to make clear the algorithm in the general case, which can be stated as follows:

2.6.5. Algorithm (Pushdown algorithm). If X, Y, Z is a triad on V and $v \in V$, v may be split into its components $v_{\ell h} \in V_{\ell h}$ by the following steps:

1. Determine the height k of v by calculating $Yv, Y^2v, \ldots$ until $Y^k v = 0$. Save the results, numbered in reverse, as $q_1 = Y^{k-1}v, q_2 = Y^{k-2}v, \ldots, q_k = v$.

2. Let $r_1 = s_1 = q_1$. Since $r_1 \in V_1$, it can be split using Algorithm 2.6.2 or 2.6.3.

3. When r_j and s_j are defined and split, with j still less than k, let

$$t_{j+1} = Y^\dagger s_j,$$
$$r_{j+1} = q_{j+1} - t_{j+1},$$
$$s_{j+1} = r_{j+1} + t_{j+1}.$$

Since $r_{j+1} \in V_1$, it can be split by Algorithm 2.6.2 or 2.6.3, and the split form should be used in s_{j+1}. (Remark: There is no t_1, since this step is used first with $j = 1$.)

4. Repeat step 3 until $j = k$. Then s_k is the split expression for v.

An Example

The pushdown algorithm can be illustrated with the same example used at the end of Section 2.5. (For another explicit example, see the end of Section 3.5.) Consider the element $v = y^2 \in V = \mathcal{F}_2^5$; we will use Algorithms 2.6.5

and 2.6.2 to split y^2 into its components in the $V_{\ell h}$, and therefore into its components in $V = \operatorname{im} \mathcal{Y} \oplus \ker \mathcal{X}$. Step 1 is to apply $\mathcal{Y}$ repeatedly:

$$v = y^2 = q_3,$$
$$\mathcal{Y}v = 4xy = q_2,$$
$$\mathcal{Y}^2 v = 8x^2 = q_1,$$
$$\mathcal{Y}^3 v = 0.$$

Therefore, the height of y^2 is 3, and the results can be labeled q_3, q_2, q_1. In the following calculations the notation $(\ )_{\ell h}$ will be used to label terms whose position in the $V_{\ell h}$ splitting has been identified. First, we know that $8x^2$ belongs to $V_1 = \ker \mathcal{Y}$. Since its weight, by (2.5.18), is $w = -4$, its chain length is $c = 1 - w = 5$, and we have step 2 of the pushdown algorithm:

$$r_1 = s_1 = q_1 = \left(8x^2\right)_{51}.$$

The first iteration of step 3 goes like this, using the fact that application of $\mathcal{Y}^\dagger$ or $\mathcal{X}$ raises h by one:

$$t_2 = \frac{1}{1(5-1)} \mathcal{X}(8x^2)_{51} = (4xy)_{52},$$

$$r_2 = q_2 - t_2 = 4xy - (4xy)_{52} = 0,$$

$$s_2 = r_2 + t_2 = (4xy)_{52}.$$

In this case, $r_2 = 0$, and there is no need to apply Algorithm 2.6.2 to split it. The next iteration begins

$$t_3 = \frac{1}{2(5-2)} \mathcal{X}(4xy)_{52} = \left(\frac{2}{3}y^2 + \frac{2}{3}xy\right)_{53},$$

$$r_3 = q_3 - t_3 = y^2 - \left(\frac{2}{3}y^2 + \frac{2}{3}xz\right) = \frac{1}{3}y^2 - \frac{2}{3}xz.$$

We know that $r_3 \in V_1$, so it can be split by Algorithm 2.6.2. Using (2.5.18) it is seen that both terms have weight $w = 0$, so they belong together, with $\ell = 1 - w = 1$, and this iteration of step 3 continues:

$$r_3 = \left(\frac{1}{3}y^2 - \frac{2}{3}xz\right)_{11},$$

$$s_3 = r_3 + t_3 = \left(\frac{1}{3}y^2 - \frac{2}{3}xz\right)_{11} + \left(\frac{2}{3}y^2 + \frac{2}{3}xz\right)_{53}.$$

Now, s_3 is the same as the original $v = y^2$, only now split into its components in $V_{\ell h}$. In view of (2.6.3), the splitting of y^2 into $\operatorname{im} \mathcal{Y} \oplus \ker \mathcal{X}$

is

$$\frac{2}{3}y^2 + \frac{2}{3}xz \in \operatorname{im} \mathcal{Y},$$

$$\frac{1}{3}y^2 - \frac{2}{3}xz \in \ker \mathcal{X}.$$

As a check of the calculation of the splitting of y^2 given above by the pushdown algorithm, notice that each term identified by the notation $(\ )_{\ell h}$ in the course of that calculation can be located in the table (2.5.23) in its correct place.

Notes and References

All of the algorithms in this section derive from Sanders [99], where they are stated in a somewhat obscure notation. In particular, our notation $(\ )_{\ell h}$ replaces the use of certain tensor products whose only purpose is to encode the information contained in the numbers ℓ and h, and the definition (2.6.5) replaces certain integrals that are used because they conveniently give the correct numerical coefficients. The pushdown algorithm is called the splitting algorithm in [99].

2.7 * Obtaining sl(2) Triads

The last two sections assume that an sl(2) triad $\{X, Y, Z\}$ acting on a vector space V is given. The question of how such a triad is obtained was not addressed. This omission will now be remedied. The triads that are useful in normal form theory arise in two ways. First, a triad can be constructed "around" any given nilpotent matrix, which may take either the X or Y role in the triad. Second, these triads induce other triads acting on different spaces, by means of certain operations called Lie algebra (anti-)homomorphisms.

Embedding a Nilpotent Operator into a Triad

The simplest way to construct a triad is to use the structure of triads given by Theorem 2.5.2, illustrated in examples such as (2.5.10) and (2.5.11). The following theorem shows how to build a triad containing a given nilpotent operator N. Unfortunately, the construction given by this theorem requires a knowledge of the Jordan chain structure of N. A second construction, which does not require this knowledge, is given below in Algorithm 2.7.2. (If this second construction is used, Theorem 2.5.2 can then be used to obtain the Jordan structure of N if desired.)

2.7.1. Theorem. Let $N : V \to V$ be a nilpotent operator. Then there exists at least one sl(2) triad $\{X, Y, Z\}$ with $X = N$, and at least one triad with $Y = N$. If $A = S + N$ is the semisimple/nilpotent decomposition of A as in Theorem 2.4.7 or Algorithm E.1.2, so that N commutes with S, then there exist such triads in which X, Y, and Z commute with S.

Proof. We will construct a triad with $Y = N$. (The construction with $X = N$ is similar.) Since N is nilpotent, V has a basis consisting of upper Jordan chains for N. Let $v_1, \ldots, v_k$ be the chain tops and $\ell_1, \ldots, \ell_j$ the chain lengths. Let $Y = N$, and define $\mathrm{wt}(Y^i v_j)$ by (2.5.7), p_{ij} by (2.5.8), X by (2.5.9), and Z by $Z(Y^i v_j) = \mathrm{wt}(Y^i v_j) Y^i v_j$. Then X, Y, and Z have matrices of the form discussed in Section 2.5, and it can be checked (directly by matrix multiplication) that $\{X, Y, Z\}$ satisfies the bracket relations for an sl(2) triad. In the case that $A = S + N$, the construction is the same except that we use Jordan chains for A (which are also Jordan chains for N). When expressed in the resulting chain basis, S will be diagonal and will have equal eigenvalues in each block of the Jordan structure. Since X, Y, and Z all respect the same block structure, they all commute with S. $\square$

In practice, the proof of Theorem 2.7.1 translates into the following procedure: Put N (or A) into upper Jordan form and set $Y = N$. Then construct matrices X and Z as in (2.5.10) or (2.5.11): For each Jordan block of Y, fill in the diagonal elements of Z with appropriate weights forming a string of integers separated by 2 and symmetrical around zero (either all odd or all even). For each string of weights, calculate the corresponding pressures and enter them in the correct Jordan block of X. Most often, we will wish to leave these matrices in Jordan form. If it is desired to have X, Y, Z expressed in some other basis than the Jordan chain basis for N (for instance, in the original basis in which N was given), the matrices can be changed to this basis by applying a similarity transformation.

An Alternative Embedding Algorithm

There are times when it is convenient to have an algorithm for embedding N into a triad that does not require computing the Jordan form of N. The following algorithm meets this requirement; it requires solving systems of linear equations of size $n^2 \times n^2$ (if N is $n \times n$), but for large n this is less work than finding the Jordan form. (We rely on Theorem 2.7.1 for a proof that solutions of these systems exist.) The algorithm is formulated to give $\{X, Y, Z\}$ with $X = N$, but it is easy to modify it to produce $Y = N$.

2.7.2. Algorithm. Given X, to produce Y and Z so that $\{X, Y, Z\}$ is an sl(2) triad.

1. Find any particular solution Y_p of the inhomogeneous linear equation $[[X, Y], X] = 2X$.

2. Let $Z = [X, Y_p]$ and observe that $[Z, X] = 2X$, as required for a triad.

3. Find a solution Y_0 of the inhomogeneous linear equation $[Z, Y_0] + 2Y_0 = -[Z, Y_p] - 2Y_p$ lying in the subspace consisting of Y_0 such that $[X, Y_0] = 0$. (That is, solve the first equation regarded as a system of equations on the subspace defined by the second equation. As remarked above, a solution exists because of Theorem 2.7.1.)

4. Let $Y = Y_0 + Y_p$ and observe that $[Z, Y] = -2Y$, as required for a triad.

5. Observe that $[X, Y] = [X, Y_p] + [X, Y_0] = Z + 0 = Z$, so that $[X, Y] = Z$, as required for a triad.

2.7.3. Remark. This construction does not prove that Z is diagonalizable (property 2 in our definition of triad), but it has already been mentioned that property 2 is redundant. It could of course be verified, in any particular case, that Z is diagonalizable.

2.7.4. Remark. It sometimes happens that N belongs to a particular class of matrices and it is desired to embed N in a triad belonging to that class. This will not necessarily be possible, but it is possible if the given class of matrices constitutes what is called a *reductive Lie algebra*. We will not give the technical definition of reductive, but when this term is encountered in the normal form literature it can be understood to mean a class of matrices closed under addition and commutator bracket, such that every nilpotent element can be embedded in a triad.

2.7.5. Remark. As in Theorem 2.7.1, it is often required that the triad commute with a given matrix S, given that X commutes with S. The construction given in Algorithm 2.7.2 can be carried out so as to satisfy this requirement. Let W be the vector space of matrices that commute with S. Given that $X \in W$, the operator $Y \mapsto [[X, Y], X]$ maps W to itself, so the equation to be solved in step 1 of the algorithm can be regarded as an equation in W rather than in the space of all $n \times n$ matrices. Then Y_p will lie in W, and so will Z in step 2. Then in step 3 we work in the space of matrices commuting with both X and S.

Induced Triads

Any triad $\{X, Y, Z\}$ of $n \times n$ matrices (or operators on $\mathbb{R}^n$) induces other triads operating on other vector spaces. Three such induced triads will arise in this book. These will be defined again when they are needed, but for convenience the definitions will be collected here as well.

Let $\mathrm{gl}(n)$ denote the real vector space of $n \times n$ real matrices. (If it is desired to consider complex matrices, simply replace every reference to real numbers or to $\mathbb{R}^n$ by complex numbers or $\mathbb{C}^n$ in the following.) If

$A \in \mathrm{gl}(n)$, define the *Lie operator*

$$\mathbb{L}_A : \mathrm{gl}(n) \to \mathrm{gl}(n)$$

by

$$\mathbb{L}_A B = [B, A] = BA - AB. \tag{2.7.1}$$

Let $\mathcal{F}_{**}^n$ denote the vector space of formal power series in $x = (x_1, \ldots, x_n)$ with real coefficients. Define

$$\mathcal{D}_A : \mathcal{F}_{**}^n \to \mathcal{F}_{**}^n$$

by

$$(\mathcal{D}_A f)(x) = f'(x)Ax = (Ax) \cdot \nabla f(x). \tag{2.7.2}$$

Here $f'(x)$ is the row vector of partial derivatives of f. Let $\mathcal{V}_{**}^n$ be the vector space of vector-valued formal power series in x, or equivalently, the space of n-tuples of (scalar) formal power series. Define

$$\mathsf{L}_A : \mathcal{V}_{**}^n \to \mathcal{V}_{**}^n$$

by

$$(\mathsf{L}_A v)(x) = v'(x)Ax - Av(x), \tag{2.7.3}$$

where $v'(x)$ is the $n \times n$ matrix of partial derivatives.

2.7.6. Theorem. If $\{X, Y, Z\}$ is an sl(2) triad, define

$$\mathbb{X} = \mathbb{L}_Y, \qquad\qquad \mathbb{Y} = \mathbb{L}_X, \qquad\qquad \mathbb{Z} = \mathbb{L}_Z;$$
$$\mathcal{X} = \mathcal{D}_Y, \qquad\qquad \mathcal{Y} = \mathcal{D}_X, \qquad\qquad \mathcal{Z} = \mathcal{D}_Z;$$
$$\mathsf{X} = \mathsf{L}_Y, \qquad\qquad \mathsf{Y} = \mathsf{L}_X, \qquad\qquad \mathsf{Z} = \mathsf{L}_Z.$$

Then $\{\mathbb{X}, \mathbb{Y}, \mathbb{Z}\}$, $\{\mathcal{X}, \mathcal{Y}, \mathcal{Z}\}$, and $\{\mathsf{X}, \mathsf{Y}, \mathsf{Z}\}$ are sl(2) triads.

Proof. The most important part of the proof is to check that

$$[\mathbb{L}_P, \mathbb{L}_Q] = \mathbb{L}_{[Q,P]} \tag{2.7.4}$$

(with P and Q reversed), and similarly for $\mathcal{D}$ and L. The calculations for L and $\mathcal{D}$ are routine; we present the details for L:

$$\mathsf{L}_P \mathsf{L}_Q v(x) = \mathsf{L}_P(v'(x)Qx - Qv(x))$$
$$= (v'(x)Qx - Qv(x))'Px - P(v'(x)Qx - Qv(x))$$
$$= v''(x)(Qx, Px) + v'(x)QPx - Qv'(x)Px$$
$$\quad - Pv'(x)Qx + PQv(x),$$

where $v''(x)(a, b)$ is the symmetric vector-valued bilinear form with coefficients $\partial^2 v_i / \partial x_j \partial x_k$. Interchanging P and Q, we obtain

$$\mathsf{L}_Q \mathsf{L}_P v(x) = v''(x)(Px, Qx) + v'(x)PQx - Pv'(x)Qx$$
$$\quad - Qv'(x)Px + QPv(x).$$

Subtracting these (and using the symmetry of v'') leaves $v'(x)[Q,P]x - [Q,P]v(x)$.

Equation (2.7.4), together with the fact that $\mathbb{L}_A$ is linear in A, is expressed by saying that $\mathbb{L}$ is a *Lie algebra antihomomorphism* (and similarly for $\mathcal{D}$ and L). From this it is easy to check the commutation rules of a triad:

$$[\mathbb{X}, \mathbb{Y}] = [\mathbb{L}_Y, \mathbb{L}_X] = \mathbb{L}_{[X,Y]} = \mathbb{L}_Z = \mathbb{Z},$$
$$[\mathbb{Z}, \mathbb{X}] = [\mathbb{L}_Z, \mathbb{L}_Y] = \mathbb{L}_{[Y,Z]} = -\mathbb{L}_{[Z,Y]} = -\mathbb{L}_{-2Y} = 2\mathbb{L}_Y = 2\mathbb{X},$$
$$[\mathbb{Z}, \mathbb{Y}] = [\mathbb{L}_Z, \mathbb{L}_X] = \mathbb{L}_{[X,Z]} = -\mathbb{L}_{[Z,X]} = -\mathbb{L}_{2X} = -2\mathbb{L}_X = -2\mathbb{Y}.$$

It remains to check that $\mathbb{Z}$, $\mathcal{Z}$, and Z are diagonalizable. This is done in Lemmas 3.3.1 and 4.5.2. (Remember that this is actually redundant, and is necessary for us only because we did not prove it in general. See Remark 2.5.1.) $\qquad\square$

2.7.7. Remark. Notice the reversal of X and Y in the definitions of these triads. This can be avoided by changing the definitions of $\mathbb{L}_A$, $\mathcal{D}_A$, and L_A to the negative of the definitions given here. In place of $\mathbb{L}_A$, it is customary in the theory of Lie algebras to define the *adjoint operator* ad_A:

$$\mathrm{ad}_A B = [A, B] = -\mathbb{L}_A B.$$

Then if $\{X, Y, Z\}$ is a triad, so is $\{\mathrm{ad}_X, \mathrm{ad}_Y, \mathrm{ad}_Z\}$, with no interchange of the first two operators. Our choice of definitions is based on what seems most natural for our applications, although it seems unnatural here.

2.7.8. Remark. See Definition 4.1.1 for more detailed definitions of $\mathcal{F}^n_{**}$ and $\mathcal{V}^n_{**}$. These spaces are infinite-dimensional, so the triad theory developed in Sections 2.5 and 2.6 does not apply to them directly. But each subspace containing homogeneous polynomials of a given degree (or vectors of such polynomials) is finite-dimensional, and is closed under the action of the respective triads. So each of these subspaces has a chain-weight basis with finitely many chains.

Notes and References

Algorithm 2.7.2 is a special case of the Jacobson–Morosov (or Morozov) lemma. Our version is based on Helgason [57] but is much more concrete. (The original is for abstract Lie groups, ours is just for matrices.) Another proof is given in Chriss and Ginzburg [30].

3
Linear Normal Forms

3.1 Perturbations of Matrices

Consider a smooth (real or complex) matrix-valued function $A(\varepsilon)$ of a (real) small parameter ε, having formal power series

$$A(\varepsilon) \sim A_0 + \varepsilon A_1 + \varepsilon^2 A_2 + \cdots. \tag{3.1.1}$$

How do the eigenvectors (or generalized eigenvectors) and eigenvalues of such a matrix vary with ε? This question arises, for instance, in studying the stability of the linear system of differential equations $\dot{x} = A(\varepsilon)x$. Or the nonlinear system $\dot{x} = f(x, \varepsilon)$ may have a rest point $x^*(\varepsilon)$ whose stability depends on the matrix $A(\varepsilon) = f_x(x^*(\varepsilon), \varepsilon)$. The same question arises for different reasons in quantum mechanics; in this case $A(\varepsilon)$ is Hermitian, hence diagonalizable, and the interest focuses on the splitting (for $\varepsilon \neq 0$) of eigenvalues that are equal when $\varepsilon = 0$. (These eigenvalues can be, for example, the spectral lines of an atom, which can split in the presence of an external field.)

In order to see the variety of possibilities that can occur, it is already interesting to examine the 2×2 case, in which the eigenvalues can be found by the quadratic formula. Here is a list of the possible cases:

1. A_0 has distinct eigenvalues (and so is diagonalizable).

2. A_0 has a double eigenvalue but is diagonalizable anyway.

    a. The eigenvalues split for $\varepsilon \neq 0$ (so $A(\varepsilon)$ remains diagonalizable).

   b. The eigenvalues do not split for $\varepsilon \neq 0$, but $A(\varepsilon)$ remains diagonalizable anyway.
   c. The eigenvalues do not split for $\varepsilon \neq 0$ and $A(\varepsilon)$ becomes nondiagonalizable.

3. A_0 has a double eigenvalue and is not diagonalizable.
   a. The eigenvalues split regularly (see below) and $A(\varepsilon)$ becomes diagonalizable.
   b. The eigenvalues split irregularly (see below) and $A(\varepsilon)$ becomes diagonalizable.
   c. The eigenvalues do not split, and $A(\varepsilon)$ remains nondiagonalizable.

Before giving examples of each possibility, notice that there is one seeming possibility that is not listed: item 3d, "the eigenvalues do not split, but $A(\varepsilon)$ becomes diagonalizable anyway." This cannot occur, because if $A(\varepsilon)$ for $\varepsilon \neq 0$ is diagonalizable with a double eigenvalue, then $A(\varepsilon) = \lambda(\varepsilon)I$, and by continuity $A(0) = \lambda(0)I$, which implies case 2b.

Case 1 is trivial: If $A(0)$ has distinct eigenvalues, then so does $A(\varepsilon)$ for small ε, with the eigenvalues depending smoothly on ε, so that $A(\varepsilon)$ remains diagonalizable. An easy example is

$$A(\varepsilon) = \begin{bmatrix} 1 & \varepsilon \\ \varepsilon & 2 \end{bmatrix}$$

with eigenvalues

$$\frac{3 \pm \sqrt{1 + 4\varepsilon^2}}{2}.$$

Notice that these eigenvalues are smooth (even analytic) in ε and can be expanded in power series.

Cases 2a,b,c are illustrated by the following matrices:

$$\begin{bmatrix} 1 & 0 \\ 0 & 1+\varepsilon \end{bmatrix}, \quad \begin{bmatrix} 1+\varepsilon & 0 \\ 0 & 1+\varepsilon \end{bmatrix}, \quad \begin{bmatrix} 1 & \varepsilon \\ 0 & 1 \end{bmatrix}.$$

An interesting variant of the last example is

$$\begin{bmatrix} 1 & \delta(\varepsilon) \\ 0 & 1 \end{bmatrix}$$

with $\delta(\varepsilon)$ being flat (but not zero). This matrix is clearly not diagonalizable, belonging to case 2c, and yet its formal power series (3.1.1) is diagonalizable (in fact, already diagonal) to all orders. This example points out that working with the power series of a smooth function may not reveal the full behavior of the function.

Case 3 is perhaps even more surprising in its range of possibilities. Consider perturbations of the matrix

$$\begin{bmatrix} 1 & 1 \\ 0 & 1 \end{bmatrix},$$

which is nondiagonalizable with double eigenvalue 1. The examples

$$\begin{bmatrix} 1+\varepsilon & 1 \\ 0 & 1+\varepsilon \end{bmatrix} \quad \text{and} \quad \begin{bmatrix} 1 & 1+\varepsilon \\ 0 & 1 \end{bmatrix}$$

remain nondiagonalizable for small $\varepsilon \neq 0$, and so belong to case 3c. (In the first example $A(\varepsilon)$ is in Jordan form for all ε, and the Jordan form changes with ε; in the second, the Jordan form of $A(\varepsilon)$ is always equal to $A(0)$ as long as $|\varepsilon| < 1$.) The example

$$\begin{bmatrix} 1 & 1 \\ 0 & 1+\varepsilon \end{bmatrix}$$

belongs to case 3a: The eigenvalues are 1 and $1+\varepsilon$, which are unequal for $\varepsilon \neq 0$ and are smooth functions of ε. (This smoothness is part of what we mean by "regular" splitting of the eigenvalues; see below for further discussion.) Since the eigenvalues are distinct for $\varepsilon \neq 0$, the matrix becomes diagonalizable. In fact, the eigenvectors may be taken to be $(1,0)$ and $(1,\varepsilon)$, so that the diagonalizing matrix is

$$T(\varepsilon) = \begin{bmatrix} 1 & 1 \\ 0 & \varepsilon \end{bmatrix}.$$

With this choice of eigenvectors, $T(\varepsilon)$ is defined and smooth for all ε, and fails to be invertible when $\varepsilon = 0$. (This helps to explain why $A(0)$ is not diagonalizable.) Turning to the case 3b example

$$\begin{bmatrix} 1 & 1 \\ \varepsilon & 1 \end{bmatrix},$$

the eigenvalues are $1 \pm \sqrt{\varepsilon}$, which again split, but now are not smooth functions of ε. The diagonalizing matrix may be taken to be

$$T(\varepsilon) = \begin{bmatrix} 1 & 1 \\ \sqrt{\varepsilon} & -\sqrt{\varepsilon} \end{bmatrix},$$

which is defined and continuous for all ε, smooth and invertible for $\varepsilon \neq 0$, but neither invertible nor smooth for $\varepsilon = 0$. The lack of smoothness in the eigenvalues and eigenvectors is one kind of "irregular" behavior that we place into case 3b. A more subtle kind of "irregularity" will be illustrated shortly, but first we pause to draw some general conclusions from the examples that have been given so far.

When faced with an $n \times n$ matrix $A(\varepsilon)$, one approach would be to consider the characteristic equation

$$f(\lambda, \varepsilon) = \det(A(\varepsilon) - \lambda I) = 0. \tag{3.1.2}$$

This is a polynomial in λ depending smoothly on ε. If the dependence on ε is analytic, the *theory of algebraic functions* implies that the n roots can be expanded in convergent *Puiseux series*, which are power series in a fractional power of ε. (The expansion process is usually carried out with the assistance of *Newton diagrams*, introduced in Section 6.1 below in another context.) Of the examples above, only the last actually required a fractional power, namely $\varepsilon^{1/2}$. Once the eigenvalues have been expanded, the eigenvectors can be determined as power series in the same fractional powers. If the dependence of $A(\varepsilon)$ on ε is not analytic, the same procedures work formally to give fractional power series that are not necessarily convergent. For several reasons discussed below, we are not going to pursue these methods in this book, but they do serve to guide our intuition by showing that a complete solution to the type of problem we are considering must sometimes take us out of the class of smooth functions, but will not require anything worse than fractional powers.

The phenomenon of fractional power series can also arise in a disguised form that is of considerable practical significance. Here is a 2×2 example. The matrix

$$A(\varepsilon) = \begin{bmatrix} 0 & 1 \\ 0 & 0 \end{bmatrix} + \varepsilon \begin{bmatrix} 1 & 0 \\ 0 & 1 \end{bmatrix} + \varepsilon^2 \begin{bmatrix} 0 & 0 \\ a^2 & 0 \end{bmatrix} \tag{3.1.3}$$

has eigenvalues $\varepsilon(1 \pm a)$. The curious feature is that although a enters the matrix only in the ε^2 term, it enters the eigenvalue at order ε. Applying the quadratic formula to the characteristic equation shows that the square root merely removes the square from ε^2, and the only evidence of any "irregularity" is that a appears in a term of the "wrong" order. Although the eigenvalue and the eigenvectors remain smooth in ε, we choose to classify this example under case 3b because of the peculiar effects that result from such examples in important applications.

Suppose, for instance, that the stability of the differential equation $\dot{x} = A(\varepsilon)x$ is being studied, where $A(\varepsilon)$ is the matrix (3.1.3). For a first attempt, one might omit the ε^2 term and examine the eigenvalues, which are both ε, leading to the conclusion that the origin is stable for $\varepsilon < 0$, unstable for $\varepsilon > 0$. This result seems to be definitive, since the rest point is hyperbolic and has no "neutral stability" that could be upset by the omitted higher-order term. Nevertheless, the result is wrong, because when the ε^2 term is included, the eigenvalues are $\varepsilon(1 \pm a)$, and if $a > 1$, the system is unstable for all $\varepsilon \neq 0$, having a saddle point at the origin. This example illustrates the need for a theory of "k-determined hyperbolicity," that is, a way of telling when the hyperbolicity of a rest point is determined by the truncation at order k of the linear part.

Although the 2×2 examples above can easily be studied by means of the characteristic equation $\det(A(\varepsilon) - \lambda I) = 0$, finding the eigenvalues first is not in general a fruitful method for approaching this kind of problem. It is difficult even to calculate the coefficients of the characteristic equation when the matrices are of any size, because it is difficult to evaluate determinants. More importantly, passing to the characteristic equation discards much information that is contained in the matrix. For instance, if $A(\varepsilon)$ is symmetric or Hermitian, the eigenvalues will be real, they will always be smooth functions of ε (so that fractional powers are not required), and the matrix will always be diagonalizable (even when eigenvalues are repeated). But all of this information disappears when we pass to the characteristic equation of $A(\varepsilon)$; for instance, there is no way to tell, from the characteristic equation, that its roots are real, apart from actually solving it. For these reasons, in quantum mechanics (where the matrices are Hermitian) the standard method is to seek the eigenvalues and eigenvectors simultaneously in the form of power series. That is, one substitutes series

$$\lambda(\varepsilon) = \lambda_0 + \varepsilon\lambda_1 + \cdots, \qquad (3.1.4)$$
$$v(\varepsilon) = v_0 + \varepsilon v_1 + \cdots, \qquad (3.1.5)$$

into the equation $A(\varepsilon)v = \lambda v$ and solves recursively for the coefficients. This method is not sufficient for our purposes, because our matrices are not guaranteed either to be diagonalizable or to have their eigenvalues depend smoothly on ε.

The procedure to be developed in this chapter is a "whole-matrix" version of the eigenvalue/eigenvector method described above. If $A(\varepsilon)$ is smoothly diagonalizable (that is, without using fractional powers), and we package its eigenvectors $v(\varepsilon)$ together as the columns of a matrix $T(\varepsilon)$, then

$$B(\varepsilon) = T(\varepsilon)^{-1}A(\varepsilon)T(\varepsilon) \qquad (3.1.6)$$

is the diagonal matrix of eigenvalues $\lambda(\varepsilon)$. We can therefore seek simultaneous power series expansions of $T(\varepsilon)$ and $B(\varepsilon)$ satisfying (3.1.6). Aside from the efficiency of treating all eigenvalues and eigenvectors at once, this method has the added advantage that it can work even when $A(\varepsilon)$ is not diagonalizable or not smoothly diagonalizable, provided that $B(\varepsilon)$ is no longer required to be diagonal, but only to be some kind of "normal form," or simplest (in some sense) matrix similar to $A(\varepsilon)$. Since normalization is essentially a perturbation procedure using power series, it will be possible to carry out only a finite number of steps; that is, we can only expect to obtain a $B(\varepsilon)$ whose power series is in normal form up to some finite order k, which should be specified in advance of the calculations. The 2×2 examples above make it clear that we cannot expect the normal form to be the Jordan canonical form, since the Jordan form is often discontinuous in ε. Since the normalization process does not introduce fractional powers, we cannot expect normalization by itself to produce complete results; in some cases it will be only a first step, to be followed by finding a *hypernormal form* or a *metanormal form* that brings increased simplification. (A hypernormal form is one that exploits free choices in the normalization process to yield a result that is in some sense the best among the possible normal forms. A metanormal form introduces fractional powers.)

Notes and References

The standard reference for perturbations of linear operators is Kato [63], but this is mostly devoted to the infinite-dimensional case (and is therefore functional-analytic in character). The method mentioned in (3.1.2), while it is not recommended, could be carried out using the perturbation techniques for roots of polynomials given in Murdock [88], Chapter 1, or by the Newton diagram and Puiseux series methods developed for different purposes in Section 6.1 below.

3.2   An Introduction to the Five Formats

Our aim is to simplify, or "normalize," the matrix series (3.1.1) by applying a similarity transformation of the form (3.1.6). In doing this, it is helpful to select a systematic "bookkeeping method" for keeping track of the calculations. There are five natural approaches, which we refer to as *formats*. We will identify these formats by the numbers 1a, 1b, 2a, 2b, and 2c; the following table is helpful in identifying the characteristics of each format:

1a	direct	iterative	
1b	direct	recursive	
2a	generated	iterative	
2b	generated	recursive	(Hori)
2c	generated	recursive	(Deprit)

In this section, the five formats will be introduced briefly and the formulas of each format given through third order. Most of this book uses only formats 1a and 2a, and it is sufficient to concentrate on these for a first reading. From a purely mathematical standpoint, there are strong reasons to regard format 2b as the most natural and most powerful; see Remark 3.2.3 below. Format 2c is of interest mainly because of its wide use among applied mathematicians. Formats 2b and 2c have the advantage of having recursive formulations, which facilitate calculations to higher order. The reader interested in a complete treatment of formats 2b and 2c should turn to Appendices C and D after studying Section 3.6 below. In Notes and References we briefly mention the notion of quadratic convergence, which can be thought of as a format in its own right (although it uses format 2b).

The first step in simplifying $A(\varepsilon) = A_0 + \varepsilon A_1 + \cdots$ is to bring A_0 into Jordan canonical form (unless another form, such as real canonical form, is more appropriate in a given application). If T_0 is any invertible matrix, then

$$T_0^{-1}(A_0 + \varepsilon A_1 + \cdots)T_0 = (T_0^{-1}A_0T_0) + \varepsilon(T_0^{-1}A_1T_0) + \cdots, \qquad (3.2.1)$$

and T_0 can always be chosen so that $T_0^{-1}A_0T_0$ is in canonical form. This will, of course, modify the higher-order terms in uncontrollable ways, and we can have no expectation that they will be simplified at this stage. *We will assume that this preliminary step has already been carried out* and that A_0 is already in the desired canonical form. This allows us to focus our attention on the higher-order terms.

In each of the five formats, the matrix series

$$A(\varepsilon) \sim A_0 + \varepsilon A_1 + \varepsilon^2 A_2 + \varepsilon^3 A_3 + \cdots \qquad (3.2.2)$$

is transformed into a new series

$$B(\varepsilon) \sim B_0 + \varepsilon B_1 + \varepsilon^2 B_2 + \varepsilon^3 B_3 + \cdots \qquad (3.2.3)$$

(in which $B_0 = A_0$) by means of a near-identity similarity transformation of the form

$$B(\varepsilon) = T(\varepsilon)^{-1} A(\varepsilon) T(\varepsilon), \qquad (3.2.4)$$

with $T(0) = I$. The differences between the formats lie in the way in which the transformation $T(\varepsilon)$ is developed or expressed. Because of convergence difficulties, it is usually possible to normalize only to some finite order in ε; for our initial, simple presentation, we take this order to be 3.

- In **format 1a**, the coordinate changes $I + \varepsilon S_1, I + \varepsilon^2 S_2, \ldots$ are applied sequentially, so that when the method is applied to third order, the complete transformation is

$$T(\varepsilon) = (I + \varepsilon S_1)(I + \varepsilon^2 S_2)(I + \varepsilon^3 S_3). \qquad (3.2.5)$$

- In **format 1b**, the job is done instead by a single transformation, expanded as

$$T(\varepsilon) = I + \varepsilon T_1 + \varepsilon^2 T_2 + \varepsilon^3 T_3. \qquad (3.2.6)$$

(Notice that the transformation here cannot be exactly the same as the one in format 1a, because when (3.2.5) is multiplied out, it produces terms of all orders up to ε^6.)

- **Format 2a** is like 1a in that several transformations are applied sequentially, but they are generated in exponential form:

$$T(\varepsilon) = e^{\varepsilon U_1} e^{\varepsilon^2 U_2} e^{\varepsilon^3 U_3}. \qquad (3.2.7)$$

- There are two natural ways to modify format 2a in a manner resembling 1b. The first, which we call **format 2b**, is to set

$$T(\varepsilon) = e^{\varepsilon V_1 + \varepsilon^2 V_2 + \varepsilon^3 V_3}. \qquad (3.2.8)$$

- **Format 2c** generates $T(\varepsilon)$ by solving the matrix differential equation

$$\frac{dT}{d\varepsilon} = (W_1 + \varepsilon W_2 + \varepsilon^2 W_3) T \qquad (3.2.9)$$

with initial condition $T(0) = I$. To understand the relationship between 3.2.8 and 3.2.9, observe that they are equal when taken only to first order: $e^{\varepsilon V_1}$ is the solution of $dT/d\varepsilon = V_1 T$ with $T(0) = I$. They are no longer equal at higher orders because (3.2.9) is a nonautonomous equation and cannot be solved by exponentials. Notice that in (3.2.9) the power of ε lags one behind the index on W; this is because solving the differential equations introduces another factor of ε.

The "1" formats (1a and 1b) work with the transformation $T(\varepsilon)$ directly; the "2" formats (2a, 2b, and 2c) work with generators of the transformation (denoted by U, V, or W) instead. The "a" formats (1a and 2a) are *iterative* in character, in that several transformations are applied in succession, while formats 1b, 2b, and 2c are *recursive* in the sense that only a single transformation is applied, but its coefficients are determined recursively.

In the literature, what we call a format is often referred to as a "method," and there is much duplication of terminology. For instance, when applied to differential equations (as we do in Chapter 4 below), format 2a can be called the "method of Lie series" or the "method of Dragt and Finn," format 2b is "Hori's method" or the "method of Mersman," and format 2c is (except for changes in the placement of factorials) the "method of Lie transforms" or "Deprit's method."

In the actual application of the formats to a normal form problem, the transformation to be used is unknown and must be determined at the same time as the transformation is applied. (One must also have a normal form *style* in mind, and select the transformation so as to achieve this style.) It is simplest to present the formats initially under the assumption that the transformation (or its generator) is *given*, and it is only necessary to compute the effect of the transformation on a given matrix series. In this way all issues connected with normal form styles can be set aside for later discussion in order to concentrate on the structure of each format.

Format 1a

Assuming, then, that S_1, S_2, and S_3 in (3.2.5) are given, we can carry out the calculation of $B(\varepsilon)$ to third order in three successive steps. The calculations are based on the geometric series

$$(I + \varepsilon X)^{-1} = I - \varepsilon X + \varepsilon^2 X^2 - \varepsilon^3 X^3 + \cdots , \qquad (3.2.10)$$

which is actually convergent (in any matrix norm) for small ε. Letting $\equiv$ denote equality through terms of order ε^3, the first step is

$$(I + \varepsilon S_1)^{-1}(A_0 + \varepsilon A_1 + \varepsilon^2 A_2 + \varepsilon^3 A_3)(I + \varepsilon S_1) \qquad (3.2.11)$$
$$\equiv (I - \varepsilon S_1 + \varepsilon^2 S_1^2 - \varepsilon^3 S_1^3)(A_0 + \varepsilon A_1 + \varepsilon^2 A_2 + \varepsilon^3 A_3)(I + \varepsilon S_1)$$
$$\equiv A_0 + \varepsilon A_1' + \varepsilon^2 A_2' + \varepsilon^3 A_3',$$

with

$$A'_1 = A_1 + A_0 S_1 - S_1 A_0, \tag{3.2.12}$$
$$A'_2 = A_2 + A_1 S_1 - S_1 A_1 + S_1^2 A_0 - S_1 A_0 S_1,$$
$$A'_3 = A_3 + A_2 S_1 - S_1 A_2 + S_1^2 A_1 - S_1 A_1 S_1 + S_1^2 A_0 S_1 - S_1^3 A_0.$$

Notice that there is no A'_0, because the constant term is not changed. Next,

$$(I + \varepsilon^2 S_2)^{-1}(A_0 + \varepsilon A'_1 + \varepsilon^2 A'_2 + \varepsilon^3 A'_3)(I + \varepsilon^2 S_2) \tag{3.2.13}$$
$$\equiv (I - \varepsilon^2 S_2)(A_0 + \varepsilon A'_1 + \varepsilon^2 A'_2 + \varepsilon^3 A'_3)(I + \varepsilon^2 S_2)$$
$$\equiv A_0 + \varepsilon A'_1 + \varepsilon^2 A''_2 + \varepsilon^3 A''_3,$$

with

$$A''_2 = A'_2 + A_0 S_2 - S_2 A_0, \tag{3.2.14}$$
$$A''_3 = A'_3 + A'_1 S_2 - S_2 A'_1.$$

There is no A''_1 because A'_1 is unchanged. Finally,

$$(I + \varepsilon^3 S_3)^{-1}(A_0 + \varepsilon A'_1 + \varepsilon^2 A''_2 + \varepsilon^3 A''_3)(I + \varepsilon^3 S_3) \tag{3.2.15}$$
$$\equiv (I - \varepsilon^3 S_3)(A_0 + \varepsilon A'_1 + \varepsilon^2 A''_2 + \varepsilon^3 A''_3)(I + \varepsilon^3 S_3)$$
$$\equiv A_0 + \varepsilon A'_1 + \varepsilon^2 A''_2 + \varepsilon^3 A'''_3$$
$$\equiv B_0 + \varepsilon B_1 + \varepsilon^2 B_2 + \varepsilon^3 B_3,$$

with

$$A'''_3 = A''_3 + A_0 S_3 - S_3 A_0. \tag{3.2.16}$$

These are all the formulas needed for format 1a (to third order) if S_1, S_2, and S_3 are given. But, as mentioned above, in an actual normalization problem these are not known in advance, and must be selected so as to achieve a desired normal form style. For this purpose, the first equation of each of the sets (3.2.12), (3.2.14), and (3.2.16) may be rewritten as follows using the *commutator bracket* $[P, Q] = PQ - QP$:

$$A'_1 = A_1 + [A_0, S_1],$$
$$A''_2 = A'_2 + [A_0, S_2],$$
$$A'''_3 = A''_3 + [A_0, S_3].$$

These equations all have the same form, and the left-hand sides are simply B_1, B_2, and B_3. They may therefore be written as

$$B_j = K_j + [A_0, S_j], \tag{3.2.17}$$

where the notation K_j is chosen to indicate a quantity that is "known" at the jth stage of the calculation, and could, if desired, be expressed entirely in terms of the original coefficients A_i and those S_i that have been calculated so far:

$$K_1 = A_1, \qquad\qquad\qquad (3.2.18)$$
$$K_2 = A_2' = A_2 + A_1 S_1 - S_1 A_1 + S_1^2 A_0 - S_1 A_0 S_1,$$
$$K_3 = A_3'' = A_3 + A_2 S_1 - S_1 A_2 + S_1^2 A_1 - S_1 A_1 S_1$$
$$+ S_1^2 A_0 S_1 - S_1^3 A_0 + A_1 S_2 + A_0 S_1 S_2 - S_1 A_0 S_2 - S_2 A_1$$
$$- S_2 A_0 S_1 - S_2 S_1 A_0.$$

Because $[P, Q] = -[Q, P]$, we can rewrite (3.2.17) as

$$[S_j, A_0] = K_j - B_j.$$

Define the linear *homological operator* $\pounds$ on the vector space of $n \times n$ matrices by

$$\pounds S = [S, A_0], \qquad\qquad (3.2.19)$$

or simply

$$\pounds = [\ , A_0].$$

Then

$$\pounds S_j = K_j - B_j. \qquad\qquad (3.2.20)$$

It is important to realize here that $\pounds$ is a linear operator on matrices, not a matrix itself. It is common to denote the vector space of all $n \times n$ real matrices by $\mathrm{gl}(n, \mathbb{R})$, and the space of $n \times n$ complex matrices by $\mathrm{gl}(n, \mathbb{C})$; when we wish to refer to one or the other of these, as needed, without specifying the field, we will write $\mathrm{gl}(n)$. Thus, $\pounds$ is a linear operator on $\mathrm{gl}(n)$:

$$\pounds : \mathrm{gl}(n) \to \mathrm{gl}(n).$$

(Of course, $\pounds$ could be represented as a matrix with respect to some basis for $\mathrm{gl}(n)$, but then it would be an $n^2 \times n^2$ matrix, not $n \times n$.)

Equation (3.2.20) is a *homological equation* of the same form as (1.1.7), (1.1.12), or, most closely, (1.1.30) in Chapter 1. This is exactly the type of problem to which Chapter 2 was devoted: One must select a *normal form style* by choosing a subspace of $\mathrm{gl}(n)$ complementary to im $\pounds$. Then, taking B_j to be the projection of K_j into this subspace, $K_j - B_j$ lies in im $\pounds$ and S_j can be found. The details depend strongly on the form of A_0, and since further discussion would take us away from the present subject of formats, it will be deferred to later sections of this chapter.

Format 1b

To expand the inverse of $T(\varepsilon)$ as given in (3.2.6), the geometric series (3.2.10) can be applied with $\varepsilon X = \varepsilon T_1 + \varepsilon^2 T_2 + \varepsilon^3 T_3$; the result is

$$T(\varepsilon)^{-1} \equiv I - \varepsilon T_1 + \varepsilon^2 (T_1^2 - T_2) - \varepsilon^3 (T_1^3 - T_1 T_2 - T_2 T_1 + T_3).$$

After some tedious but straightforward calculation, it is found that

$$T(\varepsilon)^{-1} A(\varepsilon) T(\varepsilon) \equiv B_0 + \varepsilon B_1 + \varepsilon^2 B_2 + \varepsilon^3 B_3$$

with

$$B_j = [A_0, T_j] + K_j, \tag{3.2.21}$$

where

$$K_1 = A_1,$$
$$K_2 = A_2 + [A_1, T_1] + T_1[T_1, A_0],$$
$$K_3 = A_3 + [A_2, T_1] + T_1[T_1, A_1] + T_1[T_2, A_0] + T_2[T_1, A_0] + T_1^2[A_0, T_1].$$

(The brackets here do not arise naturally, but are introduced by inspection to shorten the expressions that result from the calculation. In the generated formats (2a, 2b, 2c), the corresponding expressions appear completely in terms of brackets, and these will arise naturally.) Notice that there are no intermediate stages, such as A_j', in format 1b. The equations (3.2.21) can be converted into

$$\pounds T_j = K_j - B_j, \tag{3.2.22}$$

where $\pounds$ is again defined by (3.2.19). Again B_j must be the projection of K_j into a chosen normal form style, so that $K_j - B_j \in \operatorname{im} \pounds$ and T_j can be found. The only thing that needs to be checked is that in fact K_j is "known" at the jth stage, but this is true because $K_1 = A_1$ is known from the beginning, so that T_1 can be found; K_2 depends only on T_1 and the A_j, so T_2 can be found; and K_3 depends only on T_1, T_2, and the A_j.

Format 2a

Format 2a is based on the matrix exponential

$$e^X = I + X + \frac{1}{2}X^2 + \frac{1}{6}X^3 + \cdots + \frac{1}{j!}X^j + \cdots, \tag{3.2.23}$$

which is a convergent series (in any matrix norm) for all $X \in \operatorname{gl}(n)$ and satisfies most formulas that are true for the ordinary exponential. (The formula $e^X e^Y = e^{(X+Y)}$ holds only if X and Y commute.) The most essential fact is that the inverse of e^X is

$$e^{-X} = I - X + \frac{1}{2}X^2 - \frac{1}{6}X^3 + \cdots + \frac{(-1)^j}{j!}X^j + \cdots.$$

Thus, according to (3.2.7) the first step in format 2a, corresponding to (3.2.11) in format 1a, calls for the computation of $e^{-\varepsilon U_1}(A_0 + \varepsilon A_1 + \varepsilon^2 A_2 + \varepsilon^3 A_3)e^{\varepsilon U_1} \equiv A_0 + \varepsilon A_1' + \varepsilon^2 A_2' + \varepsilon^3 A_3'$. The symmetry in the treatment of the transformation and its inverse, to be contrasted with the asymmetry present in (3.2.11), leads to expressions for the A'_j that can be reduced entirely to iterated brackets. In fact, one finds that

$$A'_1 = A_1 + A_0 U_1 - U_1 A_0, \qquad (3.2.24)$$

$$A'_2 = A_2 + A_1 U_1 - U_1 A_1 + \frac{1}{2}\left(A_0 U_1^2 - 2U_1 A_0 U_1 + U_1^2 A_0\right),$$

$$A'_3 = A_3 + A_2 U_1 - U_1 A_2 + \frac{1}{2}\left(A_1 U_1^2 - 2U_1 A_1 U_1 + U_1^2 A_1\right)$$

$$+ \frac{1}{6}\left(A_0 U_1^3 - 3U_1 A_0 U_1^2 + 3U_1^2 A_0 U_1 - U_1^3 A_0\right),$$

and the expression for A'_3 (which also shows how the others go) can be rewritten as

$$A'_3 \equiv A_3 + [A_2, U_1] + \frac{1}{2}[[A_1, U_1], U_1] + \frac{1}{6}[[[A_0, U_1], U_1], U_1].$$

Rather than proceeding in this way, it is better to prove a theorem that will allow the bracket structure of format 2a to emerge in a natural way. For any Y in gl(n), define the *Lie operator* $\mathbb{L}_Y : \text{gl}(n) \to \text{gl}(n)$ by

$$\mathbb{L}_Y X = [X, Y] = XY - YX, \qquad (3.2.25)$$

or simply

$$\mathbb{L}_Y = [\ , Y].$$

(The operator $\mathcal{L}$ defined in (3.2.19) is just $\mathbb{L}_{A_0}$.)

3.2.1. Theorem (Fundamental theorem of Lie series). For any Y and X in gl(n),

$$e^{-Y} X e^{Y} = e^{\mathbb{L}_Y} X.$$

Proof. We will prove the theorem in the form

$$e^{-sY} X e^{sY} = e^{s\mathbb{L}_Y} X, \qquad (3.2.26)$$

where s is a real parameter. (The theorem itself is recovered by setting $s = 1$.) Since $\mathbb{L}_Y$ is a linear operator (and can be viewed as an $n^2 \times n^2$ matrix), the exponential $e^{s\mathbb{L}_Y}$ is not a new kind of object, but is defined exactly as in (3.2.23):

$$e^{s\mathbb{L}_Y} X = \sum_{j=0}^{\infty} \frac{s^j}{j!} \mathbb{L}_Y^j X.$$

The right-hand side is called a *Lie series* and is convergent for all s. Thus, both sides of (3.2.26) are real analytic functions of s, and it suffices to prove that the right-hand side is the Taylor expansion of the left-hand side, in other words, that

$$\left. \frac{d^j}{ds^j} e^{-sY} X e^{sY} \right|_{s=0} = \mathbb{L}_Y^j X$$

for each j. To show this, we first calculate

$$\frac{d}{ds}e^{-sY}Xe^{sY} = e^{-sY}(XY - YX)e^{sY} = e^{-sY}\mathbb{L}_Y Xe^{sY}.$$

That is, to differentiate any product of the form e^{-sY} times a constant matrix times e^{sY}, it suffices to apply $\mathbb{L}_Y$ to the constant matrix. Applying this to itself j times gives

$$\frac{d^j}{ds^j}e^{-sY}Xe^{sY} = e^{-sY}\mathbb{L}_Y^j Xe^{sY},$$

and the result follows by setting $s = 0$.   □

3.2.2. Remark. A second proof of this theorem, one that is conceptual rather than computational in character (making use of Lie group ideas), will be given in Section 3.6. Both proofs are important. The argument given here will be repeated in Chapter 4 in a setting in which the series are not convergent, but only formal and asymptotic. The conceptual argument fails in that case (because the Lie group is infinite-dimensional).

To develop format 2a by means of Theorem 3.2.1, define

$$\mathbb{L}_j = \mathbb{L}_{U_j} \tag{3.2.27}$$

and observe that

$$e^{-\varepsilon U_1}\left(A_0 + \varepsilon A_1 + \varepsilon^2 A_2 + \varepsilon^3 A_3\right)e^{\varepsilon U_1} \equiv e^{\varepsilon \mathbb{L}_1}\left(A_0 + \varepsilon A_1 + \varepsilon^2 A_2 + \varepsilon^3 A_3\right),$$
$$\tag{3.2.28}$$

where

$$e^{\varepsilon \mathbb{L}_1} \equiv I + \varepsilon \mathbb{L}_1 + \frac{1}{2}\varepsilon^2 \mathbb{L}_1^2 + \frac{1}{6}\varepsilon^3 \mathbb{L}_1^3.$$

This multiplies out to give (3.2.24) again, in the form

$$A_1' = A_1 + \mathbb{L}_1 A_0, \tag{3.2.29}$$

$$A_2' = A_2 + \mathbb{L}_1 A_1 + \frac{1}{2}\mathbb{L}_1^2 A_0,$$

$$A_3' = A_3 + \mathbb{L}_1 A_2 + \frac{1}{2}\mathbb{L}_1^2 A_1 + \frac{1}{6}\mathbb{L}_1^3 A_0.$$

The next step is

$$e^{-\varepsilon^2 U_2}\left(A_0 + \varepsilon A_1' + \varepsilon^2 A_2' + \varepsilon^3 A_3'\right)e^{\varepsilon^2 U_2} \tag{3.2.30}$$

$$\equiv e^{\varepsilon^2 \mathbb{L}_2}\left(A_0 + \varepsilon A_1' + \varepsilon^2 A_2' + \varepsilon^3 A_3'\right)$$

$$\equiv \left(I + \varepsilon^2 \mathbb{L}_2\right)\left(A_0 + \varepsilon A_1' + \varepsilon^2 A_2' + \varepsilon^3 A_3'\right)$$

$$\equiv A_0 + \varepsilon A_1' + \varepsilon^2 A_2'' + \varepsilon^3 A_3'',$$

with

$$A_2'' = A_2' + \mathbb{L}_2 A_0, \tag{3.2.31}$$
$$A_3'' = A_3' + \mathbb{L}_2 A_1'.$$

Finally,

$$e^{-\varepsilon^3 U_3} \left(A_0 + \varepsilon A_1' + \varepsilon^2 A_2'' + \varepsilon^3 A_3'' \right) e^{\varepsilon^3 U_3} \tag{3.2.32}$$
$$\equiv A_0 + \varepsilon A_1' + \varepsilon^2 A_2'' + \varepsilon^3 A_3'''$$
$$\equiv B_0 + \varepsilon B_1 + \varepsilon^2 B_2 + \varepsilon^3 B_3,$$

with

$$A_3''' = A_3'' + \mathbb{L}_3 A_0 = A_3'' + A_0 U_3 - U_3 A_0. \tag{3.2.33}$$

These equations are sufficient if U_1, U_2, and U_3 are known, but in the application to normal forms they are not. But the first equation of each set (3.2.29), (3.2.31), and (3.2.33) can be written as

$$B_j = K_j + \mathbb{L}_j A_0, \tag{3.2.34}$$

where K_j is known at the time each equation is to be used. Because

$$\mathbb{L}_j A_0 = [A_0, U_j] = -[U_j, A_0] = -\pounds U_j,$$

this takes the form

$$\pounds U_j = K_j - B_j. \tag{3.2.35}$$

Then, as before, B_j is taken to be the projection of K_j into a particular complement $\mathcal{C}$ of im $\pounds$, that is, into a particular normal form style. Then (3.2.35) can be solved for U_j. It is important to observe that there is a proper order in which to work through these equations to avoid circularity. Thus, the first equation of (3.2.29) is used in the form (3.2.35) with $j = 1$, to determine first B_1 and then U_1. When U_1 is known, so is the operator $\mathbb{L}_1 = \mathbb{L}_{U_1}$, so the remaining equations of (3.2.29) can be used to evaluate A_2' and A_3'. Then K_2 is available for the next application of (3.2.35), and so forth.

Format 2b

Formats 2b and 2c will be treated in a more sketchy manner here, because a full section will be devoted to each of these formats in Appendices C and D. Here we will give the formulas of each format to third order and indicate how these formulas fit into recursive "triangular" algorithms that will be developed fully in these appendices.

To expand (3.2.4) where $T(\varepsilon)$ is given by (3.2.8) we invoke Theorem 3.2.1 and write

$$B(\varepsilon) = e^{\mathbb{L}(\varepsilon)} A(\varepsilon),$$

where

$$\mathbb{L}(\varepsilon) \equiv \mathbb{L}_{\varepsilon V_1 + \varepsilon^2 V_2 + \varepsilon^3 V_3} \equiv \varepsilon \mathbb{L}_{V_1} + \varepsilon^2 \mathbb{L}_{V_2} + \varepsilon^3 \mathbb{L}_{V_3} \equiv \varepsilon \mathbb{L}_1 + \varepsilon^2 \mathbb{L}_2 + \varepsilon^3 \mathbb{L}_3,$$

and we have set

$$\mathbb{L}_j = \mathbb{L}_{V_j}.$$

Since the series for $\mathbb{L}(\varepsilon)$ begins with a term of order ε, $\mathbb{L}(\varepsilon)^j$ begins with a term of order ε^j, and it follows that $\mathbb{L}(\varepsilon)^4 \equiv 0$. Therefore,

$$e^{\mathbb{L}(\varepsilon)} \equiv I + \mathbb{L}(\varepsilon) + \frac{1}{2}\mathbb{L}(\varepsilon)^2 + \frac{1}{6}\mathbb{L}(\varepsilon)^3.$$

The easiest way to calculate $e^{\mathbb{L}(\varepsilon)} A(\varepsilon)$ is to apply $\mathbb{L}(\varepsilon)$ repeatedly to $A(\varepsilon)$ and save the results to be added later, with the coefficients $\frac{1}{2}$ and $\frac{1}{6}$ added. Thus,

$$A(\varepsilon) \equiv A_0 + \varepsilon A_1 + \varepsilon^2 A_2 + \varepsilon^3 A_3,$$
$$\mathbb{L}(\varepsilon)A(\varepsilon) \equiv \varepsilon \mathbb{L}_1 A_0 + \varepsilon^2 (\mathbb{L}_1 A_1 + \mathbb{L}_2 A_0) + \varepsilon^3 (\mathbb{L}_1 A_2 + \mathbb{L}_2 A_1 + \mathbb{L}_3 A_0),$$
$$\mathbb{L}(\varepsilon)^2 A(\varepsilon) \equiv \varepsilon^2 (\mathbb{L}_1^2 A_0) + \varepsilon^3 (\mathbb{L}_1^2 A_1 + \mathbb{L}_1 \mathbb{L}_2 A_0 + \mathbb{L}_2 \mathbb{L}_1 A_0),$$
$$\mathbb{L}(\varepsilon)^3 A(\varepsilon) \equiv \varepsilon^3 (\mathbb{L}_1^3 A_0).$$

Carrying out the addition gives

$$B_0 = A_0,$$
$$B_1 = A_1 + \mathbb{L}_1 A_0,$$
$$B_2 = A_2 + \mathbb{L}_1 A_1 + \mathbb{L}_2 A_0 + \frac{1}{2}\mathbb{L}_1^2 A_0,$$
$$B_3 = A_3 + \mathbb{L}_1 A_2 + \mathbb{L}_2 A_1 + \mathbb{L}_3 A_0$$
$$+ \frac{1}{2}(\mathbb{L}_1^2 A_1 + \mathbb{L}_1 \mathbb{L}_2 A_0 + \mathbb{L}_2 \mathbb{L}_1 A_0) + \frac{1}{6}\mathbb{L}_1^3 A_0.$$

By the now-familiar remark that $\mathbb{L}_j A_0 = -\pounds V_j$, each of these equations can be put in the form

$$\pounds V_j = K_j - B_j, \tag{3.2.36}$$

where K_j consists of all terms on the right-hand side of the previous equations except for the term $\mathbb{L}_j A_0$. As in format 1b, each K_j is known at the time it is needed.

It is convenient to set

$$A_j^{(0)} = A_j$$

and to define $A_j^{(i)}$ by

$$\mathbb{L}(\varepsilon)^i A(\varepsilon) = \varepsilon^i A_0^{(i)} + \varepsilon^{i+1} A_1^{(i)} + \varepsilon^{i+2} A_2^{(i)} + \cdots. \tag{3.2.37}$$

These quantities can then be arranged into the following triangular pattern:

$$A_0^{(0)} \quad A_1^{(0)} \quad A_2^{(0)} \quad A_3^{(0)}$$
$$A_0^{(1)} \quad A_1^{(1)} \quad A_2^{(1)}$$
$$A_0^{(2)} \quad A_1^{(2)} \qquad\qquad \text{(3.2.38)}$$
$$A_0^{(3)}$$

Recursive procedures for the construction of the entries in this triangle to any order will be developed in Section C.1. Once the triangle is built, the B_j can be obtained as weighted sums down the columns of the triangle:

$$B_j = \sum_{i=0}^{j} \frac{1}{i!} A_{j-i}^{(i)}. \qquad (3.2.39)$$

3.2.3. Remark. From a mathematical standpoint, format 2b is entirely natural. One simply has a graded Lie algebra, in this case the Lie algebra of formal power series in ε with matrix coefficients, with commutator bracket as the product. The subset of the Lie algebra consisting of series with no constant term acts on the whole algebra by the exponential of the adjoint; in our notation this amounts to the "generator" $\varepsilon v_1 + \varepsilon^2 v_2 + \cdots$ acting on $a_0 + \varepsilon a_1 + \cdots$ to produce $e^{\mathbb{L}_{\varepsilon v_1 + \varepsilon^2 v_2 + \cdots}}(a_0 + \varepsilon v_1 + \cdots)$. (The restriction to no constant term in the exponential is necessary to avoid infinite series occurring within each coefficient of the resulting formal series.) The Campbell–Baker–Hausdorff formula

$$e^u e^v = e^{u+v+\frac{1}{2}[u,v]+\frac{1}{12}[u,[u,v]]+\cdots}$$

may be used (with u and v replaced by $\mathbb{L}_u$ and $\mathbb{L}_v$) to compute the effect of two such generators applied in turn. Format 2a can be viewed as the successive application of generators having only a single term; the Campbell–Baker–Hausdorff formula shows how to convert such a sequence into the action of a single generator, and thus relates formats 2a and 2b.

Format 2c

The main idea of format 2c is not to develop $B(\varepsilon)$ directly from (3.1.6) and (3.2.9), but rather to develop a differential equation for $B(\varepsilon)$, just as (3.2.9) is a differential equation for $T(\varepsilon)$, and then solve this differential equation recursively. The details will be developed in Section D.1; for now we will only state the result of that discussion in a way that enables the formulas to be calculated. Let

$$\mathbb{L}_j = \mathbb{L}_{W_j},$$

and let $\mathbb{D}$ be the formal operator

$$\mathbb{D} = \frac{d}{d\varepsilon} + \mathbb{L}_1 + \varepsilon\mathbb{L}_2 + \varepsilon^2\mathbb{L}_3 + \cdots.$$

Notice that the powers on ε lag behind the subscripts on $\mathbb{L}$; the term $d/d\varepsilon$ should be regarded as having order -1, because when it operates on a term it decreases the order by one. The main result is that

$$B_j = \frac{1}{j!}\mathbb{D}^j A(\varepsilon)|_{\varepsilon=0}.$$

That is, first apply $\mathbb{D}$ repeatedly to $A = A(\varepsilon)$ up to the desired order; then (after this is finished) set $\varepsilon = 0$ and divide by $j!$. Each time $\mathbb{D}$ is applied, the number of terms that can be considered "known" decreases, but if we start with terms of a given order, say 3, it is possible to compute up to that order. Let $\equiv_j$ denote equality through terms of order j. Then we have

$$\begin{aligned}
A \equiv_3 \ & A_0 + \varepsilon A_1 + \varepsilon^2 A_2 + \varepsilon^3 A, \\
\mathbb{D}A \equiv_2 \ & (A_1 + \mathbb{L}_1 A_0) + \varepsilon(2A_2 + \mathbb{L}_1 A_1) \\
& + \varepsilon^2(3A_3 + \mathbb{L}_1 A_2 + \mathbb{L}_2 A_1 + \mathbb{L}_3 A_0), \\
\mathbb{D}^2 A \equiv_1 \ & (2A_2 + \mathbb{L}_1 A_1 + \mathbb{L}_2 A_0) \\
& + \varepsilon(6A_3 + 4\mathbb{L}_1 A_2 + 3\mathbb{L}_2 A_1 + 2\mathbb{L}_3 A_0 + \mathbb{L}_1^2 A_1 \\
& + \mathbb{L}_1\mathbb{L}_2 A_0 + \mathbb{L}_2\mathbb{L}_1 A_0), \\
\mathbb{D}^3 A \equiv_0 \ & 6A_3 + 6\mathbb{L}_1 A_2 + 3\mathbb{L}_2 A_1 + 2\mathbb{L}_3 A_0 + 4\mathbb{L}_1 A_2 \\
& + \mathbb{L}_2\mathbb{L}_1 A_0 + 2\mathbb{L}_1^2 A_1 + 2\mathbb{L}_1\mathbb{L}_2 A_0.
\end{aligned}$$

Setting $\varepsilon = 0$ in these equations and prefixing the correct factorial yields

$$\begin{aligned}
B_0 &= A_0, \\
B_1 &= A_1 + \mathbb{L}_1 A_0, \\
B_2 &= A_2 + \frac{1}{2}(\mathbb{L}_1 A_1 + \mathbb{L}_2 A_0), \\
B_3 &= A_3 + \mathbb{L}_1 A_2 + \frac{1}{2}\mathbb{L}_2 A_1 + \frac{1}{3}\mathbb{L}_3 A_0 + \frac{2}{3}\mathbb{L}_1 A_2 + \frac{1}{6}\mathbb{L}_2\mathbb{L}_1 A_0 \\
&\quad + \frac{1}{3}\mathbb{L}_1^2 A_1 + \frac{1}{3}\mathbb{L}_1\mathbb{L}_2 A_0.
\end{aligned}$$

Since $\mathbb{L}_j A_0 = -\pounds W_j$, these can be arranged as

$$\pounds W_j = K_j - B_j. \tag{3.2.40}$$

The rest proceeds as usual.

It is convenient to set

$$A_j^{(0)} = A_j$$

and to define $A_j^{(i)}$ by

$$\mathbb{D}^{(i)} A = A_0^{(i)} + \varepsilon A_1^{(i)} + \varepsilon^2 A_2^{(i)} + \cdots. \qquad (3.2.41)$$

Then these quantities can be arranged into the following triangular pattern:

$$
\begin{array}{cccc}
A_0^{(0)} & & & \\
A_1^{(0)} & A_0^{(1)} & & \\
A_2^{(0)} & A_1^{(1)} & A_0^{(2)} & \\
A_3^{(0)} & A_2^{(1)} & A_1^{(2)} & A_0^{(3)}
\end{array}
\qquad (3.2.42)
$$

Recursive procedures for constructing this triangle to any order will be developed in Section D.1. Once the triangle is constructed, the B_j are given by

$$B_j = \frac{1}{j!} A_0^{(j)}. \qquad (3.2.43)$$

That is, except for weighting factors the B_j are given by the top diagonal. In the usual formulation of format 2c (the Deprit triangle), the B_j are exactly the top diagonal entries, but many unnecessary factorials and binomial coefficients occur elsewhere in the formulas. For the reader's convenience, the formulas in both notations will be set out in Section D.1.

Notes and References

I obtained each of the so-called "formats" introduced in this section by applying a "method," originally developed for one of the settings discussed in Chapter 4 below, to the matrix situation. These things have been redis-covered so many times that it is probably impossible to sort out the history correctly. Since my main goal in developing this material was to obtain a "clean" version, there is no reference available that quite resembles the way it is done here; at the same time, nothing here is at all new. All, or most, of these "methods" can be found, in one form or another, in Nayfeh [93]. For instance, format 1b resembles the "generalized method of averaging" in Section 5.2.3 of [93], and format 2c is called the method of Lie transforms, Section 5.7. Format 2b is Hori's method in the version of Mersman [78]; see also Kirchgraber [65]. (In view of Remark 3.2.3, it is clear that Hori only introduced this method to applied mathematicians. It has often been used entirely independently of Hori.) Format 2a was used by Dragt and Finn in the setting of Hamiltonian systems (Section 4.9 below); see Finn [45] for further references. The ideas behind the generated formats are probably more clearly explained in the abstract theory of Lie groups than in any of the applied literature. See Notes and References for Section 3.6. The Campbell–Baker–Hausdorff formula can be found in [56].

All of the formats in this section proceed one degree in ε at a time, in the sense that (whether the method is iterative or recursive) it is necessary to solve a new homological equation for each degree. It is possible to organize the calculations in such a way that in the first step A_1 is normalized; in the second step A_2 and A_3 are normalized (with the solution of a single generalized homological equation); in the third step A_4 through A_7 are normalized; and so forth. At the kth step, 2^k terms are normalized at once. The idea is to incorporate the idea of Newton's method, or iteration with quadratic convergence, using a formula for the derivative of the exponential mapping in a Lie algebra. See [99] for hints about how to proceed.

3.3 Normal and Hypernormal Forms When A_0 Is Semisimple

The task for putting a matrix series (3.1.1) into normal form always comes down to the solution of a homological equation such as

$$\pounds U_j = K_j - B_j,$$

where

$$\pounds = \mathbb{L}_{A_0} : \mathrm{gl}(n) \to \mathrm{gl}(n),$$

K_j is known at the jth stage of the calculation, and B_j and U_j are to be determined. (The notation U_j implies that the homological equation reproduced here is the one for format 2a, (3.2.35). The others are (3.2.19), (3.2.22), (3.2.36), and (3.2.40). The important point is that $\pounds$ is the same operator in every format.) Once a homological equation has been obtained, the steps that must be carried out to "solve" the equation and arrive at a normal form are as follows:

1. Choose a subspace $\mathcal{C}$ of $\mathrm{gl}(n)$ complementary to $\mathrm{im}\,\pounds$, so that

   $$\mathrm{gl}(n) = \mathrm{im}\,\pounds \oplus \mathcal{C}.$$

   We have referred to this step as *choosing a normal form style*. The subspace $\mathcal{C}$ itself is called the *normal form space* (of the specified style). Spelling out the requirements for a matrix to belong to $\mathcal{C}$ is called *solving the description problem* for that style.

2. Let

   $$\mathbb{P} : \mathrm{gl}(n) \to \mathrm{gl}(n)$$

   be the projection onto $\mathcal{C}$ associated with the splitting described above, and let $\mathbb{I} : \mathrm{gl}(n) \to \mathrm{gl}(n)$ be the identity mapping (whose matrix is the $n^2 \times n^2$ identity matrix). Calculate $B_j = \mathbb{P}K_j$ and $K_j - B_j = (\mathbb{I} - \mathbb{P})K_j$.

3. "Invert $\mathcal{L}$ on its image," and apply this *partial inverse* (more precisely, a right inverse of the restriction of $\mathcal{L}$ to its image) to $K_j - B_j$ to obtain U_j. This right inverse mapping, and U_j itself, are not unique (since ker $\mathcal{L} \neq \{0\}$).

The object of this section and the next two is to work out these steps in the two cases A_0 semisimple and A_0 nonsemisimple. The main focus will be on the description problem, that is, the question "what does the normal form look like?" The other questions to be addressed are "how do you compute the projections?" and "how do you find the partial inverse of $\mathcal{L}$?" Notice that *none of these questions depends on the format that is being used*. Once the projection operator and right inverse operator are known, the remaining steps of the calculation will be determined by the format being followed.

One of our concerns in this section is to provide computation methods that avoid the use of $n^2 \times n^2$ matrices. The most important operators ($\mathcal{L} = \mathbb{L}_{A_0}$, the projection $\mathbb{P}$, and any other operators written in similar "blackboard bold" type) are of this size if written as matrices. If n is fairly small, it may not be too inconvenient to use these larger matrices. But in Chapter 4, the corresponding matrices can have sizes much larger than n^2, and avoiding these matrices can be essential.

The Semisimple Normal Form Style

The key to the treatment of the semisimple case is the following lemma, which implies that $\mathcal{L}$ is a semisimple operator whenever A_0 is semisimple:

3.3.1. Lemma. If S is semisimple, then $\mathbb{L}_S$ is semisimple. More precisely, if S is real and is diagonalizable over $\mathbb{R}$ or over $\mathbb{C}$, then $\mathbb{L}_S$ (which is also real) is diagonalizable over the same field. If S is complex and diagonalizable (over $\mathbb{C}$), so is $\mathbb{L}_S$.

Proof. First, assume that S is already diagonal ($S = \Lambda$ with diagonal entries $(\lambda_1, \ldots, \lambda_n)$). Let E_{rs} be the matrix with one in the (r, s) position and zero elsewhere. Then

$$\mathbb{L}_S E_{rs} = (\lambda_s - \lambda_r) E_{rs}, \qquad (3.3.1)$$

as seen in the 2×2 example

$$\begin{bmatrix} 0 & 1 \\ 0 & 0 \end{bmatrix} \begin{bmatrix} \lambda_1 & 0 \\ 0 & \lambda_2 \end{bmatrix} - \begin{bmatrix} \lambda_1 & 0 \\ 0 & \lambda_2 \end{bmatrix} \begin{bmatrix} 0 & 1 \\ 0 & 0 \end{bmatrix} = \begin{bmatrix} 0 & \lambda_2 - \lambda_1 \\ 0 & 0 \end{bmatrix}.$$

It follows that $\mathbb{L}_S$ has n^2 linearly independent eigenvectors E_{rs}, and it is semisimple, as claimed.

If S is semisimple but not diagonal, then there is a matrix T such that $T^{-1}ST = \Lambda$ is diagonal. Then the required eigenvectors of $\mathbb{L}_S$ are $TE_{rs}T^{-1}$:

$$\mathbb{L}_S(TE_{rs}T^{-1}) = [TE_{rs}T^{-1}, T\Lambda T^{-1}] = T[E_{rs}, \Lambda]T^{-1} = (\lambda_s - \lambda_r)TE_{rs}T^{-1}.$$

These eigenvectors will be real if S and T are real, and again form a full set of eigenvectors, proving diagonalizability (over the appropriate field).   □

3.3.2. Corollary. If A_0 is semisimple with eigenvalues λ_r, then $\pounds$ is semisimple with eigenvalues $\lambda_s - \lambda_r$. If A_0 is diagonal, then

$$\pounds E_{rs} = (\lambda_s - \lambda_r) E_{rs}. \tag{3.3.2}$$

Since $\pounds$ is semisimple, the problem of finding a complement to its image falls under the case covered in Section 2.1, and by Theorem 2.1.3,

$$\mathrm{gl}(n) = \mathrm{im}\,\pounds \oplus \ker \pounds. \tag{3.3.3}$$

Thus, $\ker \pounds$ constitutes a very natural normal form space when A_0 is semisimple, and defines the normal form style that we call the *semisimple normal form* (since no other normal form style is commonly used in the semisimple case). Since a matrix belongs to $\ker \pounds$ if and only if it commutes with A_0, the semisimple normal form can be defined as follows:

3.3.3. Definition. A matrix series (3.1.1) is in semisimple normal form (to order k, or to all orders) if and only if

1. the leading term A_0 is semisimple; and

2. each term A_j (for $j \leq k$, or for all j) commutes with the leading term: $[A_j, A_0] = 0$, or $A_j \in \ker \pounds$.

Regarded as a formal power series, $A(\varepsilon)$ is in semisimple normal form to all orders if and only if A_0 is semisimple and $[A(\varepsilon), A_0] = 0$ (the zero formal power series). Frequently we say (redundantly, but for emphasis) that $A(\varepsilon)$ is in semisimple normal form *with respect to A_0*. Notice that *only* the leading term is required to be semisimple.

Although this section is strictly limited to the case that A_0 is semisimple, it will be useful in later sections to have on hand the following generalization of semisimple normal form.

3.3.4. Definition. A matrix series is in *extended semisimple normal form* if

1. $A_0 = S + N$ with S semisimple, N nilpotent, and $[S, N] = 0$; and

2. $[A_j, S] = 0$ for $j \geq 0$ (or only for $1 \leq j \leq k$).

Unless the word "extended" is included, the statement that $A(\varepsilon)$ is in semisimple normal form implies that A_0 is semisimple. In the literature, extended semisimple normal form is often called "normal form with respect to S."

> **3.3.5. Remark.** The extended semisimple normal form is not a normal form style. That is, $\ker \mathbb{L}_S$ is not a complement to $\mathrm{im}\,\pounds = \mathrm{im}\,\mathbb{L}_{A_0}$

when $N \neq 0$. The dimension of $\ker \mathbb{L}_S$ is too large; additional simplifications of $A(\varepsilon)$ can be performed to bring $A(\varepsilon)$ into one of the normal form styles described in Sections 3.4 and 3.5.

It is often useful to know whether or not a given normal form style is preserved under linear coordinate changes. This is a question that must be posed for each style separately, and receives quite different answers in the different styles. For the semisimple style the answer is always yes:

3.3.6. Theorem. If $A(\varepsilon)$ is in semisimple normal form with respect to A_0, and T is any invertible $n \times n$ matrix, then $\mathbb{S}_T A(\varepsilon) = T^{-1}A(\varepsilon)T$ is in semisimple normal form with respect to $\mathbb{S}_T A_0 = T^{-1}A_0 T$.

Proof. $[T^{-1}A(\varepsilon)T, T^{-1}A_0 T] = T^{-1}[A(\varepsilon), A_0]T = 0.$ □

The Description Problem for the Semisimple Normal Form

To complete the solution of the *description problem* for the semisimple normal form style it is necessary to determine which matrices commute with A_0, that is, to calculate $\ker \mathcal{L}$. Remember that the "zeroth step" in the normalizing process, given by (3.2.1), is to put A_0 into canonical form, and that we assume that this step has already been carried out. If A_0 is real and semisimple, there are two natural canonical forms that might be used, the diagonal form (which might include complex eigenvalues in conjugate pairs), and the real semisimple canonical form (see Definition 2.1.2), in which conjugate pairs of eigenvalues are replaced by 2×2 blocks of the form (2.1.2). If A_0 is complex to begin with, and semisimple, there is only one natural canonical form, the diagonal form. All of these possibilities are covered if we consider two cases: A_0 diagonal, and A_0 in real semisimple canonical form.

Consider first the case in which A_0 is diagonal. (Its entries may be real or complex, and if complex, they may be in conjugate pairs or not.) Then (3.3.2) shows that the kernel of $\mathcal{L}$, which is its zero eigenspace, is spanned by those E_{rs} for which $\lambda_r = \lambda_s$. If the eigenvalues of A_0 are distinct, this span is simply the space of diagonal matrices. The semisimple normal form is then described by saying that $B_1, \ldots, B_k$ are diagonal.

If A_0 is diagonal but the eigenvalues of A_0 are not all distinct, it is easiest to describe the normal form if we assume that all repeated eigenvalues are placed next to one another in the diagonal. If A_0 contains a string of equal eigenvalues in its diagonal, the smallest submatrix of A_0 containing this diagonal string will be called the *block subtended by the string*. (A nonrepeated eigenvalue constitutes a string of length one, and the block it subtends is just that entry itself.) Then if r and s are integers such that the (r, s) position in A_0 lies in a block subtended by a string, it follows that $\lambda_r = \lambda_s$, and therefore $\mathcal{L}E_{rs} = 0$. We have immediately the following theorem:

3.3.7. Theorem. If A_0 is a diagonal $n \times n$ matrix, with equal eigenvalues adjacent, then ker $\mathcal{L}$ is the set of block diagonal $n \times n$ matrices, with blocks subtended by the strings of equal eigenvalues of A_0. A matrix series with such an A_0 as leading term is in semisimple normal form (up to degree k) if and only if the remaining terms (up to degree k) are in this block diagonal form.

An example of a 3×3 matrix series in semisimple normal form through order 2 is

$$\begin{bmatrix} 1 & & \\ & 1 & \\ & & 2 \end{bmatrix} + \varepsilon \begin{bmatrix} -3 & 6 & \\ -8 & 11 & \\ & & 7 \end{bmatrix} + \varepsilon^2 \begin{bmatrix} 5 & 1 & \\ 9 & \pi & \\ & & 17.5 \end{bmatrix} + \varepsilon^3 \begin{bmatrix} 96 & 0.023 & \sqrt{2} \\ 9 & 0 & 25 \\ 1 & 2 & 3 \end{bmatrix}.$$

(3.3.4)

Here A_0 contains two strings, of length two and one. The blocks subtended by these strings are the 2×2 block in the upper left corner, and the 1×1 block in the lower right. The kernel of $\mathcal{L}$ is spanned by $E_{11}, E_{12}, E_{21}, E_{22}, E_{33}$. The ε^3 term is not in block form because the series has not been normalized this far.

Next we turn to the case where the semisimple matrix A_0 is in real semisimple canonical form (see Definition 2.1.2) and contains at least one 2×2 block. We assume that the entire series $A(\varepsilon)$ is real, because if it is not, there is no reason to prefer the real form for A_0 over its complex diagonal form.

Rather than treat this case by entirely real arguments, it is simplest to calculate temporarily in complex form, and return to real form at the end. It is easy to find a (complex) similarity transformation carrying A_0 to diagonal form, and once this has been done, the normalization (with respect to $T^{-1}A_0T$) goes as before. Theorem 3.3.6 then implies that the inverse similarity carries the normalized series back to one that is in normal form with respect to A_0. But this theorem by itself does not show that this final series is real. This requires a closer look at T and the determination of the "reality condition" satisfied by a "real matrix in complex form."

We begin by constructing a matrix T that transforms A_0 into complex diagonal form by similarity. Consider first the case of a single block,

$$A_0 = \begin{bmatrix} \alpha & -\beta \\ \beta & \alpha \end{bmatrix}.$$

Letting

$$T = \frac{1}{2} \begin{bmatrix} 1 & 1 \\ -i & i \end{bmatrix}, \qquad T^{-1} = \begin{bmatrix} 1 & i \\ 1 & -i \end{bmatrix}, \tag{3.3.5}$$

it follows that

$$T^{-1}A_0T = \begin{bmatrix} \alpha + i\beta & 0 \\ 0 & \alpha - i\beta \end{bmatrix}.$$

(This choice of T is not unique, but is selected to work well in our applications; see for instance the discussion surrounding equation (4.5.36).) This is easily generalized to larger matrices in real semisimple canonical form. Thus, if

$$A_0 = \begin{bmatrix} \lambda & & & \\ & \alpha_1 & -\beta_1 & \\ & \beta_1 & \alpha_1 & \\ & & & \alpha_2 & -\beta_2 \\ & & & \beta_2 & \alpha_2 \end{bmatrix}$$

with $\beta_1 \neq 0$, $\beta_2 \neq 0$, and all entries real, then

$$T = \begin{bmatrix} 1 & & & \\ & 1/2 & 1/2 & \\ & -i/2 & i/2 & \\ & & & 1/2 & 1/2 \\ & & & -i/2 & i/2 \end{bmatrix}, \quad T^{-1} = \begin{bmatrix} 1 & & & \\ & 1 & i & \\ & 1 & -i & \\ & & & 1 & i \\ & & & 1 & -i \end{bmatrix},$$

and

$$T^{-1}A_0T = \begin{bmatrix} \lambda & & & \\ & \alpha_1 + i\beta_1 & & \\ & & \alpha_1 - i\beta_1 & \\ & & & \alpha_2 + i\beta_2 & \\ & & & & \alpha_2 - i\beta_2 \end{bmatrix}.$$

(Notice that if $\alpha_1 = \alpha_2$ and $\beta_1 = \beta_2$, equal eigenvalues are not adjacent in this matrix, and Theorem 3.3.7 does not apply. We will have more to say about this case after the proof of Theorem 3.3.9 below.) Now for any matrix $X \in \mathrm{gl}(n, \mathbb{R})$, define X' by

$$X' = \mathbb{S}_T(X) = T^{-1}XT,$$

where

$$\mathbb{S}_T : \mathrm{gl}(n, \mathbb{R}) \to \mathrm{gl}(n, \mathbb{C}).$$

This map is not onto, but its image is a subset $\mathcal{R} \subset \mathrm{gl}(n, \mathbb{C})$ that, *viewed as a real vector space* (that is, allowing multiplication only by real scalars), is isomorphic to $\mathrm{gl}(n, \mathbb{R})$. The set $\mathcal{R}$ is called the *reality subspace* of $\mathrm{gl}(n, \mathbb{C})$, and its elements (which are complex matrices) are said to *satisfy the reality condition* determined by T (or by $\mathbb{S}_T$). Matrices satisfying this condition can be returned to real form by applying $\mathbb{S}_T^{-1}$, where

$$\mathbb{S}_T^{-1}(Y) = TYT^{-1}.$$

3.3.8. Theorem. Suppose that

$$A(\varepsilon) = A_0 + \varepsilon A_1 + \cdots.$$

is a real matrix series in which A_0 is semisimple and is in real semisimple canonical form. Let T be defined as above, and let

$$A'(\varepsilon) = \mathbb{S}_T(A(\varepsilon)) = A'_0 + \varepsilon A'_1 + \cdots.$$

Then it is possible to choose a semisimple normal form

$$B'(\varepsilon) = A'_0 + \varepsilon B'_1 + \cdots$$

for the complex series $B'(\varepsilon)$ such that the series

$$B(\varepsilon) = \mathbb{S}_T^{-1}(B'(\varepsilon)) = A_0 + \varepsilon B_1 + \cdots$$

is a real matrix series in normal form with respect to A_0.

Proof. Each term of $A'(\varepsilon)$ is a complex matrix lying in $\mathcal{R}$. The computation of the normal form $B'(\varepsilon)$, using any format, can then be carried out entirely within the reality subspace $\mathcal{R}$. We give the argument for format 2a. Each homological equation for the normalization of $A'(\varepsilon)$ has the form

$$\mathcal{L}'U'_k = K'_k - B'_k, \tag{3.3.6}$$

where $\mathcal{L}' = [\ , A'_0]$; see (3.2.35), and note that we put primes on U_k and K_k. We claim that each K'_k and U'_k can be taken to lie in $\mathcal{R}$. This is true for $K'_1 = A'_1$; assume (inductively) that $K'_k \in \mathcal{R}$. Observe that $\mathcal{L}'$ maps $\mathcal{R}$ into $\mathcal{R}$; any element of $\mathcal{R}$ may be written $X' = T^{-1}XT$ for $X \in \mathrm{gl}(n, \mathbb{R})$, so that $\mathcal{L}'X' = [T^{-1}XT, T^{-1}A_0T] = T^{-1}[X, A_0]T \in \mathcal{R}$; more generally, $\mathcal{R}$ is closed under the commutator bracket. (In the language of Section 3.6 below, $\mathcal{R}$ is a Lie subalgebra of $\mathrm{gl}(n, \mathbb{C})$ regarded as a Lie algebra over $\mathbb{R}$.) Therefore, equation (3.3.6) may be interpreted as an equation in $\mathcal{R}$, and we may choose $B'_k \in \mathcal{R}$ in such a way that $K'_k - B'_k$ belongs to the image of $\mathcal{L}' : \mathcal{R} \to \mathcal{R}$. Then U'_k can be chosen to lie in $\mathcal{R}$, and since K'_{k+1} is expressed (via the fundamental theorem of Lie series, Theorem 3.2.1) using brackets of elements of $\mathcal{R}$, it lies in $\mathcal{R}$ itself, completing the induction. Finally, since each B'_k is in $\mathcal{R}$, it follows that each B_k is in $\mathrm{gl}(n, \mathbb{R})$. $\square$

Theorem 3.3.8 leads to an easy characterization of the real normal form in the case that the eigenvalues of A_0 are distinct.

3.3.9. Corollary. If A_0 is a semisimple matrix in real semisimple canonical form, and if all of the 1×1 and 2×2 diagonal blocks are distinct, then the semisimple normal form will be in real semisimple canonical form with the same block sizes.

Proof. In the notation of Theorem 3.3.8, A'_0 will be a diagonal matrix with distinct eigenvalues. Theorem 3.3.7 then implies that the $B'j$ are diagonal. Theorem 3.3.8 implies that the terms of this complex normal form satisfy the reality condition, and that applying $\mathbb{S}_T^{-1}$ will produce the normalized real series. Finally, it is easy to check that when $\mathbb{S}_T^{-1}$ is applied to a diagonal matrix satisfying the reality condition, the result is a matrix in real semisimple canonical form. $\square$

When there are repetitions among the 1×1 or 2×2 blocks in A_0, it is not as easy to describe the real normal form. It is easy to see that the normalized series is zero outside of the blocks subtended by the strings of repeated blocks in A_0, but beyond that, there are further restrictions that we will not attempt to classify. For instance, suppose that $\alpha_1 = \alpha_2$ and $\beta_1 = \beta_2$ in (3.3.5). Then, by a different choice of T (for which Theorem 3.3.8 is still valid), the complex diagonal form of A_0 can be arranged so as to make equal eigenvalues adjacent:

$$
A_0 = \begin{bmatrix} \alpha + i\beta & & & \\ & \alpha + i\beta & & \\ & & \alpha - i\beta & \\ & & & \alpha - i\beta \end{bmatrix}.
$$

Now by Theorem 3.3.7 the complex normal form will contain 2×2 blocks. When transformed back to real form this block structure will not be maintained, and it will appear that the real normal form has no particular visible structure (since the block subtended by the string of repeated blocks of A_0 is the whole matrix). But there will be a concealed normal form structure expressing the fact that in complex coordinates there are 2×2 blocks. The nature of this normal form structure could be worked out if needed.

The Computation Problem

As explained in the introduction to this section, the details of the computation problem for any normal form are mostly contained in the formulas governing the format that is being used for the calculation, but the two most important parts of the computation problem are independent of format: the projections, and the partial inverse of $\mathcal{L}$. We will explain how these are done for the semisimple normal form in two cases: when A_0 is diagonal, and when it is not necessarily diagonal (but is semisimple). The method for diagonal matrices is suitable for hand calculation when n is small enough. The general method is suitable for use with computer algebra systems.

When A_0 is diagonal, the projections of $gl(n)$ into $\text{im } \mathcal{L}$ and $\ker \mathcal{L}$ given by the splitting (3.3.3) are easy to compute. Theorem 3.3.7 identifies the kernel of $\mathcal{L}$ as the block diagonal matrix with blocks subtended by the strings of A_0; that is, the kernel consists of matrices that are zero outside of these blocks. The same reasoning used in that theorem shows that the image consists of those matrices that are zero inside these blocks (and arbitrary elsewhere): Indeed, (3.3.1) shows that E_{rs} belongs to $\text{im } \mathcal{L}$ whenever $\lambda_r \neq \lambda_s$. So to project any given matrix X into the image and the kernel it is enough to separate it into a sum of two matrices, one containing the entries of X outside these blocks (and zeros inside), the other the reverse. For instance, with A_0 as in (3.3.4), the projections of an arbitrary matrix are given by

$$\begin{bmatrix} a & b & c \\ d & e & f \\ g & h & i \end{bmatrix} = \begin{bmatrix} a & b & \\ d & e & \\ & & i \end{bmatrix} + \begin{bmatrix} & & c \\ & & f \\ g & h & \end{bmatrix}.$$

Calculation of these projections does not require the use of $n^2 \times n^2$ matrices.

The partial inversion of $\mathcal{L}$ is equally easy. If $X = (x_{rs})$ is a matrix in im $\mathcal{L}$, we can write

$$X = \sum x_{rs} E_{rs},$$

where the sum is taken over those (r, s) for which $\lambda_r \neq \lambda_s$. Setting

$$Y = \sum \frac{x_{rs}}{\lambda_s - \lambda_r} E_{rs},$$

it follows from (3.3.1) that

$$\mathcal{L}Y = X.$$

In other words, to obtain an inverse image of X under $\mathcal{L}$, assuming that its entries in the blocks subtended by the strings of A_0 are zero, divide each nonzero entry by the difference of the appropriate pair of eigenvalues. Thus, to solve $\mathcal{L}U_j = K_j - B_j$, B_j will be the part of K_j lying inside the blocks, and one possible U_j will be obtained from the part of K_j outside the blocks by dividing by eigenvalue differences. This choice of U_j will have zeros inside the blocks. The effect of nonuniqueness in U_j will be studied in the next subsection.

For the general algorithm (suitable for computer algebra systems, and not requiring that A_0 be already diagonal), it is best to number the eigenvalues of A_0 (which are assumed known) as $\lambda_{(1)}, \ldots, \lambda_{(r)}$, without repetitions (so that the multiplicity of λ_i is m_i). Let f_i be as in Corollary 2.1.5, so that $P_i = f_i(A_0) : \mathbb{C}^n \to E_i$ (or $\mathbb{R}^n \to E_i$) is the projection into the eigenspace of A_0 with eigenvalue $\lambda_{(i)}$. The eigenvalues of $\mathcal{L} = \mathbb{L}_{A_0}$ are $\lambda_{(i)} - \lambda_{(j)}$; there may be repetitions among this list. The following lemma computes the projections into the eigenspaces of $\mathcal{L}$ (which are subspaces of gl(n)). For the sake of later applications the lemma is stated for a semisimple matrix S and its associated operator $\mathbb{L}_S$; in our present application these are A_0 and $\mathcal{L}$. (The method of proof of this lemma is interesting in itself and will be used again in Chapter 4 in a different setting.)

3.3.10. Lemma. Let S be a semisimple matrix. The projection of a matrix R into the eigenspace of $\mathbb{L}_S$ with eigenvalue μ is

$$\mathbb{P}_\mu(R) = \sum_{\{(i,j):\lambda_{(i)} - \lambda_{(j)} = \mu\}} P_j R P_i.$$

Proof. It follows from the definition of eigenvalue that

$$S = \sum_{i=1}^{r} \lambda_{(i)} P_i.$$

Notice that for any vector v,

$$e^{\sigma S} v = \sum_{i=1}^{r} e^{\sigma \lambda_{(i)}} P_i(v), \qquad (3.3.7)$$

where σ is a parameter. According to Theorem 3.2.1,

$$e^{\sigma \mathbb{L}_S} R = e^{-\sigma S} R e^{\sigma S}$$

$$= \left(\sum_{j} e^{-\sigma \lambda_{(l)}} P_j \right) R \left(\sum_{i=1}^{r} e^{\sigma \lambda_{(i)}} P_i \right)$$

$$= \sum_{\mu} e^{\sigma \mu} \left(\sum_{\{(i,j) : \lambda_{(i)} + \lambda_{(j)} = \mu\}} P_j R P_i \right).$$

This last expression is a sum of the same type as (3.3.7), with S replaced by $\mathbb{L}_S$ and v replaced by R. Notice that (3.3.7) could be "read in reverse"; that is, it could be used to determine the projections $P_i(v)$ rather than, given the projections, to compute $e^{\sigma S} v$. In the same way, it follows that the coefficient of $e^{\sigma \mu}$ in the last displayed equation is the projection of R into the eigenspace of $\mathbb{L}_S$ with eigenvalue μ. $\qquad \square$

It is perhaps worthwhile to illustrate the idea of the proof of the last lemma (rather than the lemma itself) by an example. If

$$S = \begin{bmatrix} 1 & 1 \\ -2 & 4 \end{bmatrix},$$

it can be calculated (for instance, by diagonalizing S) that

$$e^{\sigma S} = \begin{bmatrix} 2e^{2\sigma} - e^{3\sigma} & -e^{2\sigma} + e^{3\sigma} \\ 2e^{2\sigma} - 2e^{3\sigma} & -e^{2\sigma} + 2e^{3\sigma} \end{bmatrix}.$$

To project

$$R = \begin{bmatrix} 1 & 0 \\ 0 & 0 \end{bmatrix},$$

into the eigenspaces of $\mathbb{L}_S$, we can compute

$$e^{-\sigma S} \begin{bmatrix} 1 & 0 \\ 0 & 0 \end{bmatrix} e^{\sigma S} = e^{-\sigma} \begin{bmatrix} -2 & 1 \\ -4 & 2 \end{bmatrix} + e^{0\sigma} \begin{bmatrix} 5 & -3 \\ 6 & -4 \end{bmatrix} + e^{\sigma} \begin{bmatrix} -2 & 2 \\ -2 & 2 \end{bmatrix}.$$

The three terms give the components in the eigenspaces with eigenvalues $-1, 0, 1$ respectively.

The next algorithm follows at once from this lemma and the previous discussion:

3.3.11. Algorithm. If A_0 is semisimple, the homological equation

$$\pounds U_j = K_j - B_j$$

can be solved by setting

$$B_j = \mathbb{P}_0(K_j)$$

and

$$U_j = \sum_{\mu \neq 0} \frac{1}{\mu} \mathbb{P}_\mu(K_j - B_j),$$

where $\mathbb{P}_\mu$ is constructed as in Lemma 3.3.10.

Hypernormal Forms in the Semisimple Case

In the last subsection, we have seen that when A_0 is diagonal, the most natural solution of $\pounds U_j = K_j - B_j$ has zeros in the blocks subtended by strings of equal eigenvalues in A_0. But the value of U_j is not unique. One could add to this U_j anything in the kernel of $\pounds$, which means that the positions inside the blocks could be filled in arbitrarily. This would have no effect on the computation of the "current" normal form term B_j, but would modify the values of later K_j and so would have an effect on later terms in the normal form. Thus, the normal form of a given $A(\varepsilon)$ is not unique, even within a given normal form style (such as the "ker $\pounds$ style" we are using here for the semisimple case). This phenomenon occurs in every branch of normal form theory, and suggests that there might be a "simplest" normal form (within a given style). This subject is sometimes called "normalizing beyond the normal form" or finding a "unique normal form." We call it a *hypernormal form.*

The first point at which nonuniqueness in the normal form arises is the "zeroth" step described in (3.2.1): There is more than one possible choice for the similarity transformation T_0 used to diagonalize a given semisimple matrix A_0. The choice made at this time will affect all later stages in the calculation of the normal form. It would be nice to make the best possible choice from the beginning, but this is hardly possible, because we do not know how the later calculations will go. Or we could use the "general" choice for T_0, containing arbitrary parameters to be chosen later, but this makes the calculations very complicated. Instead, the best approach is to make the initial choice arbitrarily (or in the easiest manner), calculate a normal form, and then go back to the first step to see whether an improvement can be made.

Suppose, for example, that a matrix series has been normalized through order ε^2 and the result is (3.3.4). Here A_0 is diagonal, but A_1 has a 2×2 block where the repeated eigenvalue of A_0 occurs. Has the normalization been done in the best way, or can further simplification (hypernormalization) be achieved? Examining the 2×2 block in A_1, we see that it is diagonalizable:

$$\begin{bmatrix} 1 & 3 \\ 1 & 4 \end{bmatrix}^{-1} \begin{bmatrix} -3 & 6 \\ -8 & 11 \end{bmatrix} \begin{bmatrix} 1 & 3 \\ 1 & 4 \end{bmatrix} = \begin{bmatrix} 3 & 0 \\ 0 & 5 \end{bmatrix}.$$

Since the matrix

$$S = \begin{bmatrix} 1 & 3 & \\ 1 & 4 & \\ & & 1 \end{bmatrix}$$

commutes with the first term of (3.3.4), applying S as a similarity leaves A_0 unchanged but diagonalizes the block in A_1 to produce

$$\begin{bmatrix} 1 & & \\ & 1 & \\ & & 2 \end{bmatrix} + \varepsilon \begin{bmatrix} 3 & 0 & \\ 0 & 5 & \\ & & 7 \end{bmatrix} + \varepsilon^2 \begin{bmatrix} * & * & \\ * & * & \\ & & 17.5 \end{bmatrix} + \cdots . \qquad (3.3.8)$$

This is still in normal form through quadratic terms, but the term of order ε has been simplified. Applying S in this manner is equivalent to choosing a different similarity in "step zero" at the beginning, but the best choice could not have been known until the term of order ε was computed.

We may continue with this example and ask whether the term of order ε^2 can be improved. The appropriate question to ask at this point is, What similarities of the form $I + \varepsilon S$ can be applied that will not affect the first two terms, but will simplify the ε^2 term? (On the "first pass" through the normalization process, similarities of the form $I + \varepsilon S$ were used to simplify the ε term. On this "second pass," at each stage we use similarities of the same form that do not affect their usual "target term," in an effort to simplify the following term.)

With matrices, there is a simple way to answer this question. (There is no comparable approach to the hypernormalization problems of Chapter 4.) Ignoring terms of order ε^3 and higher, (3.3.8) can be separated into two matrix series, using the block structure. The series formed from the 2×2 blocks is

$$\begin{bmatrix} 1 & \\ & 1 \end{bmatrix} + \varepsilon \begin{bmatrix} 3 & 0 \\ 0 & 5 \end{bmatrix} + \varepsilon^2 \begin{bmatrix} * & * \\ * & * \end{bmatrix},$$

and the 1×1 series is $2 + 7\varepsilon + 17.5\varepsilon^2$. If the first term of the 2×2 series is deleted and the rest divided by ε, the result is

$$\begin{bmatrix} 3 & 0 \\ 0 & 5 \end{bmatrix} + \varepsilon \begin{bmatrix} * & * \\ * & * \end{bmatrix}.$$

This can be viewed as a new normalization problem with a new A_0 term, which is diagonal and has distinct eigenvalues. According to the results of this section, the term of order ε can be diagonalized by a similarity of the form $I + \varepsilon S$, where I and S are 2×2. Replacing I by the 3×3 identity matrix and making S into a 3×3 matrix by adding zero entries gives a similarity $I + \varepsilon S$ that when applied to (3.3.8) diagonalizes the ε^2 term.

This can obviously be carried out to any order. The general procedure is as follows. Suppose that A_0 in (3.1.1) is diagonal, with repeated eigenvalues adjacent. Choose an integer k. Normalize the series up to order k and truncate. The result will be block diagonal in all terms, with blocks subtended by the strings of repeated eigenvalues. Separate the blocks into new matrix series of smaller dimension, delete the first term of each, and divide by ε. If the leading terms of each such series are semisimple, diagonalize them. If the eigenvalues are distinct at this point, then everything diagonalizes up to order k; otherwise, each series will block diagonalize with new (smaller) block sizes. Split the series according to these new blocks, delete the first term, divide by ε, and normalize again. The only obstruction that can occur will be if one of the new leading terms is not semisimple, in which case methods from the next two sections must be applied.

There is one important case in which the obstruction cannot arise, namely when the original matrix series is symmetric (or else Hermitian) to all orders. Of course, a symmetric matrix is always diagonalizable, but this does not guarantee that the diagonalization can be done smoothly in ε (or, more or less equivalently, as a power series in ε). It will be shown in Section 3.6 (see the discussion surrounding equation (3.6.16)) that by using a generated format with skew-symmetric (or skew-Hermitian) generators, a symmetric series can be normalized in such a way that the terms remain symmetric. Then, when the hypernormalization is attempted, each new leading term will be symmetric and therefore diagonalizable. This proves the following theorem:

3.3.12. Theorem. A symmetric or Hermitian matrix series can be diagonalized to any order by hypernormalization.

Notes and References

There is nothing new in this section, except possibly for the method used to prove Theorem 3.3.12, which is mostly contained in Section 3.6. On the other hand, it is hard to think of references to cite. One reference that does contain almost everything in this section (in very sketchy form) is Section 1.3 of Bogaevsky and Povsner [17]. I did not use this reference in preparing the section; I merely applied the well-known methods of the nonlinear semisimple case (Section 4.5 below) to the linear case and derived everything "from scratch."

3.4   Inner Product and Simplified Normal Forms

In this section and the next, we will develop three normal form styles for $A(\varepsilon)$ in the case that A_0 is not necessarily semisimple. The first, which we call the *inner product normal form*, is developed by the methods of

Section 2.2, which depend upon the existence of a suitable inner product for gl(n). The second, *simplified normal form*, is derived by modifying the inner product normal form. The third, based on starred Sections 2.5–2.7, is given in Section 3.5.

The Inner Product Normal Form Style

To construct a normal form space is to find a complement to im $\mathcal{L}$. According to Section 2.2, one such complement is given by ker $\mathcal{L}^*$, where * is the adjoint with respect to some inner product on gl(n). In order for this to be effective, there must be an inner product on gl(n) for which $\mathcal{L}^*$ is easily computable. The following lemma shows that this is the case. The lemma is stated for the complex case; there is no change if everything is real, except that the complex conjugates may be omitted (so that, for instance, A_0^* becomes $A_0^\dagger$). Two inner products are needed, one on $\mathbb{C}^n$ and another on gl($n, \mathbb{C}$). For the former, the *standard inner product* is defined by

$$\langle x, y \rangle = \sum_i x_i \bar{y}_i;$$

this induces the *standard adjoint* for matrices (or linear maps) $A : \mathbb{C}^n \to \mathbb{C}^n$, namely the *conjugate transpose*

$$A^* = \overline{A}^\dagger.$$

The inner product on gl($n, \mathbb{C}$) is the *Frobenius inner product*

$$\langle A, B \rangle = \sum_{rs} a_{rs} \bar{b}_{rs}.$$

The next lemma computes the adjoint of $\mathcal{L}$ with respect to the Frobenius inner product:

3.4.1. Lemma. For any matrix $A \in$ gl($n, \mathbb{C}$),

$$(\mathbb{L}_A)^* = \mathbb{L}_{(A^*)}.$$

In particular, the $\mathcal{L}$ operator for a given matrix series (3.1.1) has an adjoint given by

$$\mathcal{L}^* = \mathbb{L}_{(A_0^*)}. \tag{3.4.1}$$

Proof. What is needed is to show that

$$\langle \mathbb{L}_{(A^*)} B, C \rangle = \langle B, \mathbb{L}_A C \rangle,$$

so that $\mathbb{L}_{(A^*)}$ satisfies the defining conditions for the adjoint of $\mathbb{L}_A$ with respect to the inner product of matrices. First note that for any matrices P, Q, R,

$$\langle PQ, R \rangle = \sum_{ijk} p_{ij} q_{jk} \bar{r}_{ik} = \langle Q, P^* R \rangle$$

(the second equality follows from the double conjugation and transpose of P), and similarly,

$$\langle PQ, R \rangle = \langle P, RQ^* \rangle.$$

Then

$$\begin{aligned}
\langle \mathbb{L}_{(A^*)}B, C \rangle &= \langle BA^* - A^*B, C \rangle \\
&= \langle BA^*, C \rangle - \langle A^*B, C \rangle \\
&= \langle B, CA \rangle - \langle B, AC \rangle \\
&= \langle B, \mathbb{L}_A C \rangle.
\end{aligned}$$

$\square$

3.4.2. Definition. A matrix series $A(\varepsilon)$ with leading term A_0 is in *inner product normal form* (with respect to A_0) if its terms A_j for $j \geq 1$ belong to ker $\mathcal{L}^*$, that is, commute with A_0^*. (The total series $A(\varepsilon)$ need not commute with A_0^*, because A_0 need not commute with A_0^*.)

> **3.4.3. Remark.** Notice that we *do not* say that $A(\varepsilon)$ is in normal form with respect to A_0^*. Those who are familiar with only the semisimple normal form may be in the habit (based on Definition 3.3.3) of thinking that "in normal form with respect to" means the same as "commutes with." However, we *always* think of a series in normal form as normalized with respect to its own linear term. The normal form style that is used will determine what the terms of the series commute with.

As mentioned in connection with Theorem 3.3.6, different normal form styles behave differently under linear transformations. For the inner product normal form, the result is as follows:

3.4.4. Definition. A matrix T is *unitary* if $T^* = T^{-1}$, or equivalently, if T preserves the inner product (meaning that $\langle Tu, Tv \rangle = \langle u, v \rangle$ for all u and v); for real matrices, "unitary" is the same as "orthogonal." We will call a matrix T *weakly unitary* if

$$T^* = cT^{-1} \tag{3.4.2}$$

for some constant $c \neq 0$; this implies that T preserves orthogonality ($u \perp v$ if and only if $Tu \perp Tv$).

An example of a weakly unitary transformation is (3.3.5). This example could be made unitary by putting $1/\sqrt{2}$ in both T and T^{-1}, but since weakly unitary transformations are sufficient for the next theorem, it is more convenient to leave it as it is.

3.4.5. Theorem. If $A(\varepsilon)$ is in inner product normal form with respect to A_0, and T is weakly unitary, then $\mathbb{S}_T A(\varepsilon)$ is in inner product normal form with respect to $\mathbb{S}_T A_0$.

Proof. Let $j \geq 1$. Then because $T^* = cT^{-1}$,

$$
\begin{aligned}
[T^{-1}A_jT, (T^{-1}A_0T)^*] &= [T^{-1}A_jT, T^*A_0^*(T^{-1})^*] \\
&= [T^{-1}A_jT, cT^{-1}A_0^*\frac{1}{c}T] \\
&= T^{-1}[A_j, A_0^*] \\
&= 0.
\end{aligned}
$$

$\square$

3.4.6. Remark. Another approach to the effect of a linear map T is to allow nonunitary maps but change the inner product. Just as the inner product on $\mathrm{gl}(n)$ was built on the standard inner product on $\mathbb{C}^n$, it would be possible to create new inner products based on nonstandard inner products on $\mathbb{C}^n$. Then $\mathbb{S}_T A(\varepsilon)$ would be in inner product normal form with respect to the inner product induced by T, which will differ from the Frobenius inner product if T is not unitary.

Since the inner product normal form theory works over $\mathbb{R}$ or over $\mathbb{C}$, a real matrix series has a real inner product normal form. (Regarded as a complex matrix series, it will also have inner product normal forms that are not real.) In most cases, for real series, it is natural to take A_0 to be in real canonical form. But as in Theorem 3.3.8 (for the semisimple normal form), it is often desirable to calculate in complex form and return to real form at the end. It is necessary to be careful that the linear transformation from real to complex form preserves the inner product normal form, and also that the normalization process preserves the reality conditions.

3.4.7. Corollary. Suppose that $A(\varepsilon)$ is a real matrix series with A_0 in real canonical form. Then $A(\varepsilon)$ can be brought into inner product normal form by first performing a (complex) similarity so that the leading term takes Jordan form, then bringing the resulting complex series into inner product normal form, and, finally, performing the reverse similarity. It is possible to do this in such a way that the resulting series is real.

Proof. To prove the first part of the corollary, it suffices (by Theorem 3.4.5) to show that if A_0 is in real canonical form, there exists a weakly unitary matrix T such that $T^{-1}A_0T$ is in Jordan canonical form. This is easiest to see by way of an example. If

$$
A_0 = \begin{bmatrix} \alpha & -\beta & 1 & 0 \\ \beta & \alpha & 0 & 1 \\ 0 & 0 & a & -\beta \\ 0 & 0 & \beta & \alpha \end{bmatrix}
$$

and

$$
T = \begin{bmatrix} 1 & 1 & & \\ -i & i & & \\ & & 1 & 1 \\ & & -i & i \end{bmatrix} \begin{bmatrix} 1 & 0 & 0 & 0 \\ 0 & 0 & 1 & 0 \\ 0 & 1 & 0 & 0 \\ 0 & 0 & 0 & 1 \end{bmatrix},
$$

then

$$
T^{-1}A_0T = \begin{bmatrix} \alpha + i\beta & 1 & & \\ & \alpha + i\beta & & \\ & & \alpha - i\beta & 1 \\ & & & \alpha - i\beta \end{bmatrix}.
$$

The first factor in T is based on (3.3.5), which is weakly unitary and diagonalizes the semisimple part of A_0. The second factor merely re-orders the basis vectors to bring equal eigenvalues together, incidentally bringing the off-diagonal ones into the correct position. The same approach works for larger matrices. To prove the second part of the theorem (that with correct choices, the final series will be real) one follows the pattern of Theorem 3.3.8: Define the *reality subspace* of $gl(n, \mathbb{C})$ to be the image of $gl(n, \mathbb{R})$ under $\mathbb{S}_T$ (with T constructed as above), and check that the homological equations can be solved within $\mathcal{R}$, so that after applying $\mathbb{S}_T^{-1}$ the result will be real.   □

Splitting the Description Problem into Semisimple and Nilpotent Parts

Recall that according to Theorem 2.4.7, every linear operator can be written uniquely as a sum of commuting semisimple and nilpotent parts. Therefore, the leading term A_0 of an arbitrary matrix series can be written as $A_0 = S + N$. The operators $\mathcal{L} = \mathbb{L}_{A_0}$ and $\mathcal{L}^* = \mathbb{L}_{A_0^*}$ can also be decomposed into commuting semisimple and nilpotent parts. The following lemma states that these decompositions are given by

$$
\mathcal{L} = \mathbb{L}_S + \mathbb{L}_N \quad \text{and} \quad \mathcal{L}^* = \mathbb{L}_{S^*} + \mathbb{L}_{N^*}.
$$

(These equations themselves are trivial, because $\mathbb{L}_X$ is linear in X. What is significant is that these give the desired decompositions.) The lemma also states that the inner product normal form space for A_0 is the *intersection* of the inner product normal form spaces for S and N separately. In other words, the presence of S in A_0 allows certain simplifications of $A_1, A_2, \ldots$, and the presence of N allows other simplifications; when both are present together, both kinds of simplification can be achieved:

3.4.8. Lemma. Suppose that $A_0 = S + N$, where S is semisimple, N is nilpotent, and S and N commute. Then $\mathbb{L}_S$ is semisimple, $\mathbb{L}_N$ is nilpotent, and $\mathbb{L}_S$ and $\mathbb{L}_N$ commute. Also, $\mathbb{L}_{S^*}$ is semisimple, $\mathbb{L}_{N^*}$ is nilpotent, and these commute. Finally,

$$\ker \mathcal{L}^* = \ker \mathbb{L}_S^* \cap \ker \mathbb{L}_N^*. \qquad (3.4.3)$$

Proof. $\mathbb{L}_S$ is semisimple by Lemma 3.3.1. To see that $\mathbb{L}_N$ is nilpotent, suppose first that $N^2 = 0$ and let X be arbitrary. Then $\mathbb{L}_N^2 X = [[X, N], N] = XN^2 - 2NXN + N^2 X = -2NXN$ and $\mathbb{L}_N^3 X = -2N^2 XN + 2NXN^2 = 0$. This illustrates the pattern: If $N^k = 0$, then when $L_N^{2k-1} X$ is expanded, each term will contain $2k - 1$ factors of N, each of which must occur on one side or the other of X. One side or the other must contain at least k factors, since if both sides contained fewer than k, the total would not be $2k - 1$. Therefore, each term must vanish separately, and $\mathbb{L}_N^{2k-1} = 0$. The Jacobi identity for commutator brackets states that

$$[[X, S], N] + [[S, N], X] + [[N, X], S] = 0;$$

the middle term vanishes because S and N commute, so $\mathbb{L}_N \mathbb{L}_S X = [[X, S], N] = [[X, N], S] = \mathbb{L}_S \mathbb{L}_N X$. In other words, $\mathbb{L}_S$ and $\mathbb{L}_N$ commute. Since the conjugate transposes S^* and N^* are also commuting semisimple and nilpotent operators, the same arguments apply to $\mathbb{L}_S^* = \mathbb{L}_{(S^*)}$ and $\mathbb{L}_N^* = \mathbb{L}_{(N^*)}$. Finally, choose a basis for $gl(n)$ such that with respect to this basis, $\mathcal{L}^* : gl(n) \to gl(n)$ is in upper Jordan form. Because of the uniqueness of the semisimple/nilpotent decomposition, $\mathbb{L}_{S^*}$ will be the diagonal part of this matrix, and $\mathbb{L}_{N^*}$ will be the off- diagonal part. The kernel of any matrix in upper Jordan form is the intersection of the kernels of its diagonal and off-diagonal parts, because the kernel of the diagonal part consists of vectors having nonzero entries only in rows where the eigenvalue is zero, and the kernel of the off-diagonal part consists of vectors having nonzero entries only in positions corresponding to bottom rows of Jordan blocks.   □

3.4.9. Remark. An alternative proof that $\mathbb{L}_N$ is nilpotent, for those reading the starred sections, goes as follows. By Theorem 2.7.1, there is a triad $\{X, Y, Z\}$ with $X = N$. Then $\mathbb{L}_Y, \mathbb{L}_X, \mathbb{L}_Z$ is a triad. (Note the change of order, due to the fact that $\mathbb{L}_P = [\ , P]$ acts on the right, and note that since Z is semisimple, so is $\mathbb{L}_Z$.) Then $\mathbb{L}_N = \mathbb{L}_X$ is nilpotent by Theorem 2.5.2. This proof does not give a bound on the index of nilpotence, as does the proof above.

Seminormal Matrices

It is natural to ask whether the inner product normal form coincides with the semisimple normal form when A_0 is semisimple. Surprisingly, perhaps, this is not always the case. But it is the case if A_0 is diagonal. The question can be generalized: Under what circumstances will it be true that a matrix series in inner product normal form (so that its terms commute with A_0^*) is also in extended semisimple normal form (so that its terms commute with S, the semisimple part of A_0)? (See Definition 3.3.4.) For most purposes, it is enough to know that this is true if A_0 is either in Jordan or real canonical form. We will prove this here, together with a generalization to the case that A_0 is "seminormal." (The issue can be avoided by using the sl(2) normal form instead of the inner product normal form; an sl(2) normal form is always an extended semisimple normal form.)

The definition of "normal" given below is standard; "seminormal" is not. (Fortunately, since "normal" is easily confused with "normal form," the term will seldom be needed outside of this subsection.)

3.4.10. Definition. A matrix A is *normal* if $AA^* = A^*A$, or equivalently, if $[A, A^*] = 0$. A matrix is *seminormal* if its semisimple part is normal $(A = S + N, [S, S^*] = 0)$.

> **3.4.11. Remark.** That a normal matrix is semisimple is a well-known generalization of the spectral theorem; in fact, if A is normal, there exists a unitary matrix T such that $T^{-1}AT = T^*AT$ is diagonal. If $A = S + N$ is seminormal, there exists a unitary matrix T such that T^*ST is diagonal, but not necessarily one for which T^*AT is in Jordan form. The role of seminormal matrices in the theory of inner product normal forms is closely related to the fact that unitary matrices preserve inner product normal form (Theorem 3.4.5).

It is easy to check that a matrix in Jordan or real canonical form is seminormal. An example of a matrix that is not seminormal is

$$A = \begin{bmatrix} 1 & 2 \\ 3 & 1 \end{bmatrix}.$$

$A = S$ is semisimple (because its eigenvalues are distinct), but it does not commute with $S^* = S^\dagger$.

3.4.12. Lemma. If A is normal, then $\ker A = \ker A^*$. If A is seminormal, $\ker S = \ker S^*$.

Proof. Let A be normal. By Theorem 2.2.1, $\operatorname{im} A$ and $\ker A^*$ are complementary. Suppose $A^*Ax = 0$; then Ax belongs to both $\operatorname{im} A$ and $\ker A^*$, so $Ax = 0$. Therefore, $\ker A^*A = \ker A$; similarly, $\ker AA^* = \ker A^*$. Finally, $\ker A = \ker A^*A = \ker AA^* = \ker A^*$. If A is seminormal, then S is normal, and the same argument applies to S. □

The following theorem implies that when the leading term is both semisimple and normal, the inner product and semisimple normal forms coincide. This will be the case if A_0 is diagonal, or is in real semisimple canonical form:

3.4.13. Theorem. Let $A(\varepsilon)$ be in inner product normal form with respect to A_0. If A_0 is seminormal, then $A(\varepsilon)$ is in extended semisimple normal form (Definition 3.3.4). If A_0 is both semisimple and normal, then $A(\varepsilon)$ is in semisimple normal form.

Proof. If $A_0 = S + N$ is seminormal, then

$$[\mathbb{L}_S, \mathbb{L}_S^*] = [\mathbb{L}_S, \mathbb{L}_{S^*}] = \mathbb{L}_{[S^*, S]} = 0,$$

so $\mathbb{L}_S$ is normal. It follows from Lemma 3.4.12 that

$$\ker \mathbb{L}_S^* = \ker \mathbb{L}_S.$$

According to Lemma 3.4.8,

$$\ker \mathcal{L}^* \subset \ker \mathbb{L}_S^* = \ker \mathbb{L}_S.$$

If $A(\varepsilon)$ is in inner product normal form, it belongs to $\ker \mathcal{L}^*$ and therefore to $\ker \mathbb{L}_S$, so it is in extended semisimple normal form. If A_0 is semisimple and normal, the word "extended" can be deleted.   □

The Description Problem for a Single Nilpotent Jordan Block

It is best to begin the study of the inner product normal form by considering the extreme opposite of the semisimple case, namely, the case of A_0 nilpotent. Among nilpotent matrices, it is easiest to consider those having only one Jordan block. The example

$$A_0 = N = N_3 = \begin{bmatrix} 0 & 1 & 0 \\ 0 & 0 & 1 \\ 0 & 0 & 0 \end{bmatrix}$$

will reveal the structure of the inner product normal form under these circumstances. To determine the kernel of $\mathcal{L}^*$, let

$$X = \begin{bmatrix} a & b & c \\ d & e & f \\ g & h & i \end{bmatrix}$$

and calculate

$$\mathcal{L}^* X = X N^* - N^* X$$

$$= \begin{bmatrix} a & b & c \\ d & e & f \\ g & h & i \end{bmatrix} \begin{bmatrix} 0 & 0 & 0 \\ 1 & 0 & 0 \\ 0 & 1 & 0 \end{bmatrix} - \begin{bmatrix} 0 & 0 & 0 \\ 1 & 0 & 0 \\ 0 & 1 & 0 \end{bmatrix} \begin{bmatrix} a & b & c \\ d & e & f \\ g & h & i \end{bmatrix}$$

$$= \begin{bmatrix} b & c & 0 \\ e & f & 0 \\ h & i & 0 \end{bmatrix} - \begin{bmatrix} 0 & 0 & 0 \\ a & b & c \\ d & e & f \end{bmatrix}$$

$$= \begin{bmatrix} b & c & 0 \\ e - a & f - b & -c \\ h - d & i - e & -f \end{bmatrix}.$$

It follows that $X \in \ker \mathcal{L}^*$ if and only if b, c, and f vanish, $a = e = i$, and $h = d$, that is, provided that

$$X = \begin{bmatrix} a & 0 & 0 \\ d & a & 0 \\ g & d & a \end{bmatrix}. \tag{3.4.4}$$

3.4.14. Definition. A *simple striped matrix* is a square matrix $C = (c_{ij})$ such that $c_{ij} = 0$ for $j > i$ and $c_{ij} = c_{kl}$ whenever $i - j = k - l$. That is, the entries above the main diagonal are zero, and the entries within any diagonal on or below the main diagonal are equal.

3.4.15. Lemma. If $A_0 = N$ is a nilpotent $n \times n$ matrix in upper Jordan form with one Jordan block, then $\ker \mathcal{L}^*$ is the set of simple striped $n \times n$ matrices. A matrix series with such an A_0 as leading term is in inner product normal form (up to a given order) if and only if the succeeding terms (up to that order) are simple striped matrices.

Proof. If X is an arbitrary $n \times n$ matrix, XN^* will equal X with its columns shifted to the left (dropping the first column and filling the last with zeros), and N^*X will equal X with its rows shifted down. As a result, $\ker \mathbb{L}_{(N^*)}$ will consist of simple striped matrices. $\qquad\square$

According to Theorem 2.2.1, the splitting associated with the inner product normal form is orthogonal:

$$\mathrm{gl}(n, \mathbb{C}) = \mathrm{im}\, \mathcal{L} \oplus_\perp \ker \mathcal{L}^*. \tag{3.4.5}$$

Therefore, to compute the projection $\mathbb{P} : \mathrm{gl}(n) \to \ker \mathcal{L}^*$ it suffices to produce an orthonormal basis for the kernel and apply Remark 2.2.4. In the example of N_3, a basis for the kernel consists of the "basic striped matrices"

$$\begin{bmatrix} 1 & 0 & 0 \\ 0 & 1 & 0 \\ 0 & 0 & 1 \end{bmatrix}, \quad \begin{bmatrix} 0 & 0 & 0 \\ 1 & 0 & 0 \\ 0 & 1 & 0 \end{bmatrix}, \quad \begin{bmatrix} 0 & 0 & 0 \\ 0 & 0 & 0 \\ 1 & 0 & 0 \end{bmatrix},$$

which are already orthogonal with respect to the inner product for matrices, but are not of unit length. An orthonormal basis is then

$$U_1 = \begin{bmatrix} 1/\sqrt{3} & 0 & 0 \\ 0 & 1/\sqrt{3} & 0 \\ 0 & 0 & 1/\sqrt{3} \end{bmatrix}, \quad U_2 = \begin{bmatrix} 0 & 0 & 0 \\ 1/\sqrt{2} & 0 & 0 \\ 0 & 1/\sqrt{2} & 0 \end{bmatrix},$$

$$U_3 = \begin{bmatrix} 0 & 0 & 0 \\ 0 & 0 & 0 \\ 1 & 0 & 0 \end{bmatrix}.$$

It follows that the projection of an arbitrary X into the kernel of $\mathcal{L}^*$ is

$$\mathbb{P}(X) = \langle X, U_1 \rangle U_1 + \langle X, U_2 \rangle U_2 + \langle X, U_3 \rangle U_3$$

$$= \begin{bmatrix} (a+e+i)/3 & 0 & 0 \\ (d+h)/2 & (a+e+i)/3 & 0 \\ g & (d+h)/2 & (a+e+i)/3 \end{bmatrix}.$$

(Notice that the radicals disappear.) Thus, we have the following lemma:

3.4.16. Lemma. Under the circumstances of Lemma 3.4.15, the projection into normal form is obtained by "averaging the stripes" (and setting the entries outside the stripes equal to zero).

An example of using Lemma 3.4.16 to split a particular matrix in accordance with 3.4.5 is

$$\begin{bmatrix} 6 & 3 & 9 \\ 1 & 1 & 4 \\ 7 & 3 & 2 \end{bmatrix} = \begin{bmatrix} 3 & 3 & 9 \\ -1 & -2 & 4 \\ 0 & 1 & -1 \end{bmatrix} + \begin{bmatrix} 3 & 0 & 0 \\ 2 & 3 & 0 \\ 7 & 2 & 3 \end{bmatrix}. \tag{3.4.6}$$

Notice that the component in im $\mathcal{L}$ has the property that the sum of the entries in the main diagonal, or any diagonal below the main diagonal, is zero. These sums are called *stripe sums*, because they are the sums of the elements in the same diagonals used to define striped matrices.

3.4.17. Lemma. Under the circumstances of Lemma 3.4.15, a matrix belongs to im $\mathcal{L}$ if and only if its stripe sums are zero.

Proof. This follows immediately from Lemma 3.4.16 and (3.4.5). Another proof uses the following calculation for the case of $N = N_3$:

$$\mathcal{L} \begin{bmatrix} a & b & c \\ d & e & f \\ g & h & i \end{bmatrix} = \begin{bmatrix} -d & a-e & b-f \\ -g & d-h & e-i \\ 0 & g & h \end{bmatrix} = \begin{bmatrix} p & q & r \\ s & t & u \\ v & w & x \end{bmatrix}.$$

It is clear that $v = 0$, $s + w = 0$, and $p + t + x = 0$, showing that the image has zero stripe sums. Conversely, if $p, \ldots, x$ are given subject to the condition that the stripe sums are zero, the determination of $a, \ldots, i$ is easy: Clearly, $g = w$, $h = x$, and $d = -p$; by taking $i = 0$ it follows that $e = u$ and $a = q + u$; by taking $f = 0$ it follows that $b = r$; and c can be taken to vanish. $\square$

The proof of this lemma also completes the solution of the computation problem for the current example by showing how to "invert $\mathcal{L}$ on its image." The important point is that even though we are actually solving a system of n^2 equations, the system is in fact already presented as a collection of disjoint subsystems in row echelon form, ready to be solved by back substitution, and there is almost no work to do.

The Description Problem for Several Nilpotent Jordan Blocks

Most of these results continue to hold if A_0 is nilpotent but has more than one Jordan block, provided that the notions of striped matrix and stripe sum are suitably generalized. We will work out the case of two Jordan blocks explicitly.

Suppose

$$A_0 = N = \begin{bmatrix} N' & \\ & N'' \end{bmatrix}.$$

Then

$$\mathcal{L}^* \begin{bmatrix} A & B \\ C & D \end{bmatrix} = \begin{bmatrix} AN'^* - N'^*A & BN''^* - N'^*B \\ CN'^* - N''^*C & DN''^* - N''^*D \end{bmatrix}.$$

The diagonal blocks, $[A, N'^*]$ and $[D, N''^*]$, are commutators of the same form that occurred in the previous case of a single Jordan block, so the kernel of $\mathcal{L}^*$ has simple striped matrices in these positions. If $N' = N''$, the off-diagonal blocks (which in this case are square) are also commutators, and the kernel will have simple striped matrices in these positions as well. If N' and N'' are of different sizes, it is easy to check that the kernels in these positions are *nonsquare striped blocks*: The entries are constant along stripes, which are now defined as lines of "slope minus one" joining an entry on the left edge of the block to an entry on the bottom; the rest of the block is zero. For instance, if

$$N_{2,3} = \begin{bmatrix} 0 & 1 & & & \\ 0 & 0 & & & \\ & & 0 & 1 & 0 \\ & & 0 & 0 & 1 \\ & & 0 & 0 & 0 \end{bmatrix},$$

then ker $\mathcal{L}^*$ consists of matrices having the form

$$\begin{bmatrix} b & 0 & | & d & 0 & 0 \\ a & b & | & c & d & 0 \\ - & - & | & - & - & - \\ 0 & 0 & | & i & 0 & 0 \\ f & 0 & | & h & i & 0 \\ e & f & | & g & h & i \end{bmatrix}. \tag{3.4.7}$$

A matrix series having $A_0 = N_{2,3}$ will be in inner product normal form if and only if its succeeding terms have the form (3.4.7). These results are included in Theorem 3.4.21 below.

The Description Problem in the General Case

It follows from (3.4.3) that when $A_0 = S + N$, the inner product normal form space ker $\mathcal{L}^*$ can be determined in two stages:

1. Find the kernel of $\mathbb{L}_{S*}$ on $gl(n)$.

2. Find the kernel of $\mathbb{L}_{N*}$ on the subspace $\ker \mathbb{L}_{S*} \subset gl(n)$.

For the remainder of this section, we impose the condition that A_0 be seminormal (Definition 3.4.10), and in most cases the stronger assumption that A_0 is in Jordan form. If A_0 is seminormal, Theorem 3.4.13 implies that $\ker \mathbb{L}_{S*} = \ker \mathbb{L}_S$. This kernel is described by Theorem 3.3.7, and consists of block-diagonal matrices with blocks subtended by the strings of equal eigenvalues in S; this block structure is just given by the large blocks of the Jordan structure of A_0. The second step is then to find the kernel of $\mathbb{L}_{N*}$ on the subspace $\ker \mathbb{L}_S$. This is a purely nilpotent problem, and amounts to imposing a stripe structure (using the small blocks of A_0) onto each of these large blocks. This is most easily seen through an example.

If

$$
A_0 = \begin{bmatrix} 3 & & & & \\ & 3 & 1 & & \\ & 0 & 3 & & \\ & & & 5 & 1 \\ & & & 0 & 5 \end{bmatrix},
\tag{3.4.8}
$$

then the large block structure associated with A_0 is

$$
\begin{bmatrix} P & 0 \\ 0 & Q \end{bmatrix} = \begin{bmatrix} * & * & * & & \\ * & * & * & & \\ * & * & * & & \\ & & & * & * \\ & & & * & * \end{bmatrix}.
\tag{3.4.9}
$$

That is, $\ker \mathbb{L}_S$ consists of matrices having this block structure. So it remains to find the kernel of $\mathbb{L}_N^*$ acting on matrices of this form. But $\mathbb{L}_N^*$ acts on these blocks independently. That is, if we write

$$
N' = \begin{bmatrix} 0 & & \\ & 0 & 1 \\ & 0 & 0 \end{bmatrix}, \qquad N'' = \begin{bmatrix} 0 & 1 \\ 0 & 0 \end{bmatrix}, \qquad N = \begin{bmatrix} N' & 0 \\ 0 & N'' \end{bmatrix},
$$

then

$$
\mathbb{L}_N^* \begin{bmatrix} P & 0 \\ 0 & Q \end{bmatrix} = \begin{bmatrix} \mathbb{L}_{N'}^* P & 0 \\ 0 & \mathbb{L}_{N''}^* Q \end{bmatrix}.
$$

But the kernels of $\mathbb{L}_{N'}^*$ and $\mathbb{L}_{N''}^*$ are just the normal form spaces for the purely nilpotent matrices N' and N'', so these have already been determined: Each large block is gridded into square and (in the case of N')

nonsquare striped matrices. The result is

$$
\begin{bmatrix}
a & | & b & 0 & | & & \\
- & | & - & - & | & & \\
0 & | & e & 0 & | & & \\
c & | & d & e & | & & \\
- & - & - & - & - & - & - \\
& & & | & g & 0 \\
& & & | & f & g
\end{bmatrix}. \tag{3.4.10}
$$

The general rules to construct a normal form, such as (3.4.10), from a leading term A_0 in Jordan form, such as (3.4.8), are as follows:

1. Beginning with an empty $n \times n$ matrix (the size of A_0), draw the vertical and horizontal lines that form the boundaries of the large blocks of A_0. (A large blocks contains all Jordan blocks having a specific eigenvalue.) Only the diagonal large blocks will be "active" for the rest of the construction. The off-diagonal blocks (of this large block structure) receive zero entries.

2. Inside each diagonal large block, draw the vertical and horizontal lines that form the boundaries of the small blocks (Jordan blocks) of A_0. Continue each of these lines until it reaches the boundary of the large block. Within each large block, both the diagonal and off-diagonal blocks bounded by these lines will be "active." Do not enter any zeros at this stage.

3. Draw lines at 45 degrees (slope -1) joining the left edge of each active block to its bottom edge. These lines are the "stripes." In the diagonal active blocks, which are square, the first such stripe will be the main diagonal, and the others will be the subdiagonals. In the off-diagonal active blocks, which may not be square, the first stripe will *either* begin at the upper left entry *or* end at the lower right entry (but not both, unless the block is square); the other stripes will lie below this one. No stripe should ever be drawn that touches the top or right side of a block (except at a corner).

4. After all stripes have been drawn, each entry that is not on a stripe should be set equal to zero, and the entries in any one stripe should be set equal to one another (using a different letter or symbol to denote the value on each stripe).

The following definitions and lemmas will be useful later:

3.4.18. Definition.

1. A matrix obtained by the rules described above will be called a *striped matrix* having the *stripe structure defined by* A_0. The number of stripes in the stripe structure is called the *codimension* of A_0.

(The reason for the name "codimension" is that it equals n^2 minus the dimension of the set of matrices in $gl(n)$ similar to A_0.)

2. The stripe structure defined by A_0 can be "applied" to any matrix of the same size as A_0, whether or not that matrix is striped. If C is any $n \times n$ matrix, a *stripe* in C is the set of entries in C lying on one of the lines of the stripe structure of A_0, even though these entries are not equal.

3. A *stripe sum* for such a matrix C is the sum of the entries in a particular stripe.

4. The *projection into striped matrices* is the map

$$\mathbb{P} : gl(n) \to gl(n)$$

defined as follows: Let $X \in gl(n)$. Then $\mathbb{P}(X)$ is the unique striped matrix having the same stripe sums as X. In other words, each entry of X that is not in a stripe is replaced by 0, and each entry that does belong to a stripe is replaced by the average of the entries in the stripe. It is easy to check that $\mathbb{P}$ is linear and $\mathbb{P}^2 = \mathbb{P}$.

5. Suppose that a particular stripe in the stripe structure defined by A_0 has been selected. The *stripe subspace* associated with that stripe is the subspace of $gl(n)$ consisting of matrices having zero entries outside of the chosen stripe. The matrices E_{rs}, as (r, s) ranges over the indices of positions in the stripe, constitutes a natural basis for the stripe subspace.

3.4.19. Lemma. If A_0 is in upper Jordan form, arranged so that Jordan blocks with equal eigenvalues are adjacent, then the projection $\mathbb{P} : gl(n) \to \ker \mathcal{L}^*$ preserves stripe sums.

Proof. Since each element of a stripe is replaced by the average of all the elements in the stripes, the sum of the elements in each stripe is unchanged by the projection. □

3.4.20. Lemma. Under the same conditions, $im \mathcal{L}$ is the set of matrices whose stripe sums are zero.

Proof. $\mathbb{I} - \mathbb{P}$ is the projection into $im \mathcal{L}$ associated with (3.4.5). Since $\mathbb{P}$ preserves stripe sums, elements of the image of $\mathbb{I} - \mathbb{P}$ have stripe sums equal to zero. If the stripe sums of X are zero, then $\mathbb{P}(X) = 0$ and $(\mathbb{I} - \mathbb{P})(X) = X$; therefore, $\mathbb{I} - \mathbb{P}$ maps onto the space of matrices with zero stripe sums. □

The same arguments leading to (3.4.10) establish the following theorem:

3.4.21. Theorem. If A_0 is in upper Jordan canonical form, arranged so that Jordan blocks with equal eigenvalues are adjacent, then a matrix series is in inner product normal form if and only if the succeeding terms are striped matrices (with respect to the stripe structure of A_0).

The Computation Problem for the Inner Product Normal Form (General Case)

There is an "elementary" way to solve the computation problem for the inner product normal form, and a more "sophisticated" way. Although the elementary way is best for hand calculation of simple problems, the "sophisticated" way is probably best for computer algebra systems, and uses ideas that will occur again in Chapter 4.

Suppose that $A_0 = S + N$ is in Jordan form (as described in Theorem 3.4.21), and that we are required to find B_j and U_j such that

$$\pounds U_j = K_j - B_j \tag{3.4.11}$$

with $\pounds = \mathbb{L}_{S+N}$ and with $B_j \in \ker \pounds^* = \ker \mathbb{L}_S \cap \ker \mathbb{L}_{N^*}$. The elementary approach is to set

$$B_j = \mathbb{P}(K_j), \tag{3.4.12}$$

where $\mathbb{P} : \mathrm{gl}(n) \to \ker \pounds^*$ is the projection into striped matrices (see Definition 3.4.18). It follows that $K_j - B_j = (\mathbb{I} - \mathbb{P})K_j \in \mathrm{im}\,\pounds$, and the system (3.4.11) of n^2 equations for the n^2 entries of U_j is solvable. When A_0 is nilpotent ($S = 0$) this system of equations is not difficult to solve, because, when written out, it is already in row echelon form. This is no longer the case when $S \neq 0$, but the system is still not far from row echelon form, and the solution is not difficult if n is not too large.

The more sophisticated approach is a four-step procedure that can be outlined as follows. It is assumed that $A_0 = S + N$ is the semisimple/nilpotent splitting of A_0. It is not required that these be in Jordan form, but we do assume that A_0 is seminormal, so that $\ker \mathbb{L}_{S^*} = \ker \mathbb{L}_S$.

3.4.22. Algorithm. To solve equation (3.4.11):

1. Find $\widehat{U}_j$ and $B'_j \in \ker \mathbb{L}_S$ such that

$$\mathbb{L}_S \widehat{U}_j = K_j - B'_j. \tag{3.4.13}$$

2. Convert $\widehat{U}_j$ into a solution U'_j of the equation

$$\pounds U'_j = K_j - B'_j. \tag{3.4.14}$$

The method for doing this will be described below.

3. Regarding $\mathbb{L}_N$ as a map of $\ker \mathbb{L}_S \to \ker \mathbb{L}_S$ (or more precisely, restricting $\mathbb{L}_N$ to $\ker \mathbb{L}_S$), solve the homological equation

$$\mathbb{L}_N U''_j = B'_j - B_j, \tag{3.4.15}$$

requiring that $B_j \in \ker \mathbb{L}_{N^*}$. This is a nilpotent problem, and (automatically) $B_j \in \ker \pounds^*$.

4. Let $U_j = U'_j + U''_j$. Then it follows that (3.4.11) holds (with the B_j obtained in the previous step).

We will now work through these steps in more detail.

Step 1 is to ignore the nilpotent part of A_0 and solve the equation (3.4.13). Since S is semisimple, this can be handled by Algorithm 3.3.11. If A_0 is in Jordan form, the hand calculation method from Section 3.3 can be used. In this case B'_j will be block diagonal with the large block structure of A_0 (i.e., blocks subtended by the repeated eigenvalue strings of S). Then $\mathbb{L}_S$ can be inverted on its image by "dividing the nonzero eigenspaces of $\mathbb{L}_S$ by their eigenvalues." Notice that $\widehat{U}_j$ *cannot* be used to generate a transformation that will be applied for normalization purposes, because (3.4.13) is not a valid homological equation for our problem: It does not contain $\mathcal{L}$.

Step 2 is intended to rectify this problem. What is needed is to convert $\widehat{U}_j$ into a solution of (3.4.14). This is accomplished by means of the following algorithm (which does not require that A_0 be in Jordan form):

3.4.23. Algorithm. If

$$\mathbb{L}_S P = Q,$$

and if R is defined by

$$R = \sum_{r=0}^{k-1} (-1)^r (\mathbb{L}_S^{-1})^r \mathbb{L}_N^r P,$$

where k is the index of nilpotence of $\mathbb{L}_N$ and $\mathbb{L}_S^{-1}$ denotes any right inverse of $\mathbb{L}_S$ on its image, then

$$\mathbb{L}_{S+N} R = Q.$$

Proof. Apply $\mathbb{L}_{S+N} = \mathbb{L}_S + \mathbb{L}_N$ to R and observe the cancellation. Notice that the summation might as well be over $r = 0, \dots, \infty$, since $\mathbb{L}_N$ is nilpotent: The point is to calculate until the terms become zero.   □

> **3.4.24. Remark.** Since (3.4.14) is a valid homological equation containing $\mathcal{L}$, it is possible to use U'_j as a generator: *If it is so desired, we can stop at step 2.* That is, we can be content with the transformation generated by U'_j and obtain a normalized series with jth term B'_j, in block diagonal form. The result will be a matrix series in extended semisimple normal form (Definition 3.3.4). But the presence of the nilpotent part makes it possible to simplify B'_j still further to obtain striped matrices B_j (with respect to the stripe structure defined by A_0), by continuing with steps 3 and 4.

Step 3 begins by considering the equation

$$\mathcal{L} U''_j = B'_j - B_j, \tag{3.4.16}$$

where B'_j is the intermediate normal form already found, and B_j is the final normal form that is still to be found. The main point now is that every term in (3.4.16) can be considered to belong to $\ker \mathbb{L}_S$: B'_j already lies in this kernel, B_j will lie in it by Lemma 3.4.8, and we are free to look for a solution U_j'' in $\ker \mathbb{L}_S$. But, in this kernel,

$$\mathcal{L} U_j'' = \mathbb{L}_{S+N} U_j'' = \mathbb{L}_S U_j'' + \mathbb{L}_N U_j'' = \mathbb{L}_N U_j''.$$

So to solve (3.4.16) it is sufficient to solve (3.4.15) in the vector space $\ker \mathbb{L}_S$. This is a purely nilpotent problem. If A_0 is in Jordan form, $\ker \mathbb{L}_S$ consists of block-diagonal matrices. In that case it is easy to find B_j as the orthogonal projection of B_j' into $\ker \mathbb{L}_{N^*} \cap \ker \mathbb{L}_S$, and the equations for the entries of U_j'' will automatically be in row echelon form. (See the proof of Lemma 3.4.17.) If A_0 is not in Jordan form, there is no good way of solving (3.4.16) except by brute force as an $n^2 \times n^2$ system of equations. It is just at this point that the sl(2) normal form, treated in the next section, provides assistance: The pushdown algorithm of Section 2.6 provides an easy algorithmic solution of the equivalent of (3.4.16).

> **3.4.25. Remark.** A further advantage of the sl(2) normal form at this point is to eliminate the need to assume that A_0 is seminormal. Without this assumption, $\mathbb{L}_S$ must sometimes be replaced by $\mathbb{L}_{S^*}$ in the previous discussion, and step 3 no longer works, at least as stated here. Specifically, (3.4.16) is now to be solved in $\ker \mathbb{L}_{S^*}$, but with $U_j'' \in \ker \mathbb{L}_{S^*}$ it is no longer true that $\mathbb{L}_S U_j'' = 0$. With the sl(2) normal form, $\mathbb{L}_{S^*}$ does not arise.

Finally, step 4 is to add (3.4.14) and (3.4.16) to obtain $\mathcal{L}(U_j' + U_j'') = K_j - B_j$. The solution of (3.4.11) is

$$U_j = U_j' + U_j''.$$

The Simplified Normal Form

The *simplified normal form style* is a simplification of the inner product normal form, having (in general) many more zero entries than does the inner product normal form. Its disadvantages include the fact that it is defined only when A_0 is in Jordan canonical form, and that its normal form space is not orthogonal to im $\mathcal{L}$.

3.4.26. Definition. If A_0 is in upper Jordan form, arranged so that Jordan blocks with equal eigenvalues are adjacent, then a matrix series with leading term A_0 is in *simplified normal form* if its succeeding terms have nonzero entries only in the bottom position of each stripe in the stripe structure defined by A_0.

Thus, the simplified normal form of (3.4.10) is

$$
\begin{bmatrix}
a' & b' & 0 & & \\
0 & 0 & 0 & & \\
c' & d' & e' & & \\
& & & 0 & 0 \\
& & & f' & g'
\end{bmatrix},
\tag{3.4.17}
$$

and that of (3.4.7) is

$$
\begin{bmatrix}
0 & 0 & 0 & 0 & 0 \\
a' & b' & c' & d' & 0 \\
0 & 0 & 0 & 0 & 0 \\
0 & 0 & 0 & 0 & 0 \\
e' & f' & g' & h' & i'
\end{bmatrix}.
\tag{3.4.18}
$$

3.4.27. Theorem. The space of matrices with nonzero entries only at the bottom of each stripe is in fact a normal form space (a complement of im $\mathcal{L}$). A matrix series in simplified normal form is also in extended semisimple normal form (Definition 3.3.4).

Proof. Let $\widehat{\mathbb{P}}$ be the projection that replaces each matrix entry by zero except the entries at the bottom of each stripe; these are replaced by the stripe sum. Thus, for the matrix in (3.4.6) we have

$$
\widehat{\mathbb{P}}\left(\begin{bmatrix} 6 & 3 & 9 \\ 1 & 1 & 4 \\ 7 & 3 & 2 \end{bmatrix}\right) = \begin{bmatrix} 0 & 0 & 0 \\ 0 & 0 & 0 \\ 7 & 4 & 9 \end{bmatrix}.
\tag{3.4.19}
$$

To see that $\widehat{\mathbb{P}}$ is a normal form projection, it is enough to check that its image has the correct dimension for a complement to im $\mathcal{L}$, and that the complementary projection $\mathbb{I} - \widehat{\mathbb{P}}$ maps into im $\mathcal{L}$. The former is true because $\dim \operatorname{im} \widehat{\mathbb{P}} = \dim \operatorname{im} \mathbb{P}$ (both equalling the number of stripes); the latter follows from Lemma 3.4.20, because $(\mathbb{I} - \widehat{\mathbb{P}})(X)$ has stripe sums equal to zero for any X. In the example,

$$
(\mathbb{I} - \widehat{\mathbb{P}})\left(\begin{bmatrix} 6 & 3 & 9 \\ 1 & 1 & 4 \\ 7 & 3 & 2 \end{bmatrix}\right) = \begin{bmatrix} 6 & 3 & 9 \\ 1 & 1 & 4 \\ 0 & -1 & -7 \end{bmatrix}.
$$

Finally, if $A(\varepsilon)$ is a matrix series in simplified normal form, the nonzero entries of A_j for $j \geq 1$ belong to the stripes, and hence to the large block structure determined by A_0 (which is in Jordan form). Therefore, these matrices commute with the diagonal part of A_0, and $A(\varepsilon)$ is in extended semisimple normal form. □

Although it is the simplest normal form, in the sense of having the most zero entries, the simplified normal form is not a "natural" complement to im $\mathcal{L}$, because it does not arise from a natural "mathematical structure" such as the inner product structure that defines the inner product normal form or the Lie algebraic structure involved in the sl(2) normal forms of Section 3.5 below. By the same token, the projections into the simplified normal form style are not orthogonal projections (as in the inner product style), nor do they follow from the pushdown algorithm (as in the triad style). Another way to realize the "unnaturalness" of the simplified normal form is to notice that its very definition is coordinate-dependent. The inner product and sl(2) normal forms are meaningful whether or not A_0 is in Jordan form; the inner product normal form space, for instance, is just ker $\mathcal{L}^*$. It is true that the convenient description of the inner product normal form using stripe structures is valid only when A_0 is in Jordan form, but this description is not the way the normal form is *defined*. For the simplified normal form style, on the contrary, the *definition* of the normal form space depends on the stripe structure and hence on the assumption that A_0 is in Jordan form. In addition, the *proof* that the simplified normal form space is a complement of im $\mathcal{L}$ depends on Lemma 3.4.20. Thus, the simplified normal form cannot stand by itself as a complete theory, but is subordinate to the theory of inner product normal forms. (It can also be obtained from the sl(2) normal form in a similar way.)

Notes and References

The inner product used here occurs in parts of classical invariant theory and mathematical methods of quantum mechanics, but as a normal form theory in the nonlinear setting (Section 4.6 below), it was first employed by Belitskii [15] and later rediscovered independently by Elphick et al. [41]. The linear case worked out here is an immediate application, and is not by any means new, but has not (to my knowledge) been written up before. As far as the description problem is concerned, the resulting theory coincides exactly with the well-known theory of Arnol'd unfoldings for matrices, presented in pages 240–246 of Arnol'd [6] and in pages 305–320 of Wiggins [111]. The algorithms for the computation problem derive from Sanders [99], where they were presented for the sl(2) normal form style (in an abstract version covering both linear and nonlinear settings). When transposed to the inner product normal form style, these algorithms require the assumption of seminormality of A_0.

The simplified normal form is worked out for a few special nonlinear examples in [41]. The method there is to begin with the inner product normal form and "change the projection." I treated the general nonlinear case to second order in [87], and to all orders in [89]. The first detailed treatment is contained in this book (in Section 3.4 for the linear case and Section 4.6 for the nonlinear). The references [6] and [111] given above for the Arnol'd unfolding of a matrix contain several alternative versions of this unfolding differing from the basic striped matrix version; the simplified normal form for the linear setting is in the same spirit as these alternative

Arnol'd unfoldings (although it is not exactly the same as any of them), and can be derived in the same way.

3.5 * The sl(2) Normal Form

The sl(2) normal form is based on completely different principles than the inner product normal form, namely, the sl(2) representation structure introduced in the starred Sections 2.5, 2.6, and 2.7. But in the end, the normal forms themselves are not very different. If A_0 is in Jordan form, the sl(2) normal form has the same stripe structure as the inner product normal form, except that the stripes are not constant. Instead, they are constant multiples of certain *bias factors*. The projection into normal form is not obtained by averaging the stripes, but by a *biased averaging*. The stripe sums are preserved, the total being partitioned among the stripe entries in proportion to the bias factors. (The phrase "biased average," rather than "weighted average," is used to avoid conflict with the weights, or eigenvalues of Z, in sl(2) representation theory.) The significant theoretical advantages of the sl(2) approach will not be apparent until Chapter 4, where we see the interaction between nonlinear normal form theory and classical invariant theory. In the present section the main advantage of the triad theory is that it leads to algorithms that can be used to compute the normal form when A_0 is not in Jordan form. These algorithms are not easy to use by hand, but they can be programmed in a symbolic algebra system. There are times when it is convenient to use a form other than Jordan form, for instance, when A_0 has a structure of interest that is lost in Jordan form. For instance, Jordan form is not always appropriate for the study of Hamiltonian systems (Section 4.9).

To define the sl(2) normal form associated with a given leading matrix A_0 and its associated operator $\mathcal{L} = \mathbb{L}_{A_0}$, it is necessary to introduce some matrices and operators associated with A_0. First, according to Theorem 2.4.7, A_0 can be split into its semisimple and nilpotent parts: $A_0 = S + N$ with S semisimple, N nilpotent, and $[S, N] = 0$. Also, we know from Lemma 3.4.8 that $\mathcal{L} = \mathbb{L}_S + \mathbb{L}_N$ with $\mathbb{L}_S$ semisimple, $\mathbb{L}_N$ nilpotent, and $[\mathbb{L}_S, \mathbb{L}_N] = 0$. Next, by Theorem 2.7.1, there exists a triad of matrices $\{X, Y, Z\}$ with $X = N$ that commutes with S. That is to say,

$$[X, Y] = Z, \qquad [Z, X] = 2X, \qquad [Z, Y] = -2Y, \qquad (3.5.1)$$

and also

$$[S, X] = [S, Y] = [S, Z] = 0, \qquad (3.5.2)$$

and in addition, S and Z are simultaneously diagonalizable. Finally, from Theorem 2.7.6 the operators

$$\mathbb{X} = \mathbb{L}_Y, \qquad \mathbb{Y} = \mathbb{L}_X, \qquad \mathbb{Z} = \mathbb{L}_Z \qquad (3.5.3)$$

also form a triad. That is, they satisfy

$$[\mathbb{X}, \mathbb{Y}] = \mathbb{Z}, \qquad [\mathbb{Z}, \mathbb{X}] = 2\mathbb{X}, \qquad [\mathbb{Z}, \mathbb{Y}] = -2\mathbb{Y}, \qquad (3.5.4)$$

and $\mathbb{Z}$ is diagonalizable. (Warning: Notice the crossing over of X and Y in the definitions (3.5.3).)

Let $A_0 = S + N$ as above, let $\{X, Y, Z\}$ be a triad with $X = N$, and define the *pseudotranspose* of A_0 (relative to this triad) to be

$$\widetilde{A}_0 = S + Y \qquad (3.5.5)$$

(since $\widetilde{A}_0$ plays roughly the same role in the sl(2) normal form theory that A_0^* does in the inner product theory). Notice that if A_0 is in Jordan form and $\{X, Y, Z\}$ are constructed as in Theorem (2.7.1), then A_0^* and $\widetilde{A}_0$ are both very close to the transpose of A_0: The former is the conjugate transpose; the latter is obtained from the transpose by replacing the off-diagonal ones with the numbers that we have called "pressures" in Section 2.5. (If A_0 is not in Jordan form, and the triad is constructed by Algorithm 2.7.2, there is no obvious relation between A_0^* and $\widetilde{A}_0$.)

3.5.1. Theorem. The subspace $\ker \mathbb{L}_{\widetilde{A}_0} \subset \mathrm{gl}(n)$ is a normal form space, that is, a complement to $\mathrm{im}\,\pounds$. This space can be characterized as

$$\ker \mathbb{L}_{\widetilde{A}_0} = \ker \mathbb{L}_S \cap \ker \mathbb{L}_Y = \ker \mathbb{L}_S \cap \ker \mathbb{X}. \qquad (3.5.6)$$

Proof. According to Corollary 2.4.6, a complement to $\mathrm{im}\,\pounds$ (in the whole space $\mathrm{gl}(n)$) may be found by restricting attention to the zero eigenspace of $\pounds$. Let $W \subset \mathrm{gl}(n)$ denote this eigenspace, and let $'$ denote the restriction of a linear operator to W; then $\pounds'$ maps W to itself, and a complement to $\mathrm{im}\,\pounds'$ in W is a complement to $\mathrm{im}\,\pounds$ in $\mathrm{gl}(n)$, and hence a normal form space. According to Lemma 3.4.8, the semisimple and nilpotent parts of $\pounds$ are $\mathbb{L}_S$ and $\mathbb{L}_N = \mathbb{L}_X = \mathbb{Y}$, respectively. It follows that $W = \mathbb{L}_S$ and that $\pounds' = \mathbb{Y}'$. (It may help to think of the matrix of $\pounds$ in Jordan form, as in the proof of equation (3.4.3) in Lemma 3.4.8.) Thus, a complement to $\mathrm{im}\,\mathbb{Y}'$ will be a normal form space. Observe that $\{\mathbb{X}', \mathbb{Y}', \mathbb{Z}'\}$ forms a triad on W. (The most important point to check is that these operators map W into itself, which follows because they commute with $\mathbb{L}_S$.) Finally, from (2.5.4), proved as item 6 of Theorem 2.5.2, we have $W = \mathrm{im}\,\mathbb{Y}' \oplus \ker \mathbb{X}'$. Since $\ker \mathbb{X}'$ is a complement to $\mathrm{im}\,\mathbb{Y}'$, it is a normal form space. (Alternatively, we could apply item 3 of Theorem 2.5.2 to $\{\mathbb{X}', \mathbb{Y}', \mathbb{Z}'\}$ to see that $\ker \mathbb{X}'$ is the span of the chain tops of $\mathbb{Y}' = \pounds'$, and then refer back to Corollary 2.4.6 to see that this is a normal form space.) $\qquad\square$

It is important to understand that neither $\widetilde{A}_0$ nor the normal form space $\ker \mathbb{L}_{\widetilde{A}_0}$ is unique, because they depend upon the choice of $\{X, Y, Z\}$. For this reason the definition of sl(2) normal form must include a mention of the triad that is used:

3.5.2. Definition. A matrix series $A(\varepsilon)$, in which the leading term has semisimple/nilpotent decomposition $A_0 = S + N$, is said to be *in sl(2) normal form with respect to A_0 and $\{X, Y, Z\}$* if

1. $\{X, Y, Z\}$ is a triad such that $X = N$ and X, Y, and Z commute with S; and

2. each A_j for $j \geq 0$ commutes with $\widetilde{A}_0 = S + Y$.

In contrast to the case of the inner product normal form, the next corollary does not require that A_0 be seminormal:

3.5.3. Corollary. A matrix series $A(\varepsilon)$ in sl(2) normal form (with respect to A_0 and any suitable triad) is in extended semisimple normal form. If A_0 is semisimple, $A(\varepsilon)$ is in semisimple normal form.

Proof. By (3.5.6), the sl(2) normal form space is a subspace of $\ker \mathbb{L}_S$. For the case $A_0 = S$, it must be noted that $N = 0$ implies $X = Y = Z = 0$. (This triad does not actually span a Lie algebra isomorphic to sl(2).)   □

The nonuniqueness of sl(2) normal forms is highlighted by the following result, giving the behavior of this normal form style under linear transformations. (Compare Theorems 3.3.6 and 3.4.5.)

3.5.4. Theorem. If $A(\varepsilon)$ is in sl(2) normal form with respect to A_0 and $\{X, Y, Z\}$, and T is an invertible matrix, then $\mathbb{S}_T A(\varepsilon)$ is in sl(2) normal form with respect to $\mathbb{S}_T A_0$ and $\{\mathbb{S}_T X, \mathbb{S}_T Y, \mathbb{S}_T Z\}$.

Proof. By (3.5.6) it suffices to check that $\mathbb{S}_T A_j$ commutes with $\mathbb{S}_T S$ and $\mathbb{S}_T Y$. Both calculations go as in Theorem 3.3.6.   □

An Example of the sl(2) Normal Form

As for the inner product normal form, it is easiest to begin with a purely nilpotent example having a single Jordan block, such as

$$S = 0, \qquad N = N_4 = \begin{bmatrix} 0 & 1 & 0 & 0 \\ 0 & 0 & 1 & 0 \\ 0 & 0 & 0 & 1 \\ 0 & 0 & 0 & 0 \end{bmatrix}.$$

We have chosen this example because it is the smallest in which the sl(2) normal form is distinctly different from the inner product normal form. From Section 2.7, a triad with $X = N$ is given by

$$X = \begin{bmatrix} 0 & 1 & 0 & 0 \\ 0 & 0 & 1 & 0 \\ 0 & 0 & 0 & 1 \\ 0 & 0 & 0 & 0 \end{bmatrix}, \qquad Y = \begin{bmatrix} 0 & 0 & 0 & 0 \\ 3 & 0 & 0 & 0 \\ 0 & 4 & 0 & 0 \\ 0 & 0 & 3 & 0 \end{bmatrix}, \qquad Z = \begin{bmatrix} 3 & & & \\ & 1 & & \\ & & -1 & \\ & & & -3 \end{bmatrix}.$$

$$(3.5.7)$$

The first step is to construct a weight table for the induced triad $\{\mathbb{X}, \mathbb{Y}, \mathbb{Z}\}$. (Recall that a *weight table* is a list of the nonnegative weights and their multiplicities; see the example (2.5.20). Since $\mathbb{Z}$ is diagonal, the proof of Lemma 3.3.1 shows that $\mathbb{Z}E_{rs} = (w_s - w_r)E_{rs}$. Therefore, the E_{rs} form a weight basis (a basis of weight vectors) for $\mathrm{gl}(n)$, and the following "weight matrix" (written without brackets because it has no algebraic significance as a matrix) has in its (r, s) position the weight of E_{rs}:

$$
\begin{matrix}
0 & -2 & -4 & -6 \\
2 & 0 & -2 & -4 \\
4 & 2 & 0 & -2 \\
6 & 4 & 2 & 0
\end{matrix}.
$$

Notice that the weight spaces (eigenspaces of $\mathbb{Z}$) are just the stripe subspaces of $\mathrm{gl}(4)$, according to Definition 3.4.18 (including "stripes" above the main diagonal, which are not true stripes according to the definition because in a striped matrix they must vanish). The weight matrix leads immediately to the weight table:

$$
\begin{matrix}
\text{Weight} & 6 & 4 & 2 & 0 \\
\text{Multiplicity} & 1 & 2 & 3 & 4
\end{matrix}
$$

Theorem 2.5.3 implies that there is one chain each of top weights 6, 4, 2, and 0 (chain lengths 7, 5, 3, 1). The element E_{41} of weight 6 is the only choice for the top of the longest chain. Next, we must find the kernel of $\mathbb{X}$ on the stripe subspace spanned by E_{31} and E_{42}, the elements of weight 4. The fact that

$$
\mathbb{X}(aE_{31} + bE_{42}) = 3(b - a)E_{41}
$$

can be checked by working out

$$
\begin{bmatrix} 0 & 0 & 0 & 0 \\ 0 & 0 & 0 & 0 \\ a & 0 & 0 & 0 \\ 0 & b & 0 & 0 \end{bmatrix}
\begin{bmatrix} 0 & 0 & 0 & 0 \\ 3 & 0 & 0 & 0 \\ 0 & 4 & 0 & 0 \\ 0 & 0 & 3 & 0 \end{bmatrix}
-
\begin{bmatrix} 0 & 0 & 0 & 0 \\ 3 & 0 & 0 & 0 \\ 0 & 4 & 0 & 0 \\ 0 & 0 & 3 & 0 \end{bmatrix}
\begin{bmatrix} 0 & 0 & 0 & 0 \\ 0 & 0 & 0 & 0 \\ a & 0 & 0 & 0 \\ 0 & b & 0 & 0 \end{bmatrix}.
$$

Taking $a = b = 1$ gives $E_{31} + E_{42}$ as the top of the chain of length 5. The chain of length 3 is easy, since we need a matrix of the form $aE_{21} + bE_{32} + cE_{43}$ that commutes with Y (that is, lies in $\ker \mathbb{X} = \ker \mathbb{L}_Y$), and Y itself fills that role. Finally, I commutes with Y and lies in the zero weight space, giving the chain of length 1. Now the normal form is the span of these tops of the chains, so it is

$$
\begin{bmatrix}
a & 0 & 0 & 0 \\
3b & a & 0 & 0 \\
c & 4b & a & 0 \\
d & c & 3b & a
\end{bmatrix}.
\tag{3.5.8}
$$

The resemblance to a striped matrix (the inner product normal form) is clear, as well as the difference: the presence of the bias factors 3, 4, 3 in one of the stripes. We have solved the description problem for the sl(2) normal form when $A_0 = N_4$.

The Description Problem for the sl(2) Normal Form When A_0 Is in Jordan Form

In the last example, A_0 was nilpotent and contained only one Jordan block. To solve the description problem completely for A_0 in Jordan form, we consider first the nilpotent case $A_0 = N$ with several Jordan blocks, then the general case $A_0 = S + N$. It is assumed throughout this discussion that the triad $\{X, Y, Z\}$ is the one constructed in the proof of Theorem 2.7.1.

The nilpotent case with several Jordan blocks follows the same pattern as the last example, except that the stripe structure is more complicated (as it was for the inner product normal form) and there can be several stripe subspaces with the same weight. Consider the case

$$A_0 = N = N_{3,4} = \begin{bmatrix} 0 & 1 & 0 & & & & \\ 0 & 0 & 1 & & & & \\ 0 & 0 & 0 & & & & \\ & & & 0 & 1 & 0 & 0 \\ & & & 0 & 0 & 1 & 0 \\ & & & 0 & 0 & 0 & 1 \\ & & & 0 & 0 & 0 & 0 \end{bmatrix}.$$

The triad for this is $X = N$,

$$Y = \begin{bmatrix} 0 & 0 & 0 & & & & \\ 2 & 0 & 0 & & & & \\ 0 & 2 & 0 & & & & \\ & & & 0 & 0 & 0 & 0 \\ & & & 3 & 0 & 0 & 0 \\ & & & 0 & 4 & 0 & 0 \\ & & & 0 & 0 & 3 & 0 \end{bmatrix}, \quad Z = \begin{bmatrix} 2 & & & & & & \\ & 0 & & & & & \\ & & -2 & & & & \\ & & & 3 & & & \\ & & & & 1 & & \\ & & & & & -1 & \\ & & & & & & -3 \end{bmatrix}.$$

The "weight matrix" is

$$\begin{bmatrix} 0 & -2 & -4 & 1 & -1 & -3 & -5 \\ 2 & 0 & -2 & 3 & 1 & -1 & -3 \\ 4 & 2 & 0 & 5 & 3 & 1 & -1 \\ -1 & -3 & -5 & 0 & -2 & -4 & -6 \\ 1 & -1 & -3 & 2 & 0 & -2 & -4 \\ 3 & 1 & -1 & 4 & 2 & 0 & -2 \\ 5 & 3 & 1 & 6 & 4 & 2 & 0 \end{bmatrix}.$$

Notice that the weight space with weight 2 is the direct sum of two stripe subspaces, one spanned by E_{21} and E_{32}, the other by E_{54}, E_{65}, and E_{76}.

We could make a weight table and see that there are two chain tops of weight 2 (height 3). To find them, calculate the kernel of $\mathbb{X} = \mathbb{L}_Y$ on the weight space by solving the equation

$$\mathbb{X} \begin{bmatrix} 0 & 0 & 0 \\ a & 0 & 0 \\ 0 & b & 0 \\ & & & 0 & 0 & 0 & 0 \\ & & & c & 0 & 0 & 0 \\ & & & 0 & d & 0 & 0 \\ & & & 0 & 0 & e & 0 \end{bmatrix} = 0,$$

which works out to

$$\begin{bmatrix} 0 & & 0 & 0 \\ 0 & & 0 & 0 \\ 2b - 2a & & 0 & 0 \\ & & & 0 & & 0 & 0 & 0 \\ & & & 0 & & 0 & 0 & 0 \\ & & & 3c - 4d & & 0 & 0 & 0 \\ & & & 0 & & 4d - 3e & 0 & 0 \end{bmatrix} = 0.$$

The important point is that the equations for each stripe in the weight space are decoupled into the separate systems $2b - 2a = 0$ for the first stripe and $3c - 4d = 0, 4d - 3e = 0$ for the second stripe, and each system is solvable by back substitution (beginning with arbitrary values of b and e, most conveniently $b = 1$ and $e = 4$ to obtain integer values for the bias factors) to yield two separate chain tops. Similarly, the weight space of weight 1 is the direct sum of stripe spaces $aE_{14} + bE_{25} + cE_{36}$ and $dE_{51} + eE_{62} + fE_{73}$, and the kernel of $\mathbb{X}$ is determined by the decoupled systems $3b - 2a = 0, 4c - 2b = 0$ and $2e - 4d = 0, 2f - 3e = 0$. Again, each stripe subspace contributes a chain top.

When A_0 contains a semisimple part also, the results are as follows:

3.5.5. Theorem. If $A_0 = S + N$ is in Jordan form and Jordan blocks with the same eigenvalue are adjacent, each stripe subspace in the stripe structure determined by A_0 contains one basis element for the sl(2) normal form, which spans the one-dimensional kernel of $\mathbb{X}$ on the stripe subspace. The entries in the stripe, called *bias factors*, can be taken to be integers.

Proof. The first step is to determine the chain tops of the triad $\mathbb{X}, \mathbb{Y}, \mathbb{Z}$, which span $\ker \mathbb{X}$. This proceeds exactly as if A_0 were equal to N, that is, just as in the previous examples: Each stripe *in the stripe structure associated with N* contributes a chain top. Since N has the same Jordan block structure as A_0, but has all eigenvalues equal to zero, there are in general many more stripes in the stripe structure for N than in the stripe structure for A_0, but the stripes lying within the large blocks of A_0 are the same. By Lemma 3.5.1, the normal form is spanned by those chain tops lying in $\ker \mathbb{L}_S$, and by Theorem 3.3.7, $\ker \mathbb{L}_S$ consists of matrices with nonzero entries only in the large blocks. Therefore, the normal form is spanned by the chain tops lying in the stripe subspaces of the stripe structure of A_0.   □

This completes the description problem for the sl(2) normal form when A_0 is in Jordan form.

The Computation Problem

Recall that to solve the computation problem, we must present the projection from K_j into the normal form, and the inversion of $\mathcal{L}$ on its image. The remainder of the computational details belong to the format being used for the normalization. For the inner product normal form, we gave two solutions of the computation problem, an "elementary" method suitable for hand calculation when A_0 is in Jordan form and is reasonably small (see the discussion surrounding equation (3.4.11)), and a more "sophisticated" method (beginning with (3.4.13)) suitable for computer algebra systems when A_0 is seminormal. Both of these approaches can be used with the sl(2) normal form; the second method is greatly aided by the pushdown algorithm (Algorithm 2.6.5), and seminormality is no longer required. (See Remark 3.4.25.)

The elementary approach begins with a result that resembles Lemma 3.4.19:

3.5.6. Lemma. Under the same hypotheses as Theorem 3.5.5, the projection into sl(2) normal form preserves stripe sums (for the stripe structure for A_0).

Proof. Beginning with the chain tops (one for each stripe in the stripe structure for N), the chain-weight basis for gl(n) is constructed by iterating $\mathbb{Y} = \mathbb{L}_X = \mathbb{L}_N$. Notice that all chain-weight basis elements other than the chain tops lie in $\operatorname{im} \mathbb{L}_N$, and by Lemma 3.4.20, all of their stripe sums (using the stripe structure for N) are zero. Suppose that a given matrix in gl(n) is expressed as a linear combination of elements of the chain-weight basis. The projection of this matrix into the normal form can be obtained in two stages: First drop the terms that are not chain tops; then drop the terms that are chain tops for stripes that belong to the stripe structure for N but not for A_0. By Lemma 3.4.20 the terms deleted in the first stage have stripe sums equal to zero (for all stripes), so deleting them preserves all stripe sums. The terms deleted in the second stage lie outside the stripe structure of A_0, so the stripe sums of that stripe structure are still preserved.   □

So, to project a matrix into the normal form it is only necessary to compute the stripe sums for each stripe associated with A_0, and distribute that sum among the positions in the stripe proportionally to the bias factors. The "elementary" solution of the computation problem, suitable for hand calculation in small problems, is simply to compute the projection $B_j = \mathbb{P}(K_j)$ and then solve $\pounds U_j = K_j - B_j$. The inversion of $\pounds$ on its image can be done just as for the inner product normal form, since the equations to be solved are already nearly in row echelon form.

Next we describe the procedure that must be carried out to normalize a matrix series with leading term A_0 using the sl(2) normal form without assuming that A_0 is in Jordan form. For completeness the steps are compiled into a single list, with references to the lemmas and algorithms where they first appear.

3.5.7. Algorithm.

1. A_0 must be decomposed into $S + N$. This can be done by Algorithm E.1.2 (if it is not desired to find the Jordan form of A_0, which of course gives the decomposition directly).

2. A triad $\{X, Y, Z\}$ must be constructed such that $X = N$ and X, Y, and Z commute with S. This can be done either from the Jordan form of N (as in the proof of Theorem 2.7.1) or by Algorithm 2.7.2.

3. The projections P_λ into the eigenspaces of S must be constructed, using Corollary 2.1.5 if A_0 is not in Jordan form. (These projections act on $\mathbb{R}^n$, and their matrices are $n \times n$.)

4. The projections $\mathbb{P}_\mu$ into the eigenspaces of $\mathbb{L}_S$ must be constructed, using Lemma 3.3.10 if A_0 is not in Jordan form. (These projections have matrices of size $n^2 \times n^2$, and the lemma enables us to avoid using these large matrices.)

5. The equation $\mathbb{L}_S \widehat{U}_j = K_j - B'_j$ must be solved. This can be done by Algorithm 3.3.11, using the results of step 4.

6. Using Algorithm 3.4.23, $\widehat{U}_j$ must be converted into a solution U'_j of $\pounds U'_j = K_j - B'_j$.

7. Steps 3 and 4 above must be repeated with the semisimple operator Z in place of S, so that the projections into the weight spaces (eigenspaces of $\mathbb{Z} = \mathbb{L}_Z$) are available for use in the next step. (If A_0 is in Jordan form, then Z is diagonal and this step is trivial; the E_{ij} are a weight basis, and the projections into a given weight space are obtained by deleting the entries in positions with the wrong weight.)

8. The equation $\mathbb{L}_N U''_j = B'_j - B_j$ must be solved within the vector space $\ker \mathbb{L}_S$. This is done using the triad $\mathbb{X} = \mathbb{L}_Y, \mathbb{Y} = \mathbb{L}_X, \mathbb{Z} = \mathbb{L}_Z$ with its action restricted to $\ker \mathbb{L}_S$. The pushdown algorithm (Algorithm 2.6.5) is applied to B'_j to split it into terms of various chain lengths and heights; the terms for which the height equals the chain length are in the chain tops and therefore constitute B_j, the desired normal

form. Then $\mathbb{Y}^{\dagger}$, which is a right inverse of $\mathbb{L}_N$ on its image, is applied to the remaining terms to obtain U_j''. In applying the pushdown algorithm, it is necessary to be able to project each r_j into the weight spaces. If A_0 is in Jordan form, the standard basis matrices E_{ij} form a weight basis, and this can be done by the sorting algorithm (Algorithm 2.6.2). Otherwise, it is done using the projections calculated in step 7. (The method of Algorithm 2.6.3, if applied literally, would require calculating the polynomials of Corollary 2.1.5 for matrices of size n^2. Step 7 avoids this. See Remark 2.6.4.)

9. Finally, B_j has been determined in step 8, and $U_j = U_j' + U_j''$.

We conclude this section by working out an example of the strategy outlined above. The example is chosen to illustrate step 8, which is specific to the sl(2) normal form. For simplicity we have chosen a purely nilpotent example, so that the steps pertaining to the semisimple part (which are the same in the inner product style) can be ignored. Our example is in Jordan form, for two reasons: to make the calculations simple enough to reproduce, and to allow comparison with the "elementary" method (which requires Jordan form). Remember that the calculations presented below are "not intended to be seen." That is, they will normally occur inside the computer during the execution of a symbolic computation program, and only the results matter. We spell out the details here only to illustrate the method.

The problem to be solved is the homological equation

$$\pounds U = K - B$$

when $A_0 = N_4$ and

$$K = E_{21} = \begin{bmatrix} 0 & 0 & 0 & 0 \\ 1 & 0 & 0 & 0 \\ 0 & 0 & 0 & 0 \\ 0 & 0 & 0 & 0 \end{bmatrix}.$$

(We omit the usual subscript on U, K, and B because the value of this subscript is irrelevant to the calculation.) Since $S = 0$, only steps 2, 7, and 8 need to be carried out. For step 2, the triad with $X = N$ has been given in (3.5.7). Step 7 is trivial because Z is diagonal. So all that remains is to work out step 8. We will find B, the projection of $K = E_{21}$ into the normal form given by (3.5.8), and we will compute U.

Rather than write many matrices that contain mostly zeros, it is easier to use the bracket multiplication table given by

$$[E_{ij}, E_{kl}] = \delta_{jk}E_{il} - \delta_{li}E_{kj},$$

where $\delta_{ij} = 1$ if $i = j$ and 0 otherwise. This allows calculations to be done in the following manner:

$$\mathbb{Y}E_{21} = [E_{21}, E_{12} + E_{23} + E_{34}] = E_{22} - E_{11},$$
$$\mathbb{X}E_{13} = [E_{13}, 3E_{21} + 4E_{32} + 3E_{43}] = 4E_{12} - 3E_{23}.$$

In reporting the following results, such calculations will be taken for granted. Since the E_{ij} form a weight basis for $gl(n)$, the sorting algorithm (Algorithm 2.6.2) will be used in conjunction with the pushdown algorithm (Algorithm 2.6.5). The first step is to iterate $\mathbb{Y}$ on E_{21} until zero is reached and label the results in reverse order:

$$q_5 = E_{21},$$
$$q_4 = E_{22} - E_{11},$$
$$q_3 = E_{23} - 2E_{12},$$
$$q_2 = E_{24} - 3E_{13},$$
$$q_1 = -4E_{14}.$$

The next step is to set $r_1 = s_1 = q_1$, and identify the chain length and height of the terms (or in this case, term) of r_1. Since E_{14} has weight -6, and is known to have height one, its chain length is $1 - (-6) = 7$, and we can write

$$s_1 = (-4E_{14})_{71},$$

where the subscripts outside the parentheses give (ℓ, h). Next, remembering that there is no t_1, we calculate

$$t_2 = \mathbb{Y}^\dagger s_1 = \frac{1}{(1)(7-1)}\mathbb{X}(-4E_{14})_{71} = 2(E_{24} - E_{13})_{72}.$$

Proceeding with the algorithm, we obtain

$$r_2 = q_2 - t_2 = (-E_{13} - E_{24})_{51},$$

$$s_2 = r_2 + t_2,$$

$$t_3 = \mathbb{Y}^\dagger s_2 = (E_{34} - E_{12})_{52} + \frac{1}{5}(-4E_{12} + 6E_{23} - 4E_{34})_{73}$$

$$= \frac{9}{5}E_{12} + \frac{6}{5}E_{23} + \frac{1}{5}E_{34},$$

$$r_3 = q_3 - t_3 = -\frac{1}{5}(E_12 + E_{23} + E_{34})_{31},$$

$$s_3 = r_3 + t_3,$$

$$t_4 = \frac{1}{10}(-3E_{11} - E_{22} + E_{33} + 3E_{44})_{32} + \frac{1}{2}(-E_{11} + E_{22} + E_{33} - E_{44})_{53}$$

$$+ \frac{1}{5}(-E_{11} + 3E_{22} - 3E_{33} + E_{44})_{74}$$

$$= -E_{11} + E_{22},$$

$$r_4 = 0,$$

$$s_4 = r_4 + t_4 = t_4,$$

$$t_5 = \frac{1}{10}(3E_{21} + 4E_{32} + 3E_{43})_{33} + \frac{1}{2}(E_{21} - E_{43})_{54}$$

$$+ \frac{1}{5}(E_{21} - 2E_{32} + E_{43})_{75}$$

$$= E_{21},$$

$$r_5 = q_5 - t_5 = 0,$$

$$s_5 = r_5 + t_5 = t_5 = \frac{1}{10}(3E_{21} + 4E_{32} + 3E_{43})_{33} + \frac{1}{2}(E_{21} - E_{43})_{54}.$$

The projection B of $K = E_{21}$ into the normal form is therefore the part of s_5 that lies in the span of the chain tops. But this is only the term with $(\ell, h) = (3,3)$. Thus,

$$B = \begin{bmatrix} 0 & 0 & 0 & 0 \\ \frac{3}{10} & 0 & 0 & 0 \\ 0 & \frac{4}{10} & 0 & 0 \\ 0 & 0 & \frac{3}{10} & 0 \end{bmatrix}.$$

Notice that the stripe sum is preserved: $\frac{3}{10} + \frac{4}{10} + \frac{3}{10} = 1$. The result is therefore the same as would be obtained (in the "elementary method") by distributing the stripe sum 1 proportionately to the bias factors $3, 4, 3$.

One advantage of the pushdown calculation is that a right inverse of $\mathcal{L}$ on its image is given immediately by (2.6.6): Since our example is purely nilpotent, $\mathcal{L} = \mathbb{L}_N = \mathbb{L}_X = \mathbb{Y}$, and $\mathbb{Y}^\dagger$ is a right inverse of $\mathbb{Y}$ on its image. (In the general case, $\mathbb{Y}^\dagger$ is still a right inverse of $\mathbb{Y} = \mathbb{L}_N$, but this is not a right inverse of $\mathcal{L}$. That is why $\mathbb{Y}^\dagger$ is used only in step 8.) Thus, to find U

it is only necessary to apply $\mathbb{Y}^\dagger$ to

$$K - B = s_5 - B = \frac{1}{2}(E_{21} - E_{43})_{54} + \frac{1}{5}(E_{21} - 2E_{32} + E_{43})_{75}.$$

This yields

$$U = \frac{1}{8}\mathbb{X}(E_{21} - E_{43}) + \frac{1}{50}\mathbb{X}(E_{21} - 2E_{32} + E_{43}) = \begin{bmatrix} 0 & 0 & 0 & 0 \\ 0 & 0 & 0 & 0 \\ -\frac{7}{10} & 0 & 0 & 0 \\ 0 & \frac{7}{10} & 0 & 0 \end{bmatrix}.$$

Notes and References

The sl(2) normal form has roots in classical invariant theory and representation theory. As a normal form for nonlinear differential equations (Section 4.8 below) it was introduced by Richard Cushman and Jan Sanders in a series of papers, beginning with the Hamiltonian case (Section 4.9 below); see Notes and References, Sections 4.8 and 4.9, for the references. An abstract version including the linear and nonlinear cases is worked out in Sanders [99], and a linear example is given. (This paper also contains methods for the linear case that are not presented in this book.) The exposition given in the present section is the first detailed elementary presentation of the linear case of the sl(2) normal form.

3.6   Lie Theory and the Generated Formats

In this section, we leave the subject of normal form styles and return to the study of formats, introduced in Section 3.2. This section is devoted to the generated iterative format (format 2a), and is essential for the rest of the book. The ideas of this section are also the foundation for the generated recursive formats (2b and 2c), developed in Appendices C and D.

Format 2a is based on the matrix exponential, and a deeper study of this format requires a deeper understanding of the exponential, specifically, its connection with one-parameter groups. This leads to an alternative proof of Theorem 3.2.1, the fundamental theorem of Lie series that makes format 2a convenient. The idea of one-parameter groups also leads naturally to the ideas of Lie groups and Lie algebras, which we present only in the form of concrete examples rather than abstract definitions. Roughly speaking, an abstract Lie group is a group in which it is possible to do calculus. Our examples are subsets of gl(n, $\mathbb{R}$), which is isomorphic as a vector space to $\mathbb{R}^{n^2}$, so we can "do calculus" in $\mathbb{R}^{n^2}$ without any special machinery. The reason for looking at examples of Lie groups and Lie algebras is that with these ideas it becomes possible to normalize matrix series in which the coefficients belong to a particular class of matrices, without leaving that class. This is much more easily done in the generated formats 2a, 2b, and 2c than in the "direct" formats 1a and 1b. This is illustrated by treating series of symmetric matrices, thus completing the proof of Theorem 3.3.12. The section ends with a discussion of the five algorithms that are needed in each generated format. These are very simple for format 2a, but call for considerable discussion in Appendices C and D for formats 2b and 2c.

One-Parameter Groups

A *one-parameter group* (or one-parameter subgroup) in $gl(n)$ is a mapping $T : \mathbb{R} \to gl(n)$, that is, a curve or path $T(t)$ of matrices in $gl(n)$, satisfying

$$T(s+t) = T(s)T(t) \tag{3.6.1}$$

and

$$T(0) = I. \tag{3.6.2}$$

Since it follows at once that $T(t)T(-t) = I$, each matrix $T(t)$ in the one-parameter group is invertible. Therefore, actually,

$$T : \mathbb{R} \to GL(n),$$

where $GL(n)$, called the *general linear group*, is the subset of $gl(n)$ consisting of invertible matrices. Recall that $gl(n)$ denotes either $gl(n, \mathbb{R})$ or $gl(n, \mathbb{C})$; in the same way there are two general linear groups, $GL(n, \mathbb{R})$ and $GL(n, \mathbb{C})$. The field ($\mathbb{R}$ or $\mathbb{C}$) will be omitted when the results are the same in both cases. The set $GL(n)$ is a *group* because it is closed under multiplication and inversion.

If $A \in gl(n)$ is any matrix, it is easy to see that $T(t) = e^{tA}$ is a one-parameter group. (The point to notice is that $e^P e^Q = e^{P+Q}$ only when P and Q commute, but tA and sA commute for all t and s.) The matrix A is called the *generator* of the one-parameter group. The first theorem is that every one-parameter group has a generator.

3.6.1. Theorem. If $T : \mathbb{R} \to GL(n)$ is a one-parameter group and $A = \dot{T}(0)$, where $\dot{T}$ is the derivative of T, then

$$T(t) = e^{tA}.$$

Proof. Differentiating (3.6.1) with respect to s and setting $s = 0$ gives

$$\dot{T}(t) = AT(t).$$

The solution of this matrix differential equation satisfying (3.6.2) is $T(t) = e^{tA}$. □

If $T \in GL(n)$, the *similarity by T* is the operator

$$\mathbb{S}_T : gl(n) \to gl(n)$$

defined by

$$\mathbb{S}_T X = T^{-1} X T. \tag{3.6.3}$$

Since $\mathbb{S}_T$ is an invertible linear operator, and $\mathrm{gl}(n)$ is isomorphic to $\mathbb{R}^{n^2}$, $\mathbb{S}_T$ may be represented by an invertible $n^2 \times n^2$ matrix, so $\mathbb{S}_T$ can be regarded as an element of $\mathrm{GL}(n^2)$. (Of course, this depends upon a choice of basis for $\mathrm{gl}(n)$.) In this way, Theorem 3.6.1, applied to $\mathrm{GL}(n^2)$, gives a new proof of Theorem 3.2.1:

3.6.2. Theorem (Fundamental theorem of Lie series). For any X and Y in $\mathrm{gl}(n)$,

$$e^{-Y} X e^{Y} = e^{\mathbb{L}_Y} X.$$

Proof. As in Theorem 3.2.1, we will prove the theorem in the form

$$e^{-tY} X e^{tY} = e^{t\mathbb{L}_Y} X,$$

or equivalently, suppressing X,

$$\mathbb{S}_{e^{tY}} = e^{t\mathbb{L}_Y}. \tag{3.6.4}$$

This is to be regarded as an equation in $\mathrm{GL}(n^2)$, and by Theorem 3.6.1, it will be proved if we show that $T(t) = \mathbb{S}_{e^{tY}}$ is a one-parameter group and $\dot{T}(0) = \mathbb{L}_Y$. But $T(s)T(t)X = e^{-sY}e^{-tY}Xe^{tY}e^{sY} = e^{-(s+t)Y}Xe^{(s+t)Y} = T(s+t)X$, so (3.6.1) holds, and (3.6.2) is obvious. Finally,

$$\frac{d}{dt}T(t)X = \frac{d}{dt}e^{-tY} X e^{tY} = e^{-tY}(-YX + XY)e^{tY},$$

so $\dot{T}(0)X = XY - YX = [X,Y] = \mathbb{L}_Y X$ and $\dot{T}(0) = \mathbb{L}_Y$. □

Lie Groups and Lie Algebras of Matrices

Theorem 3.6.1 indicates a relationship between $\mathrm{GL}(n)$ and $\mathrm{gl}(n)$ that can be described as follows: Every one-parameter group in $\mathrm{GL}(n)$ has a generator in $\mathrm{gl}(n)$, and every element of $\mathrm{gl}(n)$ generates a one-parameter group in $\mathrm{GL}(n)$. There are certain important subgroups of the group $\mathrm{GL}(n)$ that are related in the same way to subspaces of the vector space $\mathrm{gl}(n)$. Note that if $\mathfrak{G}$ is a subgroup of $\mathrm{GL}(n)$, then any one-parameter group in $\mathfrak{G}$ is automatically a one-parameter group in $\mathrm{GL}(n)$, and therefore by Theorem 3.6.1 it has a generator.

3.6.3. Definition. If $\mathfrak{G}$ is a subgroup of $\mathrm{GL}(n)$, and $\mathfrak{g}$ is a subspace of $\mathrm{gl}(n)$, with the properties that every one-parameter group in $\mathfrak{G}$ has its generator in $\mathfrak{g}$, and every element of $\mathfrak{g}$ generates a one-parameter group in $\mathfrak{G}$, then $\mathfrak{G}$ is called a *Lie group of matrices* (or *matrix Lie group*), and $\mathfrak{g}$ is called the *Lie algebra* of $\mathfrak{G}$.

Another common definition of a Lie algebra is a vector space together with an operation $[\cdot, \cdot]$ satisfying a list of algebraic rules, which are the same algebraic rules satisfied by a commutator bracket (or by the vector cross product on $\mathbb{R}^3$); specifically, $[\cdot, \cdot]$ is linear in both arguments and satisfies the *Jacobi identity*

$$[A, [B, C]] + [B, [C, A]] + [C, [A, B]] = 0. \tag{3.6.5}$$

It will be proved in Theorem 3.6.8 that any $\mathfrak{g}$ satisfying Definition 3.6.3 is closed under the matrix commutator bracket, and is therefore a Lie algebra in this sense as well. There is also an abstract definition of Lie group, which is not required in this book. (Briefly, a Lie group is a group that is also a manifold such that the group composition and inversion are smooth.)

There are four examples of matrix Lie groups that are the most important for this book: the *general linear group* $\mathrm{GL}(n, \mathbb{R})$ of all invertible $n \times n$ real matrices; the *special linear group* $\mathrm{SL}(n, \mathbb{R})$ of all invertible $n \times n$ real matrices with determinant equal to one; the *orthogonal group* $\mathrm{O}(n)$ of all orthogonal $n \times n$ real matrices; and the *symplectic group* $\mathrm{Sp}(m) \subset \mathrm{GL}(n)$, with $n = 2m$. The symplectic group consists of all symplectic $2m \times 2m$ real matrices; these are matrices T such that $TJT^\dagger = J$, where J is the $2m \times 2m$ matrix

$$J = \begin{bmatrix} 0 & -I \\ I & 0 \end{bmatrix}, \tag{3.6.6}$$

with 0 and I being the $m \times m$ zero and identity matrices. Observe that

$$J^{-1} = J^\dagger = -J. \tag{3.6.7}$$

It must be checked that if $S, T \in \mathrm{Sp}(m)$, then ST and T^{-1} belong to $\mathrm{Sp}(m)$. The former is easy; for the latter, notice that if $TJT^\dagger = J$, taking determinants shows that T is invertible, and it follows that $J = T^{-1}J(T^{-1})^\dagger$. (Incidentally, taking the inverse then shows that $T^\dagger JT = J$, so this is another way to define $\mathrm{Sp}(m)$.) It is often said that a symplectic matrix is one that "leaves invariant the quadratic form $Q(x) = x^\dagger Jx$ defined by J." This means that if $x = Ty$ with T symplectic, then $x^\dagger Jx = y^\dagger Jy$.

Table 3.1 lists the groups and their defining properties in the first column, the associated algebras and their defining properties in the second, and the dimension of the Lie algebra in the third. The table shows that the Lie algebra of $\mathrm{GL}(n, \mathbb{R})$ is $\mathfrak{gl}(n, \mathbb{R})$. The Lie algebra of $\mathrm{SL}(n, \mathbb{R})$, called $\mathfrak{sl}(n, \mathbb{R})$, consists of all $n \times n$ matrices A with trace zero. (The case $\mathfrak{sl}(2)$ has been introduced at the beginning of Section 2.5.) The Lie algebra of $\mathrm{O}(n)$ is the skew-symmetric matrices. The Lie algebra of the symplectic group is the set $\mathfrak{hm}(m)$ of $2m \times 2m$ *Hamiltonian matrices* A, which are characterized either as those matrices satisfying $A^\dagger J + JA = 0$, or equivalently as the image of the symmetric matrices under multiplication by J. (Because of (3.6.7), an equivalent condition is that JA be symmetric.)

Table 3.1. Matrix Lie groups and their Lie algebras.

Lie Group	Lie Algebra	Dimension
$GL(n, \mathbb{R})$	$gl(n, \mathbb{R})$	n^2
$\det T \neq 0$	A arbitrary	
$SL(n, \mathbb{R})$	$sl(n, \mathbb{R})$	$n^2 - 1$
$\det T = 1$	$\operatorname{tr} A = 0$	
$O(n)$	$sk(n)$	$n(n-1)/2$
$TT^\dagger = I$	$A^\dagger = -A$	
$Sp(m)$	$hm(m)$	$m(2m+1)$
$T^\dagger JT = J$	$A^\dagger J + JA = 0$	

3.6.4. Theorem. The Lie algebra of each Lie group, and its dimension, is correctly given in Table 3.1.

Proof. The first step is to show that if $T(t)$ is a one-parameter group in $\mathfrak{G}$, then $\dot{T}(0) \in \mathfrak{g}$. In fact, we will prove a stronger result: If $T(t)$ is any path in $\mathfrak{G}$ with $T(0) = I$, then $\dot{T}(0) \in \mathfrak{g}$. For $\mathfrak{G} = GL(n, \mathbb{R})$ there is nothing to prove, since $gl(n, \mathbb{R})$ imposes no conditions. If $\mathfrak{G} = O(n)$, the result follows by differentiating $T(t)T(t)^\dagger = I$ to obtain $\dot{T}(t)T(t)^\dagger + T(t)\dot{T}(t)^\dagger = 0$, noting that the product rule is valid for matrices if the order of factors is maintained, and that $\dot{T}(t)^\dagger$ is unambiguous because derivative and transpose commute. Setting $t = 0$ and using $T(0) = I$ gives $\dot{T}(0) + \dot{T}(0)^\dagger = 0$, or $\dot{T}(0) \in sk(n)$. For $\mathfrak{G} = Sp(m)$, differentiating $T^\dagger JT = J$ at $t = 0$ gives $\dot{T}(0)^\dagger J + J\dot{T}(0) = 0$. The case of $\mathfrak{G} = SL(n, \mathbb{R})$ is trickier. Given the path $T(t)$, define $A(t) = \dot{T}(t)T(t)^{-1}$. Then $T(t)$ satisfies the differential equation $\dot{T}(t) = A(t)T(t)$, and Wronski's formula

$$\det T(t) = [\det(T(0)] \exp \left(\int_0^t \operatorname{tr} A(s) \, ds \right) \tag{3.6.8}$$

from the theory of linear differential equations, together with the fact that $\det T(t) = 1$ is constant, says that $0 = (d/dt) \det T(t) = (\operatorname{tr} A(t))(\det T(t))$, implying $\operatorname{tr} A(t) = 0$. But $\dot{T}(0) = A(0)$, so its trace is zero.

The second step is to show that if $A \in \mathfrak{g}$, then e^{At} is in $\mathfrak{G}$. For $\mathfrak{g} = gl(n, \mathbb{R})$, it is enough to note that e^{At} is invertible (with inverse e^{-At}). For $\mathfrak{g} = sk(n)$, let $\Omega \in sk(n)$; then $(e^{\Omega t})^\dagger = e^{\Omega^\dagger t} = e^{-\Omega t} = (e^{\Omega t})^{-1}$, so $e^{\Omega t} \in O(n)$. For $\mathfrak{g} = hm(m)$, we need to prove that $e^{tA^\dagger} Je^{tA} = J$. This is clearly true at $t = 0$, and the derivative of the left-hand side is $e^{tA^\dagger}(A^\dagger J + JA)e^{tA} = 0$, so it is true for all t. For the case of $\mathfrak{g} = sl(n)$, we have (using a formula closely related to Wronski's) $\det e^{tA} = e^{\operatorname{tr}(tA)} = e^{t(\operatorname{tr} A)} = 1$.

To check the dimensions, $gl(n, \mathbb{R})$ is obvious and $sl(n, \mathbb{R})$ has only one constraint (trace zero). The dimension of $sk(n)$ is equal to the number of above-diagonal entries, which is $1 + 2 + \cdots + (n-1) = n(n-1)/2$. The elements of $hm(m)$ are in one-to-one correspondence with the symmetric $2m \times 2m$ matrices, whose dimension equals the number of diagonal and above-diagonal entries, namely $1 + 2 + \cdots + 2m = (2m)(2m+1)/2$.   □

3.6.5. Remark. It can be checked that not only $\dot{T}(0)$, but $\dot{T}(t)T(t)^{-1}$ for any t, belongs to $\mathfrak{g}$ if $T(t) \in \mathfrak{G}$. Therefore, any path $T(t)$ in $\mathfrak{G}$ satisfies a differential equation of the form $\dot{T}(t) = A(t)T(t)$ with $A(t) \in \mathfrak{g}$. Conversely, if $\dot{T} = A(t)T(t)$ with $A(t) \in \mathfrak{g}$ and $T(0) \in \mathfrak{G}$, then $T(t) \in \mathfrak{G}$. The one-parameter groups are the special case with A constant.

3.6.6. Remark. An important illustration of these ideas, not related to normal forms, is the topic of "infinitesimal rotations" in space. The orthogonal group $O(3)$ has two connected components, with determinant ± 1; the part $SO(3)$ with positive determinant is the *special orthogonal group*, or group of rotations. The motion of a rigid body with one fixed point can be represented by a matrix $T(t) \in SO(3)$: For each t, $T(t)$ is the unique matrix that rotates the body from its initial position to its position at time t. Then it follows from Remark 3.6.5 that $\dot{T}(t) = \Omega(t)T(t)$ with $\Omega(t) \in sk(3)$, an equation that is usually written in mechanics texts as $\dot{r}(t) = \omega(t) \times r(t)$, where $r(t)$ is any vector fixed in the body (and so rotating with respect to the space frame) and $\omega = (\omega_1, \omega_2, \omega_3)$ with

$$\Omega(t) = \begin{bmatrix} 0 & -\omega_3 & \omega_2 \\ \omega_3 & 0 & -\omega_1 \\ -\omega_2 & \omega_1 & 0 \end{bmatrix}.$$

The map $\Omega \mapsto \omega$ is an isomorphism of $sk(3)$ that sends commutator bracket to cross product.

Conjugation, Similarity, and the Lie Operator

In order to show that the Lie algebras $\mathfrak{g}$ are closed under commutator bracket, it would suffice to check each Lie algebra separately. But it is better to derive the bracket operation from the relation between Lie groups and their algebras, because this gives a single argument valid for all Lie groups and algebras. This argument will play an important role in Chapter 4 as well. We begin with an operation called *conjugation*, which involves $\mathfrak{G}$ only. By differentiation we deduce from this an operation called *similarity* involving $\mathfrak{G}$ and $\mathfrak{g}$, and, by a second differentiation, we deduce the *Lie operator*, which is equivalent to the bracket operation on $\mathfrak{g}$.

Let $T \in \mathfrak{G}$. *Conjugation* by T is the group homomorphism

$$\mathbb{C}_T : \mathfrak{G} \to \mathfrak{G}$$

defined for $S \in \mathfrak{G}$ by

$$\mathbb{C}_T S = T^{-1} S T. \tag{3.6.9}$$

($\mathbb{C}$ here does not refer to the complex numbers, but conforms to our convention that operators on matrices are written in blackboard bold.) Replacing S by e^{tA} with $A \in \mathfrak{g}$ gives

$$\mathbb{C}_T e^{tA} = T^{-1} e^{tA} T.$$

It is easy to check that this is a one-parameter group in $\mathfrak{G}$, so its generator belongs to $\mathfrak{g}$; this generator is just the derivative at $t = 0$, which is $T^{-1}AT$. In other words, we have shown that if $T \in \mathfrak{G}$ and $A \in \mathfrak{g}$, then $T^{-1}AT \in \mathfrak{g}$. This defines the *similarity* operation, which (for any $T \in \mathfrak{G}$) is the linear map

$$\mathbb{S}_T : \mathfrak{g} \to \mathfrak{g}$$

given by

$$\mathbb{S}_T A = T^{-1} A T. \tag{3.6.10}$$

The only difference between conjugation and similarity is that one acts on $\mathfrak{G}$ and the other on $\mathfrak{g}$.

3.6.7. Remark. When $\mathfrak{G} = \mathrm{GL}(n)$ and $\mathfrak{g} = \mathrm{gl}(n)$ we have $\mathfrak{G} \subset \mathfrak{g}$, and so $\mathbb{C}_T$ is a restriction of $\mathbb{S}_T$ to $\mathfrak{G}$. It is not true in general that $\mathfrak{G} \subset \mathfrak{g}$, so $\mathbb{C}_T$ is in general a completely distinct operation from $\mathbb{S}_T$, not a restriction of it. In the setting developed in Chapter 4, the conjugation and similarity operations do not resemble one another. See (4.4.5) and (4.4.9).

Substituting e^{sB} for T in the definition of similarity gives

$$\mathbb{S}_{e^{sB}} A = e^{-sB} A e^{sB}.$$

This is a path in $\mathfrak{g}$, and since $\mathfrak{g}$ is a vector space, its derivative at $s = 0$ also belongs to $\mathfrak{g}$. (It is not a one-parameter group.) This derivative is

$$\mathbb{L}_B A = [A, B] = AB - BA. \tag{3.6.11}$$

We have simultaneously defined the *Lie operator*

$$\mathbb{L}_B : \mathfrak{g} \to \mathfrak{g}$$

for any $B \in \mathfrak{g}$, and proved the following result:

3.6.8. Theorem. The Lie algebra $\mathfrak{g}$ of any Lie group $\mathfrak{G}$ is closed under commutator bracket.

3.6.9. Remark. In Lie group theory it is customary to define the conjugation, similarity, and Lie operator as TST^{-1}, TAT^{-1}, and $[B, \cdot]$, respectively, rather than $T^{-1}ST$, $T^{-1}AT$, and $[\cdot, B]$ as we have done. The common notations for similarity and Lie operator in this setting are

$$\mathrm{Ad}_T\, A = T^{-1}AT \quad \text{and} \quad \mathrm{ad}_B\, A = [B, A].$$

These are called the *adjoint operators*. The advantage of ad_B over $\mathbb{L}_B$ is that the former is a Lie algebra homomorphism, the latter an antihomomorphism. For us this is a minor nuisance given that $\mathbb{L}_B$ is more natural for the normal forms application.

Format 2a, Algorithmic Version

In Section 3.2, format 2a was developed to third order using the fundamental theorem of Lie series, and it was clear that the procedure could be carried out to any order. As stated there, the method at the jth stage calls for the solution of a homological equation of the form

$$\pounds U_j = K_j - B_j,$$

where K_j is a matrix that is computable at the point when it is needed. Formulas were given for K_1, K_2, and K_3. Here we present a slightly different packaging of the same method, suitable for algorithmic implementation, that does not require the use of such formulas; instead, the meaning of A_j is changed at each iteration.

Suppose that a matrix series is given, truncated at some degree k in ε,

$$A(\varepsilon) \equiv A_0 + \varepsilon A_1 + \cdots + \varepsilon^k A_k, \tag{3.6.12}$$

and that it is desired to normalize this series to order k. (Here $\equiv$ means equal modulo ε^{k+1}.) Assume that the series has already been normalized to order $j - 1$, and we are ready to do order j. (For the first step, $j = 1$.) Let $\varepsilon^j U_j$ be the next generator to be used, with U_j as yet unknown, and write $\mathbb{L}_j$ for $\mathbb{L}_{U_j}$. According to the fundamental theorem,

$$e^{-\varepsilon^j U_j} A(\varepsilon) e^{\varepsilon^j U_j} \equiv e^{\varepsilon^j \mathbb{L}_j} A(\varepsilon). \tag{3.6.13}$$

Since $e^{\varepsilon^j \mathbb{L}_j} = I + \varepsilon^j \mathbb{L}_j + \cdots$, the jth-order term of the right-hand side is $A_j + \mathbb{L}_j A_0$, which we temporarily call B_j. Now, $\mathbb{L}_j A_0 = [A_0, U_j] = -[U_j, A_0] = -\pounds U_j$, so

$$\pounds U_j = A_j - B_j. \tag{3.6.14}$$

This is a homological equation to be solved for B_j and U_j; the procedure for doing this will depend on the normal form style that is adopted, and is not part of format 2a. Once U_j has been determined, we go back to (3.6.13) and compute the right-hand side completely (up to order k). The result is renamed $A(\varepsilon)$, and we return to (3.6.12) with the next value of j.

In this formulation, the meaning of A_j changes at the end of each stage of the calculation, and it must be clearly understood that A_j in the homological equation (3.6.14) is not the original A_j given at the beginning (except in the case $j = 1$). In fact, A_j in (3.6.14) is exactly what we called K_j (for format 2a) in Section 3.2, and to avoid confusion we will continue to write the homological equation as $\mathcal{L}U_j = K_j - B_j$ except when explicitly using the algorithmic form stated here.

Normalizing Within a Class of Matrices

The principal *theoretical* advantage of the generated formats 2a, 2b, and 2c (as distinct from their *computational* advantages) is that they make it easy to normalize matrix series whose coefficients belong to certain specific classes of matrices, without leaving the class. For instance, suppose that

$$A(\varepsilon) = A_0 + \varepsilon A_1 + \cdots \tag{3.6.15}$$

and all A_j belong to a Lie algebra $\mathfrak{g}$. It is desired to simplify this series as much as possible subject to the constraint that the new coefficients also belong to $\mathfrak{g}$. Since $\mathbb{S}_T$ maps $\mathfrak{g}$ to itself if $T \in \mathfrak{G}$, any such similarity can be applied to (3.6.15). Since $e^{\varepsilon^j U_j} \in \mathfrak{G}$ if $U_j \in \mathfrak{g}$, generators from $\mathfrak{g}$ should be used at each stage in the normalization process. At the jth stage, the homological equation

$$\mathcal{L}U_j = K_j - B_j$$

must be solved. Since K_j (in format 2a) is constructed from A_j and from brackets of the A_i and U_i for $i < j$, all of which belong to $\mathfrak{g}$, it follows that $K_j \in \mathfrak{g}$, since $\mathfrak{g}$ is closed under bracket. Since $\mathcal{L} = \mathbb{L}_{A_0}$ and $A_0 \in \mathfrak{g}$,

$$\mathcal{L} : \mathfrak{g} \to \mathfrak{g}.$$

So the normalization process proceeds as usual: Select a complement of im $\mathcal{L}$ in $\mathfrak{g}$ (a normal form style) and project K_j into this complement to obtain B_j; then $K_j - B_j \in$ im $\mathcal{L}$ and $U_j \in \mathfrak{g}$ exists.

For example, if the A_j are Hamiltonian matrices, the series can be normalized within the class of Hamiltonian matrices using symplectic similarities with Hamiltonian generators. We will return to this example in Section 4.9. But the technique is not limited to cases in which the class of matrices to be preserved is a Lie algebra $\mathfrak{g}$. More generally, the class could be any subspace of gl(n) that is invariant under the action of $\mathbb{S}_T$ for T in a Lie group $\mathfrak{G} \subset$ GL(n). (Such a subspace is called a *representation space* of $\mathfrak{G}$, and $\mathfrak{G}$ is said to *act on* the space by similarity; see Remark 2.5.1.) We turn now to an example that is of this more general type.

Normalizing a Symmetric or Hermitian Matrix Series

The class sym(n) of real symmetric $n \times n$ matrices is not a Lie algebra. Nevertheless, it is often of interest to normalize a series (3.6.20) in which the A_j are symmetric, preserving the symmetry. The important fact is that

$T^{-1}AT = T^{\dagger}AT$ is symmetric if A is symmetric and T is orthogonal; that is,

$$\mathbb{S}_T : \text{sym}(n) \to \text{sym}(n)$$

if $T \in O(n)$. Since by the spectral theorem of linear algebra any symmetric matrix can be diagonalized by an orthogonal similarity, at the "zeroth stage" (3.2.1) we can prepare the series so that A_0 is diagonal, while preserving the symmetry of the remaining matrices. To normalize A_1 in format 2a we apply a similarity by $e^{\varepsilon U_1}$; this is orthogonal, so that the similarity preserves symmetry if $U_1 \in \text{sk}(n)$, the Lie algebra of $O(n)$. It is then necessary to solve the homological equation

$$\pounds U_1 = A_1 - B_1.$$

At first sight there seems to be a difficulty, because A_1 and B_1 are to be symmetric, while U_1 is to be skew-symmetric. But, in fact,

$$\pounds : \text{sk}(n) \to \text{sym}(n), \tag{3.6.16}$$

because if U is skew-symmetric and $A_0 = \Lambda$ is diagonal, then

$$(\pounds U)^{\dagger} = (U\Lambda - \Lambda U)^{\dagger} = \Lambda^{\dagger}U^{\dagger} - U^{\dagger}\Lambda^{\dagger} = \Lambda(-U) - (-U)\Lambda = \pounds U.$$

Unlike all previous examples (and in fact all subsequent examples in this book), this $\pounds$ map is not an operator (or endomorphism), that is, it does not map a vector space into itself, so it does not make sense to ask whether this $\pounds$ is or is not semisimple. Therefore, to find a complement to $\text{im}\,\pounds$ it is necessary to calculate the image directly. For example, if

$$A_0 = \begin{bmatrix} \lambda & & \\ & \lambda & \\ & & \mu \end{bmatrix},$$

with two equal eigenvalues λ and $\mu \neq \lambda$, then

$$\pounds \begin{bmatrix} 0 & a & b \\ -a & 0 & c \\ -b & -c & 0 \end{bmatrix} = \begin{bmatrix} 0 & 0 & (\mu-\lambda)b \\ 0 & 0 & (\mu-\lambda)c \\ (\mu-\lambda)b & (\mu-\lambda)c & 0 \end{bmatrix}.$$

In the general case, $\text{im}\,\pounds$ consists of those symmetric matrices that have zero entries in the blocks subtended by the strings of equal eigenvalues in A_0. Thus, a complement of $\text{im}\,\pounds$ consists of those symmetric block-diagonal matrices whose blocks are subtended by the strings. This coincides with the semisimple normal form (see Theorem 3.3.7), except for the additional condition of symmetry. The projection of A_1 into B_1 is obtained by separating A_1 into block and out-of-block parts; for instance, in our $n = 3$ example,

$$\begin{bmatrix} a & b & c \\ b & d & e \\ c & e & f \end{bmatrix} = \begin{bmatrix} a & b & \\ b & d & \\ & & f \end{bmatrix} + \begin{bmatrix} & & c \\ & & e \\ c & e & \end{bmatrix}.$$

The first term is B_1, the second is $A_1 - B_1$, and it is guaranteed that there will exist a U_1 in $\mathrm{sk}(n)$. The higher-order terms in the series will be modified, but will remain symmetric, and the process can be continued to higher order. It is only necessary to observe that in the higher-order homological equations $\mathcal{L}U_j = K_j - B_j$, each K_j is symmetric. But in the notation of (3.2.29)–(3.2.35), $K_2 = A'_2$, $K_3 = A''_3$, and so forth. That is, each K_j is just the first term in the series that has not yet been normalized, and we have just seen that the terms of the series remain symmetric at each step.

The same arguments work for a series of Hermitian matrices, with † replaced by *, the conjugate transpose. The orthogonal group should be replaced by the unitary group, and skew-symmetric matrices by skew-Hermitian ones. (The unitary group is a Lie group consisting of complex matrices with $T^* = T^{-1}$, and the skew-Hermitian matrices form its Lie algebra.)

This discussion completes the proof of Theorem 3.3.12, which states that a symmetric or Hermitian matrix series can be diagonalized to any order by hypernormalization. The algorithm for the diagonalization consists of three steps applied repeatedly. In the symmetric case these are:

1. Diagonalize the leading term by an orthogonal similarity.

2. Normalize the series to the desired order using skew-symmetric generators.

3. Drop the first term, divide by ε, and separate the series into several series of smaller matrices.

4. Return to step 1 with the new series.

Diffeomorphisms and Vector Fields

It is easy to confuse Lie groups with Lie algebras when both consist of matrices. Another way to picture them, which will be crucial in the next chapter, is to think of elements of a matrix Lie group $\mathfrak{G}$ as *diffeomorphisms* of $\mathbb{R}^n$, and elements of its Lie algebra as *vector fields* on $\mathbb{R}^n$. (A diffeomorphism is a smooth mapping with a smooth inverse. In this chapter, the diffeomorphisms and vector fields are all linear.)

Any invertible matrix $S \in \mathrm{GL}(n, \mathbb{R})$ defines a diffeomorphism $\psi : \mathbb{R}^n \to \mathbb{R}^n$ in an obvious manner:

$$\widetilde{x} = \psi(x) = Sx. \tag{3.6.17}$$

Any (not necessarily invertible) matrix $A \in \mathrm{gl}(n, \mathbb{R})$ defines a vector field v by $v(x) = Ax$, which may be viewed as a differential equation:

$$\dot{x} = Ax. \tag{3.6.18}$$

The "solution operator" of (3.6.18), the map φ^t such that $x = \varphi^t(a)$ is the solution of (3.6.18) with initial condition $x(0) = a$, is given by

$$\varphi^t(x) = e^{tA}x, \qquad (3.6.19)$$

or $\varphi^t = e^{tA}$, which is a one-parameter group of diffeomorphisms. In other words, the exponential map from the Lie algebra to the Lie group simply assigns to each (linear) differential equation its solution.

A second element $T \in \mathrm{GL}(n, \mathbb{R})$ may be regarded as a change of coordinates

$$x = Ty, \qquad (3.6.20)$$

which can be applied both to the diffeomorphism (3.6.17) and to the vector field (3.6.18). The results are

$$\tilde{y} = T^{-1}STy = \mathbb{C}_T(S)y \qquad (3.6.21)$$

and

$$\dot{y} = T^{-1}ATy = \mathbb{S}_T(S)y. \qquad (3.6.22)$$

Thus, conjugation and similarity, $\mathbb{C}_T$ and $\mathbb{S}_T$, are the natural coordinate change operations for linear diffeomorphisms and linear vector fields respectively. The generalized conjugation and similarity operations in Chapter 4 are the natural coordinate change operations for nonlinear diffeomorphisms and vector fields.

The Five Algorithms of a Generated Format

Throughout this chapter, we have been concerned with applying a transformation $T(\varepsilon)$ to a matrix series $A(\varepsilon)$ to obtain a new matrix series $B(\varepsilon) = T(\varepsilon)^{-1}A(\varepsilon)T(\varepsilon) = \mathbb{S}_{T(\varepsilon)}A(\varepsilon)$. In view of (3.6.22), this is equivalent to applying a coordinate change $x = T(\varepsilon)y$ to a differential equation $\dot{x} = A(\varepsilon)x$ to obtain $\dot{y} = B(\varepsilon)y$. It is natural to consider, along with this, the inverse transformation $y = T(\varepsilon)^{-1}x$, which carries $\dot{y} = B(\varepsilon)y$ into $\dot{x} = A(\varepsilon)x$. This suggests that in each of the generated formats (2a, 2b, and 2c) there should exist five algorithms to carry out the following operations:

I. Given a generator, find the transformation (or coordinate change) $T(\varepsilon)$ that it generates, expressed as a power series in ε.

II. Given a generator, find the inverse of the transformation it generates.

III. Given a generator and a matrix series $A(\varepsilon)$, find the transformed matrix series $B(\varepsilon)$.

IV. Given a generator and a matrix series $B(\varepsilon)$, find the inversely transformed matrix series $A(\varepsilon)$.

V. The normal form problem: Given $A(\varepsilon)$, find simultaneously the generator (or generators) needed to transform $A(\varepsilon)$ into the simplest possible $B(\varepsilon)$, and also find $B(\varepsilon)$.

We will conclude our study of format 2a by stating the five algorithms needed for that format.

3.6.10. Algorithm (Algorithm I for format 2a). Given a generator $\varepsilon^j U_j$, the transformation it generates is

$$T(\varepsilon) = e^{\varepsilon^j U_j} = I + \varepsilon^j U_j + \frac{1}{2}\varepsilon^{2j} U_j^2 + \cdots .$$

3.6.11. Algorithm (Algorithm II for format 2a). Given a generator $\varepsilon^j U_j$, the inverse of the transformation it generates is

$$T(\varepsilon)^{-1} = e^{-\varepsilon^j U_j} = I - \varepsilon^j U_j + \frac{1}{2}\varepsilon^{2j} U_j^2 + \cdots .$$

That is, the inverse transformation is generated by $-\varepsilon^j U_j$.

3.6.12. Algorithm (Algorithm III for format 2a). Given $\varepsilon^j U_j$, $A(\varepsilon)$ is transformed into

$$B(\varepsilon) = \mathbb{C}_{e^{\varepsilon^j U_j}} A(\varepsilon) = e^{\varepsilon^j \mathbb{L}_{U_j}} A(\varepsilon).$$

That is, writing $\mathbb{L}_j = \mathbb{L}_{U_j}$,

$$B(\varepsilon) = \left(I + \varepsilon^j \mathbb{L}_j + \frac{1}{2}\varepsilon^{2j} \mathbb{L}_j^2 + \cdots\right)(A_0 + \varepsilon A_1 + \varepsilon^2 A_2 + \cdots).$$

3.6.13. Algorithm (Algorithm IV for format 2a). Given the generator $\varepsilon^j U_j$ and the "transformed" series $B(\varepsilon)$, the "original" series is

$$A(\varepsilon) = e^{-\varepsilon^j \mathbb{L}_j} B(\varepsilon).$$

3.6.14. Algorithm (Algorithm V for format 2a). Suppose that $A(\varepsilon)$ has already been normalized up to order $j - 1$. Using the desired normal form style, choose U_j and B_j to solve the homological equation

$$\mathcal{L} U_j = A_j - B_j,$$

where $\mathcal{L} = \mathbb{L}_{A_0}$, and then apply Algorithm III. The resulting series will agree with $A(\varepsilon)$ up to order $j-1$, A_j will be replaced by B_j, and the higher-order terms will be modified. Rename the new series $A(\varepsilon)$ and repeat the process at the next order $j + 1$.

Notes and References

Lie theory is usually treated as an advanced subject, and most treatments begin with the abstract definition of a Lie group as a manifold with a group structure such that the group operations are smooth. The treatment given here is limited to matrix Lie groups, and is influenced by the elementary article by Howe [61]. I would suggest that an absolute beginner in Lie theory should start with [61] and then proceed to Part I of Hausner and Schwartz [56], which introduces manifolds but is still relatively concrete. The fundamental theorem of Lie series is Lemma 4, page 65, in [56]. A more advanced solid introduction to Lie theory is Chapter 3 of Warner [110].

3.7   Metanormal Forms and k-Determined Hyperbolicity

Until now we have attempted to simplify a matrix series $A(\varepsilon)$ by using transformations $T(\varepsilon)$ that are smooth in ε. But it was seen in Section 3.1 that the eigenvalues of $A(\varepsilon)$ can involve fractional powers of ε when A_0 is nonsemisimple, and fractional powers are not smooth. So it is natural to ask whether a matrix series $A(\varepsilon)$ that is already in normal form can be simplified still further, perhaps even diagonalized or brought into Jordan form, by performing transformations that use fractional powers. The final result of such a process will be called a *metanormal form*. At the end of the section, metanormal forms will be applied to the problem of k-*determined hyperbolicity*. This is the problem of deciding, given the k-jet of a smooth matrix function $A(\varepsilon)$, whether or not $A(\varepsilon)$ is hyperbolic for small $\varepsilon > 0$. (There are two possible answers: yes and undecided. The answer "no" cannot be obtained from a finite jet, because eigenvalues on the imaginary axis could be pushed off by higher-order terms.)

Shearing and Transplanting

A *shearing transformation* is a linear transformation defined by a *shearing matrix* of the form

$$
T(\varepsilon) = \begin{bmatrix} \varepsilon^{\nu_1} & & & \\ & \varepsilon^{\nu_2} & & \\ & & \ddots & \\ & & & \varepsilon^{\nu_n} \end{bmatrix}, \tag{3.7.1}
$$

where $\nu_1, \ldots, \nu_n$ are rational numbers (possibly negative or zero). A shearing matrix is to be regarded as defined for $\varepsilon > 0$; at $\varepsilon = 0$ the matrix either reduces to zero and is not invertible (so that $T(0)$ does not define a similarity transformation) or contains entries that become infinite and so is not defined. Thus, any simplifications of $A(\varepsilon)$ obtained via shearing matrices are valid only for $\varepsilon > 0$. (The case $\varepsilon < 0$ may be treated separately by redefining the parameter, but $\varepsilon = 0$ is definitely excluded.)

If $A(\varepsilon)$ is a matrix series (using either integer or fractional powers) and $T(\varepsilon)$ is a shearing matrix, we may consider the matrix

$$B(\varepsilon) = T(\varepsilon)^{-1} A(\varepsilon) T(\varepsilon). \qquad (3.7.2)$$

Although $B(0)$ is not defined, there may be a fractional power μ such that

$$B(\varepsilon) = \delta^\mu C(\varepsilon), \qquad (3.7.3)$$

where $C(\varepsilon)$ is a fractional power series with a nonzero constant term C_0 and no negative powers, so that $C(0) = C_0$ is defined. For $\varepsilon > 0$, the transformation from $A(\varepsilon)$ to $C(\varepsilon)$ described by (3.7.2) and (3.7.3) consists of similarity and scaling, but the transformation from $A(0)$ to $C(0)$ cannot be described in this way; for instance, A_0 may be nonsemisimple and C_0 may be semisimple, or have a different Jordan block structure from A_0. The series $C(\varepsilon)$ will be called a *transplanting* of $A(\varepsilon)$. The motivation for this terminology is to picture $A(\varepsilon)$ for $\varepsilon > 0$ as a plant growing from a root placed at A_0, transplanted to a new location with root at C_0. (In contrast, the smooth near-identity transformations $T(\varepsilon)$ used in earlier sections leave the root unchanged.) What we have called the *root* of the family of matrices goes under a variety of names depending on what is being emphasized; these names include *leading term*, *unperturbed matrix*, and *organizing center*. (The term *organizing center* is most common in families having several parameters, and will be used in Chapter 6 in the context of unfoldings.)

The simplest example of this construction is the case

$$A(\varepsilon) = A_0 + \varepsilon A_1 = \begin{bmatrix} 0 & 1 \\ 0 & 0 \end{bmatrix} + \varepsilon \begin{bmatrix} a & 0 \\ b & a \end{bmatrix} \qquad (3.7.4)$$

with $b \neq 0$; notice that A_1 is in inner product normal form with respect to A_0. Taking

$$T(\varepsilon) = \begin{bmatrix} 1 & 0 \\ 0 & \varepsilon^{1/2} \end{bmatrix}, \qquad T(\varepsilon)^{-1} = \begin{bmatrix} 1 & 0 \\ 0 & \varepsilon^{-1/2} \end{bmatrix}$$

yields

$$B(\varepsilon) = \varepsilon^{1/2} \begin{bmatrix} 0 & 1 \\ b & 0 \end{bmatrix} + \varepsilon \begin{bmatrix} a & 0 \\ 0 & a \end{bmatrix},$$

so that $\mu = \frac{1}{2}$ and

$$C(\varepsilon) = \begin{bmatrix} 0 & 1 \\ b & 0 \end{bmatrix} + \varepsilon^{1/2} \begin{bmatrix} a & 0 \\ 0 & a \end{bmatrix}.$$

The new "root" or leading term C_0 has distinct eigenvalues $\pm\sqrt{b}$, so it is possible to diagonalize C_0 and then to normalize the term of order $\frac{1}{2}$, using the semisimple normal form style. (Introducing a new parameter $\eta = \varepsilon^{1/2}$ eliminates the fractional power, so that the standard formalism for normalization can be used.) Since the leading term is diagonal with distinct eigenvalues, it follows from Theorem 3.3.7 that the normalized form of B_1

will be diagonal. (Normalization of B_1 will introduce higher-order terms in η; these can be normalized to any order, and all of the normalized terms will be diagonal.)

The final simplified form of a matrix series, obtained after one or more repetitions of normalization, transplanting, and normalizing again with respect to the new root, will be called a *metanormal form*. The goal of metanormalization is, when possible, to obtain a diagonal final form. If this is not possible, one seeks to bring the matrix into Jordan form. (We have already seen in Section 3.1 that the Jordan form can be discontinuous at $\varepsilon = 0$, but here that difficulty does not arise, since $\varepsilon = 0$ is already excluded.)

In the remainder of this section, we will study metanormal forms in the following situations, building up to the general case:

1. The general 2×2 case with nonsemisimple leading term.

2. The $n \times n$ case having nilpotent leading term with one Jordan block.

3. Nilpotent leading term with several Jordan blocks.

4. The general case.

The General 2×2 Case

The general case of a 2×2 matrix power series $A(\varepsilon)$, when A_0 is nonsemisimple, looks as follows in inner product normal form:

$$A(\varepsilon) = \begin{bmatrix} a_0 & 1 \\ 0 & a_0 \end{bmatrix} + \varepsilon \begin{bmatrix} a_1 & 0 \\ b_1 & a_1 \end{bmatrix} + \varepsilon^2 \begin{bmatrix} a_2 & 0 \\ b_2 & a_2 \end{bmatrix} + \cdots. \qquad (3.7.5)$$

If $b_j = 0$ for $j = 1, 2, \ldots$ (to all orders), then $A(\varepsilon)$ is already in Jordan form to all orders, and it is clearly impossible to diagonalize $A(\varepsilon)$ or to improve on the existing normalization.

3.7.1. Remark. If $A(\varepsilon)$ is a smooth matrix function and not just a formal power series, it may still be diagonalizable, if its eigenvalues differ by a flat function. But this is not detectable from its formal power series. It should also be pointed out that the procedure we are considering is algorithmic in cases where there is a $b_j \neq 0$, but not when there isn't: It is never possible to discover that all b_j are zero with a finite amount of calculation.

We therefore assume that there is a smallest integer j such that $b_j \neq 0$, so that

$$A(\varepsilon) = \widehat{a}(\varepsilon)I + \begin{bmatrix} 0 & 1 \\ 0 & 0 \end{bmatrix} + \varepsilon^j \begin{bmatrix} 0 & 0 \\ b_j & 0 \end{bmatrix} + \begin{bmatrix} \widetilde{a}(\varepsilon) & 0 \\ \widetilde{b}(\varepsilon) & \widetilde{a}(\varepsilon) \end{bmatrix}, \qquad (3.7.6)$$

where $\widehat{a}(\varepsilon) = a_0 + \varepsilon a_1 + \cdots + \varepsilon^j a_j$, and $\widetilde{a}(\varepsilon)$ and $\widetilde{b}(\varepsilon)$ contain terms of orders ε^{j+1} and higher; notice that the diagonal entries a_j are included in

$\widehat{a}(\varepsilon)$ and not in the matrix with coefficient ε^j. Apply to this the similarity by

$$T(\varepsilon) = \begin{bmatrix} 1 & 0 \\ 0 & \varepsilon^\nu \end{bmatrix}, \qquad (3.7.7)$$

where ν is to be determined; the result is

$$B(\varepsilon) = \widehat{a}(\varepsilon)I + \begin{bmatrix} 0 & \varepsilon^\nu \\ \varepsilon^{j-\nu}b_j & 0 \end{bmatrix} + \begin{bmatrix} \widetilde{a}(\varepsilon) & 0 \\ \varepsilon^{-\nu}\widetilde{b}(\varepsilon) & \widetilde{a}(\varepsilon) \end{bmatrix}.$$

Thus, the nilpotent part of A_0 is "moved up" by an amount ν to order ε^ν, while the term b_j is "moved down" an equivalent amount to order $\varepsilon^{j-\nu}$. These terms will be of the same order if $j - \nu = \nu$, or $\nu = j/2$ (which is a fraction if j is odd, an integer if j is even). In either case, the matrix now takes the form

$$B(\varepsilon) = \widehat{a}(\varepsilon)I + \varepsilon^{j/2} \begin{bmatrix} 0 & 1 \\ b_j & 0 \end{bmatrix} + \mathcal{O}\left(\varepsilon^{(j+1)/2}\right); \qquad (3.7.8)$$

it is important to note that the lowest order appearing in $\varepsilon^{-\nu}\widetilde{b}(\varepsilon)$ is $\varepsilon^{(j+1)/2}$. The term $\widehat{a}(\varepsilon)I$ will never change under any further similarities that may be applied, and may conveniently be set aside; therefore, the new (transplanted) series to be considered is

$$C(\varepsilon) = \begin{bmatrix} 0 & 1 \\ b_j & 0 \end{bmatrix} + \mathcal{O}(\varepsilon^{1/2}). \qquad (3.7.9)$$

(This is an example of a transplanting operation more complicated than (3.7.3).) Since $b_j \neq 0$, the eigenvalue splits (to $\pm\sqrt{b_j}$). As in the first example, it is possible to diagonalize C_0 and then to put the rest (up to any order) in semisimple normal form, which will be diagonal. When the initial term of (3.7.8), and the factor $\varepsilon^{j/2}$, are restored, the result is still diagonal.

Nilpotent Leading Term with One Jordan Block

Next we consider the case that A_0 is nilpotent with one Jordan block, and the rest of the series (to some order) is in inner product normal form with respect to A_0, so that it consists of simple striped matrices. This topic will be introduced by treating the example

$$A(\varepsilon) = \begin{bmatrix} 0 & 1 & 0 & 0 \\ 0 & 0 & 1 & 0 \\ 0 & 0 & 0 & 1 \\ 0 & 0 & 0 & 0 \end{bmatrix} + \varepsilon \begin{bmatrix} a & 0 & 0 & 0 \\ b & a & 0 & 0 \\ c & b & a & 0 \\ d & c & b & a \end{bmatrix}. \qquad (3.7.10)$$

After this example has been worked out in detail, it will be easy to see how the $n \times n$ case goes.

To determine the correct shearing matrix to apply to (3.7.10), begin with an unspecified shearing matrix

$$T(\varepsilon) = \begin{bmatrix} \varepsilon^{\nu_1} & & & \\ & \varepsilon^{\nu_2} & & \\ & & \varepsilon^{\nu_3} & \\ & & & \varepsilon^{\nu_4} \end{bmatrix}. \tag{3.7.11}$$

Then

$$B(\varepsilon) = T(\varepsilon)^{-1}A(\varepsilon)T(\varepsilon) = \begin{bmatrix} a\varepsilon & \varepsilon^{\nu_2-\nu_1} & 0 & 0 \\ b\varepsilon^{1+\nu_1-\nu_2} & a\varepsilon & \varepsilon^{\nu_3-\nu_2} & 0 \\ c\varepsilon^{1+\nu_1-\nu_3} & b\varepsilon^{1+\nu_2-\nu_3} & a\varepsilon & \varepsilon^{\nu_4-\nu_3} \\ d\varepsilon^{1+\nu_1-\nu_4} & c\varepsilon^{1+\nu_2-\nu_4} & b\varepsilon^{1+\nu_3-\nu_1} & a\varepsilon \end{bmatrix}. \tag{3.7.12}$$

It is natural to impose the requirement that terms containing the same coefficient a, b, c, or d should be of the same order in (3.7.12); this is already true of the terms in a. It follows that the "gaps" between the shearing powers should be equal: $\nu_2 - \nu_1 = \nu_3 - \nu_2 = \nu_4 - \nu_3$. It will prove convenient to denote this gap by 2θ, so that

$$T(\varepsilon) = \begin{bmatrix} \varepsilon^{\nu_1} & & & \\ & \varepsilon^{\nu_1+2\theta} & & \\ & & \varepsilon^{\nu_1+4\theta} & \\ & & & \varepsilon^{\nu_1+6\theta} \end{bmatrix}.$$

Notice also that (3.7.12) involves only the gaps between the powers, so the value of ν_1 is irrelevant. It will be convenient to choose ν_1 so that the powers appearing in $T(\varepsilon)$ are symmetrically spaced around zero; thus we take $\nu_1 = 3\theta$ and obtain

$$T(\varepsilon) = \begin{bmatrix} \varepsilon^{-3\theta} & & & \\ & \varepsilon^{-\theta} & & \\ & & \varepsilon^{\theta} & \\ & & & \varepsilon^{3\theta} \end{bmatrix}. \tag{3.7.13}$$

The value of θ remains to be determined. With $T(\varepsilon)$ in this form, (3.7.12) becomes

$$B(\varepsilon) = \begin{bmatrix} a\varepsilon & \varepsilon^{2\theta} & 0 & 0 \\ b\varepsilon^{1-2\theta} & a\varepsilon & \varepsilon^{2\theta} & 0 \\ c\varepsilon^{1-4\theta} & b\varepsilon^{1-2\theta} & a\varepsilon & \varepsilon^{2\theta} \\ d\varepsilon^{1-6\theta} & c\varepsilon^{1-4\theta} & b\varepsilon^{1-2\theta} & a\varepsilon \end{bmatrix}. \tag{3.7.14}$$

Notice that each stripe subspace of gl(4) becomes an eigenspace of the similarity operation $\mathbb{S}_{T(\varepsilon)}$, the eigenvalues (from the main diagonal down) being 1, $\varepsilon^{-2\theta}$, $\varepsilon^{-4\theta}$, and $\varepsilon^{-6\theta}$. (The superdiagonal receives a factor of $\varepsilon^{2\theta}$, so the "stripes" above the diagonal, which are not true stripes according to Definition 3.4.18, also define eigenspaces of the similarity.)

3.7.2. Remark. Those who have read the starred sections will recognize that (3.7.13) may be written

$$T(\varepsilon) = \varepsilon^{-\theta Z} = e^{-\theta \ln \varepsilon Z},$$

where

$$Z = \begin{bmatrix} 3 & & & \\ & 1 & & \\ & & -1 & \\ & & & -3 \end{bmatrix}$$

is the matrix obtained in (3.5.7) by embedding $A_0 = X$ into an sl(2) triad $\{X, Y, Z\}$. From the fundamental theorem of Lie series we have

$$\mathbb{S}_{T(\varepsilon)} = e^{-\theta \ln \varepsilon \mathbb{L}_Z},$$

so the eigenspaces of the similarity are just the eigenspaces, or weight spaces, of $\mathbb{Z} = \mathbb{L}_Z$, which are the stripe subspaces of gl(4). In the present section we are developing the shearing theory using the inner product normal form rather than the sl(2) normal form, so we will make no further use of this observation. But the same shearing theory works with the sl(2) normal form, and then it is possible to point out some additional connections between the two theories. See Notes and References for further remarks.

Now we use (3.7.14) to fix the value of θ in various cases, and so complete the shearing analysis for (3.7.10). First, assume that $d \neq 0$. In this case we match the exponent $1 - 6\theta$ associated with d to the exponent 2θ of the superdiagonal elements (coming from the nilpotent matrix A_0), to obtain

$$1 - 6\theta = 2\theta,$$

or $\theta = \frac{1}{8}$. With this value of θ,

$$B(\varepsilon) = \begin{bmatrix} a\varepsilon & \varepsilon^{1/4} & 0 & 0 \\ b\varepsilon^{3/4} & a\varepsilon & \varepsilon^{1/4} & 0 \\ c\varepsilon^{1/2} & b\varepsilon^{3/4} & a\varepsilon & \varepsilon^{1/4} \\ d\varepsilon^{1/4} & c\varepsilon^{1/2} & b\varepsilon^{3/4} & a\varepsilon \end{bmatrix} = \varepsilon^{1/4} \begin{bmatrix} 0 & 1 & 0 & 0 \\ 0 & 0 & 1 & 0 \\ 0 & 0 & 0 & 1 \\ d & 0 & 0 & 0 \end{bmatrix} + \mathcal{O}(\varepsilon^{1/2}).$$

The "root" of the transplanted family of matrices is then

$$C_0 = \begin{bmatrix} 0 & 1 & 0 & 0 \\ 0 & 0 & 1 & 0 \\ 0 & 0 & 0 & 1 \\ d & 0 & 0 & 0 \end{bmatrix},$$

which has four distinct eigenvalues (the fourth roots of d). Since C_0 is semisimple with distinct eigenvalues, the transplanted series can be diagonalized to any order by normalizing with respect to C_0.

3.7.3. Remark. An effective way to obtain the characteristic equation for C_0 in this and the following examples is to write the eigenvector equation $C_0 x = \lambda x$, which in this case gives $x_2 =$

λx_1, $x_3 = \lambda x_2$, $x_r = \lambda x_3$, and $dx_1 = \lambda x_4$. The first three of these equations generate the eigenvector recursively in the form $(x_1, \lambda x_1, \lambda^2 x_1, \lambda^3 x_1)$, and the last equation then becomes $dx_1 = \lambda^4 x_1$. Since $x_1 \neq 0$ (since the eigenvector cannot be zero), it may be canceled to yield $d = \lambda^4$, which is the characteristic equation. Attempts to evaluate $\det(C_0 - \lambda I)$ by minors or other methods are much less effective for this type of problem.

If $d = 0$ but $c \neq 0$, the dominant term that defines the shearing will be c. Setting $1 - 4\theta = 2\theta$ gives $\theta = \frac{1}{6}$, leading to

$$
B(\varepsilon) = \varepsilon^{1/3} \begin{bmatrix} 0 & 1 & 0 & 0 \\ 0 & 0 & 1 & 0 \\ c & 0 & 0 & 1 \\ 0 & c & 0 & 0 \end{bmatrix} + \mathcal{O}(\varepsilon^{2/3}).
$$

The eigenvalues of the leading term are 0 and the three cube roots of $2c$, which are again distinct. If $d = c = 0$ but $b \neq 0$, then $\theta = \frac{1}{4}$ and

$$
C_0 = \begin{bmatrix} 0 & 1 & 0 & 0 \\ b & 0 & 1 & 0 \\ 0 & b & 0 & 1 \\ 0 & 0 & b & 0 \end{bmatrix},
$$

which again has four distinct eigenvalues. Finally, if $d = c = b = 0$ in (3.7.10), and if the series for $A(\varepsilon)$ has no more terms, then (whether a is zero or not) $A(\varepsilon)$ is already in Jordan form and is not diagonalizable; if there are additional terms, we proceed as in (3.7.5) by passing over the diagonal terms to the first term A_j having at least one nonzero stripe below the diagonal, and basing the shearing on this term.

In this 4×4 example, the eigenvalues always split completely if there is a nonzero subdiagonal stripe, but this is not true in higher dimensions. The 5×5 case can give rise to the new leading term

$$
C_0 = \begin{bmatrix} 0 & 1 & 0 & 0 & 0 \\ 0 & 0 & 1 & 0 & 0 \\ c & 0 & 0 & 1 & 0 \\ 0 & c & 0 & 0 & 1 \\ 0 & 0 & c & 0 & 0 \end{bmatrix}, \tag{3.7.15}
$$

with characteristic equation $\lambda^2(\lambda^3 - 3c) = 0$ and double eigenvalue 0; the other three eigenvalues are distinct. Since the nullity of C_0 is one, there is a 2×2 nilpotent block in the Jordan form of C_0. It follows that the normal form of $C(\varepsilon)$ has the structure

$$
\begin{bmatrix} * & * & & & \\ * & * & & & \\ & & * & & \\ & & & * & \\ & & & & * \end{bmatrix}.
$$

This can be subjected to a further shearing transformation (see the general case of shearing, below) to split the zero eigenvalue. The next theorem shows that the general $n \times n$ problem (having nilpotent A_0 with a single Jordan block) behaves like this 5×5 example: When transplanting is possible, it produces at least some splitting of the eigenvalues, so that normalizing produces smaller block sizes and progress is made toward the goal of diagonalization.

3.7.4. Theorem. Suppose $A(\varepsilon)$ is a matrix series in inner product normal form, A_0 is nilpotent with one Jordan block, and there is a term in $A(\varepsilon)$ having a nonzero stripe below the diagonal. Then there exists a shearing transformation that transplants $A(\varepsilon)$ to a new series $C(\varepsilon)$ whose leading term C_0 has more than one eigenvalue.

Proof. Let

$$
T(\varepsilon) = \begin{bmatrix} \varepsilon^{(-n+1)\theta} & & & & \\ & \varepsilon^{(-n+3)\theta} & & & \\ & & \ddots & & \\ & & & \varepsilon^{(n-3)\theta} & \\ & & & & \varepsilon^{(n-1)\theta} \end{bmatrix}.
$$

(The coefficients of θ increase by 2 along the diagonal.) Let A_j be the first term having a nonzero stripe below the diagonal, and consider $T(\varepsilon)^{-1}(A_0 + \varepsilon^j A_j)$. (The terms between A_0 and A_j are multiples of I and will not be affected by the shearing.) Each stripe in $A_0 + \varepsilon^j A_j$ will be multiplied by a factor of the form $\varepsilon^{2m\theta}$ for various integers m; the dominant stripe will be the nonzero stripe closest to the lower left-hand corner. When θ is chosen so that this stripe and the superdiagonal stripe have the same power of ε, these two stripes will constitute the new C_0. We claim that C_0 has at least two distinct eigenvalues. Since C_0 has zero diagonal, its trace (which is the sum of its eigenvalues) is zero; therefore, if the eigenvalues are equal, they must all be zero. But this is not the case, because it is clear that the characteristic equation is not $\lambda^n = 0$. □

Nilpotent Leading Term with Multiple Jordan Blocks

When the leading term is nilpotent with one Jordan block, progress toward diagonalization is measured by the splitting of eigenvalues. If the leading term is nilpotent with more than one Jordan block, splitting of eigenvalues may or may not occur; when it does not, progress toward diagonalization still takes place, but in a different way. We will see that if all of the eigenvalues of C_0 are zero (so that it is still nilpotent), then the length of the longest Jordan chain of C_0 (which is the size of its largest Jordan block, when put in Jordan form) will be greater than the corresponding value for A_0. It follows that after repeated transplanting, if the eigenvalue never splits, the leading term will eventually become nilpotent with one block, the case treated above. The next transplanting will then split the eigenvalue.

As before, we begin with an example,

$$A(\varepsilon) = \begin{bmatrix} 0 & 1 & & & \\ 0 & 0 & & & \\ & & 0 & 1 & 0 \\ & & 0 & 0 & 1 \\ & & 0 & 0 & 0 \end{bmatrix} + \varepsilon \begin{bmatrix} a & 0 & c & 0 & 0 \\ b & a & d & c & 0 \\ 0 & 0 & g & 0 & 0 \\ e & 0 & h & g & 0 \\ f & e & i & h & g \end{bmatrix}. \tag{3.7.16}$$

The second term here is a striped matrix in inner product normal form with respect to A_0. To this we apply similarity by a matrix

$$T(\varepsilon) = \begin{bmatrix} \varepsilon^{-\theta} & & & & \\ & \varepsilon^{\theta} & & & \\ & & \varepsilon^{-2\theta} & & \\ & & & \varepsilon^{0} & \\ & & & & \varepsilon^{2\theta} \end{bmatrix}, \tag{3.7.17}$$

where θ is to be determined, obtaining

$$B(\varepsilon) = \mathbb{S}_{T(\varepsilon)}A(\varepsilon) = \begin{bmatrix} \varepsilon a & \varepsilon^{2\theta} & \varepsilon^{1-\theta}c & 0 & 0 \\ \varepsilon^{1-2\theta}b & \varepsilon a & \varepsilon^{1-3\theta}d & \varepsilon^{1-\theta}c & 0 \\ 0 & 0 & \varepsilon g & \varepsilon^{2\theta} & 0 \\ \varepsilon^{1-\theta}e & 0 & \varepsilon^{1-2\theta}h & \varepsilon g & \varepsilon^{2\theta} \\ \varepsilon^{1-3\theta}f & \varepsilon^{1-\theta}e & \varepsilon^{1-4\theta}i & \varepsilon^{1-2\theta}h & \varepsilon g \end{bmatrix}. \tag{3.7.18}$$

The shearing matrix (3.7.17) is constructed blockwise according to the block structure of A_0, using, for each block, a string of exponents separated by two and symmetrically arranged around zero. (See Remark 3.7.2; this shearing matrix is related in the same way to the matrix Z of the triad having $X = A_0$.) The eigenspaces of $\mathbb{S}_{T(\varepsilon)}$ are direct sums of stripe subspaces of $gl(5)$, as is easily seen from (3.7.18); notice that sometimes two stripes receive the same shearing exponent. The dominant term (with the lowest exponent) is i (which does not denote $\sqrt{-1}$); next are d and f, with equal dominance; then b and h; then c and e; and finally, a.

If $i \neq 0$, setting $1 - 4\theta = 2\theta$ (the exponent of the entries coming from A_0) gives $\theta = \frac{1}{6}$. The new leading term is

$$C_0 = \begin{bmatrix} 0 & 1 & 0 & 0 & 0 \\ 0 & 0 & 0 & 0 & 0 \\ 0 & 0 & 0 & 1 & 0 \\ 0 & 0 & 0 & 0 & 1 \\ 0 & 0 & i & 0 & 0 \end{bmatrix}. \tag{3.7.19}$$

Notice that the eigenvalues of C_0 do not split, as they did in the previous cases. Instead, this matrix C_0 (with $i \neq 0$) has only the eigenvalue zero, and its kernel is one-dimensional. This means that the Jordan form of C_0 consists of a single Jordan block, so that this stage of the transplanting process leaves us with a new problem of the previously studied type. When this problem is put into inner product normal form and transplanted once again, the eigenvalues will split.

If $i = 0$ but d and f are not both zero, then $\theta = \frac{1}{5}$ and the leading term is

$$C_0 = \begin{bmatrix} 0 & 1 & 0 & 0 & 0 \\ 0 & 0 & d & 0 & 0 \\ 0 & 0 & 0 & 1 & 0 \\ 0 & 0 & 0 & 0 & 1 \\ f & 0 & 0 & 0 & 0 \end{bmatrix}. \qquad (3.7.20)$$

The characteristic equation of this matrix is $\lambda^5 = df$, so if d and f are both nonzero, the eigenvalues split completely (to the five cube roots of df) and the problem is diagonalizable. If one (but not both) of d and f is zero, the eigenvalue zero does not split, but (as in the case $i \neq 0$) the kernel is one-dimensional, and the new problem has a single Jordan block. If $i = d = f = 0$ but b and h are not both zero, then $\theta = \frac{1}{4}$ and

$$C_0 = \begin{bmatrix} 0 & 1 & 0 & 0 & 0 \\ b & 0 & 0 & 0 & 0 \\ 0 & 0 & 0 & 1 & 0 \\ 0 & 0 & h & 0 & 1 \\ 0 & 0 & 0 & h & 0 \end{bmatrix}.$$

The characteristic equation is $-\lambda(\lambda^2 - b)(\lambda^2 - 2h) = 0$, and some splitting of the eigenvalues must occur; the splitting is complete if $b \neq 2h$. If $i = d = f = b = h = 0$ but c and e are not both zero, then $\theta = \frac{1}{3}$ and

$$C_0 = \begin{bmatrix} 0 & 1 & c & 0 & 0 \\ 0 & 0 & 0 & c & 0 \\ 0 & 0 & 0 & 1 & 0 \\ e & 0 & 0 & 0 & 1 \\ 0 & e & 0 & 0 & 0 \end{bmatrix}.$$

The characteristic equation is $\lambda^2(-\lambda^3 + 3ce) = 0$, and (partial) splitting of eigenvalues occurs only if both c and e are nonzero. Otherwise, the eigenvalues are all zero, and the nullity is 2, so there are two Jordan blocks; these turn out to have block sizes 4 and 1, so that (as claimed in introducing this topic), progress toward diagonalization has been made by increasing the size of the largest block. Since the length of the longest Jordan chain of C_0 equals its index of nilpotence, an easy way to see that this length is greater than 3 is to check that $C_0^3 \neq 0$.

3.7.5. Theorem. Suppose $A(\varepsilon)$ is a matrix series in inner product normal form, A_0 is nilpotent and is in Jordan form with more than one Jordan block, and there is a term in $A(\varepsilon)$ having a nonzero stripe not lying on the diagonal. Then there exists a shearing transformation that transplants

$A(\varepsilon)$ to a new series $C(\varepsilon)$ whose leading term C_0 either has more than one eigenvalue, or else is nilpotent with a Jordan chain longer than the longest chain of A_0.

Proof. Let A_j be the first term with a nontrivial off-diagonal stripe. If we construct the shearing matrix blockwise as in the example, the leading term of the transplanted matrix series will have the form $C_0 = A_0 + R$, where R is a striped matrix consisting of just those stripes of A_j that are dominant (i.e., have the smallest shearing power). Let k be the index of nilpotence of A_0; it can then be proved, from the structure of C_0, that $C_0^k \neq 0$ (see Notes and References below). From this it follows that either C_0 is not nilpotent, or it is nilpotent with index of nilpotence greater than k. If C_0 is not nilpotent, then (since the sum of the eigenvalues equals the trace, which is zero) C_0 has more than one eigenvalue; if it is, then since its index of nilpotence is greater than that of A_0, it has a longer Jordan chain.   □

The General Case

All of the ingredients necessary to handle the general case are contained in the special cases already treated. Suppose that A_0 is an arbitrary matrix in upper Jordan canonical form, and that the remainder of the terms in $A(\varepsilon)$ (at least up to some order) are in inner product normal form with respect to A_0. Let $\widehat{A}(\varepsilon)$ be the truncation of $A(\varepsilon)$ at the order to which it has been normalized. Then $\widehat{A}(\varepsilon)$ is in block-diagonal form, with blocks of the size of the large blocks of A_0. For further consideration we will separate $\widehat{A}(\varepsilon)$ into several matrix series, each consisting of one of these blocks, and attend to these series one at a time. Let $\widetilde{A}(\varepsilon)$ be one such series, and let $\widetilde{A}_j$ be the first term with a nontrivial off-diagonal stripe. Since $\widetilde{A}_0$ has only one eigenvalue a_0, it may be written $\widetilde{A}_0 = a_0 \widetilde{I} + \widetilde{N}$ where $\widetilde{N}$ is nilpotent. ($\widetilde{I}$ and $\widetilde{N}$ are marked with a tilde to emphasize that they are of the same size as the block that is being considered.) Since each $\widetilde{A}_i$ for $1 = 1, \ldots, j-1$ is diagonal, we write $\widetilde{A}_i = \widetilde{\Lambda}_i$ and

$$\widetilde{A}(\varepsilon) = N + \left(a_0 \widetilde{I} + \varepsilon \widetilde{\Lambda}_1 + \cdots + \varepsilon^{j-1}\widetilde{\Lambda}_{j-1}\right) + \varepsilon^j \widetilde{A}_j + \cdots.$$

The intermediate terms are not susceptible to shearing and may be set aside (compare the treatment of (3.7.5)), while $\widetilde{N} + \varepsilon^j \widetilde{A}_j$ may be sheared to produce a new leading term of the form $\widetilde{C}_0 = \widetilde{N} + \widetilde{R}$, where $\widetilde{R}$ contains the dominant stripes of $\widetilde{A}_j$. Either the eigenvalues of $\widetilde{C}_0$ will (at least partially) split, or $\widetilde{C}_0$ will remain nilpotent with a larger index of nilpotence than $\widetilde{A}_0$; in either case, progress has been made toward diagonalization of the series. The next step is to put $\widetilde{C}_0$ into Jordan form and normalize with respect to it. Notice that if the eigenvalues of $\widetilde{C}_0$ split, a new block structure (using large blocks of $\widetilde{C}_0$) will result; each of the separate series into which the problem has already been divided may then divide again. The stage is then set for another shearing operation (if necessary). The process terminates when the entire original series has been diagonalized (or equivalently, split into separate 1×1 series) or when it is impossible to proceed because some part of the problem has no nontrivial off-diagonal stripes.

3.7.6. Remark. Although there certainly exist formal power series $A(\varepsilon)$ that cannot be diagonalized, this fact can never be discovered by a finite amount of calculation, since it would be required to know that when $A(\varepsilon)$ is normalized *to all orders*, there are no nontrivial off-diagonal stripes. In a computational problem, then, we can stop calculating and say that a diagonal form has not been found, but not that it is impossible. In the case of smooth $A(\varepsilon)$, even computing to all orders (and finding no off-diagonal stripes) does not prove nondiagonalizability; see Remark 3.7.1.

In cases where $A(\varepsilon)$ proves to be diagonalizable by transplanting, the final result may be stated as follows: There exists a fractional power series $T(\varepsilon)$, which in general contains both positive and negative fractional powers of ε, such that

$$B(\varepsilon) = T(\varepsilon)^{-1} A(\varepsilon) T(\varepsilon) \qquad (3.7.21)$$

is diagonal and is a fractional power series in ε containing only positive fractional powers. Here $T(\varepsilon)$ is the product of all of the shearing and normalizing transformations that are used in the course of the simplification of $A(\varepsilon)$, multiplied in the correct order. If $A(\varepsilon)$ is broken down into blocks of smaller sizes during the analysis, these blocks must be reassembled to produce (3.7.21).

k-Determined Hyperbolicity

In Section 3.1, it was shown that the matrix series (3.1.3), if truncated at order ε, has eigenvalues in the right half-plane for $\varepsilon > 0$, but the full series has one eigenvalue in each half-plane (if $a > 1$). Both the truncated and full matrices are hyperbolic for $\varepsilon > 0$ (i.e., have no eigenvalues on the imaginary axis), but they are of different stability type, as indicated by the number of eigenvalues (counting multiplicity) in the left half-plane. This raises the following general question: Suppose that

$$\widehat{A}(\varepsilon) = A_0 + \varepsilon A_1 + \cdots + \varepsilon^k A_k \qquad (3.7.22)$$

is hyperbolic for all $\varepsilon > 0$; under what circumstances can we say that any smooth matrix $A(\varepsilon)$ having $\widehat{A}(\varepsilon)$ as its k-jet will be hyperbolic and have the same hyperbolicity type? Because of the restriction that $\varepsilon > 0$, the method of metanormalization suggests itself as natural. We assume that $A(\varepsilon)$ is diagonalizable as in (3.7.21) by a sequence of transplantings and normalizations. Notice that since $A(\varepsilon)$ and $B(\varepsilon)$ in (3.7.21) are similar for $\varepsilon > 0$, they will be of the same hyperbolicity type. We suppose that $B(\varepsilon)$

has been computed up to some fractional order, and that this portion $\widehat{B}(\varepsilon)$ of $B(\varepsilon)$ is hyperbolic and diagonal; the next theorem then implies that $A(\varepsilon)$ is hyperbolic, with the same stability type as $\widehat{B}(\varepsilon)$. Since the computation of $\widehat{B}(\varepsilon)$ will depend only on some k-jet $\widehat{A}(\varepsilon)$ of $A(\varepsilon)$, it follows that the hyperbolicity of $A(\varepsilon)$ is k-determined. (The value of k may not be closely related to the number of terms appearing in $\widehat{B}(\varepsilon)$. It may, for instance, be necessary to go to a high value of k in order to break down the Jordan block size enough to make $B(\varepsilon)$ diagonal, but then a small number of terms of $B(\varepsilon)$ may give a truncation $\widehat{B}(\varepsilon)$ that is hyperbolic.)

3.7.7. Theorem. Suppose that $C(\varepsilon)$ and $D(\varepsilon)$ are continuous matrix-valued functions defined for $\varepsilon > 0$, and that

$$C(\varepsilon) = \Lambda(\varepsilon) + \varepsilon^R D(\varepsilon),$$

where

$$\Lambda(\varepsilon) = \begin{bmatrix} \lambda_1(\varepsilon) & & & \\ & \ddots & & \\ & & \ddots & \\ & & & \lambda_n(\varepsilon) \end{bmatrix} = \varepsilon^{r_1}\Lambda_1 + \cdots + \varepsilon^{r_j}\Lambda_j;$$

here $r_1 < r_2 < \cdots < r_j < R$ are rational numbers, and $\Lambda_1,\ldots,\Lambda_j$ are diagonal. Then there exists $\varepsilon_0 > 0$ such that for $0 < \varepsilon < \varepsilon_0$, the eigenvalues of $C(\varepsilon)$ are approximately equal to the diagonal entries $\lambda_i(\varepsilon)$ of $\Lambda(\varepsilon)$, with error $\mathcal{O}(\varepsilon^R)$.

Proof. Assume that the basis vectors for $\mathbb{R}^n$ are numbered so that equal eigenvalues in $\Lambda(\varepsilon)$ are adjacent. Let the first string of equal eigenvalues be $\lambda_1(\varepsilon),\ldots,\lambda_p(\varepsilon)$, and call their common value $\lambda(\varepsilon)$. We claim that $C(\varepsilon)$ has p eigenvalues having the form $\lambda(\varepsilon) + \varepsilon^R\sigma_i(\varepsilon)$, for $i = 1,\ldots,p$. The theorem then follows by repeating the argument for the other strings of equal entries in $\Lambda(\varepsilon)$, or else by renumbering the basis vectors so that each string takes a turn in the first position.

After the ε-dependent change of variables $\mu \leftrightarrow \sigma$ defined by $\mu = \lambda(\varepsilon) - \varepsilon^R\sigma$, the characteristic equation $\det(C(\varepsilon) - \mu I) = 0$ takes the form

$$\det \begin{bmatrix} \varepsilon^R(D_{11}(\varepsilon) - \sigma I_p) & \varepsilon^R D_{12}(\varepsilon) \\ \varepsilon^R D_{21}(\varepsilon) & E(\varepsilon) + \varepsilon^R D_{22}(\varepsilon) \end{bmatrix} = 0,$$

where D_{11} is the upper left $p \times p$ block of D; D_{12}, D_{21}, and D_{22} are the remaining blocks of D; and $E(\varepsilon)$ is a diagonal matrix having entries $\lambda_i(\varepsilon) - \lambda(\varepsilon)$ for $i > p$. It is possible to remove a factor of ε^R from each of the first p rows of this determinant. For each $i > p$, there is a smallest power of ε, say ε^{r_k}, in which $\lambda_i(\varepsilon)$ differs from $\lambda(\varepsilon)$ in at least one of its coefficients; we may then remove a factor of ε^{r_k} from the ith row of the determinant. The resulting equation for σ has the form

$$\det \begin{bmatrix} D_{11}(\varepsilon) - \sigma I_p & D_{12}(\varepsilon) \\ o(1) & F + o(1) \end{bmatrix} = 0.$$

Here $o(1)$ is the little-oh asymptotic order symbol, denoting terms that approach zero as $\varepsilon \to 0$, and F is a diagonal matrix with nonzero diagonal entries (each entry being the difference of the lowest-order differing coefficients in $\lambda_i(\varepsilon)$ and $\lambda(\varepsilon)$). This equation can be rewritten as

$$(\det D(\varepsilon) - \sigma I_p)(\det F) + o(1) = 0,$$

which has p solutions (counting multiplicity) for σ when $\varepsilon = 0$; it follows from Rouché's theorem that there are p solutions for small $\varepsilon > 0$. $\qquad\square$

Notes and References

The treatment of shearing in this section is based on Bogaevski and Povsner [17], Chapter 1, although many of the arguments have been modified. The normal form used in [17] is what we call the sl(2) normal form (although it is developed in a different way and sl(2) is never mentioned). Our treatment is organized in such a way that it is valid for both the inner product and sl(2) normal forms, although we have used the inner product normal form in the presentation.

The missing part of the proof of Theorem 3.7.5, that $C_0^k = (A_0 + R)^k \neq 0$, is proved in Theorem 1.5.1 of [17]. The proof is purely combinatorial in character and not particularly illuminating. (It examines the products that occur in the expansion of $(A_0 + R)^k$, remembering that A_0 and R do not commute, and counts the number of vanishing stripes that can occur.) It would be nice to have a proof that relies only on the fact that R is an eigenvector of $\mathbb{L}_Z$, where $\{X, Y, Z\}$ is an sl(2) triad with $X = A_0$.

The question of k-determined hyperbolicity has no doubt arisen in various forms many times, but it appears explicitly in Murdock [82] (where it was settled in a 2×2 special case by the quadratic formula) and in subsequent joint work with Clark Robinson ([91], [90]), where a general condition was given. A more computable version was presented in Murdock [84], but the treatment here is the first that is algorithmic (in that it makes use of the metanormal form to render the question decidable). Note that what we call k-determined hyperbolicity here is equivalent to $(k + 1)$-hyperbolicity in the terminology of the earlier papers.

4
Nonlinear Normal Forms

Chapters 2 and 3 have been somewhat of a digression from the problem posed in Chapter 1, that of normalizing a system of nonlinear differential equations. In this chapter we return to that problem, equipped with the methods discussed in Chapters 2 and 3.

The material in this chapter falls under three headings: *settings, formats,* and *styles.* A *setting* is a particular context in which normal form problems can arise, such as "autonomous differential equations with a rest point," called "setting 1" below. A *format* is a set of algorithms for handling the transformations needed to normalize a system, while a *style* specifies what kind of "simplest form" is to be achieved. The formats and styles developed here are exactly parallel to those developed in Chapter 3 in the setting of matrix series, which might be called "setting 0" for normal form theory. Thus, there are formats 1a, 1b, 2a, 2b, and 2c, and there are semisimple, inner product, sl(2), and simplified styles.

4.1 Preliminaries

Before beginning, it is convenient to collect in one place the definitions of various vector spaces of functions that will be used in this chapter, and a few remarks about vector spaces of formal power series. It is assumed here that all functions and vector fields are real. In the applications it is often necessary to deal with the complex case as well; the same notations will be used (with $\mathbb{R}$ and $\mathbb{R}^n$ replaced by $\mathbb{C}$ and $\mathbb{C}^n$).

4.1.1. Definition.

1. $\mathfrak{F}^n$ is the space of smooth functions $f : \mathbb{R}^n \to \mathbb{R}$, also called *smooth scalar fields*. *Smooth* (in this book) always means infinitely differentiable.

2. $\mathfrak{F}_j^n$ is the space of polynomial functions $f : \mathbb{R}^n \to \mathbb{R}$ that are homogeneous of degree j.

3. $\mathfrak{F}_*^n$ is the space of all polynomial functions $f : \mathbb{R}^n \to \mathbb{R}$, regardless of degree. (This is the direct sum of the $\mathfrak{F}_j^n$ over all $j = 0, 1, 2, \ldots$. Recall that a direct *sum* of infinitely many spaces allows only finitely many nonzero terms.

4. $\mathfrak{F}_{**}^n$ is the space of all formal power series in $x_1, \ldots, x_n$ with real coefficients, also called *formal scalar fields*. This space is the direct *product* of the $\mathfrak{F}_j^n$ over all j (which is like a direct *sum*, but allows infinitely many nonzero terms). Note that $\mathfrak{F}_j^n$ and $\mathfrak{F}_*^n$ are subspaces of $\mathfrak{F}^n$, but $\mathfrak{F}_{**}^n$ is not, because a formal power series need not converge to a definite function. The Taylor series or ∞-jet map $j^\infty : \mathfrak{F}^n \to \mathfrak{F}_{**}^n$ is onto (by the Borel–Ritt theorem, Theorem A.3.2), and its kernel is the space of flat functions; see Remark 1.1.5.

5. $\mathcal{V}^n$ is the space of smooth vector fields $v : \mathbb{R}^n \to \mathbb{R}^n$.

6. $\mathcal{V}_j^n$ is the vector space of *polynomial vector fields* on $\mathbb{R}^n$, that is, vector fields $v : \mathbb{R}^n \to \mathbb{R}^n$ such that each component v_i $(i = 1, \ldots, n)$ is a homogeneous polynomial of degree $j + 1$, that is, $v_i \in \mathfrak{F}_{j+1}^n$. The index $j = -1, 0, 1, 2, \ldots$ will be called the *grade* of the vector field. Warning: The grade of a vector field is one less than its degree.

7. $\mathcal{V}_*^n$ is the space of polynomial vector fields on $\mathbb{R}^n$, the direct sum of $\mathcal{V}_j^n$ for $j = -1, 0, 1, 2, \ldots$.

8. $\mathcal{V}_{**}^n$ is the space of *formal power series vector fields*, or simply *formal vector fields* on $\mathbb{R}^n$. $\mathcal{V}_{**}^n$ may be viewed as the set of formal power series in $x_1, \ldots, x_n$ with vector coefficients, or as the direct product of the $\mathcal{V}_j^n$ for all j, or as the set of n-tuples of elements of $\mathfrak{F}_{**}^n$. A formal vector field is not truly a vector field, since it need not converge, so $\mathcal{V}_{**}^n$ is not a subspace of $\mathcal{V}^n$, but again the map $j^\infty : \mathcal{V}^n \to \mathcal{V}_{**}^n$ is onto, with a kernel consisting of flat vector fields. (This map j^∞ is just the "scalar" j^∞, applied componentwise.)

9. $\mathcal{V}^n(2\pi)$ is the vector space of all smooth vector fields on the space $\mathbb{R}^{n+1}$ with variables $(x_0, x_1, \ldots, x_n)$, having the special form

$$v(x_0, x_1, \ldots, x_n) = (0, v_1(x), \ldots, v_n(x)), \qquad (4.1.1)$$

and subject to the condition that each v_i is periodic in x_0 with period 2π.

10. $V_j^n(2\pi)$ is the subspace of $V^n(2\pi)$ consisting of vector fields of the form (4.1.1) for which each v_i is a polynomial of degree $j + 1$ in the variables $x_1, \ldots, x_n$ with coefficients that are smooth functions of x_0 with period 2π.

The vector spaces $\mathcal{F}_*^n$ and V_*^n are infinite-dimensional vector spaces that have countable bases in the usual sense, such that every element can be written uniquely as a *finite* linear combination of the basis elements. (The standard basis elements for these spaces are the scalar and vector monomials defined in equations (4.5.10) and (4.5.12) below.) On the other hand, the spaces $\mathcal{F}_{**}^n$ and V_{**}^n have *formal bases*. By a formal basis, we mean a countably infinite set of elements of a vector space such that each vector can be written uniquely as an infinite linear combination of the basis elements, and, conversely, every infinite linear combination of the basis elements defines a vector. (This contrasts, for instance, with the notion of basis commonly used for Hilbert spaces, in which infinite linear combinations are allowed, provided that the coefficients are square-summable.) Any basis (in the usual sense) for $\mathcal{F}_*^n$ or V_*^n is automatically also a formal basis for $\mathcal{F}_{**}^n$ or V_{**}^n.

In addition to the vector space structure, each space of functions listed above ($\mathcal{F}^n$, $\mathcal{F}_*^n$, and $\mathcal{F}_{**}^n$) admits the structure of a ring (by which we mean a commutative ring with identity), and each space of vector fields (V^n, V_*^n, and V_{**}^n) forms a module over the appropriate ring. That is, it is possible to multiply two functions (of the same type, smooth, polynomial, or formal) and to multiply a function times a vector field (again of the same type). Further information about these ring and module structures is contained in Appendices A and B. There is also a "composition structure" that is frequently used: In the smooth case, for instance, an element φ of V^n can be thought of as a mapping $\varphi : \mathbb{R}^n \to \mathbb{R}^n$ instead of as a vector field, and a function f or a vector field v may then be composed with φ to form $f \circ \varphi$ or $v \circ \varphi$. (We will use this only for mappings φ that can be understood as coordinate changes, and in the case of vector fields v we will be interested in $\varphi'(x)^{-1} v(\varphi(x))$ rather than $v \circ \varphi$ itself.)

These composition operations are self-explanatory in the smooth and polynomial cases, but require comment in the case of formal power series. The requirement for an operation to be possible with formal power series is that *each individual coefficient* of the resulting series must be computable (in principle) in a finite number of steps, without taking infinite sums or limits (since there is nothing to guarantee that such sums or limits will converge); of course, the amount of computation required to determine a coefficient will generally be unbounded as one goes farther out in the series. It follows that the computation of $f \circ \varphi$ or $v \circ \varphi$ is possible when f, v, and φ are formal power series, provided that the series φ has no constant term. (Observe that if $e^{\cos x}$ is computed by substituting the power series for $y = \cos x$ into the power series for e^y, the coefficient of each power of x is an infinite series of constants. In this example, these series are convergent, but with formal power series they need not be, and no power series at all, even formal, will result. But, when $e^{\sin x}$ is expanded in the same way, all coefficients are finite sums, because the series for $\sin x$ has no constant term.)

4.2   Settings for Nonlinear Normal Forms

All of the normal form problems that we treat in this book can be written in the following "universal" form:

$$\dot{x} = a(x, \varepsilon) \sim a_0(x) + \varepsilon a_1(x) + \varepsilon^2 a_2(x) + \cdots. \qquad (4.2.1)$$

The difference between one "setting" and another lies in the vector spaces to which the coefficients a_j belong. For instance, the problem of normalizing a matrix series

$$A(\varepsilon) \sim A_0 + \varepsilon A_1 + \varepsilon^2 A_2 + \cdots,$$

discussed in Chapter 3, is equivalent to the problem of normalizing the differential equation

$$\dot{x} \sim A_0 x + \varepsilon A_1 x + \varepsilon^2 A_2 x + \cdots, \qquad (4.2.2)$$

which falls under the form (4.2.1) with each a_j belonging to the space $\mathcal{V}_0^n$ of linear vector fields (grade 0 or degree 1) on $\mathbb{R}^n$:

$$a_j \in \mathcal{V}_0^n \quad \text{for all } j. \qquad (4.2.3)$$

If this is classified as "setting 0," the three settings described here may be listed as follows:

1. Autonomous differential equations near a rest point.

2. Periodic differential equations (method of averaging).

3. Autonomous differential equations near a periodic solution, or periodic differential equations near a rest point.

We will discuss each of these settings in turn, identifying the vector spaces to which the coefficients a_j belong in each case. The rest of this book is almost entirely concerned with setting 1.

Setting 1: Autonomous Differential Equations near a Rest Point

If the smooth autonomous differential equation

$$\dot{v} = f(v),$$

with $v \in \mathbb{R}^n$, has a rest point at $v = p$, so that $f(p) = 0$, we may introduce $u = v - p$ and expand the equation in a Taylor series around $u = 0$ as

$$\dot{u} = a(u) \sim a_0(u) + a_1(u) + a_2(u) + \cdots , \qquad (4.2.4)$$

where a_j is a vector field that is homogeneous of degree $j + 1$ and where

$$a_0(u) = Au \qquad (4.2.5)$$

with $A = f'(p)$. To bring (4.2.4) into the "universal" form (4.2.1), set

$$u = \varepsilon x. \qquad (4.2.6)$$

By homogeneity, $a_j(\varepsilon x) = \varepsilon^{j+1} a_j(x)$, so that both sides can be divided by ε. The relationship between the smooth functions $a(u)$ in (4.2.4) and $a(x, \varepsilon)$ in (4.2.1) is then

$$a(x, \varepsilon) = a(\varepsilon x)/\varepsilon. \qquad (4.2.7)$$

Setting 1 is characterized by the condition

$$a_j \in \mathcal{V}_j^n. \qquad (4.2.8)$$

Notice the convenience of the definition of the "grade" of a vector field: Although the grade does not equal the degree in x, it does equal the order in ε after this scaling.

A significant difference between "setting 0" (Chapter 3) and setting 1 is that here, the coefficients a_j belong to distinct vector spaces depending on j. This means that the homological equations are in different vector spaces for each degree, and the choice of a complement to im $\mathcal{L}$ cannot be made once and for all as in the linear case. This has already been seen in the examples of Chapter 1.

The "scaling" transformation (4.2.6) can be viewed as merely a book-keeping device, which can be eliminated at any time by setting $\varepsilon = 1$. But it is useful to view it as an actual change of coordinates, called *dilation*. For instance, it is frequently possible to prove theorems of the following type for specific systems of the form 4.2.1: "For every $r > 0$ there exists $\varepsilon_0 > 0$ such that if $\|x\| \leq r$ and $0 < \varepsilon < \varepsilon_0$, then such-and-such is true." (For instance, perhaps solutions beginning in this ball approach the origin as $t \to \infty$.) If the system in question results from (4.2.4) by way of (4.2.6), then such a result can be carried back to (4.2.4) as "for every $r > 0$ there exists $\varepsilon_0 > 0$ such that if $0 < \varepsilon < \varepsilon_0$ and $\|u\| \leq \varepsilon r$, then such-and-such is true." Since ε does not appear in (4.2.4), it is only the product εr that matters, and the statement is equivalent to "there exists $\rho > 0$ such that for $\|x\| \leq \rho$, such-and-such is true." In other words, proving a result for a *compact* neighborhood of the origin in (4.2.1) is equivalent to proving it for a *small* neighborhood of the origin in (4.2.4).

It is worth pointing out the effect of dilation on the kind of asymptotic validity that each of the expansions (4.2.1) and (4.2.4) possesses. Since

(4.2.4) is a Taylor series in u around $u = 0$, the truncation at $a_k(u)$ has an error of the order of the first omitted term:

$$a(u) = a_0(u) + a_1(u) + \cdots + a_k(u) + \mathcal{O}\left(\|u\|^{k+2}\right), \qquad (4.2.9)$$

meaning there exists a constant $c > 0$ such that

$$\|a(u) - a_0(u) - a_1(u) - \cdots - a_k(u)\| \leq c\|u\|^{k+2}.$$

(This is simply an asymptotic expansion, neither a "pointwise" nor a "uniform" expansion, because there are no parameters present. In this context a *parameter* is a variable other than those in which the function is expanded.) After dilation, this becomes

$$\|a(x, \varepsilon) - a_0(x) - \varepsilon a_1(x) - \cdots - \varepsilon^k a_k(x)\| \leq c\varepsilon^{k+1}\|x\|^{k+2}.$$

Thus, for x in any (fixed) compact set, the error committed by making the truncation is bounded by a constant times ε^{k+1}. In other words,

$$a(x, \varepsilon) = a_0(x) + \varepsilon a_1(x) + \cdots + \varepsilon^k a_k(x) + \mathcal{O}\left(\varepsilon^{k+1}\right) \qquad (4.2.10)$$

uniformly for x in compact sets. (This is now an asymptotic expansion *in ε only*, with x as a vector parameter, so the question of uniformity becomes meaningful.)

It is possible to do normal form theory for (4.2.4) directly, without introducing ε, provided that the work is done in (slight modifications of) format 1a, 1b, or 2a; see Section 4.3 below. In format 2c it is essential to have ε, since in that format the transformations are generated as solutions of differential equations in which ε is the independent variable; compare (3.2.9).

Setting 2: Periodic Differential Equations and the Method of Averaging

A system of the form

$$\dot{y} \sim \varepsilon f_1(t, y) + \varepsilon^2 f_2(t, y) + \cdots, \qquad (4.2.11)$$

where f is periodic in t with period 2π and $y \in \mathbb{R}^n$, is said to be in *standard form for the method of averaging*. The goal of averaging is to make a periodic, ε-dependent change of variables under which the system becomes autonomous (independent of t) to any specified order. We will see (in Section 4.3) that this is a normal form problem. Unlike setting 1, the small parameter ε is in the problem from the beginning, and there are no special homogeneity requirements on the coefficients: They all belong to the same vector space, the space of smooth periodic vector fields on $\mathbb{R}^n$.

To cast (4.2.11) into the form of (4.2.1) we introduce $x = (x_0, x_1, \ldots, x_n) \in \mathbb{R}^{n+1}$ and rewrite the system as

$$\dot{x}_0 = 1$$

together with

$$\begin{bmatrix} \dot{x}_1 \\ \vdots \\ \dot{x}_n \end{bmatrix} \sim \varepsilon f_1(x) + \varepsilon^2 f_2(x) + \cdots .$$

This has the form of (4.2.1), where now $x \in \mathbb{R}^{n+1}$ (written as $(x_0, \ldots, x_n)$) and where

$$a_0(x) = \begin{bmatrix} 1 \\ 0 \\ \vdots \\ 0 \end{bmatrix} \quad \text{and} \quad a_j(x) = \begin{bmatrix} 0 \\ f_j(x) \end{bmatrix} \qquad (4.2.12)$$

for $j > 0$. Each a_j for $j \geq 1$ belongs to the vector space $\mathcal{V}^n(2\pi)$ defined at the beginning of this chapter.

Setting 3: Autonomous Differential Equations near a Periodic Solution

If the smooth autonomous differential equation

$$\dot{v} = f(v),$$

with $v \in \mathbb{R}^n$, has a periodic solution $v = p(t)$ (which without loss of generality we assume to have period 2π), we may introduce $z = v - p(t)$ and expand the equation in a Taylor series around $z = 0$ as

$$\dot{z} \sim g_0(t, z) + g_1(t, z) + g_2(t, z) + \cdots , \qquad (4.2.13)$$

where g_j is a periodic vector field on $\mathbb{R}^n$ that is homogeneous of degree $j + 1$ and where

$$g_0(t, z) = f'(p(t))z. \qquad (4.2.14)$$

By Floquet theory there is a periodic linear change of variables $z = P(t)u$ that eliminates the time dependence of the linear term, and we obtain

$$\dot{u} \sim Au + h_1(t, u) + h_2(t, u) + \cdots , \qquad (4.2.15)$$

where A is a constant matrix. Finally, putting $x = (x_0, x_1, \ldots, x_n) = (t, \varepsilon u) \in \mathbb{R}^{n+1}$ leads to the form (4.2.1), with

$$a_0(x) = \begin{bmatrix} 1 \\ 0 \\ \vdots \\ 0 \end{bmatrix} + \begin{bmatrix} 0 \\ & A \end{bmatrix} x \qquad (4.2.16)$$

and $a_j(x) = (0, h_j(x))$ for $j > 0$. These coefficients satisfy

$$a_j \in \mathcal{V}_j^n(2\pi)$$

for $j \geq 1$.

If the smooth periodic differential equation

$$\dot{v} = f(t, v),$$

with $v \in \mathbb{R}^n$ and period 2π in t, has a rest point $v = p$, so that $f(t, p) = 0$ for all t, then after introducing $z = v - p$ the equations take the form (4.2.13), with (4.2.14) replaced by

$$g_0(t, z) = f_v(t, p)z. \qquad (4.2.17)$$

The reduction to (4.2.1) is the same as before.

Notes and References

For the most part, this section consists only in establishing notation, and calls for no citations. The notions of asymptotic expansion (pointwise and uniform), dilation, and the method of averaging are treated more thoroughly in Murdock [88] (and in many other sources). In particular, it is shown in Section 5.4 of [88] how to bring many common systems, such as Duffing and Van der Pol oscillators, into standard form for the method of averaging (referred to there as periodic standard form).

The method of averaging constitutes a large subject in its own right, going beyond the periodic case considered here to include quasiperiodic, almost periodic, and nonperiodic situations. For a complete overview of the method of averaging, one would have to read (at least) all of the following sources: Sanders and Verhulst [100], Lochak and Meunier [72], Hapaev [55], and Murdock [84]. Although each of these references is a survey of part of the field, there is almost no overlap between the content of any two of them. One of the most important papers on the subject of averaging is Perko [96]; ideas contained in this paper are frequently rediscovered (often in weaker form) by others.

4.3   The Direct Formats (1a and 1b)

Recall from Section 3.2 that a *format* is a set of notations and procedures for handling the coordinate transformations used in normalization. The formats for nonlinear normal forms are classified according to the same scheme introduced in Section 3.2 for the linear case. As in the linear case, formats 1a and 1b are the easiest to explain, although the "generated" formats 2a, 2b, and 2c have strong advantages and format 2a will be the format most often used. In this section we will develop format 1a to the second order for the "universal" equation (4.2.1), and it will be clear how to continue to higher orders. This will serve to introduce the operator $\mathcal{L}$, which is the same in all formats. Slight modifications in the appearance of format 1a will then be introduced so that it can be applied directly to each of the three settings described in the last section without reducing these settings to the "universal" form. That is, we will discuss setting 1 without dilation by ε, setting 2 without introducing $x_0 = t$, and setting 3 without either of these. The section ends with a brief treatment of format 1b.

The notations in this chapter for each format are designed to agree as much as possible with those in Chapter 3 for the same format. For instance, in the linear case the first transformation in format 1a is $I + \varepsilon S_1$, or $x = \xi + \varepsilon S_1 \xi$, where S_1 is a matrix. Here this becomes $x = \xi + \varepsilon s_1(\xi)$, where s_1 is a nonlinear transformation.

Format 1a in Universal Form

The "zeroth step" in the normalization of (4.2.1) is to perform a change of variables that is independent of ε and brings $a_0(x)$ into some desired form. This will depend on the setting; for instance in setting 1, where $a_0(x) = Ax$, we can bring A into either Jordan or real canonical form, just as in Chapter 3 (see equation (3.2.1)). In setting 2 the form for a_0 achieved in (4.2.12) is already the desired form. In setting 3, we can use (4.2.16) and also put A into a canonical form. As in Chapter 3, it will usually be assumed that the desired simplification of $a_0(x)$ has already been accomplished, and it is desired to normalize the higher terms (up to some order) *with respect to* the existing $a_0(x)$.

If a smooth invertible coordinate change $x = h(\xi, \varepsilon)$ is applied to the smooth differential equation $\dot{x} = a(x, \varepsilon)$, the result is

$$\dot{\xi} = h_\xi(\xi, \varepsilon)^{-1} a(h(\xi, \varepsilon), \varepsilon). \tag{4.3.1}$$

This is obtained by differentiating the coordinate change to get $\dot{x} = h_\xi(\xi, \varepsilon)\dot{\xi}$, solving for $\dot{\xi}$, using the differential equation for x, and inserting the coordinate change into $a(x, \varepsilon)$ so that the entire equation (4.3.1) is expressed in terms of the new variable ξ. A coordinate change of the form

$$x = \xi + \varepsilon^j s_j(\xi) \tag{4.3.2}$$

will be invertible on compact sets for small ε. (That is, given a compact set K there exists ε_0 such that for $0 \le \varepsilon < \varepsilon_0$, (4.3.2) can be solved for ξ as a smooth function of x for $x \in K$. This follows from the inverse function theorem and the Heine–Borel theorem.) The following lemma shows that the coordinate change leaves (4.2.1) unchanged to order $j - 1$, and gives the formula for the transformed jth term:

4.3.1. Lemma. The coordinate change (4.3.2) carries (4.2.1) into

$$\dot{\xi} = b(\xi, \varepsilon) \sim a_0(\xi) + \cdots + \varepsilon^{j-1} a_{j-1}(\xi) + \varepsilon^j b_j(\xi) + \cdots, \tag{4.3.3}$$

where

$$b_j(\xi) = a_j(\xi) + a_0'(\xi)s_j(\xi) - s_j'(\xi)a_0(\xi). \tag{4.3.4}$$

Proof. In the proof we will omit the subscript j on s_j. Differentiating the coordinate change gives $h_\xi(\xi, \varepsilon) = I + \varepsilon^j s'(\xi)$. The inverse may be expanded formally by (3.2.10) as

$$h_\xi(\xi, \varepsilon)^{-1} \sim I - \varepsilon^j s'(\xi) + \cdots.$$

To expand $a(h(\xi, \varepsilon), \varepsilon)$ we begin with the individual terms:

$$a_i(\xi + \varepsilon^j s(\xi)) = a_i(\xi) + \varepsilon^j a_i'(\xi)s(\xi) + \cdots.$$

It follows that to order ε^j we have

$$a_0(\xi + \varepsilon^j s(\xi)) = a_0(\xi) + \varepsilon^j a_0'(\xi)s(\xi) + \cdots$$

and for $1 \leq i \leq j$,

$$\varepsilon^i a_i(\xi + \varepsilon^j s(\xi)) = \varepsilon^i a_i(\xi) + \cdots.$$

Now

$$h_\xi(\xi, \varepsilon)^{-1} a(h(\xi, \varepsilon), \varepsilon) \sim (I - \varepsilon^j s'(\xi) + \cdots)$$
$$\cdot (a_0(\xi) + \cdots + \varepsilon^{j-1} a_{j-1}(\xi) + \varepsilon^j (a_j(\xi) + a_0'(\xi)s(\xi)) + \cdots).$$

Multiplying out gives (4.3.3).   □

Define the *homological operator* $\mathcal{L}$ acting on vector fields $v : \mathbb{R}^n \to \mathbb{R}^n$ as follows (see (1.1.6)):

$$(\mathcal{L}v)(x) = v'(x)a_0(x) - a_0'(x)v(x). \tag{4.3.5}$$

Notice that each term is the product of a matrix and a column vector. We have written $(\mathcal{L}v)(x)$ rather than $\mathcal{L}v(x)$ because, strictly speaking, $\mathcal{L}$ operates on the mapping v to give $\mathcal{L}v$, which is then evaluated at some (variable) point, which may be, for example, x or ξ; there is no variable, as such, in the definition of $\mathcal{L}$ (although some variable must be used for calculating). Thus, equation (4.3.4) can be written in the form

$$\mathcal{L}s_j = a_j - b_j, \tag{4.3.6}$$

which can be evaluated at ξ or, equally well, at x. It is not always convenient to adhere to the strictly correct notation $(\mathcal{L}v)(x)$, and our usage will vary to suit the context (as explained in the following remark).

> **4.3.2. Remark.** The logical point we are discussing here applies also to basic calculus, in which d/dx is the differentiation operator applied to *expressions in x*, and the symbol $'$ is the differentiation operator for *functions from $\mathbb{R}$ to $\mathbb{R}$*. When we write "if $f(x) = x^2$, then $f'(x) = 2x$," we are saying that if f is the function that sends any number to its square, then f' is the function that sends any number to its double. This function f may be written, in what is called

bound (or *dummy*) *variable notation*, as $x \mapsto x^2$, and it is correct to write $(x \mapsto x^2)' = (x \mapsto 2x)$, or even $(x \mapsto x^2)' = (t \mapsto 2t)$. But it is not, strictly speaking, correct to write $(x^2)' = 2x$, because x^2 is not a *function*, but an *expression*. One can run into paradoxes by confusing these. For instance, what is $(t^2)'$? This might be understood as $(t \mapsto t^2)'$, which is $t \mapsto 2t$, or as $(x \mapsto t^2)'$, which is $x \mapsto 0$. This difficulty is usually avoided by specifying whether $' = d/dx$ or $' = d/dt$, but more properly $'$ is an *operator on functions* and does not involve any variable at all; then $(t^2)'$ is not a correctly formulated expression. In contrast, d/dx and d/dt are *operators on expressions*, so that $(d/dt)t^2$ and $(d/dx)t^2$ are meaningful and unambiguous. Having pointed this out, we must add at once that while it is sometimes very helpful to understand this point (for instance when thinking of differentiation, or $\mathcal{L}$, as an operator on a vector space of functions), it is seldom necessary to insist on it, and we will often follow the usual practice, commonly called "abuse of notation," by which $\mathcal{L}$ can be applied to *expressions* "regarded as" functions of a particular *vector variable* (which must be made clear from the context).

To normalize (4.2.1) up to a specified order k using format 1a, we perform a sequence of k transformations of the form used in Lemma 4.3.1, with $j = 1, \ldots, k$. If the *computation problem* is to be solved, that is, if a specific system with given $a_i(x)$ is to be normalized including the explicit computation of the normalized terms, it is necessary to keep track of the transformations used. In this case (4.2.1) can be rewritten with $x^{(1)}$ in place of x, and the successive new variables can be called $x^{(2)}, \ldots, x^{(k)}$. At the end, $x^{(k)}$ can be replaced by y, and the normalized equation written as

$$\dot{y} = b(y, \varepsilon) \sim b_0(y) + \varepsilon b_1(y) + \cdots + \varepsilon^k b_k(y) + \cdots . \tag{4.3.7}$$

Of course, the function b_0 will be the same as a_0. If we are interested only in solving the *description problem*, that is, in describing the form that the normalized equation will take, it is not necessary to keep track of the transformations. We can apply a transformation of the form $x = \xi + \varepsilon s_1(\xi)$ to (4.2.1), and then rename ξ as x so that the transformed equation again takes the form (4.2.1), except that a_1 is now normalized and the higher a_i are modified. Then we can apply $x = \xi + \varepsilon^2 s_2(\xi)$ and again rename ξ as x. The final normalized equation will have the same appearance as (4.2.1), except that the a_j for $j = 1, \ldots, k$ will belong to certain appropriate vector subspaces of functions defining the specific normal form style that is being used. Since the full formulas for the computation problem are difficult in format 1a (as is already apparent from the first three steps for the linear case, given in Section 3.2), we will take the latter approach here. For the computation problem, format 2b or 2c is recommended.

After all of these remarks, and with the experience of the linear case behind us, what remains to be said is entirely anticlimactic. The first step is to put $x = \xi + \varepsilon s_1(\xi)$ into (4.2.1). By Lemma 4.3.1 and (4.3.6) we have the homological equation

$$\mathcal{L}s_1 = a_1 - b_1. \tag{4.3.8}$$

The specific form of $\mathcal{L}$ will depend on the setting, and will be discussed later in this section. Next a complement to $\operatorname{im}\mathcal{L}$ must be selected; this depends not only on the setting, but on the form of a_0, and will be treated in Sections 4.5–4.8. Then (as we have done so often in Chapter 3) a_1 must be projected into this complement to obtain b_1, after which $b_1 - a_1$ will belong to $\operatorname{im}\mathcal{L}$, and s_1 can be calculated. At this point (4.2.1) has been transformed to

$$\dot{\xi} = b_0(\xi) + \varepsilon b_1(\xi) + \varepsilon^2 \widehat{a}_2(\xi) + \varepsilon^3 \widehat{a}_3(\xi) + \cdots,$$

where $b_0 = a_0$, b_1 is as above, and $\widehat{a}_i(\xi)$ are the modified higher-order terms. (In Section 3.2 we used the notation A'_i for the corresponding terms, but $'$ applied to a_i would indicate a derivative.) Renaming ξ as x, this equation becomes

$$\dot{x} = b_0(x) + \varepsilon b_1(x) + \varepsilon^2 \widehat{a}_2(x) + \varepsilon^3 \widehat{a}_3(x) + \cdots,$$

and we can apply $x = \xi + \varepsilon^2 s_2(\xi)$ with a new homological equation $\mathcal{L}s_2 = \widehat{a}_2 - b_2$. For the description problem it is not necessary to keep track of the formula for $\widehat{a}_2$. For this reason, and to avoid having to apply several "hats" to the same letter, it is better to write the homological equation as

$$\mathcal{L}s_2 = k_2 - b_2,$$

where k_2 simply denotes a vector field that can be regarded as known at this stage of the calculation. Here again, $\mathcal{L}$ maps some vector space of functions to itself, perhaps a different space than in the first step. (This will depend on the setting.) Next k_2 is projected into a complement of $\operatorname{im}\mathcal{L}$ to obtain b_2, and the process repeats. The homological equation at the jth stage will be

$$\mathcal{L}s_j = k_j - b_j. \tag{4.3.9}$$

From the standpoint of the description problem, it is only this one equation that is significant, because it says that at each stage, b_j must belong to a complement of $\operatorname{im}\mathcal{L}$, which can be freely chosen and whose choice constitutes the selection of a normal form style.

The next item of business is to specify the operator $\mathcal{L}$ more precisely in each setting.

Format 1a in Setting 1 (near a Rest Point)

Setting 1 is characterized by the conditions $x \in \mathbb{R}^n$, $a_0(x) = Ax$ for some $n \times n$ matrix A, and $a_j \in \mathcal{V}_j^n$, the vector space of homogeneous polynomial vector fields of grade j (degree $j + 1$). Since $a'_0(x) = A$, the homological operator (4.3.5) takes the form

$$(\mathcal{L}v)(x) = v'(x)Ax - Av(x). \tag{4.3.10}$$

This will be the most frequent form for $\mathcal{L}$ used in this book. Notice that if $v(x) = Bx$ is a linear vector field, $\mathcal{L}v(x) = BAx - ABx = [B, A]x$. This is essentially the same as the $\mathcal{L}$ operator of Chapter 3, $\mathcal{L}B = [B, A_0]$.

If $v \in \mathcal{V}_j^n$, then the degree of v is $j + 1$, so $v'(x)$ is a matrix with entries homogeneous of degree j. Multiplying such a matrix by Ax produces a vector of degree $j + 1$, or grade j. Since $Av(x)$ is also of grade j, it follows that the restriction $\mathcal{L}_j$ of $\mathcal{L}$ to $\mathcal{V}_j^n$ satisfies

$$\mathcal{L}_j : \mathcal{V}_j^n \to \mathcal{V}_j^n \tag{4.3.11}$$

for each j. Therefore, when (in later sections) we turn our attention to normal form styles, it will be necessary to choose a complement $\mathcal{N}_j$ to im $\mathcal{L}_j$ in each space $\mathcal{V}_j^n$:

$$\mathcal{V}_j^n = \text{im } \mathcal{L}_j \oplus \mathcal{N}_j. \tag{4.3.12}$$

The nature of the normal form subspaces $\mathcal{N}_j$ will depend upon A, and (just as in Chapter 3) will hinge on whether A is semisimple, nilpotent, or a sum of both.

Format 1a for setting 1 can be formulated directly in the form (4.2.4), without dilation. It is convenient to use the variable x, rather than u, because the homological equations are the same as in the dilated form. Consider a vector field

$$\dot{x} = a(x) \tag{4.3.13}$$

with $a(0) = 0$ and $a \in \mathcal{V}^n$. The ∞-jet map

$$j^\infty : \mathcal{V}^n \to \mathcal{V}_{**}^n, \tag{4.3.14}$$

already mentioned in item 8 of Definition 4.1.1, assigns to a smooth vector field its Taylor series around the origin, regarded as a formal power series, which can be written as

$$j^\infty(a)(x) = Ax + a_1(x) + a_2(x) + \cdots, \tag{4.3.15}$$

where $a_j \in \mathcal{V}_j^n$. This *formal power series vector field* (which, in general, is not a true vector field, since it may not converge and therefore does not define a vector at each point x) will usually be written as a *formal power series differential equation*,

$$\dot{x} = Ax + a_1(x) + a_2(x) + \cdots, \tag{4.3.16}$$

which is not a true differential equation. Although it is convenient to work with formal power series, it is always possible to work instead with smooth vector fields expanded as Taylor polynomials to some degree, with a smooth remainder, and it is always this version of the theory that is used at the stage when dynamical consequences are being drawn from normal form theory (as in Chapters 5 and 6 below); see (5.0.1) and (5.0.2).

To bring (4.3.16) into normal form using format 1a, we perform a sequence of transformations, first $x = \xi + s_1(\xi)$ with $s_1 \in \mathcal{V}_1^n$, then (after replacing ξ by x) $x = \xi + s_2(\xi)$ with $s_2 \in \mathcal{V}_2^n$, and so on. Lemma 4.3.1 continues to hold, with $\varepsilon = 1$, and everything works out exactly as before. In particular, the operator $\mathcal{L}$ is still defined by (4.3.5) and the homological equations are (4.3.9).

The map j of (4.3.14) is defined using derivatives of a at the origin, so the formal system (4.3.16) is unchanged if a is modified away from the origin. The ultimate aim of normal form theory is to explain the dynamics of (4.3.13) *in a neighborhood of the origin* by means of the dynamics of a truncation of (4.3.16) in normal form. The language of *germs of vector fields* is sometimes used as a way of expressing the notion of "behavior near the origin." Although we will not use this language, it deserves to be described briefly.

Two vector fields v_1 and v_2 in $\mathcal{V}^n$ (or, more generally, any two mappings from $\mathbb{R}^n$ into the same space) are said to *have the same germ at the origin* if there is a neighborhood U of the origin such that $v_1(x) = v_2(x)$ for all $x \in U$. (U is not fixed in advance.) "Having the same germ" is an equivalence relation, and equivalence classes are called "germs." (It is not even necessary for a vector field to be defined on all of $\mathbb{R}^n$ for it to "have a germ" at the origin, only to be defined near the origin. But this does not increase the supply of germs, since any vector field defined near the origin has the same germ as some vector field that is defined everywhere.) The map j can be regarded as a map from germs of vector fields to formal power series vector fields. Some dynamical behaviors (such as stability of the origin) have the property that if they hold for one vector field, they hold for any vector field with the same germ. It is this kind of behavior that one may hope to predict from normal form theory in setting 1.

Format 1a in Setting 2 (Method of Averaging)

Setting 2 is characterized by (4.2.1) with $x = (x_0, x_1, \ldots, x_n) \in \mathbb{R}^{n+1}$, with $a_0(x) = (1, 0, \ldots, 0)$ and $a_j \in \mathcal{V}^n(2\pi)$. Let K be a compact set in $\mathbb{R}^n$ and let $\widehat{K} = [0, 2\pi] \times K$. Consider transformations $x = \xi + \varepsilon^j s(\xi)$ with $s \in \mathcal{V}^n(2\pi)$, so that $s(\xi)$ has first component zero and is periodic in ξ_0. (The transformation itself is not periodic in ξ_0, because of the ξ term.) Then the transformation is invertible for small ε on $\widehat{K}$, and once this is known, it is also in fact invertible on $\mathbb{R} \times K$.

Since $a_0'(x) = 0$,

$$\pounds s(x) = \pounds \begin{bmatrix} 0 \\ s_1(x) \\ \vdots \\ s_n(x) \end{bmatrix} = s'(x) \begin{bmatrix} 1 \\ 0 \\ \vdots \\ 0 \end{bmatrix} = \begin{bmatrix} 0 \\ \partial s_1/\partial x_0 \\ \vdots \\ \partial s_n/\partial x_0 \end{bmatrix} = \frac{\partial}{\partial x_0} s(x).$$

That is,

$$\pounds = \frac{\partial}{\partial x_0} : \mathcal{V}^n(2\pi) \to \mathcal{V}^n(2\pi). \tag{4.3.17}$$

This is an operator on an infinite-dimensional space that is independent of j, so the homological equations (4.3.9) take place in the same space for each j.

For setting 1, the treatment of normal form styles has been postponed to later sections of this chapter. For setting 2, it is easy to give a complete treatment at once. As a periodic function of x_0, any vector field in $\mathcal{V}^n(2\pi)$ has a Fourier series in x_0 (with coefficients that are vector functions of $x_1, \ldots, x_n$). This Fourier series may be separated into a "constant" term (the average over x_0) and a "mean-free" part. The operator $\pounds$ annihilates the constant term, and maps onto the space of mean-free fields. Therefore,

$$\mathcal{V}^n(2\pi) = \mathrm{im}\, \pounds \oplus \ker \pounds, \tag{4.3.18}$$

and $\ker \pounds$ may be chosen as the normal form space. That is, a differential equation in setting 2 is in normal form (up to degree k in ε) if and only if its terms (up to degree k) are independent of x_0. The projection into $\ker \pounds$ associated with the direct sum (4.3.18) is the *averaging operator*.

Let us restate these results in the natural coordinates for this problem, that is, in the form of (4.2.11). The near-identity transformations then take the form

$$y = \eta + \varepsilon^j s_j(t, \eta)$$

with $\eta \in \mathbb{R}^n$ and $s_j : \mathbb{R} \times \mathbb{R}^n \to \mathbb{R}^n$, periodic in t. The homological operator, calculated above, becomes simply

$$\pounds = \frac{\partial}{\partial t}.$$

A system (4.2.11) is in normal form when its terms (up to specified order) are independent of t, and the projection into normal form is the averaging operation. (Of course, this projection can only be applied one step at a time, with intermediate calculations. Thus, f_1 in (4.2.11) can be averaged with respect to t to give the order-ε term of the normalized equation, but then s_1 must be calculated and f_2 must be modified before another averaging operation may be applied.)

Format 1a in Setting 3 (near a Periodic Orbit)

Setting 3 is an important part of normal form theory, but one that (for limitations of time and space) will not be treated in depth in this book. But it is at least desirable to show how this setting fits into our general framework. Setting 3 consists of (4.2.1) with $x \in \mathbb{R}^{n+1}$, a_0 given by (4.2.16), and $a_j \in \mathcal{V}_j^n(2\pi)$ for $j \geq 1$. The transformations have the form (4.3.2) with $s_j \in \mathcal{V}_j^n(2\pi)$. The homological operator, defined by (4.3.5), works out as

$$(\mathcal{L}s)(x) = s'(x)\left(\begin{bmatrix} 1 \\ 0 \\ \vdots \\ 0 \end{bmatrix} + \begin{bmatrix} 0 & \\ & A \end{bmatrix} x\right) - \begin{bmatrix} 0 & \\ & A \end{bmatrix} s(x)$$

$$= \begin{bmatrix} 0 \\ \partial\widehat{s}/\partial x_0 + \widehat{s}'(x)A\widehat{x} - A\widehat{s}(x) \end{bmatrix},$$

where $\widehat{}$ denotes omission of the first entry of a vector (which in each case is zero). Expressed in the natural coordinates of (4.2.15), but using x in place of u, this looks as follows. The formal power series differential equations are

$$\dot{x} = Ax + a_1(t, x) + a_2(t, x) + \cdots, \tag{4.3.19}$$

where each a_j is a homogeneous polynomial vector field on $\mathbb{R}^n$ of degree $j+1$ depending 2π-periodically on t. The coordinate changes have the form

$$x = \xi + s_j(t, \xi) \tag{4.3.20}$$

with $s_j : \mathbb{R} \times \mathbb{R}^n \to \mathbb{R}^n$, homogeneous of degree $j + 1$ and 2π-periodic in t. The homological operator is

$$(\mathcal{L}s_j)(x) = \frac{\partial s_j}{\partial t} + s'(x)Ax - As(x). \tag{4.3.21}$$

The first term is like setting 2, the rest like setting 1, and the goal of normalization will be a combination of settings 1 and 2: to eliminate time dependence (up to a given order) and simplify x dependence.

Format 1b

In format 1b, (4.2.1) is normalized to (4.3.7) in one step, by a single transformation of the form

$$x = y + \varepsilon t_1(y) + \varepsilon^2 t_2(y) + \cdots + \varepsilon^k t_k(y).$$

(The notation t_i is awkward, since it conflicts with the use of t for time, but it is chosen to agree with (3.2.6). Since format 1b will not be used except in the present discussion, the notation will not pose a difficulty.)

We will carry out the calculations to the second order, $k = 2$, and (in contrast with our treatment of format 1a) we will keep track of the details, so that our work would suffice to solve the computation problem (in any specific application) and not only the description problem. The calculations here are valid for any setting (expressed in universal form (4.2.1)). Differentiating the transformation with respect to time gives

$$\dot{x} = (I + \varepsilon t_1'(y) + \varepsilon^2 t_2'(y))\dot{y}.$$

The inverse of the matrix here can be calculated by using (3.2.10) with $X = t_1'(y) + \varepsilon t_2'(y)$. The leading terms are

$$(I + \varepsilon t_1'(y) + \varepsilon^2 t_2'(y))^{-1} \equiv I - \varepsilon t_1(y) + \varepsilon^2(t_1'(y) - t_2'(y)),$$

where $\equiv$ indicates computation modulo ε^3. Working modulo ε^3, we can expand

$$a_0(y + \varepsilon t_1(y) + \varepsilon^2 t_2(y))$$

into

$$a_0(y) + a_0'(y)(\varepsilon t_1(y) + \varepsilon^2 t_2(y)) + \frac{1}{2}a_0''(y)(\varepsilon t_1(y), \varepsilon t_1(y))$$

$$\equiv a_0(y) + \varepsilon(a_0'(y)t_1(y)) + \varepsilon^2\left(a_0'(y)t_2(y) + \frac{1}{2}a_0''(y)(t_1(y), t_1(y))\right).$$

The notation here is as follows: If $v(x)$ is a vector field, and $h \in \mathbb{R}^n$, then the Taylor expansion of v around x is

$$v(x + h) = v(x) + v'(x)h + \frac{1}{2}v''(x)(h, h) + \cdots,$$

where $v''(x)(h, h)$ is the vector whose ith component is

$$\sum_{jk} \frac{\partial^2 v_i(x)}{\partial x_j \partial x_k} h_j h_k.$$

The terms εa_1 and $\varepsilon^2 a_2$ can be expanded similarly, but it is not necessary to take them as far:

$$\varepsilon a_1(y + \varepsilon t_1(y) + \varepsilon^2 t_2(y)) \equiv \varepsilon a_1(y) + \varepsilon^2 a_1'(y)t_1(y),$$
$$\varepsilon^2 a_2(y + \varepsilon t_1(y) + \varepsilon^2 t_2(y)) \equiv \varepsilon^2 a_2(y).$$

Multiplying the matrix inverse times the sum of these expanded terms, and comparing with (4.3.7), gives

$$b_0(y) = a_0(y),$$
$$b_1(y) = a_1(y) + a_0'(y)t_1(y) - t_1'(y)a_0(y),$$
$$b_2(y) = k_2(y) + a_0'(y)t_2(y) - t_2'(y)a_0(y),$$

where (suppressing the argument y in each factor)

$$k_2 = a_2 + a_1't_1 + \frac{1}{2}a_0''(t_1, t_1) - t_1'a_1 - t_1'a_0't_1 + t_1'a_0.$$

The terms that have been collected into k_2 are those that do not involve t_2, and are therefore known at the beginning of the second stage in the calculation, when t_1 has been determined but not t_2. With $k_1 = a_1$, these equations can now be written in homological form:

$$\mathcal{L}t_1 = k_1 - b_1,$$
$$\mathcal{L}t_2 = k_2 - b_2.$$

As usual, each b_j is the projection of k_j into a selected complement of im $\mathcal{L}$, and then t_j can be calculated by partially inverting $\mathcal{L}$.

Notes and References

The idea of putting systems of differential equations into normal form probably originated with Poincaré, who studied the linearization of systems using what we call format 1b (except that he assumed that his systems were analytic and required the normalizing transformation to be convergent). Poincaré's results can be found in Arnol'd [6]. After this the method was used by Birkhoff and others for Hamiltonian systems (see Section 4.9 below) and by Krylov, Bogoliubov, and Mitropolski for the method of averaging (see Nayfeh [93]), and then by so many others that it is pointless to give references.

4.4   Lie Theory and the Generated Formats

As in Chapter 3, there are three generated formats, designated as formats 2a, 2b, and 2c (or by their longer titles, generated iterative, generated recursive (Hori version), and generated recursive (Deprit version)). Again as in Chapter 3, format 2a is fundamental, and is developed in detail here. Formats 2b and 2c (in the nonlinear case) are worked out in Sections C.2 and D.2, respectively.

In Section 3.2, format 2a (for the linear case) was developed starting from the notion of the exponential of a matrix. In Section 3.6 this was connected with the notions of Lie group and Lie algebra to provide a deeper understanding. In the present nonlinear context, there is nothing that corresponds exactly with the exponential, and even the notion of Lie group does not apply exactly. Nevertheless, it is possible to build a valid theory analogous to the theory of Chapter 3. We begin with a concise restatement of the main ideas of Section 3.6:

1. The set $GL(n)$ of invertible $n \times n$ matrices forms a Lie group under matrix multiplication.

2. The set $gl(n)$ of all $n \times n$ matrices forms a Lie algebra under commutator bracket.

3. Differentiation of a path through the identity in $GL(n)$ gives an element of $gl(n)$, and exponentiation associates to each element A of $gl(n)$ the one-parameter group (or one-parameter subgroup) e^{tA} in $GL(n)$ that it generates, which is also the solution operator of the differential equation $\dot{x} = Ax$.

4. $GL(n)$ acts on itself by conjugation, $\mathbb{C}$, and on $gl(n)$ by similarity, $\mathbb{S}$; $gl(n)$ acts on itself by the Lie operator, $\mathbb{L}$. Differentiation converts conjugation to similarity and similarity to the Lie operator; the Lie operator coincides with the commutator bracket, in the sense that $\mathbb{L}_B A = [A, B]$.

5. The fundamental theorem of Lie series expresses the similarity operator defined by a one-parameter group as the exponential of the Lie operator of its generator.

6. A smooth family $A(\varepsilon)$ of matrices in $gl(n)$, expanded as a power series, may be normalized by successive similarities $\mathbb{C}_{e^{\varepsilon^j U_j}}$ generated by $U_j \in gl(n)$ for $j = 1, 2, \ldots$. The U_j are found by solving homological equations. According to the fundamental theorem of Lie series, the effect of such a similarity on the power series of $A(\varepsilon)$ may be calculated by applying $e^{\varepsilon^j \mathbb{L}_{U_j}}$. The normalization can be carried out using generators only, without calculating or using the transformations that they generate.

7. When the transformation is required, it can, of course, be calculated from its generator U_j via the (convergent) power series for the matrix exponential $e^{\varepsilon^j U_j}$.

Each item in this list corresponds to a topic that will be developed in this section for nonlinear problems. The starting point is the idea, developed briefly in Section 3.6, that elements of $gl(n)$ may be viewed as linear vector fields, and elements of $GL(n)$ as linear diffeomorphisms. These are replaced here by arbitrary vector fields and diffeomorphisms, and some conclusions must be modified or weakened:

1'. The set $\mathfrak{G} = \mathrm{Diff}(\mathbb{R}^n)$ of diffeomorphisms of $\mathbb{R}^n$ forms a group under composition, but not exactly a Lie group (because it is infinite-dimensional).

2'. The set $\mathfrak{g} = \mathcal{V}^n$ of smooth vector fields on $\mathbb{R}^n$ forms an infinite-dimensional Lie algebra under the *Lie bracket* operation
$$[a, b](x) = a'(x)b(x) - b'(x)a(x).$$

3'. The connection between one-parameter subgroups of $\mathfrak{G}$ and generators in $\mathfrak{g}$ is weaker than in the linear case. A vector field $a \in \mathfrak{g}$ may generate only a *(local) flow* φ^t in $\mathbb{R}^n$, which is weaker than a one-parameter group of diffeomorphisms. We will not use exponential notation (such as $\varphi^t = e^{at}$) to denote the relation between a generator and its flow, since the exponential of a vector field has no natural meaning in our context. (See also item 7' and Remark 4.4.1 below.)

4'. The operations of conjugation (C), similarity (S), and Lie operator (L) are defined and related to each other very much like $\mathbb{C}$, $\mathbb{S}$, and $\mathbb{L}$.

5'. The fundamental theorem of Lie series holds in a formal sense, but the series are no longer convergent.

6'. The normalization process is formally the same as in the linear case, and can again be carried out using generators without computing the transformations that they generate.

7'. The computation of a transformation from its generator cannot be done by a convergent series. (This is another reason for avoiding the exponential notation in item 3' above.) But there is a formal version of the same algorithm: If $a \in \mathfrak{g}$, the formal power series of its flow φ^t can be obtained by applying $e^{t\mathcal{D}_a}$, understood as a formal power series of operators, to the identity function x; here $\mathcal{D}_a$ is the derivative operator defined by the vector field (see the remarks below).

4.4.1. Remark. In differential geometry a vector field a is frequently identified with the differentiation operator (on smooth functions) defined by a. We distinguish these notions: A vector field is simply a column vector $a(x) = (a_1(x), \ldots, a_n(x))$, while the differentiation operator it defines is written

$$\mathcal{D}_a = a_1(x)\frac{\partial}{\partial x_1} + \cdots + a_n(x)\frac{\partial}{\partial x_n}.$$

Because we distinguish these, we also distinguish between the (local) flow φ^t of a and the exponential $e^{t\mathcal{D}_a}$. The latter is understood strictly as a formal power series in t with operator coefficients, not as a diffeomorphism. Admittedly, if we were to identity a with $\mathcal{D}_a$ and φ^t with $e^{t\mathcal{D}_a}$, the analogy between the formulas in this section and those in Section 3.6 would be closer. But it seems that clarity is best served by making the distinctions.

4.4.2. Remark. One additional comment may be useful for those familiar with differential geometry. Frequently *Lie differentiation* is defined to be an operation valid for tensors of all ranks, coinciding with our $\mathcal{D}_v$ for functions (tensors of rank zero) and with our L_v for vectors (tensors of rank one). We will avoid the custom (common in differential geometry) of using the same symbol for $\mathcal{D}_v$ and L_v. The actual distinction between $\mathcal{D}_v$ and L_v is that $\mathcal{D}_v$ should be applied to objects (whether scalar or vector) that transform like functions under coordinate changes (so that f becomes $f \circ \varphi$), while L_v should be applied to objects that transform like differential equations (so that v becomes $(\varphi')^{-1}v \circ \varphi$). An example of a vector quantity that transforms like a function is a coordinate change; we have, for instance, already spoken (in item 7') of applying $\mathcal{D}_v$ to the vector-valued identity function x. What we intend, of course, is to apply $\mathcal{D}_v$ to the components of x.

Diffeomorphisms, Vector Fields, and Flows: Items 1', 2', and 3'

A *diffeomorphism* of $\mathbb{R}^n$ with itself is a smooth invertible mapping $\varphi : \mathbb{R}^n \to \mathbb{R}^n$ with a smooth inverse. The set $\mathfrak{G} = \mathrm{Diff}(\mathbb{R}^n)$ of all diffeomorphisms of $\mathbb{R}^n$ with itself clearly forms a group under composition, and it is natural to ask whether this is (in some sense) a Lie group. It is, of course, not a *matrix* Lie group, which is the only kind of Lie group we have defined, but it makes sense to ask whether the same things that we have done with matrix Lie groups $\mathfrak{G}$ can be done with $\mathrm{Diff}(\mathbb{R}^n)$. The answer is mostly, but not entirely, yes. The most important thing that we did with a matrix Lie group $\mathfrak{G}$ was to differentiate paths through the identity to obtain the Lie algebra $\mathfrak{g}$ associated with $\mathfrak{G}$. So we ask what such a path looks like for $\mathrm{Diff}(\mathbb{R}^n)$, and what its derivative might be.

Suppose that $\varphi_t : \mathbb{R}^n \to \mathbb{R}^n$ is a smooth family of diffeomorphisms of $\mathbb{R}^n$; in other words, φ_t is a diffeomorphism for each real t (or at least each t in a neighborhood of zero) and $(t,x) \mapsto \varphi_t(x)$ is smooth as a function of $(t,x) \in \mathbb{R}^n$. Suppose also that $\varphi_0(x) = x$, so that φ_0 is the identity diffeomorphism. Then for fixed x we can define

$$\dot{\varphi}_0(x) = \frac{d}{dt}\varphi_t(x)\bigg|_{t=0}. \tag{4.4.1}$$

This may be thought of as the velocity vector of the moving point $\varphi_t(x)$ at "time" zero, and the values of $\dot{\varphi}_0(x)$ for all x clearly form a smooth vector field on $\mathbb{R}^n$:

$$\dot{\varphi}_0 \in \mathcal{V}^n.$$

This suggests that we should take $\mathfrak{g} = \mathcal{V}^n$ to be the "Lie algebra" associated with $\mathfrak{G} = \mathrm{Diff}(\mathbb{R}^n)$. It will be seen, later in this section, that $\mathcal{V}^n$ is closed under a bracket operation that satisfies the algebraic rules for a Lie algebra product.

4.4.3. Remark. Those familiar with the general theory of Lie groups might expect that we would form $\mathfrak{g}$ by making $\mathrm{Diff}(\mathbb{R}^n)$ into a manifold and passing to its tangent space. But $\mathrm{Diff}(\mathbb{R}^n)$ cannot be made into a finite-dimensional manifold, and the theory of infinite-dimensional (Banach) manifolds is difficult. We avoid this by differentiating $\varphi_t(x)$ with respect to t for fixed x, rather than attempting to differentiate φ_t with respect to t as an element of $\mathrm{Diff}(\mathbb{R}^n)$.

For the reverse construction from $\mathfrak{g}$ to $\mathfrak{G}$, let a vector field $a \in \mathcal{V}^n$ be given, and let the solution of the differential equation

$$\dot{x} = a(x)$$

with initial condition $x(0) = x_0$ be denoted by $\varphi^t(x_0)$. Then φ^t is called the *flow* (more precisely, the *local flow*) of a, and a is called the *generator* of the flow. (We will usually omit the subscript zero from the initial condition and write $\varphi^t(x)$ for the solution passing through x at $t = 0$.) It follows from the theory of differential equations that for each x there is an open interval of t around zero for which $\varphi^t(x)$ is defined, and if this interval has an endpoint, either for positive or negative t, then $\|\varphi^t(x)\| \to \infty$ as t approaches the endpoint. For $t = 0$, one has $\varphi^0(x) = x$, but for $t \neq 0$, there is no guarantee that φ^t is defined for all x. Thus, it may be that φ^t is not a diffeomorphism of $\mathbb{R}^n$ except at $t = 0$, so we cannot say that $\varphi^t \in \text{Diff}(\mathbb{R}^n)$, even for t near zero. On the other hand, the derivative $(\varphi^t)'(x)$ is defined for some interval in t for each x, and is nonsingular for t near zero (since it equals the identity matrix I when $t = 0$). Therefore, by the inverse function theorem, φ^t is at least a local diffeomorphism near each point for small t. If $U \subset \mathbb{R}^n$ is an open subset with compact closure, there is a common interval of t such that $\varphi^t(x)$ is defined for all $x \in U$ and is a diffeomorphism of U onto its image.

The flow φ^t of a vector field a satisfies the (local) *flow property*, which states that

$$\varphi^s(\varphi^t(x)) = \varphi^{s+t}(x) \tag{4.4.2}$$

whenever both sides are defined. (This is a well-known consequence of the uniqueness theorem for solutions of a differential equation.) In other words, φ^t is "nearly" a one-parameter subgroup of $\text{Diff}(\mathbb{R}^n)$.

4.4.4. Remark. It is not necessary to worry very much about the interval of existence of $\varphi^t(x)$, for the following reason. In applications we are usually concerned with the behavior of the flow φ^t in some compact set K, the closure of a bounded open set. Let $r > 0$ be a real number such that K is contained in $\|x\| < r$, and let $R > r$. Let μ be a smooth function such that $\mu(x) = 1$ for $\|x\| < r$ and $\mu(x) = 0$ for $\|x\| > R$. Then the flow of the vector field $\mu(x)a(x)$ is defined for all x and t, because solutions can never approach infinity. Also, the (modified) flow coincides with the desired flow on K. As a rule, therefore, we will ignore the problem and speak as though φ^t were defined everywhere, except when there is a need to be particularly careful.

Conjugation, Similarity, and Lie Operator: Item 4′

In Section 3.6, the three operators $\mathbb{C}$, $\mathbb{S}$, and $\mathbb{L}$ were defined and evaluated as follows:

$$\mathbb{C}_T S = T^{-1} S T, \tag{4.4.3}$$

$$\mathbb{S}_T A = \frac{d}{dt} \mathbb{C}_T e^{tA} \bigg|_{t=0} = T^{-1} A T,$$

$$\mathbb{L}_B A = \frac{d}{ds} \mathbb{S}_{e^{sB}} A \bigg|_{s=0} = [A, B].$$

In the cases of $\mathbb{S}_T A$ and $\mathbb{L}_B A$, the first expression given is the definition and the second is an evaluation. Here we define and evaluate the corresponding operators C, S, and L in the nonlinear setting, by the following similar formulas, which will be explained below:

$$\mathsf{C}_\psi \varphi = \psi^{-1} \circ \varphi \circ \psi, \tag{4.4.4}$$

$$\mathsf{S}_\psi a = \frac{d}{dt} \mathsf{C}_\psi \varphi^t \bigg|_{t=0} = \psi'(x)^{-1} a(\psi(x)),$$

$$\mathsf{L}_b a = \frac{d}{ds} \mathsf{S}_{\psi^s} a \bigg|_{s=0} = [a, b].$$

Let $\psi \in \mathfrak{G}$. *Conjugation* by ψ is the map

$$\mathsf{C}_\psi : \mathfrak{G} \to \mathfrak{G}$$

defined for $\varphi \in \mathfrak{G}$ by

$$\mathsf{C}_\psi \varphi = \psi^{-1} \circ \varphi \circ \psi, \tag{4.4.5}$$

that is,

$$(\mathsf{C}_\psi \varphi)(x) = \psi^{-1}(\varphi(\psi(x))).$$

This is to be compared with equation (3.6.9).

Again, let $\psi \in \mathfrak{G}$. *Similarity* by ψ will be a map

$$\mathsf{S}_\psi : \mathfrak{g} \to \mathfrak{g}$$

defined as follows: Let $a \in \mathfrak{g}$ be a vector field and let φ^t be its flow, so that

$$\frac{d}{dt} \varphi^t(x) \bigg|_{t=0} = a(x). \tag{4.4.6}$$

Then

$$\mathsf{C}_\psi \varphi^t = \psi^{-1} \circ \varphi^t \circ \psi$$

is a flow, and by analogy with the linear case, $\mathsf{S}_\psi a$ should be its generator. Remember that the generator of $\mathsf{C}_\psi \varphi^t$ is the vector field obtained by differentiating with respect to t at $t = 0$. Therefore, the nonlinear similarity operator is defined as follows:

$$(S_\psi a)(x) = \frac{d}{dt}(C_\psi \varphi^t)(x)\Big|_{t=0}.$$ (4.4.7)

A usable formula for $S_\psi a$ can be obtained by evaluating the right-hand side of (4.4.7). It is worthwhile to do this calculation with some care.

4.4.5. Lemma.

$$\frac{d}{dt}\psi^{-1}(\varphi^t(\psi(x)))\Big|_{t=0} = \psi'(x)^{-1}a(\psi(x)).$$

Proof. Differentiating $\psi^{-1}(\psi(x)) = x$ gives $(\psi^{-1})'(\psi(x))\psi'(x) = I$. (It is important to read such formulas carefully, to distinguish evaluations from multiplications. In this case we mean the derivative matrix of ψ^{-1}, evaluated at $\psi(x)$, multiplied by $\psi'(x)$.) Therefore,

$$(\psi^{-1})'(\psi(x)) = \psi'(x)^{-1}.$$ (4.4.8)

Now

$$\frac{d}{dt}\psi^{-1}(\varphi^t(\psi(x))) = (\psi^{-1})'(\varphi^t(\psi(x)))\frac{d}{dt}\varphi^t(\psi(x)).$$

Setting $t = 0$ and using (4.4.6) and (4.4.8) gives the result.   □

It follows from this lemma that the similarity operation, defined by (4.4.7), may be evaluated by the formula

$$(S_\psi a)(x) = \psi'(x)^{-1}a(\psi(x)).$$ (4.4.9)

(As in Remark 4.3.2, the operator is applied to a *before* it is evaluated at a particular point x, although we will usually omit the parentheses around $S_\psi a$ that help to emphasize this fact.) The "action" of $\mathfrak{G}$ on $\mathfrak{g}$ given by (4.4.9) is nothing other than the transformation of the differential equation

$$\dot{x} = a(x)$$ (4.4.10)

by the coordinate change

$$x = \psi(\xi)$$ (4.4.11)

to obtain

$$\dot{\xi} = b(\xi) = \psi'(\xi)^{-1}a(\psi(\xi)) = (S_\psi a)(\xi).$$ (4.4.12)

Written without variables, the transformed vector field is

$$b = S_\psi a.$$ (4.4.13)

Often (4.4.13) will be evaluated at x rather than ξ (often without even mentioning ξ). This is equivalent to performing the change of variables (4.4.11) and then replacing ξ by x. Notice that in contrast to the linear case, similarity in the context of vector fields and diffeomorphisms cannot easily be confused with conjugation.

Following the pattern of the linear case, the next step is to replace ψ by ψ^s in the definition of similarity, where ψ^s is the flow of a vector field $b \in \mathfrak{g}$, and differentiate with respect to s at $s = 0$. This will define the Lie operator L_b. It is convenient to write ψ_s for ψ^s in some of the following calculations, so that the derivative can be written ψ'_s rather than $(\psi^s)'$. Then the definition reads

$$(\mathsf{L}_b a)(x) = \frac{d}{ds}(\mathsf{S}_{\psi_s} a)(x)\bigg|_{s=0} = \frac{d}{ds}\psi'_s(x)^{-1}a(\psi_s(x))\bigg|_{s=0}. \qquad (4.4.14)$$

To understand what this means, remember that the vector field $a(x)$ defines a differential equation

$$\dot{x} = \frac{dx}{dt} = a(x) \qquad (4.4.15)$$

with flow φ^t. The vector field $b(x)$ also defines a differential equation, which we write as

$$\frac{dx}{ds} = b(x), \qquad (4.4.16)$$

with flow ψ^s. It is convenient to think of t as time and s as simply a parameter. Then ψ^s (for any s) may be used as a coordinate change to be applied to (4.4.15) and its flow. The transformed flow will be $\mathsf{C}_{\psi^s}\varphi^t = \psi^{-s} \circ \varphi^t \circ \psi^s$, and (according to (4.4.13)) the transformed vector field will be $\mathsf{S}_{\psi^s} a$. Then (4.4.14) will give the rate of change of the transformed vector field as the parameter s is varied. For application to format 2a, s will be replaced by ε, ε^2, etc.

4.4.6. Remark. Stated more precisely, for any compact set $K \subset \mathbb{R}^n$ there is an interval $-s_0 < s < s_0$ for which ψ^s is a valid coordinate change in a neighborhood of K and transforms a in that neighborhood as explained above. Alternatively, we can apply to b the procedure of Remark 4.4.4 and obtain a valid transformation ψ^s defined for all s and x.

4.4.7. Lemma.

$$(\mathsf{L}_b a)(x) = a'(x)b(x) - b'(x)a(x).$$

Proof. Since $\psi_s(x)$ is a solution of $dx/ds = b(x)$, we have

$$\frac{d}{ds}\psi_s(x) = b(\psi_s(x)).$$

Differentiating this with respect to x yields

$$\frac{d}{ds}\psi'_s(x) = b'(\psi_s(x))\psi'_s(x).$$

This is a matrix differential equation of the form $dX(s)/ds = B(s)X(s)$; if $X(s)$ is invertible, it follows (by differentiating $X(s)X(s)^{-1} = I$) that

$dX(s)^{-1}/ds = -X(s)^{-1}B(s)$. Therefore,

$$\frac{d}{ds}\psi_s'(x)^{-1} = -\psi_s'(x)^{-1}b'(\psi_s(x)).$$

It follows that

$$\frac{d}{ds}\psi_s'(x)^{-1}a(\psi_s(x)) = -\psi_s'(x)^{-1}b'(\psi_s(x))a(\psi_s(x)) \qquad (4.4.17)$$
$$+ \psi_s'(x)^{-1}a'(\psi_s(x))b(\psi_s(x)).$$

Setting $s = 0$ and using $\psi_0(x) = x$ and $\psi_0'(x) = I$ gives the result.   □

Recall from equation (3.6.11) that for matrix Lie algebras, $\mathbb{L}_B A = [A, B]$. Lemma 4.4.7 then suggests that we define the *Lie bracket* of two vector fields a and b to be the vector field $[a, b]$ whose value at any x is given by

$$[a, b](x) = a'(x)b(x) - b'(x)a(x). \qquad (4.4.18)$$

The "correct" notation, in view of Remark 4.3.2, is $[a, b](x)$, but by "abuse of notation" it is common to write $[a(x), b(x)]$ although the value of the bracket at x cannot be calculated from a knowledge of the values of a and b at x only. Notice that if A and B are matrices,

$$[Ax, Bx] = [A, B]x,$$

where $[A, B]$ is the usual commutator bracket for matrices. It is clear that $[a, b]$ is linear in a and b, and it will be shown later that the Jacobi identity holds:

$$[a, [b, c]] + [b, [c, a]] + [c, [a, b]] = 0. \qquad (4.4.19)$$

Therefore, $\mathfrak{g} = \mathcal{V}^n$ forms a Lie algebra under $[\cdot, \cdot]$. In view of (4.4.18), Lemma 4.4.7 states the equivalence of the Lie operator and the Lie bracket, namely,

$$\mathsf{L}_b a(x) = [a(x), b(x)]. \qquad (4.4.20)$$

The fundamental operator of normal form theory is the homological operator $\pounds$, defined in (4.3.5). This is a special case of the Lie operator:

$$\pounds v = [v, a_0] = \mathsf{L}_{a_0} v. \qquad (4.4.21)$$

Here a_0 is the leading term of the vector field whose normal form is sought. In setting 1, this is a linear vector field given by $a_0(x) = Ax$ for a matrix A. In this case, we often write L_A in place of L_{a_0}, as for instance in (4.5.5) below.

The Fundamental Theorem of Lie Series: Item 5'

For matrices, the fundamental theorem of Lie series, Theorems 3.2.1 and 3.6.2, can be written either as

$$e^{-sY}Xe^{sY} = e^{s\mathbb{L}_Y}X \qquad (4.4.22)$$

or else as

$$\mathbb{S}_{e^s Y} = e^{s\mathsf{L}_Y}. \tag{4.4.23}$$

The first form is an equality in $gl(n)$, the second in the group $GL(gl(n))$ of invertible linear operators on the vector space $gl(n)$. (In Section 3.6 we referred to $GL(gl(n))$ as $GL(n^2)$, using the fact that $gl(n)$ is isomorphic, as a vector space, to $\mathbb{R}^{n^2}$.) In the matrix context, all of these expressions are analytic functions of s; that is, they are *equal to* their convergent power series expansions.

Let us attempt to translate these equations into the context of vector fields and diffeomorphisms; this, of course, will not prove anything, but it can at least suggest conjectures. Beginning with (4.4.23), replace Y by b, e^{sY} by ψ^s (the flow of b), and $\mathbb{S}$ by S, to obtain

$$\mathsf{S}_{\psi^s} = e^{s\mathsf{L}_b}. \tag{4.4.24}$$

Applying this to a vector field a yields $\mathsf{S}_{\psi^s} a = e^{s\mathsf{L}_b} a$, which, written out in full, becomes

$$\varphi_s'(x)^{-1} a(\varphi_s(x)) = e^{s\mathsf{L}_b} a(x) = a(x) + s\mathsf{L}_b a(x) + \frac{s^2}{2}\mathsf{L}_b^2 a(x) + \cdots. \tag{4.4.25}$$

But there is a problem with this equation: The left-hand side of (4.4.25) is only a smooth function of s, not necessarily analytic. So the right-hand side, which is a power series in s, cannot be expected to be a convergent expansion, and equation (4.4.25) cannot be strictly true. We will soon see that it is true when $=$ is replaced by $\sim$; that is, the right-hand side is the formal power series expansion of the left-hand side. Equation (4.4.24) suffers from even greater difficulties. It would have to be understood as an equation in the group $GL(\mathfrak{g})$ of invertible linear operators on $\mathfrak{g}$, stating that S_{φ^s} is a one-parameter subgroup of $GL(\mathfrak{g})$ having generator L_b. One might then hope to prove it along the lines of the proof of Theorem 3.6.2, relying on the fact that one-parameter subgroups have generators. But unlike $GL(gl(n))$, which is equivalent to the matrix Lie group $GL(n^2)$, $GL(\mathfrak{g})$ is not a matrix Lie group. $GL(\mathfrak{g})$ is also not a group of diffeomorphisms. So we have no theorem stating that one-parameter subgroups of $GL(\mathfrak{g})$ have generators. (Theorem 3.6.1 says that one-parameter subgroups of matrix Lie groups have generators, and for diffeomorphism groups the generator of a flow is its associated vector field.) Therefore, there is no hope of imitating the proof of Theorem 3.6.2. It is sometimes convenient to regard (4.4.24) as true, but only as a shorthand for the formal power series version of (4.4.25), which we now state and prove precisely, following the proof method used for Theorem 3.2.1.

4.4.8. Theorem (Lie series for vector fields). The formal power series expansion of $S_{\psi^s}a(x)$ is given by

$$\psi_s'(x)^{-1}a(\psi_x(x)) \sim e^{sL_b}a(x) = a(x) + sL_b a(x) + \frac{s^2}{2}L_b^2 a(x) + \cdots.$$

Proof. The key to the proof is equation (4.4.17) in the proof of Lemma 4.4.7, which may be written as

$$\frac{d}{ds}\psi_s'(x)^{-1}a(\psi_s(x)) = \psi_s'(x)^{-1}(L_b a)(\psi_s(x)).$$

Notice that both sides of this equation contain $\psi_s'(x)^{-1}$ times a vector-valued function evaluated at $\psi_s(x)$. The equation may be interpreted as saying, to differentiate such an expression with respect to s, apply L_b to the vector-valued function. Applying this remark repeatedly gives

$$\frac{d^j}{ds^j}\psi_s'(x)^{-1}a(\psi_s(x)) = \psi_s'(x)^{-1}(L_b^j a)(\psi_s(x)).$$

Setting $s = 0$ gives

$$\frac{d^j}{ds^j}\psi_s'(x)^{-1}a(\psi_s(x))\bigg|_{s=0} = L_b^j a(x).$$

The desired formal power series now follows by Taylor's theorem.   □

Format 2a for Vector Fields: Item 6′

With the machinery developed above, it is a straightforward matter to describe format 2a for vector fields. We will follow the "algorithmic form" of format 2a, which was presented for matrices in equations (3.6.12) through (3.6.14).

Suppose that a differential equation is given in the form

$$\dot{x} = a(x,\varepsilon) \equiv a_0(x) + \varepsilon a_1(x) + \cdots + \varepsilon^k a_k(x), \qquad (4.4.26)$$

truncated at some degree k in ε, and that it is desired to normalize this series to order k. (Here $\equiv$ means equal modulo ε^{k+1}.) Remember that each a_j will belong to a specific vector space, depending upon the setting; for setting 1, $a_j \in \mathcal{V}_j^n$. Assume that the series has already been normalized to order $j-1$, and we are ready to do order j. (For the first step, $j=1$.) Let u_j be the next generator to be used (which as yet is not known, but which must belong to the appropriate vector space depending on the setting). Let ψ_j^s be the flow of the vector field u_j; that is, $\psi_j^s(\xi)$ is the solution of

$$\frac{dx}{ds} = u_j(x)$$

passing through $x = \xi$ at $s = 0$, where s is simply a parameter. To obtain the transformation that will be used to simplify a_j, we set $s = \varepsilon^j$; the change of variables is then

$$x = \psi_j^{\varepsilon^j}(\xi).$$

(In the matrix context this corresponds to setting $s = \varepsilon^j$ in e^{sU_j} to obtain $e^{\varepsilon^j U_j}$.) As usual, as soon as the coordinate change is made, we rename ξ as x (so that the occurrences of x from here on in this discussion are actually ξ). Write L_j for L_{u_j}. Then according to Theorem 4.4.8,

$$S_{\psi_j^s} a(x, \varepsilon) \equiv e^{s\mathsf{L}_j} a(x, \varepsilon),$$

and therefore

$$S_{\psi_j^{\varepsilon^j}} a(x, \varepsilon) \equiv e^{\varepsilon^j \mathsf{L}_j} a(x, \varepsilon). \qquad (4.4.27)$$

Since $e^{\varepsilon^j \mathsf{L}_j} = I + \varepsilon^j \mathsf{L}_j + \cdots$, the jth-order term of the right-hand side is $a_j + \mathsf{L}_j a_0$, which we temporarily call b_j. Using (4.4.21), $\mathsf{L}_j a_0 = [a_0, u_j] = -[u_j, a_0] = -\mathsf{L}_{a_0} u_j = -\mathcal{L} u_j$, so

$$\mathcal{L} u_j = a_j - b_j. \qquad (4.4.28)$$

This is a homological equation to be solved for b_j and u_j; the procedure for doing this will depend on the normal form style that is adopted, and is not part of format 2a. It is, however, important to remark that the homological equation must be solved within the correct vector space for the setting in question; in setting 1, for instance, a_j, b_j, and u_j will all belong to $\mathcal{V}_j^n$. Once u_j has been determined, the right-hand side of (4.4.27) must be computed completely (up to order k). The result is renamed $a(x, \varepsilon)$, and the normalization procedure begins again with the modified equation (4.4.26) and the next value of j.

As in Section 3.6, there is a possibility of confusion in this "algorithmic" version of format 2a due to the fact that the a_i change at every stage of the calculation. It is perhaps best to write the homological equation (4.4.28) in the form

$$\mathcal{L} u_j = k_j - b_j, \qquad (4.4.29)$$

to emphasize that k_j is known at the jth stage of the calculation but is not equal to the original a_j. (This matches the notation for format 2a in Section 3.2.) The final normalized system will sometimes be written

$$\dot{y} = b(y, \varepsilon) \equiv b_0(y) + \varepsilon b_1(y) + \cdots + \varepsilon^k b_k(y), \qquad (4.4.30)$$

just as in (4.3.7). Here $b_0 = a_0$, and the b_j for $j = 1, \ldots, k$ are the vector fields determined by solving (4.4.29) at each stage. The complete transformation from the original form (4.4.26) to the final form (4.4.30) is then given by

$$x = \psi_k^{\varepsilon^k} \circ \cdots \circ \psi_2^{\varepsilon^2} \circ \psi_1^{\varepsilon}(y). \qquad (4.4.31)$$

When this transformation is not needed, the normalized equation may be written in the same form (4.4.26) as the original equation.

Computing the Transformation: Item 7'

Format 2a (and the other generated formats) do not require that the transformations ψ_j be computed; it is enough to compute their generators (which fall out from the solution of the homological equation). But there are times when it is desired to know the transformations, or at least their formal power series to a certain order. There is a simple algorithm to generate these series.

To develop this algorithm we begin with the following lemma, which is quite similar to Theorem 4.4.8. A vector field b is given, with flow ψ^s. A (scalar) function $f : \mathbb{R}^n \to \mathbb{R}$ is also given. The lemma computes the transform of $f(x)$ under the coordinate change $x = \psi^s(y)$, just as Theorem 4.4.8 computes the transform of a vector field $a(x)$ under the same transformation. (Substituting $x = \psi^s(y)$ into $f(x)$ gives $f(\psi^s(y))$; the lemma computes $f(\psi^s(x))$.) Because the transformation rule for scalars is different from that for vectors (due to the absence of $\psi'_s(x)^{-1}$), L_b is replaced by $\mathcal{D}_b$. (See Remark 4.4.2.)

4.4.9. Lemma (Lie series for scalar fields). The formal power series expansion of $f(\psi^s(x))$ is given by

$$f(\psi^s(x)) \sim e^{s\mathcal{D}_b} f(x) = f(x) + s\mathcal{D}_b f(x) + \frac{s^2}{2} \mathcal{D}_b^2 f(x) + \cdots .$$

Proof. The proof begins with the observation that

$$\frac{d}{ds} f(\psi^s(x)) = (\mathcal{D}_b f)(\psi^s(x)),$$

which is just the chain rule. The rest of the proof is the same as that of Theorem 4.4.8.   □

To obtain the formal power series of the transformation $\psi^s(x)$ itself, observe that Lemma 4.4.9 remains valid if the scalar function f is replaced by a vector-valued function $f : \mathbb{R}^n \to \mathbb{R}^n$, as long as this is understood as a mapping (which transforms like a function) and not as a vector field; the operator $\mathcal{D}_b$ should be understood to be applied componentwise to f (see Remark 4.4.2). Taking f to be the identity function $f(x) = x$ now yields

$$\psi^s(x) \sim e^{s\mathcal{D}_b} x. \tag{4.4.32}$$

When this is applied to the individual generators u_j used in the normalizing process, with $\mathcal{D}_j = \mathcal{D}_{u_j}$, the result is

$$\psi_j^{\varepsilon^j}(x) \sim e^{\varepsilon^j \mathcal{D}_j} x. \tag{4.4.33}$$

Format 2a for Setting 1 Without Dilation

Format 2a can also be developed for setting 1 without dilation, that is, for a system in the form

$$\dot{x} = Ax + a_1(x) + a_2(x) + \cdots \tag{4.4.34}$$

with $a_j \in \mathcal{V}_j^n$. (Compare the discussion beginning with (4.3.13) for format 1a.) As usual let us suppose, inductively, that the system has been normalized up to a_{j-1} (that is, up to grade $j - 1$, which is degree j). The next generator will be a vector field u_j of grade j, which must be determined. This generator will be used in the following way: The differential equation

$$\frac{dx}{ds} = u_j(x) \tag{4.4.35}$$

has a flow ψ_j^s. Since $u_j(0) = 0$, solutions beginning close enough to the origin remain near the origin from time $s = 0$ to time $s = 1$ before (possibly) escaping to infinity. It follows that $\psi_j^1(x)$, the time-one map of the flow applied to x, is defined for x in some neighborhood of the origin. It is this time-one map that will be the transformation taking (4.4.34) one step further in the normalization process. Of course, the formulas themselves turn out to be exactly the same as those worked out above, with ε set equal to 1. Thus, u_j will be found by solving the usual homological equation $\mathcal{L}u_j = a_j - b_j$ (or $k_j - b_j$).

Since each ψ_j is defined only in a neighborhood of the origin, the same is true for the partially normalized equations at each stage; these neighborhoods will, in general, shrink at each stage, but for each j there will be a neighborhood in which the result is valid. This result is quite natural when stated in the language of jets and germs introduced in Section 4.3. The formal power series differential equation (4.4.34), whether truncated or taken to infinity, is a jet of some actual differential equation defined either on all of $\mathbb{R}^n$ or only on a neighborhood of the origin; any two such equations having the same germ will also have the same jet. The formal calculations involved in normalization operate only with the jet, leading from an "original" jet to a "normalized" jet. However, the transformation produced by these calculations to any finite stage is a polynomial transformation, and is well-defined (as an invertible transformation) in some neighborhood of the origin. (It is a germ of a diffeomorphism.) Therefore, it can be applied to germs of vector fields having the original jet, leading to germs having the normalized jet.

Summary: The Five Algorithms of Format 2a

As explained in Section 3.6, there are five algorithms that make up each generated format. Algorithms I and V have been developed in detail above, and in format 2a the remaining algorithms are so similar that they require only brief comments.

I. Given a generator, it is not usually possible to obtain the exact transformation that it generates, because that would require solving a system of nonlinear differential equations. But the formal power series of the transformation it generates is given by (4.4.32). For the generators used in normalization, form (4.4.33) expresses the correct dependence upon ε.

II. The negative of a generator generates the inverse transformation in format 2a. Thus, algorithm II reduces to algorithm I.

III. Given a generator, the effect of the transformation it generates is given by Theorem 4.4.8. Equation (4.4.27) is the version needed for normalization (with the correct dependence on ε).

IV. Again, in format 2a the inverse is obtained by changing the sign of the generator. No new algorithm is required.

V. The algorithm for normalization is to repeat algorithm III with a finite sequence of generators $u_1, u_2, \ldots, u_k$ ending with the desired order of normalization. Each generator is determined by solving a homological equation of the form $\mathcal{L}u_j = k_j - b_j$ (or, if care is taken to change the meaning of a after each step, $\mathcal{L}u_j = a_j - b_j$), and the flow ψ_j^s of each generator is applied with $s = \varepsilon^j$. The solution of the homological equation is guided by the normal form style that is chosen, and will be discussed thoroughly in the sections of this chapter devoted to normal form styles.

Further Study of the Lie Algebra of Vector Fields

For matrices, the bracket $[A, B] = AB - BA$ is literally the commutator of A and B, so that $[A, B] = 0$ if and only if A and B commute. For vector fields, $[a, b] = a'b - b'a$ is not a commutator bracket; that is, it cannot be written $[a, b] = ab - ba$. (The "product" ab of two column vectors does not even make sense.) Nevertheless, we have claimed that the Jacobi identity (4.4.19) holds. To complete the study of the Lie algebra of vector fields requires that we prove the Jacobi identity and explore the closely related idea of "commuting vector fields," which play an important role in the applications of normal form theory in Chapter 5 below.

It has already been pointed out, in Remark 4.4.1, that with every vector field $a(x) = (a_1(x), \ldots, a_n(x))$ there is associated a differential operator

$$\mathcal{D}_a = a_1(x)\frac{\partial}{\partial x_1} + \cdots + a_n(x)\frac{\partial}{\partial x_n} \qquad (4.4.36)$$

acting on the space $\mathcal{F}^n$ of smooth functions. Although a commutator bracket does not make sense for vector fields thought of as column vectors, it very much makes sense for the differential operators associated with these vector fields: The commutator of $\mathcal{D}_a$ and $\mathcal{D}_b$ is the operator

$$[\mathcal{D}_a, \mathcal{D}_b] = \mathcal{D}_a\mathcal{D}_b - \mathcal{D}_b\mathcal{D}_a. \tag{4.4.37}$$

A short calculation using (4.4.18) and (4.4.37) shows that this bracket is related to the Lie bracket of vector fields by the equation

$$[\mathcal{D}_a, \mathcal{D}_b] = \mathcal{D}_{[b,a]}, \tag{4.4.38}$$

which says that $\mathcal{D}$ is a *Lie algebra antihomomorphism*; that is, it reverses brackets. (In doing the calculation for (4.4.38), it is probably best to write out the expressions in summation notation with indices, so that the second derivatives that arise can be handled nicely. Part of the point is to see that the second derivatives cancel out.)

The Jacobi identity is automatic for commutator brackets. (It follows by expanding everything as differences of products and canceling terms.) That is, it is clear that

$$[\mathcal{D}_a, [\mathcal{D}_b, \mathcal{D}_c]] + [\mathcal{D}_b, [\mathcal{D}_c, \mathcal{D}_a]] + [\mathcal{D}_c, [\mathcal{D}_a, \mathcal{D}_b]] = 0.$$

Since $\mathcal{D}$ is a Lie algebra antihomomorphism, it follows (care being taken with reversals) that

$$\mathcal{D}_{[a,[b,c]]+[b,[c,a]]+[c,[a,b]]} = 0,$$

and finally, since a differential operator is zero only if its coefficients are zero, we have the Jacobi equation for vector fields,

$$[a, [b, c]] + [b, [c, a]] + [c, [a, b]] = 0. \tag{4.4.39}$$

This equation can be rewritten, using the anticommutativity of Lie brackets, as $[a, [b, c]] + [[a, c], b] - [[a, b], c] = 0$, which in turn, using (4.4.20), becomes $\mathsf{L}_{[b,c]}a + \mathsf{L}_b\mathsf{L}_c a - \mathsf{L}_b\mathsf{L}_c a = 0$, or

$$\mathsf{L}_{[b,c]} = [\mathsf{L}_c, \mathsf{L}_b]. \tag{4.4.40}$$

That is, L is also a Lie algebra antihomomorphism.

4.4.10. Definition. Two vector fields a and b are said to *commute* if $[a, b]$ is identically zero.

4.4.11. Theorem. Let a and b be vector fields with flows φ^t and ψ^t, and suppose a and b commute ($[a, b] = 0$). Then:

1. $\mathcal{D}_a$ and $\mathcal{D}_b$ commute (in the usual sense, as linear operators).

2. L_a and L_b commute.

3. The vector field a is *equivariant* under the flow of b. That is, the diffeomorphism ψ^s (for any s) transforms the vector field a into itself:

$$S_{\psi^s}a = a.$$

4. The flows of a and b commute, in the sense that $\varphi^t \circ \psi^s = \psi^s \circ \varphi^t$ for all s and t.

Proof. Items 1 and 2 follow immediately from (4.4.38) and (4.4.40). For item 3, write the condition $[a, b] = 0$ in the form $\mathsf{L}_b a = 0$. By the definition of L_b (4.4.14), this is equivalent to

$$\frac{d}{ds}\mathsf{S}_{\psi^s}a\bigg|_{s=0} = 0.$$

We claim that this derivative is in fact zero for all s. A short calculation using (4.4.9) shows that for any diffeomorphisms φ and ψ, $\mathsf{S}_{\varphi\circ\psi} = \mathsf{S}_\psi \circ \mathsf{S}_\varphi$. Then

$$\begin{aligned}
\frac{d}{ds}\mathsf{S}_{\psi^s}a &= \frac{d}{dh}\mathsf{S}_{\psi^{s+h}}a\bigg|_{h=0} \\
&= \frac{d}{dh}\mathsf{S}_{\psi^h\circ\psi^s}a\bigg|_{h=0} \\
&= \frac{d}{dh}\mathsf{S}_{\psi^s}\circ\mathsf{S}_{\psi^h}a\bigg|_{h=0} \\
&= \mathsf{S}_{\psi^s}\frac{d}{dh}\mathsf{S}_{\psi^h}a\bigg|_{h=0} \\
&= 0.
\end{aligned}$$

Therefore, $\mathsf{S}_{\psi^s}a$ is constant, and equals its value when $s = 0$, which is a; this proves item 3. For item 4, observe that according to item 3, if ψ^s (with s fixed) is applied to the vector field a as a coordinate transformation, the result is again a. When a diffeomorphism carries one vector field to another, it also carries the flow of the first vector field to the flow of the other. In the present case, then, ψ^s carries φ^t to itself, which is the content of item 4. □

Notes and References

General references for Lie theory have been given in Notes and References for Section 3.6. An account of Lie theory emphasizing what we call format 2a is given in Finn [45]. This source contains historical remarks and references; however, it assumes that the systems being normalized are Hamiltonian (as did most of the early work on the subject). For this reason it is best read in connection with Section 4.9 below, although all of the ideas are transferable to the general setting of vector fields and diffeomorphisms.

4.5   The Semisimple Normal Form

The next four sections are devoted to a detailed treatment of normal forms for vector fields with a rest point at the origin, that is, vector fields in setting 1. Such vector fields will be written in the undilated form (4.3.16), that is,

$$\dot{x} = a(x) = Ax + a_1(x) + a_2(x) + \cdots, \qquad (4.5.1)$$

with $a_j \in \mathcal{V}_j^n$, so that each a_j is a homogeneous polynomial vector field of grade j, degree $j + 1$. In other words, we work with formal power series vector fields (also called *formal vector fields*) in $\mathcal{V}_{**}^n$, without the dilation parameter ε of (4.2.6). In this section it is assumed that A is semisimple.

A *normal form style* for (4.5.1) is a family of subspaces

$$\mathcal{N}_j \subset \mathcal{V}_j^n \qquad (4.5.2)$$

for $j = 1, 2, \ldots$, such that for each j,

$$\mathcal{V}_j^n = \operatorname{im} \mathcal{L}_j \oplus \mathcal{N}_j. \qquad (4.5.3)$$

Here $\mathcal{L}_j : \mathcal{V}_j^n \to \mathcal{V}_j^n$ is the restriction to $\mathcal{V}_j^n$ of the homological operator $\mathcal{L} : \mathcal{V}^n \to \mathcal{V}^n$ defined in (4.3.5) and (4.3.10) by

$$(\mathcal{L}v)(x) = v'(x)a_0(x) - a_0'(x)v(x) = v'(x)Ax - Av(x), \qquad (4.5.4)$$

where $a_0(x) = Ax$. This operator is the same as L_{a_0} defined in (4.4.14) and evaluated in Lemma 4.4.7, and any of the following notations may be used interchangeably:

$$\mathcal{L} = \mathsf{L}_{a_0} = \mathsf{L}_{Ax} = \mathsf{L}_A. \qquad (4.5.5)$$

When a normal form style has been selected, (4.5.1) is said to be in *normal form* to order k (or to all orders) if

$$a_j \in \mathcal{N}_j \qquad (4.5.6)$$

for $j = 1, \ldots, k$ (or for all j). According to Sections 4.3, 4.4, C.2, and D.2, any system of the form (4.5.1) may be brought into normal form (of any style) by solving homological equations of the form (4.3.9), (4.4.29), or the equivalent in the other formats; these equations are solvable because of (4.5.3).

If a normal form style has been defined, so that $\mathcal{N}_j$ is specified for $j = 1, 2, \ldots$, we will write

$$\mathcal{N}_+ = \{a_1(x) + a_2(x) + \cdots \in \mathcal{V}_{**}^n : a_j \in \mathcal{N}_j\}. \qquad (4.5.7)$$

That is, $\mathcal{N}_+$ is the collection of all formal power series that begin with quadratic terms and are in normal form to all orders; this is the direct product of the spaces $\mathcal{N}_j$ for $j \geq 1$. Frequently it happens that the definition of a specific normal form style is meaningful for $j = -1$ and $j = 0$ (constant and linear vector fields) as well, even though these are not needed in normal form theory. (They will play a role in unfolding theory in Chapter 6, and in each normal form style that we discuss, $\mathcal{N}_0$ coincides with the corresponding linear normal form style studied in Chapter 3.) Thus, it is usually possible to define the *normal form space*

$$\mathcal{N} = \{a_{-1} + a_0(x) + a_1(x) + a_2(x) + \cdots \in \mathcal{V}_{**}^n : a_j \in \mathcal{N}_j\}. \qquad (4.5.8)$$

In equation (4.5.8), $a_0(x)$ no longer coincides with Ax, but is any element of $\mathcal{N}_0$; a_{-1} is an element of $\mathbb{R}^n$ and need not be written as a function of x. It will turn out that for each normal form style that we study, $\mathcal{N}$ has the structure of a module over a certain ring of (scalar-valued) functions, and the terms of grade -1 and 0 play an important role in understanding this module structure even though they are not used in the normal form itself.

The Semisimple Normal Form Style

The first steps in the semisimple theory for vector fields correspond to Lemma 3.3.1 and Corollary 3.3.2 in the linear case. Since there is no module structure in the linear theory, many of the subsequent steps do not correspond to anything in Chapter 3, although many similarities will appear.

Recall from Chapter 3 that $gl(n)$ denotes either $gl(n, \mathbb{R})$ or $gl(n, \mathbb{C})$, and that a matrix is semisimple if and only if it is diagonalizable over the complex numbers. It is useful to introduce the *multi-index* notation, in which $m = (m_1, \ldots, m_n)$ is a vector of nonnegative integers, $|m|$ is defined by

$$|m| = m_1 + \cdots + m_n, \tag{4.5.9}$$

and x^m by

$$x^m = x_1^{m_1} x_2^{m_2} \cdots x_n^{m_n}. \tag{4.5.10}$$

Although we will not use it until the next section, $m!$ is defined by

$$m! = (m_1!)(m_2!) \cdots (m_n!). \tag{4.5.11}$$

The monomials (4.5.10) having $|m| = j$ constitute the *standard basis* for $\mathcal{F}_j^n$. Let e_i, $i = 1, \ldots, n$, be the standard basis vectors for $\mathbb{R}^n$. Then the *monomial vector fields* (or simply *vector monomials*)

$$x^m e_i \quad \text{for} \quad |m| = j + 1 \tag{4.5.12}$$

make up the *standard basis* for $\mathcal{V}_j^n$. For instance, if $n = 2$ and $j = 2$, this basis includes

$$x^{(2,1)} e_1 = \begin{bmatrix} x_1^2 x_2 \\ 0 \end{bmatrix}$$

and

$$x^{(0,3)} e_2 = \begin{bmatrix} 0 \\ x_2^3 \end{bmatrix}.$$

4.5.1. Lemma. The dimensions of the vector spaces $\mathcal{F}_j^n$ and $\mathcal{V}_j^n$ are given by the binomial coefficients

$$\dim \mathcal{F}_j^n = \binom{n+j-1}{j}$$

and

$$\dim \mathcal{V}_j^n = n \binom{n+j}{j+1}.$$

Proof. It is easy to see that the number of ways of placing $n-1$ dots and j dashes in a row (as if to form an imaginary Morse code letter) is $\binom{n+j-1}{j}$. (This is the same as the number of ways of drawing all of the marbles, counting order but without replacement and without distinguishing between marbles of the same color, from a bag containing $n-1$ black marbles and j white marbles.) There is a one-to-one correspondence of such Morse code letters with monomials x^m having $|m| = j$: Starting at the beginning of the Morse code letter, replace each dash with x_1 until the first dot is reached, then replace each dash with x_2 until the next dot is reached, and so forth. (If the first entry is a dot, begin at once with x_2; if several dots occur in a row, skip the appropriate number of variables.) Finally, ignore the dots and simplify the product. This proves the formula for $\dim \mathcal{F}_j^n$. The second formula follows at once from $\dim \mathcal{V}_j^n = n \dim \mathcal{F}_{j+1}^n$.   □

4.5.2. Lemma (cf. Lemma 3.3.1). If $S \in \mathrm{gl}(n)$ is semisimple, then the operator

$$\mathsf{L}_S = \mathsf{L}_{Sx} : \mathcal{V}_j^n \to \mathcal{V}_j^n$$

given by

$$\mathsf{L}_S v(x) = v'(x)Sx - Sv(x)$$

is semisimple for each j.

Proof. Suppose first that S is diagonal with eigenvalues $\lambda_1, \ldots, \lambda_n$. Then an easy calculation shows that

$$\mathsf{L}_S x^m e_i = (\langle m, \lambda \rangle - \lambda_i) x^m e_i, \tag{4.5.13}$$

where

$$\langle m, \lambda \rangle = m_1 \lambda_1 + \cdots + m_n \lambda_n. \tag{4.5.14}$$

Therefore, the standard basis for $\mathcal{V}_j^n$ consists of eigenvectors of L_S, which is therefore semisimple.

If S is semisimple but not diagonal, there exists a linear coordinate change $x = Ty$ such that $T^{-1}ST = \Lambda$ is diagonal and has the same eigenvalues $\lambda_1, \ldots, \lambda_n$ as S. If these are numbered in the order that they occur in the diagonal of Λ, then by (4.5.13), $y^m e_i$ is an eigenvector of L_Λ with eigenvalue $\langle m, \lambda \rangle - \lambda_i$. We claim that, when transformed back to the variable x according to the correct transformation law for vector fields, $y^m e_i$ gives an eigenvector for L_S with the same eigenvalue. When this is proved we are finished, because L_S has a full set of eigenvectors.

It is worthwhile to do the required calculation in somewhat greater generality. Let $A \in \mathrm{gl}(n)$ and $T \in \mathrm{GL}(n)$. Any vector field $a(x)$ transforms under $x = Ty$ into $b(y) = T^{-1}a(Ty)$. Then

$$
\begin{aligned}
(\mathsf{L}_{T^{-1}AT})b(y) &= b'(y)(T^{-1}AT)y - (T^{-1}AT)b(y) \qquad\qquad (4.5.15)\\
&= (T^{-1}a'(Ty)T)(T^{-1}AT)y - (T^{-1}AT)(T^{-1}a(Ty))\\
&= T^{-1}(a'(Ty)A(Ty) - Aa(Ty))\\
&= T^{-1}(\mathsf{L}_A a)(Ty),
\end{aligned}
$$

which is nothing other than the vector field $\mathsf{L}_A a(x)$ expressed in y. It follows that if $a(x)$ is an eigenvector of L_A, then $b(y)$ is an eigenvector of $\mathsf{L}_{T^{-1}AT}$ with the same eigenvalue, and conversely; this does not require that A be semisimple or that $T^{-1}AT$ be diagonal. When $A = S$ is semisimple, $T^{-1}AT = \Lambda$ is diagonal, and $b(y) = y^m e_i$, this specializes to the required result for the lemma. $\qquad\qquad\qquad\qquad\qquad\qquad\qquad\qquad\qquad\square$

4.5.3. Corollary (cf. Corollary 3.3.2). If A in (4.5.1) is semisimple with eigenvalues λ_i, then $\mathcal{L} = \mathsf{L}_A : \mathcal{V}^n_{**} \to \mathcal{V}^n_{**}$ is semisimple with eigenvalues $\langle m, \lambda \rangle - \lambda_i$; more specifically, the eigenvalues of $\mathcal{L}_j : \mathcal{V}^n_j \to \mathcal{V}^n_j$ are those for which $|m| = j + 1$. If A is diagonal, then

$$
\mathcal{L}x^m e_i = (\langle m, \lambda \rangle - \lambda_i)x^m e_i. \qquad\qquad (4.5.16)
$$

It follows at once from Section 2.1 that

$$
\mathcal{V}^n_j = \operatorname{im} \mathcal{L}_j \oplus \ker \mathcal{L}_j, \qquad\qquad (4.5.17)
$$

and we can define the *semisimple normal form style* by setting

$$
\mathcal{N}_j = \ker \mathcal{L}_j. \qquad\qquad (4.5.18)
$$

This is well defined for all $j \geq -1$, so both $\mathcal{N}_+$ and $\mathcal{N}$ can be defined, as in (4.5.7) and (4.5.8).

4.5.4. Definition (cf. Definition 3.3.3). A formal power series vector field $a(x)$ is in *semisimple normal form* with respect to a_0 (to all orders, or to grade k) if

1. $a_0(x) = Ax$ with A semisimple; and

2. each homogeneous vector field $a_j(x)$ (for $j \geq 1$, or for $1 \leq j \leq k$) commutes with the vector field Ax (that is, $a_j \in \ker \mathsf{L}_A$).

Occasionally, it is useful to extend the definition of semisimple normal form for vector fields whose linear part is not semisimple. As in Chapter 3, unless the word "extended" is used, "semisimple normal form" implies that A is semisimple; as in Remark 3.3.5, extended semisimple normal form is not a true normal form style (in our technical sense).

4.5.5. Definition (cf. Definition 3.3.4). We say that $a(x)$ is in *extended semisimple normal form* if

1. $a_0(x) = Ax$, with $A = S + N$ being the semisimple/nilpotent decomposition of A; and

2. the terms a_j for $j \geq 1$ commute with Sx $(a_j \in \ker \mathsf{L}_S)$.

The next theorem states that semisimple normal form is preserved by all linear coordinate changes:

4.5.6. Theorem (cf. Theorem 3.3.6). If $a(x)$ is in semisimple normal form and T is an invertible linear transformation (or matrix), then $S_T a(x) = T^{-1}a(Tx)$ is in semisimple normal form.

Proof. Notice that for linear transformations, the formula $(S_\varphi a)(x) = \varphi'(x)^{-1}a(\varphi(x))$ becomes $(S_T a)(x) = T^{-1}a(Tx)$; this is helpful because it follows that

$$(S_T a)'(x) = T^{-1}a'(Tx)T,$$

which does not introduce second derivatives of φ. We actually prove the following generalization of Theorem 4.5.6: If a and b are vector fields with $[a, b] = 0$, then $[S_T a, S_T b] = 0$. We have

$$
\begin{aligned}
[S_T a, S_T b](x) &= (S_T a)'(x)S_T b(x) - (S_T b)'(x)S_T a(x) \\
&= (T^{-1}a'(Tx)T)(T^{-1}b(Tx)) - (T^{-1}b'(Tx)T)(T^{-1}a(Tx)) \\
&= T^{-1}(a'(Tx)b(Tx) = b'(Tx)a(Tx)) \\
&= T^{-1}[a, b](Tx) \\
&= 0.
\end{aligned}
$$

$\square$

Equivariants, Invariants, and the Semisimple Normal Form Module

Recall (see Remarks 4.4.1 and 4.4.2) that if v is a (smooth) vector field and f a (smooth) scalar field, then

$$\mathcal{D}_{v(x)}f(x) = f'(x)v(x) = (v(x) \cdot \nabla)f(x)$$

is a (smooth) scalar field, called the derivative of f along (the flow of) v. We write $\mathcal{D}_A$, an abbreviation of $\mathcal{D}_{Ax}$, for the derivative along the linear vector field Ax, viewed as a map $\mathcal{D}_A : \mathcal{F}^n \to \mathcal{F}^n$. A function (or scalar field) f is called an *invariant* of (the flow of) Ax if

$$\frac{d}{dt}f(e^{At}x) = 0, \tag{4.5.19}$$

or equivalently, $\mathcal{D}_A f = 0$ or $f \in \ker \mathcal{D}_A$. Since

$$\mathcal{D}_A(f + g) = \mathcal{D}_A f + \mathcal{D}_A g$$

and

$$\mathcal{D}_A(fg) = f\mathcal{D}_A g + g\mathcal{D}_A f, \tag{4.5.20}$$

it follows that if f and g are invariants, then $f + g$ and fg are invariants also; that is, $\ker \mathcal{D}_A$ is both a vector space over $\mathbb{R}$, and also a subring of $\mathcal{F}^n$, known as the *ring of (smooth) invariants*. The map $\mathcal{D}_A$ can also be applied to formal power series $f(x) = \sum_j f_j(x) \in \mathcal{F}^n_{**}$ by applying $\mathcal{D}_A$ to each $f_j(x) \in \mathcal{F}^n_j$ separately; that is, $\mathcal{D}_A : \mathcal{F}^n \to \mathcal{F}^n$ can be viewed as a map $\mathcal{D}_A : \mathcal{F}^n_{**} \to \mathcal{F}^n_{**}$, denoted by the same symbol. The kernel of this map,

$$\ker \mathcal{D}_A \subset \mathcal{F}^n_{**}, \tag{4.5.21}$$

is the *ring of formal invariants*. (It is also a vector space.) The map j^∞ carries the ring of smooth invariants to the ring of formal invariants, and it is seldom necessary to specify which ring is intended. It is also possible to view $\mathcal{D}_A$ as a map of $\mathcal{F}^n_*$ to itself, the kernel being the *ring of polynomial invariants*. Just as every basis for $\mathcal{F}^n_*$ (as a vector space over $\mathbb{R}$) is also a formal basis (allowing infinite linear combinations) for $\mathcal{F}^n_{**}$, so every basis for the space of polynomial invariants is a formal basis for the space of formal invariants. (See the discussion following Definition 4.1.1.)

In Definition 3.3.3, it was said that a matrix series with semisimple leading term is in normal form if each succeeding term commutes with the leading term. For vector fields, the notion of "commutativity" has been discussed in Section 4.4 (Definition 4.4.10 and Theorem 4.4.11). If $a(x)$ is in semisimple normal form, then each term $a_j(x)$ commutes with $a_0(x) = Ax$, and each a_j is *equivariant* under the flow $\psi^s = e^{As}$ of a_0. As with $\mathcal{D}_A$ above, we write L_A as an abbreviation for $\mathsf{L}_{Ax} = \mathsf{L}_{a_0}$, and regard L_A as a mapping of either $\mathcal{V}^n$, $\mathcal{V}^n_*$, or $\mathcal{V}^n_{**}$ to itself; the kernels of these maps are the spaces of *smooth equivariants, polynomial equivariants*, and *formal equivariants*, respectively. Each of these is a vector space, and any basis for the space of polynomial equivariants is a formal basis for the space of formal equivariants.

4.5.7. Lemma. Let $A \in \mathrm{gl}(n)$ be semisimple. Then the (smooth, polynomial, or formal) normal form space of A, that is, the space of (smooth, polynomial, or formal) equivariants $\mathcal{N} = \ker \mathcal{L} = \ker \mathsf{L}_A$, is a module over the ring of (smooth, polynomial, or formal) invariants $\ker \mathcal{D}_A$.

Proof. The important point is that if f is an invariant and v an equivariant (so that $\mathcal{D}_A f = 0$ and $\mathsf{L}_A v = 0$), then fv is an equivariant. This is an immediate consequence of the formula

$$\mathsf{L}_A(fv) = (\mathcal{D}_A f)v + f(\mathsf{L}_A v), \tag{4.5.22}$$

which follows from the definitions of L_A and $\mathcal{D}_A$ by a simple calculation. In order to be a module, $\ker \mathsf{L}_A$ must be closed under addition and under multiplication by elements of $\ker \mathcal{D}_A$, and must satisfy the same axioms as a vector space (with the field, $\mathbb{R}$ or $\mathbb{C}$, replaced by the ring $\ker \mathcal{D}_A$). These are merely restatements of familiar facts of the algebra of formal power series. $\qquad \square$

4.5.8. Remark. The notions of invariant and equivariant are properly defined in terms of group representations (see Remark 2.5.1). Let $\mathfrak{G}$ be any group of diffeomorphisms; then $\varphi \in \mathfrak{G}$ acts on scalar fields by sending f to $f \circ \varphi$, and on vector fields by sending v to $(\varphi')^{-1} v \circ \varphi$. (These are actually antirepresentations, since applying first φ, then ψ, is equivalent to applying $\varphi \circ \psi$ rather than $\psi \circ \varphi$.) A scalar field fixed by the action of $\mathfrak{G}$ on scalar fields is an *invariant* of $\mathfrak{G}$, whereas a vector field fixed by the action on vector fields is an *equivariant*. See also the discussion of invariants and equivariants in Section 1.1, following equation (1.1.26).

The Description Problem

The *description problem* for the semisimple normal form is the problem of describing what terms may be present in each $a_j(x)$ in (4.5.1) when it is in normal form. It is possible to solve the description problem by merely identifying the elements of $\mathcal{N} = \ker \mathsf{L}_A$, without any mention of the invariants or the module structure. This may be done by giving a set (usually infinite) of basis elements of $\mathcal{N}$ as a vector space over $\mathbb{R}$. But it is often helpful to describe the ring $\ker \mathcal{D}_A$ of invariants first, and then present a list of basic equivariants that generate $\ker \mathsf{L}_A$ as a module over the invariants, since this list of generators will be finite. The simplest example has already been worked out in Section 1.1. Rather than attempt a general result for arbitrary semisimple A (which, to a considerable extent, we were able to do in the linear case, in Section 3.3), we will concentrate on several examples. We will work out two examples for 3×3 real matrices A that are diagonal over the reals, and an important family of 4×4 real matrices whose eigenvalues are pure imaginary. This case, the *double center*, is a natural generalization of the (single) center treated in Section 1.1.

For all of our examples, we assume $A \in \mathrm{gl}(n, \mathbb{R})$, which is a natural assumption for differential equation applications. Since A is semisimple, there will exist a matrix T (which may be real or complex) such that $T^{-1}AT = \Lambda$ is diagonal (either real or complex). If T and Λ are real, we assume that $A = \Lambda$, in other words, that A has already (in "step zero") been brought into diagonal form. If T and Λ contain (strictly) complex entries, we assume that A is in real semisimple canonical form (Definition 2.1.2). In either case, the following proposition (which is an immediate corollary of the results and definitions above) provides the starting point for the solution of the description problem.

4.5.9. Corollary. If Λ is a real (respectively, complex) diagonal matrix with diagonal entries $\lambda_1, \ldots, \lambda_n$, then the space $\ker \mathcal{D}_\Lambda$ of polynomial invariants is spanned (as a vector space over $\mathbb{R}$ or $\mathbb{C}$, respectively) by the monomials x^m such that $\langle m, \lambda \rangle = 0$; these will be called the *invariant monomials*. The space $\ker \mathsf{L}_\Lambda$ of polynomial equivariants is spanned (over $\mathbb{R}$ or $\mathbb{C}$) by the vector monomials $x^m e_i$ such that $\langle m, \lambda \rangle - \lambda_i = 0$; these will be called the *equivariant monomials*. The same invariant and equivariant monomials provide a formal basis (allowing infinite linear combinations) for the spaces of formal invariants and equivariants.

Proof. Equation (4.5.13) shows that the eigenvectors of L_Λ are $x^m e_i$, and the associated eigenvalues are $\langle m, \lambda \rangle - \lambda_i$. The kernel of a linear transformation is spanned by its eigenvectors of eigenvalue zero. This proves the result for the equivariants, and the result for the invariants follows in the same way from

$$\mathcal{D}_\Lambda x^m = \langle m, \lambda \rangle x^m. \tag{4.5.23}$$

□

4.5.10. Remark. The term *resonance* is frequently used in the literature to refer to either of the conditions $\langle m, \lambda \rangle = 0$ or $\langle m, \lambda \rangle - \lambda_i = 0$, and then x^m and $x^m e_i$, respectively, are referred to as *resonant monomials*. We will generally avoid these usages, because the terms "invariant" and "equivariant" are more precise (distinguishing the scalar and vector cases), and also because this notion of resonance is easily confused with the more common notion of resonance between frequencies of oscillation. In cases (such as the double center, below) where pure imaginary eigenvalues occur in conjugate pairs, this more natural notion comes into play. In that problem, there are four eigenvalues, $\pm i\omega_1$ and $\pm i\omega_2$, but only two frequencies, ω_1 and ω_2; a resonance between these frequencies is a relation $k_1\omega_1 + k_2\omega_2 = 0$, where k_1 and k_2 are integers (not necessarily positive). A resonance relation of the form $q\omega_1 - p\omega_2 = 0$ is usually written as $\omega_1 : \omega_2 = p : q$.

Two Hyperbolic Examples

Most introductory treatments of normal form theory for differential equations emphasize the case where all eigenvalues of A lie on the imaginary axis. It is indeed true (as we will see in Chapter 6) that this is the most important case from the standpoint of bifurcation theory, but there is actually considerable interest in hyperbolic cases as well; normal forms in hyperbolic systems can be used to compute "strong stable and unstable manifolds" (stable and unstable manifolds with rate conditions). The two examples that we present here are given not because of any intrinsic importance that they have in applications, but because they illustrate the variety of behaviors that can be found in the module structure of the equivariants (over the invariants).

The first example is

$$A = \begin{bmatrix} 6 & & \\ & 2 & \\ & & 3 \end{bmatrix}. \tag{4.5.24}$$

Since $A = \Lambda$ is real and already diagonal, Corollary 4.5.9 tells us that the invariants are spanned by the monomials

$$x^m = x_1^{m_1} x_2^{m_2} x_3^{m_3}$$

for which

$$6m_1 + 2m_2 + 3m_3 = 0.$$

Since the powers m_i must be nonnegative, the only solution of this equation is $(m_1, m_2, m_3) = (0, 0, 0)$, so the only invariant monomial is 1. It follows that the ring of invariants is

$$\ker \mathcal{D}_A = \mathbb{R}. \tag{4.5.25}$$

This has a natural geometric interpretation: The flow of the vector field Ax consists of orbits that approach the origin as $t \to -\infty$. Since any invariant must be constant on the orbits (see (4.5.19)), any *continuous* invariant must be equal everywhere to its value at the origin, a limit point of all the orbits; thus any continuous invariant is constant. Our "ring of invariants" is the ring of *formal* invariants, but any truncation of a formal power series invariant is a polynomial invariant, which is continuous and must reduce to a constant; therefore, the formal power series can contain only a constant term.

Since the ring of invariants is the field $\mathbb{R}$, the module structure of the space of equivariants is the same as its vector space structure. According to Corollary 4.5.9, a vector space basis for the equivariants consists of those vector monomials $x^m e_i$ for which

$$6m_1 + 2m_2 + 3m_3 - \lambda_i = 0.$$

In other words,

$$\begin{bmatrix} x_1^{m_1} x_2^{m_2} x_3^{m_3} \\ 0 \\ 0 \end{bmatrix} \qquad \text{with } 6m_1 + 2m_2 + 3m_3 - 6 = 0,$$

$$\begin{bmatrix} 0 \\ x_1^{m_1} x_2^{m_2} x_3^{m_3} \\ 0 \end{bmatrix} \qquad \text{with } 6m_1 + 2m_2 + 3m_3 - 2 = 0,$$

$$\begin{bmatrix} 0 \\ 0 \\ x_1^{m_1} x_2^{m_2} x_3^{m_3} \end{bmatrix} \qquad \text{with } 6m_1 + 2m_2 + 3m_3 - 3 = 0.$$

It is not hard to see that there are only finitely many solutions to these equations, namely,

$$\begin{bmatrix} x_1 \\ 0 \\ 0 \end{bmatrix}, \quad \begin{bmatrix} 0 \\ x_2 \\ 0 \end{bmatrix}, \quad \begin{bmatrix} 0 \\ 0 \\ x_3 \end{bmatrix}, \quad \begin{bmatrix} x_3^2 \\ 0 \\ 0 \end{bmatrix}, \quad \begin{bmatrix} x_2^3 \\ 0 \\ 0 \end{bmatrix}. \tag{4.5.26}$$

The module of equivariants, or normal form space $\mathcal{N}$, is simply the five-dimensional real vector space spanned by these equivariant monomials. The portion $\mathcal{N}_+$ (see(4.5.7)) of $\mathcal{N}$ that is actually used in the normal form is spanned only by the last two of these monomials. In other words, a system of differential equations with linear part (4.5.24) is in normal form (to all orders) if and only if it has the form

$$\dot{x}_1 = 6x_1 + c_1 x_3^2 + c_2 x_2^3,$$
$$\dot{x}_2 = 2x_2,$$
$$\dot{x}_3 = 3x_3.$$

The next example differs from the last one only in the sign of one entry, but its normal form module has a very different structure.

$$A = \begin{bmatrix} 6 & & \\ & 2 & \\ & & -3 \end{bmatrix}. \tag{4.5.27}$$

According to Corollary 4.5.9, the invariants are spanned (as a vector space over $\mathbb{R}$) by x^m for which

$$6m_1 + 2m_2 - 3m_3 = 0.$$

Writing this as $2m_2 = 3m_3 - 6m_1$, we see that the right-hand side is divisible by 3, which implies $m_2 = 3\ell$ for some integer ℓ, which must be nonnegative. It follows that $m_3 - 2m_1 = 2\ell$; if we write $m_1 = k$, then $m_3 = 2k + 2\ell$, or $m = (k, 3\ell, 2k + 2\ell)$, for $k, \ell \geq 0$. Thus, the invariant monomials are

$$x_1^k x_2^{3\ell} x_3^{2k+2\ell} = (x_1 x_3^2)^k (x_2^3 x_3^2)^\ell.$$

An infinite linear combination of such monomials is simply a formal power series in the *basic invariants* $I_1 = x_1 x_3^2$ and $I_2 = x_2^3 x_3^2$, so we have identified the ring of invariants as

$$\ker \mathcal{D}_A = \mathbb{R}[[x_1 x_3^2, x_2^3 x_3^2]] = \mathbb{R}[[I_1, I_2]]. \tag{4.5.28}$$

($\mathbb{R}[[y_1, \ldots, y_k]]$ is a standard notation for the ring of formal power series in the variables $y_1, \ldots, y_k$, just as $\mathbb{R}[y_1, \ldots, y_k]$ denotes the ring of polynomials in the same variables.) The quantities $x_1 x_3^2$ and $x_2^3 x_3^2$ are *algebraically independent*, meaning that no polynomial in these quantities (other than the zero polynomial) can be identically zero. (In fact, they are *functionally independent*, which is stronger. This is clear because the second contains a variable, x_2, that is not contained in the first, so that the two quantities vary independently and cannot satisfy any functional relation that is not identically zero.) As a consequence of the algebraic independence of these quantities, each element of $\ker \mathcal{D}_A$ has a *unique* expression as a power series in the form (4.5.28).

Again, according to Corollary 4.5.9, the equivariants are spanned (over $\mathbb{R}$) by those $x^m e_i$ for which $6m_1 + 2m_2 - 3m_3$ equals 6, 2, or -3 (according as $i = 1, 2$, or 3). There are two "smallest" m giving a value of 6, namely $m = (1, 0, 0)$ and $m = (0, 3, 0)$, and one each giving 2 and -3, $m = (0, 1, 0)$ and $m = (0, 0, 1)$; all other solutions equal one of these plus one of the values of m found above that give a value of zero, $m = (k, 3\ell, 2k + \ell)$ for $k, \ell \geq 0$. In other words, each equivariant monomial is equal to an invariant monomial times one of the following *basic equivariants*:

$$v_1 = \begin{bmatrix} x_1 \\ 0 \\ 0 \end{bmatrix}, \quad v_2 = \begin{bmatrix} 0 \\ x_2 \\ 0 \end{bmatrix}, \quad v_3 = \begin{bmatrix} 0 \\ 0 \\ x_3 \end{bmatrix}, \quad v_4 = \begin{bmatrix} x_2^3 \\ 0 \\ 0 \end{bmatrix}. \tag{4.5.29}$$

(The equivariant monomial with $i = 1$ and $m = (0, 3, 0)$ has been placed last so that the basic equivariants will be grouped according to their degree $|m|$.) Another way to say this is that the vector fields $v_1, \ldots, v_4$ in (4.5.29) *generate* ker L_A as a module over the ring ker $\mathcal{D}_A$ given by (4.5.28): Every formal power series equivariant is a linear combination of $v_1, \ldots, v_4$ with coefficients that are formal power series in I_1, I_2. However, the representation of equivariants in this manner is not unique, the simplest counterexample being that

$$x_2^3 x_3^2 \begin{bmatrix} x_1 \\ 0 \\ 0 \end{bmatrix} = x_1 x_3^2 \begin{bmatrix} x_2^3 \\ 0 \\ 0 \end{bmatrix} \quad \text{or} \quad I_2 v_1 - I_1 v_4 = 0. \tag{4.5.30}$$

The relation (4.5.30) is an example of what is known as a *syzygy*, and illustrates the difference between a module and a vector field: Although the generating set (4.5.29) is not "linearly independent" over the coefficient ring (because of the syzygy), it is not possible to reduce the generating set to a smaller one (called a "basis" in the vector space case) that is linearly independent. If I_1 and I_2 were members of a field, we could solve (4.5.30) for $v_4 = (I_2/I_1)v_1$ and thus eliminate v_4 from the generating set, but in the ring ker $\mathcal{D}_A$ it is not possible to divide by I_1. ($I_2/I_1 = x_2^3/x_1$, which cannot be expressed as a formal power series in x_1, x_2, x_3.)

4.5.11. Remark. The technical definition of a syzygy is as follows: Let M be a finitely generated module over a ring R, with generators $v_1, \ldots, v_k$, so that every element of M may be written (although not necessarily uniquely) as $r_1 v_1 + \cdots + r_k v_k$, with $r_i \in R$. Let $u_1, \ldots, u_k$ be a set of arbitrary symbols, and let M' be the set of expressions $r_1 u_1 + \cdots + r_k u_k$, with $r_i \in R$; M' is called the *free module* over R with generators $u_1, \ldots, u_k$. Let $h : M' \to M$ be the module homomorphism defined by $h(\sum r_i u_i) = \sum r_i v_i$. Then the kernel of h, which is a submodule of M', is the *module of syzygies* among $v_1, \ldots, v_k$. In our example (4.5.30), the syzygy, properly speaking, would be the element $I_2 u_1 - I_1 u_4$ of M', but it is common to represent this by the equation $I_2 v_1 - I_1 v_4 = 0$, which is equivalent to $h(I_2 u_1 - I_1 u_4) = 0$, or $I_2 u_1 - I_1 u_4 \in \ker h$. (We cannot say that $I_2 v_1 - I_1 v_4$ is the syzygy, because this is simply zero and it would not do to say that zero is the syzygy; this would convey no information.) Expressing the syzygy in terms of the u_i, which do not in fact "satisfy the syzygy" as do the v_i, enables us to regard the syzygy as an *element* of some set, rather than as an *equation*.

Although it is not possible to reduce (4.5.30) to a basis (in the vector space sense) for the module of equivariants, there does exist a method of restoring uniqueness to the expressions for the equivariants. Namely, we can block the use of I_1 in the coefficient of v_4, allowing v_1, v_2, v_3 to be multiplied by arbitrary elements of ker $\mathcal{D}_A$ but permitting v_4 to be multiplied only by formal power series in I_2. This at once eliminates the syzygy (in the sense that the equivariant $I_2 v_1$ cannot also be written as $I_1 v_4$). A little thought will show that every formal power series equivariant can be expressed in a unique way in the form

$$f_1(I_1, I_2)v_1 + f_2(I_1, I_2)v_2 + f_3(I_1, I_2)v_3 + f_4(I_2)v_4, \qquad (4.5.31)$$

where each f_i is a formal power series in its arguments (and it is crucial that f_4 is a function only of I_2). Another way to express this fact is that

$$\mathcal{N} = \ker \mathsf{L}_A = \mathbb{R}[[I_1, I_2]]v_1 \oplus \mathbb{R}[[I_1, I_2]]v_2 \oplus \mathbb{R}[[I_1, I_2]]v_3 \oplus \mathbb{R}[[I_2]]v_4.$$
$$(4.5.32)$$

Either of the expressions (4.5.31) or (4.5.32) is called a *Stanley decomposition* of the normal form module. It should be noted that although the expression of an equivariant in the form (4.5.31) is unique, the form itself (the Stanley decomposition) is not unique; we could equally well have eliminated the syzygy by blocking I_2 from the coefficient of v_1, obtaining a second Stanley decomposition of the same module.

4.5.12. Remark. The arguments used to obtain (4.5.32) and other Stanley decompositions of both rings and modules in this chapter are special cases of general arguments occurring in the theory of *Gröbner bases*, introduced in Section A.5. In general, given a module of syzygies or an ideal of relations (see Remark 4.5.15), it is necessary to construct a Gröbner basis before obtaining a Stanley decomposition. When the module of syzygies (or ideal of relations) has a single generator, the generator already constitutes a Gröbner basis, and it is often possible to obtain the Stanley decomposition in an ad hoc manner. A characteristic feature of a Gröbner basis is the important role played by the *leading terms* of the Gröbner basis elements. The two Stanley decompositions mentioned above result from different choices of the "leading" term in (4.5.30).

The Double Center

Any system of the form (4.5.1) with $x \in \mathbb{R}^4$ and

$$A = \begin{bmatrix} 0 & -\omega_1 & & \\ \omega_1 & 0 & & \\ & & 0 & -\omega_2 \\ & & \omega_2 & 0 \end{bmatrix} \qquad (4.5.33)$$

is referred to as a *double center*; this is a natural generalization of the *single center* studied in Section 1.1, and the first of a larger class of *multiple centers*, in which x has even dimension and A has additional 2×2 blocks of the same form as in (4.5.33). Our goal in the next few subsections is to solve the description problem for the normal form of the double center. It will be convenient in parts of the following discussion to assume that the vector field a in (4.5.1) is a *polynomial* vector field rather than a formal power series, so that we can speak of actual vectors $a(x)$ attached at points $x \in \mathbb{R}^4$; although this facilitates visualization, all of the results are valid for formal power series as well. We will use the following notation:

1. $\mathcal{V}$ denotes $\mathcal{V}_*^4$, the space of real polynomial vector fields on $\mathbb{R}^4$.

2. $\mathcal{F}$ denotes $\mathcal{F}_*^4$, the space of real-valued polynomial functions on $\mathbb{R}^4$.

3. $\mathcal{L}$ denotes L_A, regarded as an operator on $\mathcal{V}$.

4. $\mathcal{D}$ denotes $\mathcal{D}_A$, acting on $\mathcal{F}$.

The normal form module that we wish to describe is $\ker \mathcal{L}$, regarded as a module over $\ker \mathcal{D}$; in words, this will be called the module of *real equivariants* over the ring of *real invariants*. In the course of the following discussion it will be convenient to introduce two additional sets of (complex) spaces and operators, denoted by $\mathcal{V}', \mathcal{F}', \mathcal{L}', \mathcal{D}'$ and $\mathcal{V}'', \mathcal{F}'', \mathcal{L}'', \mathcal{D}''$, and two additional (complex) modules, $\ker \mathcal{L}'$ (a module over $\ker \mathcal{D}'$) and $\ker \mathcal{L}''$ (a module over $\ker \mathcal{D}''$). These will be defined when they arise.

As in (3.3.5), the matrices

$$T = \frac{1}{2} \begin{bmatrix} 1 & 1 & & \\ -i & i & & \\ & & 1 & 1 \\ & & -i & i \end{bmatrix}, \qquad T^{-1} = \begin{bmatrix} 1 & i & & \\ 1 & -i & & \\ & & 1 & i \\ & & 1 & -i \end{bmatrix} \qquad (4.5.34)$$

diagonalize A:

$$T^{-1}AT = \Lambda = \begin{bmatrix} i\omega_1 & & & \\ & -i\omega_1 & & \\ & & i\omega_2 & \\ & & & -i\omega_2 \end{bmatrix}. \qquad (4.5.35)$$

For our present purposes, it is best to regard the matrix T^{-1} as defining a map $T^{-1} : \mathbb{R}^4 \to \mathbb{C}^4$ by $\zeta = T^{-1}x$, with $x = (x_1, x_2, x_3, x_4)$ and $\zeta =$

(z_1, w_1, z_2, w_2); written out, this map appears as

$$z_1 = x_1 + ix_2, \qquad (4.5.36)$$
$$w_1 = x_1 - ix_2,$$
$$z_2 = x_3 + ix_4,$$
$$w_2 = x_3 - ix_4.$$

The image of this map is the *reality subspace*

$$\mathcal{R} = \{(z_1, w_1, z_2, w_2) \in \mathbb{C}^4 : w_1 = \overline{z}_1, w_2 = \overline{z}_2\}. \qquad (4.5.37)$$

Then the matrix T can be regarded as defining a map $T : \mathcal{R} \to \mathbb{R}^4$. (If T is applied to all of $\mathbb{C}^4$, it maps into $\mathbb{C}^4$ rather than $\mathbb{R}^4$.) Notice that $\mathcal{R}$ is not a complex subspace of $\mathbb{C}^4$ (that is, it is not closed under multiplication by complex numbers), but is a "real vector space" in the sense that (although its elements are complex) it is closed under multiplication by real numbers.

Now suppose that the change of coordinates $\zeta = T^{-1}x$ is applied to the differential equation (4.5.1), with A given by (4.5.33). The result is the system

$$\dot{\zeta} = T^{-1}a(T\zeta). \qquad (4.5.38)$$

At first sight, this is defined only on $\mathcal{R}$, and not on all of $\mathbb{C}^4$, because $a(T\zeta)$ is defined only if $T\zeta$ is real. But if we consider that $a(x)$ is expressed as a polynomial, it is actually possible to evaluate $a(T\zeta)$ for all ζ simply by multiplying out the polynomial expressions, as one might do without thinking. (In other words, the vector field $a(T\zeta)$ defined on $\mathcal{R}$ has a unique extension to $\mathbb{C}^4$ if we insist that the extension be a complex polynomial vector field.) The resulting system can be written out as

$$\dot{z}_1 = i\omega_1 z_1 + f_1(z_1, w_1, z_2, w_2), \qquad (4.5.39)$$
$$\dot{w}_1 = -i\omega_1 w_1 + g_1(z_1, w_1, z_2, w_2),$$
$$\dot{z}_2 = i\omega_2 z_2 + f_2(z_1, w_1, z_2, w_2),$$
$$\dot{w}_2 = -i\omega_2 w_2 + g_2(z_1, w_1, z_2, w_2).$$

The actual expressions for f_1, g_1, f_2, g_2 in terms of $a(x)$ can be rather complicated. Each monomial $x^m e_s$ in $a(x)$ (we avoid writing e_i because in the present context $i = \sqrt{-1}$) can give rise to several monomials, appearing in different places in (4.5.39). For instance, a monomial $x^m e_1$ will contribute to f_1 and g_1, and the number of terms produced will depend on m. However, each of these terms will have the same degree as the original term from which it came.

At this point, it is possible to apply normal form theory directly to the system (4.5.39). To do so, we define the following spaces and operators:

1. $\mathcal{V}'$ denotes the space of complex polynomial vector fields $v : \mathbb{C}^4 \to \mathbb{C}^4$.

2. $\mathcal{F}'$ denotes the space of complex-valued polynomial functions $f : \mathbb{C}^4 \to \mathbb{C}$.

3. $\mathcal{L}'$ denotes L_Λ, regarded as an operator on $\mathcal{V}'$. (Note that the matrix of the linear part of (4.5.39) is Λ, defined in (4.5.35).

4. $\mathcal{D}'$ denotes $\mathcal{D}_\Lambda$, acting on $\mathcal{F}'$.

Notice that all of these spaces are complex. The associated normal form space is the module $\ker \mathcal{L}'$ of *complex equivariants* over the ring $\ker \mathcal{D}'$ of *complex invariants*.

However, although normal form theory can be applied in this way to the complex system, doing so does not give exactly the result that we want for the original real system. The difficulty is that if we normalize the system carelessly, we may produce a complex system that does not transform back into a real system when the transformation from x to ζ is reversed. In fact, since $a(x)$ is a *real* vector field, (4.5.39) is not an arbitrary complex vector field on $\mathbb{C}^4$, but belongs to a restricted class of such vector fields. One way to see this is that since $\mathbb{R}^4$ is invariant under the flow of (4.5.1), and $\mathbb{R}^4$ maps to $\mathcal{R} \subset \mathbb{C}^4$, $\mathcal{R}$ must be invariant under the flow of (4.5.39); in other words, the vectors of this vector field, when evaluated at points of $\mathcal{R}$, must be tangent to $\mathcal{R}$. Analytically, the condition is

$$g_s(\zeta) = \overline{f}_s(\zeta) \quad \text{for} \quad \zeta \in \mathcal{R}. \qquad (4.5.40)$$

To see this, observe that the right-hand side of (4.5.38) can be written out as $T^{-1}a = (a_1 + ia_2, a_1 - ia_2, a_3 + ia_4, a_3 - ia_4)$, and the a_s are real on $\mathcal{R}$. Conversely, any complex vector field on $\mathbb{C}^4$ satisfying the condition (4.5.40), when restricted to $\mathcal{R}$, can be transformed back into a real vector field in the variable x. For this reason (4.5.40) is known as the *reality condition* for complex vector fields. We define the following spaces and operators:

1. $\mathcal{V}''$ denotes the space of complex polynomial vector fields on $\mathbb{C}^4$ that satisfy the reality condition (4.5.40).

2. $\mathcal{F}''$ denotes the space of complex-valued polynomial functions on $\mathbb{C}^4$ that take real values on $\mathcal{R}$. (This is the natural "reality condition" for scalar fields, since a scalar field from $\mathcal{F}''$ multiplied by a vector field from $\mathcal{V}''$ gives a vector field in $\mathcal{V}''$; compare the fact that $\mathcal{R}$ is a vector space over $\mathbb{R}$ but not $\mathbb{C}$.)

3. $\mathcal{L}''$ denotes L_Λ, regarded as an operator on $\mathcal{V}''$. Thus, $\mathcal{L}''$ is the restriction of $\mathcal{L}'$ to the subspace $\mathcal{V}'' \subset \mathcal{V}$.

4. $\mathcal{D}''$ denotes $\mathcal{D}_\Lambda$, acting on $\mathcal{F}''$; it is a restriction of $\mathcal{D}'$.

Now, at last, we have found the correct setting in which to compute the normal form module that we want, which is $\ker \mathcal{L}''$ regarded as a module over $\ker \mathcal{D}''$. This module will be isomorphic to the purely real module $\ker \mathcal{L}$ over $\ker \mathcal{D}$, and will transform into this module under the coordinate change $x = T\zeta$.

4.5.13. Remark. It is common to write (4.5.39) in the compressed form

$$\dot{z}_1 = i\omega_1 z_1 + f_1(z_1, \overline{z}_1, z_2, \overline{z}_2), \qquad (4.5.41)$$
$$\dot{z}_2 = i\omega_2 z_2 + f_2(z_1, \overline{z}_1, z_2, \overline{z}_2).$$

Since w_s has been replaced by $\overline{z}_s$, these expressions are valid only on $\mathcal{R}$. We will not follow this practice here, because the version of normal form theory that we have developed cannot be applied directly to (4.5.41). It is possible to develop a version suited to this abridged notation, but we prefer to apply the machinery developed already to the full system (4.5.39). The discussion that follows is somewhat lengthy, but at the end it will be seen how to obtain the final result just as quickly as if we had used (4.5.41). See also Remark 4.5.14.

Since the linear part Λ of (4.5.39) is diagonal, the vector field (f_1, g_1, f_2, g_2) is equivariant if and only if it contains only equivariant monomials. It is convenient to write vector monomials in $\zeta = (z_1, w_1, z_2, w_2)$ in the form $z_1^{m_1} w_1^{n_1} z_2^{m_2} w_2^{n_2} e_s$. The condition for this monomial to be equivariant (which in the usual notation would be $\langle m, \lambda \rangle = \lambda_s$) is then

$$m_1(i\omega_1) + n_1(-i\omega_1) + m_2(i\omega_2) + n_2(-i\omega_2) = \begin{cases} +i\omega_1 & \text{if } s = 1, \\ -i\omega_1 & \text{if } s = 2, \\ +i\omega_2 & \text{if } s = 3, \\ -i\omega_2 & \text{if } s = 4. \end{cases}$$
$$(4.5.42)$$

In other words, the monomial $z_1^{m_1} w_1^{n_1} z_2^{m_2} w_2^{n_2}$ can appear in (f_1, g_1, f_2, g_2) only if

$$(m_1 - n_1)\omega_1 + (m_2 - n_2)\omega_2 = \begin{cases} \omega_1 & \text{for } f_1, \\ -\omega_1 & \text{for } g_1, \\ \omega_2 & \text{for } f_2, \\ -\omega_2 & \text{for } g_2. \end{cases} \qquad (4.5.43)$$

When a vector field is known to satisfy the reality condition (4.5.40), the second and fourth of these equations follow from the first and third and can be omitted.

If it is required to describe the normal form space only up to some finite degree, without finding the module structure, these equations are sufficient. We will illustrate this by determining the cubic normal form for the 1 : 2 resonance. Then we will turn to the harder problem of describing the normal form to all orders by finding its module structure, first in the nonresonant case and then for the general $p : q$ resonance.

The Cubic Normal Form for the $1:2$ *Resonance*

The $1:2$ *resonant double center* is the case in which $\frac{\omega_1}{\omega_2} = \frac{1}{2}$, or

$$\omega_1 = \omega, \qquad \omega_2 = 2\omega. \tag{4.5.44}$$

(See Remark 4.5.10 for the word *resonance*.) For this example, we will work carefully through the determination of the quadratic and cubic parts of ker $\mathcal{L}'$, ker $\mathcal{L}''$, and ker $\mathcal{L}$ (as defined above). This is rather tedious, and once the ideas are understood it is possible to reach the final result much more quickly by skipping some of the steps. But we are aiming here at complete conceptual clarity rather than speed.

For the $1:2$ resonance the first and third lines of (4.5.43) become

$$(m_1 - n_1) + 2(m_2 - n_2) = \begin{cases} 1 & \text{for } f_1, \\ 2 & \text{for } f_2. \end{cases} \tag{4.5.45}$$

By making a table of (m_1, n_1, m_2, n_2) totalling to 2 or 3 (for quadratic and cubic terms), it can be seen that the solutions for f_1 are $(m_1, n_1, m_2, n_2) = (0, 1, 1, 0)$, $(2, 1, 0, 0)$, and $(1, 0, 1, 1)$. For f_2 they are $(2, 0, 0, 0)$, $(1, 1, 1, 0)$, $(0, 0, 2, 1)$. Thus, the equivariant monomials of the form $(f_1, 0, 0, 0)$ and $(0, 0, f_2, 0)$, up to cubic terms, can be listed as follows:

$$(w_1 z_2, 0, 0, 0), \qquad (z_1^2 w_1, 0, 0, 0), \qquad (z_1 z_2 w_2, 0, 0, 0), \tag{4.5.46}$$
$$(0, 0, z_1^2, 0), \qquad (0, 0, z_1 w_1 z_2, 0), \qquad (0, 0, z_2^2 w_2, 0).$$

Six more equivariant monomials of the form $(0, g_1, 0, 0)$ and $(0, 0, 0, g_2)$ can be obtained from the second and fourth lines of (4.5.43), but these will not be needed in the sequel. Together, these twelve vector monomials form a basis (over $\mathbb{C}$) for the vector space of complex equivariants of degrees 2 and 3, that is, for the quadratic and cubic part of ker $\mathcal{L}'$. Our goal is to use this information to determine, first the quadratic and cubic part of ker $\mathcal{L}''$ (that is, the complex quadratic and cubic equivariants that satisfy the reality condition), and finally the quadratic and cubic part of ker $\mathcal{L}$ (the real equivariants).

Just as the reality subspace $\mathcal{R}$ is a *real* subspace of the complex vector space $\mathbb{C}^4$, so ker $\mathcal{L}''$ is a real subspace of ker $\mathcal{L}'$; that is, it is a subspace of ker $\mathcal{L}'$ when ker $\mathcal{L}'$ is *viewed as a vector space over* $\mathbb{R}$. Viewed in this way, the cubic and quadratic part of ker $\mathcal{L}'$ has dimension 24 rather than 12, and has a basis (over $\mathbb{R}$) consisting of the 12 equivariants mentioned in the last paragraph together with these same equivariants multiplied by i. In particular, to the 6 equivariants listed in (4.5.46) should be added 6 more, namely,

$$(iw_1 z_2, 0, 0, 0), \qquad (iz_1^2 w_1, 0, 0, 0), \qquad (iz_1 z_2 w_2, 0, 0, 0), \tag{4.5.47}$$
$$(0, 0, iz_1^2, 0), \qquad (0, 0, iz_1 w_1 z_2, 0), \qquad (0, 0, iz_2^2 w_2, 0),$$

and there will be twelve others involving g_1 and g_2. We do not need these, because for equivariants satisfying the reality condition, g_1 and g_2 are determined by (4.5.40). Therefore, we can determine a basis (over $\mathbb{R}$) for ker $\mathcal{L}''$ by "completing" each of the vector fields in (4.5.46) and (4.5.47) so that each of them satisfies (4.5.40). For instance, if $f_1 = w_1 z_2$, then g_1 must be a polynomial in (z_1, w_1, z_2, w_2) that equals $\overline{w}_1 \overline{z}_2$ on $\mathcal{R}$; but on $\mathcal{R}$, $\overline{w}_1 = z_1$ and $\overline{z}_2 = w_2$; so g_1 should be $z_1 w_2$. Thus, we obtain

$$(w_1 z_2, z_1 w_2, 0, 0), \qquad (z_1^2 w_1, w_1^2 z_1, 0, 0), \qquad (z_1 z_2 w_2, w_1 w_2 z_2, 0, 0),$$
$$(0, 0, z_1^2, w_1^2), \qquad (0, 0, z_1 w_1 z_2, w_1 z_1 w_2), \qquad (0, 0, z_2^2 w_2, w_2^2 z_2),$$
$$(i w_1 z_2, -i z_1 w_2, 0, 0), \quad (i z_1^2 w_1, -i w_1^2 z_1, 0, 0), \quad (i z_1 z_2 w_2, -i w_1 w_2 z_2, 0, 0),$$
$$(0, 0, i z_1^2, -i w_1^2), \qquad (0, 0, i z_1 w_1 z_2, -i w_1 z_1 w_2), \quad (0, 0, i z_2^2 w_2, -i w_2^2 z_2).$$
$$(4.5.48)$$

The normal form of (4.5.37) through cubic terms in the $1:2$ resonance case is now obtained by taking for (f_1, g_1, f_2, g_2) the real linear combinations of the 12 vector fields listed in (4.5.48).

Since this normal form satisfies the reality condition (4.5.40), it follows that when restricted to $\mathcal{R}$, the equations for w_1 and w_2 are the complex conjugates of the equations for z_1 and z_2. This can also be seen from (4.5.48): For instance, a linear combination, with real coefficients α and β, of the first and seventh of these vector fields gives

$$\alpha \begin{bmatrix} w_1 z_1 \\ z_1 w_2 \\ 0 \\ 0 \end{bmatrix} + \beta \begin{bmatrix} i w_1 z_2 \\ -i z_1 w_2 \\ 0 \\ 0 \end{bmatrix} = \begin{bmatrix} (\alpha + i\beta) w_1 z_2 \\ (\alpha - i\beta) z_1 w_2 \\ 0 \\ 0 \end{bmatrix},$$

and on $\mathcal{R}$ this becomes $(a \overline{z}_1 z_1, \overline{a z}_1 z_1, 0, 0)$. Notice that *as far as the f_1 and f_2 terms are concerned*, the *real* linear combinations of (4.5.48) coincide with the *complex* linear combinations of (4.5.46), although this is not true for the g_1 and g_2 terms. It follows that through cubic terms, the normal form for the z_1 and z_2 equations of (4.5.37), restricted to $\mathcal{R}$, is

$$\dot{z}_1 = i\omega_1 z_1 + a_1 \overline{z}_1 z_2 + a_2 z_1^2 \overline{z}_1 + a_3 z_1 z_2 \overline{z}_2, \qquad (4.5.49)$$
$$\dot{z}_2 = i\omega_2 z_2 + a_4 z_1^2 + a_5 z_1 \overline{z}_1 z_2 + a_6 z_2^2 \overline{z}_2, \;.$$

where the a_j are complex constants. As we will see in a moment, these equations contain all the information required to write down the real normal form (or equivalently, ker $\mathcal{L}$) quite easily, either in the original variables (x_1, x_2, x_3, x_4) or in polar coordinates $(r_1, \theta_1, r_2, \theta_2)$. Having worked through this painstaking derivation, we now see clearly that (4.5.49) can be obtained immediately from (4.5.45): One simply solves this to obtain (4.5.46), then takes complex linear combinations to get (4.5.49), always writing $\overline{z}_i$ in place of w_i and ignoring g_1 and g_2 altogether.

4.5.14. Remark. Equation (4.5.49) is the same result that would be obtained if we had developed the normal form theory suited to the notation of (4.5.39). In other words, we adopt this notation *after* normalizing, even though in Remark 4.5.13 we declined to adopt it *before* normalizing.

To obtain the real normal form from (4.5.49), it is necessary only to substitute $z_1 = x_1 + ix_2$, $z_2 = x_3 + ix_4$ (from (4.5.36)), set $a_s = \alpha_s + i\beta_s$ (with α_s and β_s real), multiply out, and separate real and imaginary parts. For instance, taking only the portion $\dot{z}_1 = i\omega_1 z_1 + a_1 z_1 z_2$, we have $\dot{x}_1 + i\dot{x}_2 = i\omega_1(x_1 + ix_2) + (\alpha_1 + i\beta_1)(x_1 - ix_2)(x_3 + ix_4)$, which separates into

$$\dot{x}_1 = -\omega_1 x_2 + \alpha_1(x_1 x_3 + x_2 x_4) - \beta_1(x_1 x_4 - x_2 x_3),$$
$$\dot{x}_2 = +\omega_1 x_1 + \alpha_1(x_1 x_4 - x_2 x_3) + \beta_1(x_1 x_3 + x_2 x_4).$$

To obtain the polar form, write

$$z_1 = r_1 e^{i\theta_1}, \quad \bar{z}_1 = r_1 e^{-i\theta_1}, \quad z_2 = r_2 e^{i\theta_2}, \quad \bar{z}_2 = r_2 e^{-i\theta_2}. \quad (4.5.50)$$

Since $\dot{z}_1 = \dot{r}_1 e^{i\theta_1} + ir_1\dot{\theta}_1 e^{i\theta_1}$, and similarly for $\dot{z}_2$, (4.5.49) becomes

$$\dot{r}_1 e^{i\theta_1} + ir_1\dot{\theta}_1 e^{i\theta_1} = i\omega r_1 e^{i\theta_1} + a_1 r_1 r_2 e^{i(-\theta_1+\theta_2)} + a_2 r_1^3 e^{i\theta_1} + a_3 r_1 r_2^2 e^{i\theta_1},$$
$$(4.5.51)$$
$$\dot{r}_2 e^{i\theta_2} + ir_2\dot{\theta}_2 e^{i\theta_2} = 2i\omega r_2 e^{i\theta_2} + a_4 r_1^2 e^{2i\theta_1} + a_5 r_1^2 r_2 e^{i\theta_2} + a_6 r_2^3 e^{i\theta_2}.$$

A complex constant times a polar complex number produces real and imaginary parts that are linear combinations of sines and cosines:

$$(a + ib)\rho e^{i\varphi} = \rho(a\cos\varphi - b\sin\varphi) + i(b\cos\varphi + a\sin\varphi).$$

Once the products in (4.5.51) have been computed in terms of the real and imaginary parts of the arbitrary complex constants $a_1, \ldots, a_6$, these may be replaced by twelve arbitrary real constants $c_1, \ldots, c_{12}$ to fill the appropriate positions. Separating the real and imaginary parts of (4.5.51) then gives

$$\dot{r}_1 = [c_1 \sin(2\theta_1 - \theta_2) + c_2 \cos(2\theta_1 - \theta_2)]r_1 r_2 + c_3 r_1^3 + c_4 r_1 r_2^2, \quad (4.5.52)$$
$$\dot{r}_2 = [c_5 \sin(2\theta_1 - \theta_2) + c_6 \cos(2\theta_1 - \theta_2)]r_1^2 + c_7 r_1^2 r_2 + c_8 r_2^3,$$
$$\dot{\theta}_1 = \omega + [c_1 \sin(2\theta_1 - \theta_2) - c_2 \cos(2\theta_1 - \theta_2)]r_2 + c_9 r_1^2 + c_{10} r_2^2,$$
$$\dot{\theta}_2 = 2\omega + [c_5 \sin(2\theta_1 - \theta_2) + c_6 \cos(2\theta_1 - \theta_2)]\frac{r_1^2}{r_2} + c_{11} r_1^2 + c_{12} r_2^2.$$

Although the θ_2 equation seems to be singular at the origin (because of the coefficient r_1^2/r_2), it must be remembered that polar coordinates are not valid at the origin. In Cartesian coordinates, the system is regular.

Invariants and Equivariants in the Nonresonant Case

As the example of $1:2$ resonance shows, solving the description problem for the double center to any finite order is simply a computational task. A more serious question is to solve the description problem to all orders by identifying the ring of invariants, and the module of equivariants over this ring, by means of generators and relations. There are two cases according to whether ω_1/ω_2 is irrational (the *nonresonant case*) or rational (the *resonant case*). For each case we must consider the following items, using notations introduced earlier:

1. The ring $\ker \mathcal{D}'$ of complex invariants. This is the set of monomials $z_1^{m_1} w_1^{n_1} z_2^{m_2} w_2^{n_2}$ such that

$$(m_1 - n_1)\omega_1 + (m_2 - n_2)\omega_2 = 0. \qquad (4.5.53)$$

2. The ring $\ker \mathcal{D}''$ of complex invariants satisfying the reality condition. Recall that for scalar fields the reality condition is that they take real values on $\mathcal{R}$.

3. The ring $\ker \mathcal{D}$ of real invariants, which is isomorphic to $\ker \mathcal{D}''$.

4. The module $\ker \mathcal{L}'$ of complex equivariants (f_1, g_1, f_2, g_2) over $\ker \mathcal{D}'$. But as we have learned by studying the $1:2$ resonance, it is not actually necessary to determine this module; it is enough to determine the set of (f_1, f_2) satisfying

$$(m_1 - n_1)\omega_1 + (m_2 - n_2)\omega_2 = \begin{cases} \omega_1 & \text{for } f_1, \\ \omega_2 & \text{for } f_2. \end{cases} \qquad (4.5.54)$$

The set of such (f_1, f_2) also forms a module over $\ker \mathcal{D}'$, which we call the module of *restricted complex equivariants*.

5. The module $\ker \mathcal{L}''$ of complex equivariants (f_1, g_1, f_2, g_2) satisfying the reality condition, which for vector fields is (4.5.40); this is a module over the ring $\ker \mathcal{D}''$. It is easy to determine this module from the module of restricted complex invariants; they are almost the same, after changing the ring from $\ker \mathcal{D}'$ to $\ker \mathcal{D}''$. Let (f_1, f_2) be a generator (over $\ker \mathcal{D}'$) for the module of restricted complex equivariants; this generator must be replaced by the two generators (f_1, f_2) and (if_1, if_2) if the ring is changed to $\ker \mathcal{D}''$. (This is exactly parallel to adjoining the basis elements (4.5.47) to (4.5.46), except that now we are working with modules rather than vector spaces.) Each of these generators could then be "completed" to give generators $(f_1, *, f_2, *)$ and $(if_1, *, if_2, *)$ for $\ker \mathcal{L}''$ over $\ker \mathcal{D}''$, where the entries denoted by "$*$" are determined by (4.5.40); however, it is not necessary to do this calculation, because (f_1, f_2) and (if_1, if_2) already contain the information needed for the next step.

6. The module ker $\mathcal{L}$ of real equivariants over the real invariants ker $\mathcal{D}$. The generators of this module are again obtained from the generators (f_1, f_2) of the module of restricted complex equivariants: For each such generator, we convert the two associated generators (f_1, f_2) and (if_1, if_2) into real vector fields, either in Cartesian or polar coordinates, exactly as we did for the 1 : 2 resonance. This will be worked out in detail below.

Suppose first that ω_1/ω_2 is irrational. Then the only solutions of (4.5.53) have $m_1 = n_1$ and $m_2 = n_2$, so the complex invariant monomials have the form

$$(z_1 w_1)^{m_1}(z_2 w_2)^{m_2}. \tag{4.5.55}$$

These automatically satisfy the reality condition (for scalars), since on $\mathcal{R}$, $z_s w_s = z_s \bar{z}_s = |z_s|^2 \in \mathbb{R}$. Therefore, ker $\mathcal{D}'$ is spanned over $\mathbb{C}$ by the monomials (4.5.55), while ker $\mathcal{D}''$ is spanned over $\mathbb{R}$ by the same monomials. We need no longer distinguish between ker $\mathcal{D}''$ and ker $\mathcal{D}$, and will refer to both as *real invariants*. The generating monomials, also called the *basic invariants* of the nonresonant double center, may be regarded as defined either on $\mathcal{R}$ or on $\mathbb{R}^4$, and may be written in complex, real, and polar coordinates as

$$I_1 = z_1 \bar{z}_1 = x_1^2 + x_2^2 = r_1^2, \tag{4.5.56}$$
$$I_2 = z_2 \bar{z}_2 = x_3^2 + x_4^2 = r_2^2.$$

Any real invariant can be written as

$$I = \varphi(I_1, I_2) = \varphi(z_1 \bar{z}_1, z_2 \bar{z}_2) = \varphi(x_1^2 + x_2^2, x_3^2 + x_4^2) = \varphi(r_1^2, r_2^2), \tag{4.5.57}$$

where φ is a polynomial (or, for formal invariants, a formal power series) in two variables *with real coefficients*. The invariants I_1 and I_2 are algebraically independent, and the expression of any invariant in the form (4.5.57) is unique, so we have now completely determined the invariant ring in the nonresonant case.

Continuing with the nonresonant case, the only solutions of the equivariance condition (4.5.54) have $m_1 = n_1 + 1$ and $m_2 = n_2$ (for f_1), or $m_1 = n_1$ and $m_2 = n_2 + 1$ (for f_2). Thus, f_1 can contain only monomials of the form

$$z_1(z_1 w_1)^{n_1}(z_2 w_2)^{n_2} = Iz_1,$$

where I is an invariant monomial, and f_2 can contain monomials of the form Iz_2. So there are two pairs (f_1, f_2), namely,

$$(f_1, f_2) = (z_1, 0), \ (0, z_2),$$

that serve to generate the module of restricted complex equivariants over the complex invariants. When these are subjected to the "doubling process" described in items 5 and 6 above, we obtain four generators $(z_1, 0)$, $(iz_1, 0)$,

$(0, z_2)$, and $(0, iz_2)$, which give rise to the following four real vector fields, the *basic equivariants* for the nonresonant double center:

$$v_1 = (x_1, x_2, 0, 0), \qquad (4.5.58)$$
$$v_2 = (-x_2, x_1, 0, 0),$$
$$v_3 = (0, 0, x_3, x_4),$$
$$v_4 = (0, 0, -x_4, x_3).$$

These are linear vector fields corresponding to radial expansion and rotation in the x_1, x_2 and x_3, x_4 coordinate planes, exactly the result that should be expected considering the corresponding result for the single center obtained in Section 1.1. The module $\mathcal{N}$ of real equivariants is generated by (4.5.58) over the ring of real invariants (4.5.57) with no syzygies. That is, the nonresonant double center in real normal form to all orders appears (uniquely) as

$$\dot{x} = Ax + \varphi_1(I_1, I_2)v_1 + \varphi_2(I_1, I_2)v_2 + \varphi_3(I_1, I_2)v_3 + \varphi_4(I_1, I_2)v_4,$$
$$(4.5.59)$$

where A is given by (4.5.33) and the φ_s are formal power series in I_1 and I_2 beginning with linear terms (because only $\mathcal{N}_+$ is used in the normal form; that is, the terms after Ax must be at least quadratic).

In polar coordinates, the basic equivariants, written as $v = (\dot{r}_1, \dot{\theta}_1, \dot{r}_2, \dot{\theta}_2)$, are

$$v_1 = (r_1, 0, 0, 0), \qquad (4.5.60)$$
$$v_2 = (0, 1, 0, 0),$$
$$v_3 = (0, 0, r_2, 0),$$
$$v_4 = (0, 0, 0, 1).$$

The differential equations in normal form then appear as

$$\dot{r}_1 = \varphi_1(r_1^2, r_2^2)r_1, \qquad (4.5.61)$$
$$\dot{\theta}_1 = \omega_1 + \varphi_2(r_1^2, r_2^2),$$
$$\dot{r}_2 = \varphi_3(r_1^2, r_2^2)r_2,$$
$$\dot{\theta}_2 = \omega_2 + \varphi_4(r_1^2, r_2^2),$$

where once again the φ_s begin with linear terms in their arguments (hence quadratic terms in r_1 and r_2). This may be compared with (1.1.25); notice that the powers of r_s are odd in the radial equations, even in the angular equations. As in Chapter 1, the stability of the origin can be determined from these equations unless the formal power series φ_1 and φ_3 are zero. If both $\dot{r}_1$ and $\dot{r}_2$ are negative to the lowest nonvanishing order, the origin is stable; otherwise (unless both vanish to all orders), it is unstable. The case $\varphi_1 = \varphi_3 = 0$ is undecided because the original smooth system (before passing to formal power series) may contain flat functions; see Remark 1.1.7.

Invariants and Equivariants in the Resonant Case

Turning to the resonant case, assume that

$$\frac{\omega_1}{\omega_2} = \frac{p}{q} \tag{4.5.62}$$

in lowest terms. In this case the invariance condition (4.5.53) can be replaced by

$$(m_1 - n_1)\frac{p}{q} + (m_2 - n_2) = 0, \tag{4.5.63}$$

with solutions

$$m_1 - n_1 = kq, \qquad m_2 - n_2 = -kp \tag{4.5.64}$$

for integers k, subject to the condition that the m_s and n_s are nonnegative. For simplicity we will assume that p and q are positive; this involves no loss of generality, since in the present context (as opposed to the Hamiltonian case treated in Section 4.9) it is always possible to change coordinates so that ω_1 and ω_2 are positive, as in Remark 1.1.1. Then (4.5.64) may be split into two sets of equations involving nonnegative terms only. For $k \geq 0$,

$$m_1 = n_1 + kq, \qquad n_2 = m_2 + kp,$$

with n_1, m_2, and k arbitrary (but nonnegative), leading to invariant monomials of the form

$$z_1^{n_1+kq} w_1^{n_1} z_2^{m_2} w_2^{m_2+kp} = (z_1^q w_2^p)^k (z_1 w_1)^{n_1} (z_2 w_2)^{m_2}.$$

For $k < 0$, write $k = -\ell$ with $\ell > 0$; then

$$n_1 = m_1 + \ell q, \qquad m_2 = n_2 + \ell p,$$

giving

$$z_1^{m_1} w_1^{m_1+\ell q} z_2^{n_2+\ell p} w_2^{n_2} = (w_1^q z_2^p)^\ell (z_1 w_1)^{m_1} (z_2 w_2)^{n_2}.$$

The collection of (infinite) linear combinations of these monomials can be described as the ring of formal power series in the *basic invariants*

$$I_1 = z_1 w_1, \quad I_2 = z_2 w_2, \quad \widetilde{I}_3 = z_1^q w_2^p, \quad \widetilde{I}_4 = w_1^q z_2^p. \tag{4.5.65}$$

Here I_1 and I_2 are real (on $\mathcal{R}$) and coincide with the invariants I_1 and I_2 that were discovered previously in the nonresonant case. The invariants $\widetilde{I}_3$ and $\widetilde{I}_4$ are complex, and will eventually be replaced by real invariants denoted I_3 and I_4. However, we will temporarily study the full complex ring of invariants, and impose the reality condition afterward.

The expression of invariants as formal power series in the quantities (4.5.65) is not unique, because the basic invariants are not algebraically independent, but satisfy the relation

$$\widetilde{I}_3 \widetilde{I}_4 = I_1^q I_2^p. \tag{4.5.66}$$

Given any formal power series in the basic invariants (4.5.65), this relation may be used to eliminate all monomials containing both $\widetilde{I}_3$ and $\widetilde{I}_4$, so that the terms remaining fall into two classes: monomials in I_1, I_2, and $\widetilde{I}_3$; and monomials in I_1, I_2, and $\widetilde{I}_4$ that contain at least one factor of $\widetilde{I}_4$. (The last condition is included so that any monomial containing only I_1 and I_2 will be placed into the first class. This is an arbitrary decision, but it is important to make a well-defined sorting of the terms.) Therefore, every invariant may be expressed *uniquely* in the form

$$\varphi\left(I_1, I_2, \widetilde{I}_3\right) + \psi\left(I_1, I_2, \widetilde{I}_4\right)\widetilde{I}_4, \tag{4.5.67}$$

where φ and ψ are (complex) formal power series in their arguments. That is to say, the complex ring of invariants is

$$\ker \mathcal{D}_\Lambda = \mathbb{C}\left[\left[I_1, I_2, \widetilde{I}_3\right]\right] \oplus \mathbb{C}\left[\left[I_1, I_2, \widetilde{I}_4\right]\right]\widetilde{I}_4. \tag{4.5.68}$$

Either (4.5.67) or (4.5.68) is called a *Stanley decomposition* of the (complex) ring $\ker \mathcal{D}_\Lambda$.

4.5.15. Remark. The *relation* (4.5.66) is different in nature from a *syzygy*, such as (4.5.30), defined in Remark 4.5.11; one is a relation among elements of the ring of invariants, the other among elements of the module of equivariants. (In spite of the distinction, it is not uncommon to find relations referred to as syzygies, and it seems to be traditional to do so in invariant theory. We will adhere to the distinction.) The technical definition of relations among generators of a ring is as follows. Let $p_1, \ldots, p_r$ be polynomials in $\mathbb{R}[x_1, \ldots, x_n]$, and consider the ring $\mathcal{R} = \mathbb{R}[[p_1, \ldots, p_r]]$ of formal power series in these polynomials. We will refer to $p_1, \ldots, p_r$ as *algebraic generators* of $\mathcal{R}$. (Although we are interested in formal power series, we will never need anything but polynomials as generators.) These *algebraic generators* are to be distinguished from *module generators*, because powers and products of the p_i are formed in creating ring elements; no powers or products are taken in creating module elements from their generators, only linear combinations using coefficients from the ring. Consider also the ring $\mathcal{R}' = \mathbb{R}[[t_1, \ldots, t_r]]$ of formal power series in independent variables $t_1, \ldots, t_r$. Define a ring homomorphism $h : \mathcal{R}' \to \mathcal{R}$ defined by $h(f(t_1, \ldots, t_r)) = f(p_1, \ldots, p_r)$. The kernel of h is the *ideal of relations* satisfied by $p_1, \ldots, p_r$. As in Remark 4.5.15, a "relation" is an *element* of one set ($\mathcal{R}'$), corresponding to an *equation* that holds in a different set ($\mathcal{R}$). In case the kernel contains only zero, the elements $p_1, \ldots, p_r$ are called *algebraically independent*. In our case, the kernel (or ideal of relations) is the principal ideal generated by the single element $t_1 t_2 - t_3 t_4 = 0$, corresponding to (4.5.66).

In the language of Section A.5, this generator constitutes a Gröbner basis for the ideal, and the two classes of monomials described just before equation (4.5.74) make up the standard monomials with respect to this ideal. For further discussion, including the reason why Gröbner basis arguments intended for rings of polynomials apply also to formal power series rings in the present situation, see Sections A.5 and A.6.

To find the real basic invariants I_3 and I_4 mentioned above, it suffices to take the real and imaginary parts of

$$\tilde{I}_3 = z_1^q \bar{z}_2^p = (x_1 + ix_2)^q (x_3 - ix_4)^p = \left(r_1 e^{i\theta_1}\right)^q \left(r_2 e^{-i\theta_2}\right)^p, \qquad (4.5.69)$$

which are the same (up to sign) as the real and imaginary parts of $\tilde{I}_4$ on $\mathcal{R}$. It is awkward to write these out in the x coordinates, but it is simple to express them in polar coordinates:

$$I_3 = r_1^q r_2^p \cos \Phi, \qquad (4.5.70)$$
$$I_4 = r_1^q r_2^p \sin \Phi,$$

with

$$\Phi = q\theta_1 - p\theta_2. \qquad (4.5.71)$$

The relation satisfied by these invariants, in place of (4.5.66), is

$$I_3^2 + I_4^2 = r_1^{2q} r_2^{2p} = I_1^q I_2^p. \qquad (4.5.72)$$

Every (real formal power series) invariant can be expressed (nonuniquely) as a formal power series in $I_1, \ldots, I_4$. The relation (4.5.72) may then be used to eliminate all powers of I_4 (or, alternatively, of I_3), so that each invariant can be expressed uniquely as

$$\varphi(I_1, I_2, I_3) + \psi(I_1, I_2, I_3)I_4, \qquad (4.5.73)$$

where φ and ψ are real formal powers series in their arguments. That is, a Stanley decomposition of the real invariant ring $\ker \mathcal{D}_A$ can be written

$$\ker \mathcal{D}_A = \mathbb{R}[[I_1, I_2, I_3]] \oplus \mathbb{R}[[I_1, I_2, I_3]]I_4. \qquad (4.5.74)$$

4.5.16. Remark. Notice that (4.5.74) reveals that the ring $\ker \mathcal{D}_A$ may be viewed as a module over its subring $\mathbb{R}[[I_1, I_2, I_3]]$ with generators 1 and I_4. These are *module* generators; that is, no powers of I_4 are taken. Furthermore, 1 and I_4 satisfy no syzygies over $\mathbb{R}[[I_1, I_2, I_3]]$, although $I_1, \ldots, I_4$ do satisfy a relation. This may help to clarify the distinction between a syzygy and a relation.

The equivariants are determined by the condition (4.5.43). The first of these equations is equivalent (in the resonant case) to

$$(m_1 - n_1 - 1)p + (m_2 - n_2)q = 0,$$

with solutions

$$m_1 - n_1 - 1 = kq, \qquad m_2 - n_2 = -kp.$$

Since we are assuming $p > 0$ and $q > 0$, we have

$$m_1 = kq + n_1 + 1, \qquad n_2 = m_2 + kp$$

for arbitrary $n_1 \geq 0$, $m_2 \geq 0$, and $k \geq 0$, and also

$$n_1 = m_1 + \ell q - 1, \qquad m_2 = n_2 + \ell p$$

for arbitrary $m_1 \geq 0$, $n_2 \geq 0$, and $\ell = -k \geq 1$. These two cases lead to monomials of the forms

$$z_1^{kq+n_1+1} w_1^{n_1} z_2^{m_2} w_2^{m_2+kp} = (z_1 w_1)^{n_1} (z_2 w_2)^{m_2} (z_1^q w_2^p)^k z_1$$
$$= I_1^{n_1} I_2^{m_2} I_3^k z_1 = I z_1$$

and

$$z_1^{m_1} w_1^{m_1+\ell q-1} z_2^{n_2+\ell p} w_2^{n_2} = (z_1 w_1)^{m_1} (z_2 w_2)^{n_2} (w_1^q z_2^p)^{\ell-1} w_1^{q-1} z_2^p$$
$$= I w_1^{q-1} z_2^p,$$

where I denotes an invariant monomial. The expressions we have just calculated are those monomials that can appear in f_1 when (4.5.39) is in normal form. A similar analysis of the second equivariance condition shows that f_2 contains terms of the form $I z_2$ and $I z_1^q w_2^{p-1}$. Thus, the following four pairs $u = (f_1, f_2)$ generate the module of restricted complex equivariants over the complex invariants:

$$u_1 = (z_1, 0), \quad u_2 = (w_1^{q-1} z_2^p, 0), \quad u_3 = (0, z_2), \quad u_4 = (0, z_1^q w_2^{p-1}).$$
$$(4.5.75)$$

The following syzygies hold among the (f_1, f_2):

$$\tilde{I}_3 u_2 = I_1^{q-1} I_2^p u_1, \quad \tilde{I}_4 u_4 = I_1^q I_2^{p-1} u_3. \qquad (4.5.76)$$

Using (4.5.74), the Stanley decomposition of this module is found to be

$$(\mathbb{C}[[I_1, I_2, \tilde{I}_3]] \oplus \mathbb{C}[[I_1, I_2, \tilde{I}_4]] \tilde{I}_4) u_1 \oplus \mathbb{C}[[I_1, I_2, \tilde{I}_4]] u_2 \qquad (4.5.77)$$
$$\oplus (\mathbb{C}[[I_1, I_2, \tilde{I}_3]] \oplus \mathbb{C}[[I_1, I_2, \tilde{I}_4]] \tilde{I}_4) u_3 \oplus \mathbb{C}[[I_1, I_2, \tilde{I}_3]] u_3.$$

To obtain the real module, the four two-dimensional vector fields $u_1, \ldots, u_4$ should be subjected to the "doubling process" (items 5 and 6 of the list of steps given above) to obtain eight real vector fields that will generate the module of real equivariants over the ring of real invariants. This results in the four basic equivariants $v_1, \ldots, v_4$ of (4.5.60) found in the nonresonant case, together with the following four additional basic equivariants, with Φ as in (4.5.71):

$$v_5 = (r_1^{q-1} \cos \Phi, -r_1^{q-2} r_2^p \sin \Phi, 0, 0), \qquad (4.5.78)$$
$$v_6 = (r_1^{q-1} r_2^p \sin \Phi, r_1^{q-2} r_2^p \cos \Phi, 0, 0),$$
$$v_7 = (0, 0, r_1^q r_2^{p-1} \cos \Phi, r_1^q r_2^{p-2} \sin \Phi),$$
$$v_8 = (0, 0, -r_1^q r_2^{p-1} \sin \Phi, r_1^q r_2^{p-2} \cos \Phi).$$

As in the $1 : 2$ resonance, some of these appear to be singular if $p = 1$ or $q = 1$, but they are regular in rectangular coordinates. These equivariants are subject to the following syzygies:

$$I_3 v_5 + I_4 v_6 = I_1^{q-1} I_2^p v_1, \qquad (4.5.79)$$
$$I_3 v_7 - I_4 v_8 = I_1^q I_2^{p-1} v_3.$$

The Stanley decomposition resulting from this can be described in words as follows: All eight of the basic equivariants can receive coefficients from the ring (4.5.74), but for a unique representation either I_3 should be excluded from v_5 and v_7, or I_4 should be excluded from v_6 and v_8.

4.5.17. Remark. The equivariants u_2 and u_4, and their real counterparts, have degree $p + q - 1$. We have assumed $p > 0$ and $q > 0$; in the general case the degree is $|p| + |q| - 1$. Therefore, if $|p| + |q| \geq 4$, the normal form through cubic terms will be the same as if the double center were nonresonant. Since it turns out that the cubic terms are sufficient to determine much of the dynamical behavior of the system (and even of its unfolding, discussed in Section 6.7), a double center satisfying this condition is said to have *no strong resonances*, and is sometimes even called nonresonant.

The $1 : 1$ Resonance

The case of $1 : 1$ resonance ($p = q = 1$, or $\omega_1 = \omega_2$) is somewhat special, because, having repeated eigenvalues, it is capable of having both semisimple and nonsemisimple versions. By a change of time unit we can assume $\omega_1 = \omega_2 = 1$. It is helpful to rewrite the semisimple system (4.5.39) in the form

$$\dot{z}_1 = iz_1 + f_1(z_1, z_2, w_1, w_2), \qquad (4.5.80)$$
$$\dot{z}_2 = iz_2 + f_2(z_1, z_2, w_1, w_2),$$
$$\dot{w}_1 = -iw_1 + g_1(z_1, z_2, w_1, w_2),$$
$$\dot{w}_2 = -iw_1 + g_2(z_1, z_2, w_1, w_2).$$

This makes no change in the previous calculations of the normal form (except for the trivial rearrangement of arguments in f_s and g_s), but causes the matrix of the linear part to have its equal eigenvalues adjacent:

$$A = \begin{bmatrix} i & & & \\ & i & & \\ & & -i & \\ & & & -i \end{bmatrix}. \qquad (4.5.81)$$

The nonsemisimple version corresponding to this has linear part

$$A = \begin{bmatrix} i & 1 & & \\ 0 & i & & \\ & & -i & 1 \\ & & 0 & -i \end{bmatrix}. \tag{4.5.82}$$

Of course, the normal form for this system cannot be obtained using the semisimple normal form style we are studying in this section, and will be treated in Section 4.7.

The Computation Problem

As in the linear case (Section 3.3), the details of the computation procedures for reducing a specific system to normal form are mostly contained in the formulas of the various formats, but the calculation of the projections (into im $\mathcal{L}$ and its chosen complement) will depend on the normal form style being used (and hence on the matrix A). Likewise, the partial inversion of $\mathcal{L} = \mathsf{L}_A$ required to solve the homological equation depends on A. These questions will now be addressed in the case where A is semisimple and the semisimple normal form style is used.

Again as in Section 3.3, there are two ways to answer these questions, one valid when A is diagonal, and another requiring only that A be semisimple. The first is simple enough to use for hand calculation in small problems, and the second is complicated enough that it is usually feasible to use it only with symbolic processing systems.

For convenience, we assume that format 2a is used, so that the homological equation has the form

$$\mathcal{L}u_j = k_j - b_j. \tag{4.5.83}$$

If A is diagonal with entries $\lambda = (\lambda_1, \ldots, \lambda_n)$, each vector monomial $x^m e_i$ is an eigenvector of $\mathcal{L}$, with eigenvalue $\langle m, \lambda \rangle - \lambda_i$. The semisimple normal form style requires that $b_j \in \ker \mathcal{L}$, so it is necessary only to select from k_j (written as a linear combination of vector monomials) those terms for which $\langle m, \lambda \rangle - \lambda_i = 0$; these terms constitute

$$b_j = \mathsf{P}_0 k_j, \tag{4.5.84}$$

the projection of k_j into the zero eigenspace of $\mathcal{L}$. Then

$$k_j - b_j = (\mathsf{I} - \mathsf{P}_0)k_j = \sum_{mi} c_{mi} x^m e_i \tag{4.5.85}$$

is a sum of monomials for which $\langle m, \lambda \rangle - \lambda_i \neq 0$, and a possible solution of (4.5.83) is

$$u_j = \sum_{mi} \frac{c_{mi}}{\langle m, \lambda \rangle - \lambda_i} x^m e_i, \tag{4.5.86}$$

with the sum extended over the same set of m and i as in (4.5.85). This solution is not unique; others can be found by adding any element of ker $\mathcal{L}$ to this u_j.

In the case that A is semisimple but not necessarily diagonal, the projections P_μ of $\mathcal{V}_j^n$ onto the eigenspace of $\mathcal{L} = \mathsf{L}_A$ with eigenvalue μ can be computed by the following lemma. This lemma is similar to Lemma 3.3.10 in the linear case, but it is not possible to extract an explicit formula for P_μ here, because of the nonlinearity of the vector field v. The notation is as follows: The distinct eigenvalues of A (assumed known) are $\lambda_{(1)}, \ldots, \lambda_{(r)}$, with multiplicities $m_1 + \cdots + m_r = n$, and P_i is the matrix giving the projection of $\mathbb{C}^n$ onto the eigenspace E_i of A with eigenvalue $\lambda_{(i)}$.

4.5.18. Lemma. Let $v \in \mathcal{V}_j^n$. Suppose that the expression

$$e^{\sigma \mathsf{L}_A} v(x) = \left(\sum_{i=1}^r e^{-\sigma \lambda_{(i)}} P_i \right) v \left(\sum_{k=1}^r e^{\sigma \lambda_{(k)}} P_k x \right) \tag{4.5.87}$$

is expanded in the form

$$e^{\sigma \mathsf{L}_A} v(x) = \sum_\mu e^{\sigma \mu} v_\mu(x) \tag{4.5.88}$$

(using whatever values of μ arise in expanding the right-hand side of (4.5.87)). Then each such μ will equal some eigenvalue of L_A, and the projection of v into the eigenspace with eigenvalue μ is

$$\mathsf{P}_\mu v = v_\mu. \tag{4.5.89}$$

Proof. Equation (4.5.87) follows from the fundamental theorem of Lie series, Theorem 4.4.24, which says that if $b(x)$ is a vector field with flow φ^σ, then $e^{\sigma \mathsf{L}_b} = \mathsf{S}_{\varphi^\sigma}$. Take $b(x) = Ax$, so that $\mathsf{L}_b = \mathsf{L}_A$, and observe that $\varphi^\sigma(x) = e^{\sigma A}$, so that

$$e^{\sigma \mathsf{L}_A} v(x) = \mathsf{S}_{e^{\sigma A}} v(x) = e^{-\sigma A} v(e^{\sigma A} x).$$

Since $A = \sum_{i=1}^r \lambda_{(i)} P_i$,

$$e^{\sigma A} = \sum_{i=1}^r e^{\sigma \lambda_{(i)}} P_i,$$

and similarly for $e^{-\sigma A}$; (4.5.87) follows. On the other hand,

$$\mathsf{L}_A = \sum_\mu \mu \mathsf{P}_\mu,$$

where μ ranges over the eigenvalues of L_A, and P_μ are the projections into the corresponding eigenspaces. Therefore, for any σ (this is most clear for $\sigma = 1$),

$$e^{\sigma \mathsf{L}_A} = \sum_{\sigma \mu} e^\mu \mathsf{P}_\mu.$$

The remainder of the lemma follows at once.   □

Although the procedure for computing P_μ given in the last lemma does not produce an explicit formula, it can still be implemented using a symbolic algebra system, making the following algorithm possible. (See Algorithm 3.3.11.)

4.5.19. Algorithm. If A is semisimple, the homological equation

$$\mathcal{L}u_j = k_j - b_j$$

can be solved by setting

$$b_j = P_0(k_j)$$

and

$$u_j = \sum_{\mu \neq 0} \frac{1}{\mu} P_\mu(k_j - b_j),$$

where the P_μ are calculated by Lemma 4.5.18.

Notes and References

Introductory treatments of the semisimple normal form focusing on concrete applications are given in Nayfeh [94] and Kahn and Zarmi [62].

A treatment of the description problem for (what we call) extended semisimple normal forms, when A is in Jordan form, is given in Bruno (or Brjuno) [20], Theorem 1, page 153. This theorem essentially says that the normal form contains only monomials that are equivariant (or "resonant") with respect to the flow of the semisimple part of A. It does not address the module structure of the normal form space. The reference goes on to discuss integrability of the normal form (which we address in Chapter 5) and questions of convergence and divergence of the normal form (which we ignore).

The semisimple normal form is sometimes regarded as being essentially equivalent to the method of averaging. In particular, the resonant double center problem (except for the nonsemisimple 1 : 1 resonance) can be easily handled by the method of averaging: Taking the problem in the dilated form (4.2.1), with $a_0(x) = Ax$, and making the substitution $x = ye^{At}$, leads to an equation that (after scaling time to adjust the period to 2π) is in standard form for the method of averaging (4.2.11). This calculation depends upon the fact that e^{At} is periodic when A is the matrix of a resonant double center (or, more generally, if the eigenvalues of A are pure imaginary, are integer multiples of a fixed imaginary number, and A is semisimple). Problems that can be treated by averaging can also be treated by a variety of multiple scale methods. However, there are many semisimple problems for which the method of averaging is not adequate. The nonresonant double center leads to a problem in multifrequency averaging, which is more difficult than periodic averaging, and systems with a hyperbolic part are not amenable to averaging at all. (Some authors use the name "method of averaging" to refer to semisimple normal forms, confusing the issue still further.)

The results presented here for the double center can be generalized to triple and higher-order multiple centers, but the treatment of resonances is more complicated. See Murdock [81] and [84] for a related problem from the standpoint of averaging, and see Section B.1 below for more information. There is a pure submodule (over the ring of integers) of the module of integer vectors representing resonance relations that hold among the frequencies, and the rank of this submodule is the "number of resonances" that hold, ranging from zero (nonresonance) to the complete resonance case in which $\omega_1 : \omega_2 : \cdots : \omega_n$ is a ratio of integers. Passing to polar coordinates, there is a unimodular transformation that separates the angles into "fast" and "slow" angles; the projection into normal form is equivalent to averaging over the fast angles. It is easy to describe the equivariant (or "resonant") monomials that appear in the normal form, but the Stanley decomposition of the module of equivariants has not been worked out.

4.6 The Inner Product and Simplified Normal Forms

When A is semisimple, a suitable normal form style can be defined by $\mathcal{N} = \ker \mathcal{L} = \ker \mathsf{L}_A$, as shown in the last section. When A is not semisimple, neither is $\mathcal{L} = \mathsf{L}_A$, and the splitting $\mathcal{V} = \operatorname{im} \mathcal{L} \oplus \ker \mathcal{L}$ no longer holds. It will be shown that for any matrix A,

$$\mathcal{V} = \operatorname{im} \mathcal{L} \oplus \ker \mathcal{L}^* = \operatorname{im} \mathsf{L}_A \oplus_\perp \ker \mathsf{L}_{A^*}, \qquad (4.6.1)$$

so that the normal form style can be taken to be

$$\mathcal{N} = \ker \mathcal{L}^* = \ker \mathsf{L}_{A^*},$$

where A^* is the transpose (or conjugate transpose, if A is complex) of the matrix A. Since (4.6.1) is proved by the use of an inner product, the normal form $\mathcal{N}$ will be called the *inner product normal form*. A second normal form style, derived from the inner product normal form in the special case that A is in Jordan form, will be called the *simplified normal form*. It is characterized (in part) by the fact that vector fields in simplified normal form have nonzero entries only in those positions corresponding to bottom rows of Jordan blocks of A. The linear part $\mathcal{N}_0$ of $\mathcal{N}$, for both the inner product and simplified normal forms, reduces to the linear inner product and simplified normal forms of Section 3.4. (Of course, these linear terms do not form part of the actual normal form of a vector field, because only $\mathcal{N}_+$ is used in this way. But the linear terms will reappear in the context of unfoldings in Chapter 6.)

Both the inner product and simplified normal form spaces $\mathcal{N}$ can be viewed as modules over the ring of functions invariant under a certain group action; the group action, and hence the ring, are the same for both normal form styles. Normalized vector fields in the inner product normal form style are equivariant under the group action, but in the simplified style they are not.

Inner Products for $\mathcal{F}_*^n$ and $\mathcal{V}_*^n$

The goal of this subsection is to prove (4.6.1). This will be done through a sequence of lemmas and definitions. Two inner products will be defined, one, denoted by $\langle \mid \rangle$, on the vector space $\mathcal{F}_*^n$ of polynomials on $\mathbb{R}^n$, and the other, denoted by $\langle \, , \, \rangle$, on the space $\mathcal{V}_*^n$ of polynomial vector fields. (See Definition 4.1.1.) The inner products are defined by specifying the inner product of any two monomials, and extending the definition by linearity in each factor (or, when working with complex numbers, by linearity in the first factor and conjugate linearity in the second). These vector spaces are infinite-dimensional, but each polynomial contains finitely many terms, so the inner product of any two polynomials is a finite sum and is well-defined. If we tried to define the inner products on the corresponding spaces $\mathcal{F}_{**}^n$ and $\mathcal{V}_{**}^n$ of formal power series, the inner products would become infinite series of real numbers, which usually would not converge. (Of course, the formal power series themselves do not converge, but the monomials serve to distinguish the terms. The inner product of two formal power series cannot be regarded as a formal power series, since there is no variable present to identify the terms; it is just a series of constants.) So the inner products will be defined only for polynomials, but we will nevertheless regard two formal power series as orthogonal if their truncations at any degree are orthogonal, or equivalently, if their homogeneous terms of each degree are orthogonal.

The inner product for $\mathcal{F}_*^n$ will be defined first in the real case. The standard basis for $\mathcal{F}_*^n$ consists of the monomials x^m. (For notation, see (4.5.9), (4.5.10), (4.5.11).) Define

$$\langle x^\ell \mid x^m \rangle = \begin{cases} m! & \text{if } \ell = m, \\ 0 & \text{if } \ell \neq m. \end{cases} \tag{4.6.2}$$

If $p, q \in \mathcal{F}_*^n$ are given by

$$p(x) = \sum_\ell p_\ell x^\ell, \qquad q(x) = \sum_m q_m x^m,$$

with only finitely many of the p_ℓ and q_m unequal to zero, the inner product of p and q will then be

$$\langle p \mid q \rangle = \sum_m m! p_m q_m. \tag{4.6.3}$$

Let the operator $p(\partial)$ be defined by

$$p(\partial) = \sum_\ell p_\ell \left(\frac{\partial}{\partial x_1}\right)^{\ell_1} \cdots \left(\frac{\partial}{\partial x_n}\right)^{\ell_n}. \tag{4.6.4}$$

4.6.1. Lemma. If $p, q \in \mathcal{F}_*^n$, then

$$\langle p|q\rangle = (p(\partial)q)(0) = p(\partial)q(x)\big|_{x=0}.$$

Proof. Since $p(\partial)q$ is clearly bilinear in p and q, it is enough to check that

$$(\partial^\ell x^m)\big|_{x=0} = \begin{cases} m! & \text{if } \ell = m \\ 0 & \text{if } \ell \neq m. \end{cases}$$

This is left to the reader. $\qquad\qquad\square$

4.6.2. Lemma. If $T : \mathbb{R}^n \to \mathbb{R}^n$ is any linear transformation, and $T^* = T^\dagger$ is the adjoint or transpose of T (with respect to the standard inner product on $\mathbb{R}^n$), then

$$\langle p \circ T | q \rangle = \langle p | q \circ T^* \rangle.$$

Proof. Let the matrix elements of T with respect to the standard basis of $\mathbb{R}^n$ be t_{ij}, and consider the change of coordinates $y = T^*x$, or $y_j = \sum_i t_{ij} x_i$. Then

$$\frac{\partial}{\partial x_i} = \sum_j \frac{\partial y_j}{\partial x_i}\frac{\partial}{\partial y_j} = \sum_j t_{ij}\frac{\partial}{\partial y_j}.$$

Now

$$\langle p|q\circ T^*\rangle = (p(\partial)(q\circ T^*))(0)$$

$$= \sum_\ell p_\ell \left(\frac{\partial}{\partial x_1}\right)^{\ell_1} \cdots \left(\frac{\partial}{\partial x_n}\right)^{\ell_n} q(T^*x)\Bigg|_{x=0}$$

$$= \sum_\ell p_\ell \left(\sum_j t_{1j}\frac{\partial}{\partial y_j}\right)^{\ell_1} \cdots \left(\sum_j t_{nj}\frac{\partial}{\partial y_j}\right)^{\ell_n} q(y)\Bigg|_{y=0}$$

$$= ((p\circ T)(\partial)q)(0)$$

$$= \langle p\circ T|q\rangle.$$

$$\square$$

The discussion so far must be modified slightly when $\mathcal{F}_*^n$ is regarded as a vector space over $\mathbb{C}$. The basic definition (4.6.2) is unchanged, but since an inner product for a complex vector space must be conjugate linear in the second factor, (4.6.3) is replaced by

$$\langle p|q\rangle = \sum_m m! p_m \bar{q}_m. \tag{4.6.5}$$

Lemma 4.6.1 becomes

$$\langle p|q \rangle = (p(\partial)\bar{q})(0), \tag{4.6.6}$$

where

$$\bar{q}(z) = \sum_m \bar{q}_m z^m, \tag{4.6.7}$$

z being the complex variable that replaces x. Lemma 4.6.2 remains true exactly as stated, except that T^* is now the conjugate transpose of T; the coordinate change $y = T^*x$ in the proof is replaced by $w = T^*z$, with $w_j = \sum_i \bar{t}_{ij} z_i$. Notice that we work with $\bar{q}(z)$, not $\overline{q(z)}$; therefore, the variables $\bar{z}_i$ never appear, and the differentiation in the proof is the usual complex differentiation of analytic functions of several complex variables. The use of complex variables in this context is not very significant, because it will be shown later that the calculation of normal forms for general matrices A can be separated into two parts by decomposing A into semisimple and nilpotent parts $(A = S+N)$; if A is in real normal form, the semisimple part can be handled by the methods of the last section, and the nilpotent part does not require complex variables. For the remainder of this discussion of inner products, the theorems will be formulated in a manner that is correct for both real and complex cases, but only the real case will be proved.

4.6.3. Lemma. The adjoint of the operator $\mathcal{D}_A$, with respect to the inner product $\langle \,|\, \rangle$ on $\mathcal{F}_*^n$, is given by

$$\mathcal{D}_A^* = \mathcal{D}_{A^*}.$$

In other words,

$$\langle \mathcal{D}_A p|q \rangle = \langle p|\mathcal{D}_{A^*}q \rangle.$$

Proof. Putting $T = e^{tA}$ in Lemma 4.6.2 gives

$$\langle p(e^{At}x)|q(x) \rangle = \langle p(x)|q(e^{A^*t}x) \rangle.$$

Differentiating with respect to t and setting $t = 0$ gives the result. $\square$

To define an inner product for $\mathcal{V}_*^n$, consider the basis (in the sense discussed above) consisting of vector monomials $x^\ell e_i$ (where ℓ is a multi-index), and define

$$\langle x^\ell e_i, x^m e_j \rangle = \begin{cases} m! & \text{if } \ell = m \text{ and } i = j, \\ 0 & \text{otherwise.} \end{cases} \tag{4.6.8}$$

If $v = (v_1, \ldots, v_n) = \sum_i v_i e_i$ and $w = \sum_j w_j e_j$ are vector fields, the inner product defined by (4.6.8) is related to that defined above for scalar polynomials by

$$\langle v, w \rangle = \sum_i \langle v_i|w_i \rangle. \tag{4.6.9}$$

4.6.4. Lemma. Let $A \in \text{gl}(n)$ be regarded as a linear transformation of $\mathcal{V}_*^n$ to itself. Then

$$\langle Av, w \rangle = \langle v, A^*w \rangle.$$

That is, the adjoint of A with respect to the inner product $\langle \, , \, \rangle$ is A^*, the transpose (in the real case) or conjugate transpose (in the complex case) of A, the same as its adjoint with respect to the standard inner product on $\mathbb{R}^n$ (or $\mathbb{C}^n$).

Proof. Assuming the complex case for generality, the calculation goes as follows:

$$\langle Av, w \rangle = \sum_i \langle (Av)_i | w_i \rangle$$

$$= \sum_i \left\langle \sum_j a_{ij} v_j \Big| w_i \right\rangle$$

$$= \sum_i \sum_j a_{ij} \langle v_j | w_i \rangle$$

$$= \sum_j \left\langle v_j \Big| \sum_i \bar{a}_{ij} w_i \right\rangle$$

$$= \sum_j \langle v_j | (A^*w)_j \rangle$$

$$= \langle v, A^*w \rangle. \qquad \square$$

4.6.5. Lemma. Let $T \in \text{GL}(n)$ and let $\mathsf{S}_T : \mathcal{V}_*^n \to \mathcal{V}_*^n$ be the similarity operator

$$(\mathsf{S}_T v)(x) = T^{-1} v(Tx).$$

(This is the special case of (4.4.9) for a linear map $\varphi = T$.) Then the adjoint of S_T with respect to $\langle \, , \, \rangle$ is S_{T^*}; that is,

$$\langle T^{-1} v(Tx), w(x) \rangle = \langle v(x), T^{*-1} w(T^*x) \rangle.$$

Proof. This is an immediate combination of Lemmas 4.6.4 and 4.6.2, extended to vector fields via (4.6.9). $\qquad \square$

4.6.6. Lemma. Let $A \in \text{gl}(n)$. Then the adjoint of L_A with respect to $\langle \, , \, \rangle$ is L_{A^*}. That is,

$$(\mathsf{L}_A)^* = \mathsf{L}_{A^*},$$

or

$$\langle \mathsf{L}_A v, w \rangle = \langle v, \mathsf{L}_{A^*} w \rangle.$$

Proof. Put $T = e^{At}$ into Lemma 4.6.5 and differentiate, as in the proof of Lemma 4.6.3. □

Finally, (4.6.1) follows from Lemma 4.6.6 and Theorem 2.2.1.

The Inner Product Normal Form Module

On the strength of (4.6.1), we can now define the inner product normal form style.

4.6.7. Definition (cf. Definition 3.4.2). A formal power series vector field $a \in \mathcal{V}^n_{**}$ with linear part Ax is in *inner product normal form* if its terms $a_j \in \mathcal{V}^n_j$ satisfy $a_j \in \ker \mathcal{L}^*$, or more precisely,

$$a_j \in \mathcal{N}_j = \ker(\mathcal{L}^* : \mathcal{V}^n_j \to \mathcal{V}^n_j)$$

for $j = 1, 2, \ldots$. In view of (4.5.7), this may be written

$$a_1 + a_2 + \cdots \in \mathcal{N}_+,$$

where

$$\mathcal{N} = \ker \mathcal{L}^* = \ker \mathsf{L}_{A^*}.$$

It follows from Theorem 4.4.11 that the terms a_j of a vector field in inner product normal form are equivariants of the flow of the linear vector field A^*x. Notice that this is *not* the linear part of the vector field a itself, but rather its conjugate transpose, so it is *not* correct to say (as it is for the semisimple normal form according to Lemma 4.5.7) that the nonlinear terms commute with the linear term. However, the rest of the proof of Lemma 4.5.7 goes through without change (except that A is replaced by A^*). In particular, the operator $\mathcal{D}_{A^*}$ defined by

$$\mathcal{D}_{A^*} f(x) = f'(x) A^* x = (A^* x \cdot \nabla) f(x) \tag{4.6.10}$$

satisfies

$$\mathcal{D}_{A^*}(fg) = f \mathcal{D}_{A^*} g + g \mathcal{D}_{A^*} f \tag{4.6.11}$$

and

$$\mathsf{L}_{A^*}(fv) = (\mathcal{D}_{A^*} f)v + f(\mathsf{L}_{A^*} v), \tag{4.6.12}$$

so that $\mathcal{N}$ is a module over $\ker \mathcal{D}_{A^*} \subset \mathcal{F}^n_{**}$, the *ring of (formal) invariants* of the flow $e^{A^* t}$; $\mathcal{N}$ is called the *module of (formal) equivariants*.

4.6.8. Theorem (cf. Theorem 3.4.5). If $a(x)$ is in inner product normal form with respect to $a_0(x)$, and if T is a unitary or weakly unitary matrix (see Definition 3.4.4), then $\mathsf{S}_T a(x)$ is in inner product normal form with respect to $\mathsf{S}_T a_0(x)$.

Proof. We need to check that $L_{(T^{-1}AT)} \cdot S_T a(x) = 0$. For unitary T, with $T^* = T^{-1}$, this expression is equal to

$$[T^{-1}a(Tx)]'(T^{-1}AT)^*x - (T^{-1}AT)^*(T^{-1}a(Tx))$$
$$= T^{-1}a'(Tx)A^*Tx - T^*A^*a(T(x))$$
$$= T^{-1}\{v'(Tx)A^*Tx - A^*v(Tx)\} = T^{-1}(L_{A^*} \cdot a)(Tx),$$

which equals zero because $L_{A^*} \cdot a = 0$. $\qquad\square$

As in Corollary 3.4.7, it follows that if $a(x)$ is a real vector field with linear part $a_0(x) = Ax$ with A in real canonical form, the semisimple normal form can be found by changing to (complex) coordinates in which A takes Jordan form, normalizing, and changing back. This will be illustrated later in this section for the example of the nonsemisimple double center. Because of this, most of our attention will be paid to the case where A is in Jordan form.

Splitting the Description Problem

4.6.9. Lemma (cf. Lemma 3.4.8). If $A = S + N$ is the semisimple/nilpotent splitting of A (meaning that S is semisimple, N is nilpotent, and $SN = NS$), then $L_A = L_S + L_N$ is the semisimple/nilpotent splitting of $\mathcal{L}$ on $\mathcal{V}_j^n$ for each j, $L_{A^*} = L_{S^*} + L_{N^*}$ is the semisimple/nilpotent splitting of $\mathcal{L}^*$ on $\mathcal{V}_j^n$, and

$$\ker \mathcal{L}^* = \ker L_{S^*} \cap \ker L_{N^*}.$$

Proof. Since S and S^* are semisimple, Lemma 4.5.2 implies that L_S and L_{S^*} are semisimple on each $\mathcal{V}_j^n$. The proof that L_N (and likewise L_{N^*}) are nilpotent will be given momentarily. The remainder of the proof of the lemma is exactly the same as the proof of Lemma 3.4.8. In particular, the equation for $\ker \mathcal{L}^*$ is proved by considering a basis for each finite-dimensional space $\mathcal{V}_j^n$ (rather than gl(n) in Lemma 3.4.8) for which $\mathcal{L}^*$ is in Jordan form.

Before proving that L_N is nilpotent, it is worth pointing out that this is true only on the finite-dimensional spaces $\mathcal{V}_j^n$, not on $\mathcal{V}_*^n$ or $\mathcal{V}_{**}^n$. The difficulty is that the index of nilpotence of L_N increases with j, and no power of L_N vanishes on all vector fields. The easiest proof of nilpotence (for those reading the starred sections) is the one that follows the lines of Remark 3.4.9. Alternatively, we may proceed as follows. For $j = -1$ (constant vector fields), $L_N = -N$, and the nilpotence is clear. For $j = 0$ (linear vector fields), $L_N = \mathbb{L}_N$ (more precisely, $L_N(Ax) = (\mathbb{L}_N A)x$), and the argument in Lemma 3.4.8 applies. The next case is $j = 1$ (quadratic vector fields). Let $v \in \mathcal{V}_1^n$; then

$$L_N v(x) = v'(x)Nx - Nv(x),$$

and therefore,

$$L_N^2 v(x) = (v'(x)Nx)'Nx - N(v'(x)Nx - Nv(x))$$
$$= v''(x)(Nx)(Nx) - v'(x)N^2x - Nv'(x)Nx + N^2v(x),$$

where $v''(x)(a)(b)$ is the vector-valued bilinear function of two vectors a and b defined by the three-index array of components of $v''(x)$. Since v is quadratic, $v'''(x) = 0$, and continued iteration of L_N does not create terms containing higher derivatives of v. Instead, each term of $L_N^i v(x)$ is of one of the following types:

$$N^j v''(x)(N^k x)(N^\ell x),$$
$$N^j v'(x)N^k x,$$
$$N^j v(x),$$

for various values of j, k, and ℓ, such that in each term the total number of factors of N equals i (so that for terms of the first type, for instance, $i = j + k + \ell$). Each type of term has a fixed number of "positions" in which N can occur, the maximum being three positions, for terms of the first type. Let r be the index of nilpotence of N. Then for $i > 3r$, each term in $L_N^i v(x)$ contains at least r factors of N in one of its positions, and therefore each term vanishes. The same argument works for homogeneous vector fields v of any degree, but the number of "positions" in which N can occur increases as the degree of v increases. For instance, for cubic v, there will be terms of the type

$$N^j v'''(x)(N^k x)(N^\ell x)(N^m x),$$

and we must have $i > 4r$ before the vanishing of such terms is guaranteed. ☐

4.6.10. Lemma (cf. Theorem 3.4.13). If $A = S + N$ is in Jordan canonical form or real canonical form, or more generally if A is seminormal (Definition 3.4.10), then

$$\ker L_{S^*} = \ker L_S,$$

and

$$\ker \mathcal{L}^* = \ker L_S \cap \ker L_{N^*}.$$

In particular, if A is seminormal, any vector field in inner product normal form with respect to A is in extended semisimple normal form. Similarly, $\ker \mathcal{D}_{S^*} = \ker \mathcal{D}_S$ and $\ker \mathcal{D}_A^* = \ker \mathcal{D}_S \cap \ker \mathcal{D}_{N^*}$.

Proof. The proof is identical to that of Theorem 3.4.13, with L (or $\mathcal{D}$) in place of $\mathbb{L}$. ☐

Splitting the Computation Problem

The computation problem (for the inner product normal form when $A = S + N$) can also be split into semisimple and nilpotent parts, provided that A is seminormal. The procedure follows the same steps as in the linear case. The homological equation that must be solved at the jth stage in the calculation (in the notation of format 2a) is

$$\mathcal{L}u_j = \mathsf{L}_S u_j = k_j - b_j, \tag{4.6.13}$$

with $b_j \in \ker \mathsf{L}_{A^*}$. The following four steps reduce the determination of b_j and u_j to the solution of two simpler homological equations, one semisimple and one nilpotent. The justification of these steps is the same as in Section 3.4 and will not be repeated. (As in Chapter 3, the sl(2) normal form is preferable if A is not seminormal.)

4.6.11. Algorithm (cf. Algorithm 3.4.22). Assuming that A is semi-normal, the following steps give the jth stage of the normalization of the vector field $a(x) = Ax + a_1(x) + \cdots$:

1. Find $b_j' \in \ker \mathsf{L}_S$ such that $k_j - b_j' \in \operatorname{im} \mathsf{L}_S$, and find $\widehat{u}_j$ such that

$$\mathsf{L}_S \widehat{u}_j = k_j - b_j'. \tag{4.6.14}$$

   This is a semisimple problem.

2. Convert $\widehat{u}_j$ into a solution u_j' of

$$\mathsf{L}_A u_j' = k_j - b_j', \tag{4.6.15}$$

   by taking

$$u_j' = \sum_{r=0}^{k-1} (-1)^r \left(\mathsf{L}_S^{-1} \right)^r \mathsf{L}_N^r \widehat{u}_j, \tag{4.6.16}$$

   where k is the index of nilpotence of L_N on $\mathcal{V}_j^n$ and where L_S^{-1} denotes any right inverse of L_S on its image. See Algorithm 3.4.23. As in item 2 of Algorithm 3.4.22, it is possible to stop here if all that is required is an extended semisimple normal form (Definition 4.5.5).

3. Regarding L_N as a map of $\ker \mathsf{L}_S \to \ker \mathsf{L}_S$, find $b_j \in \ker \mathsf{L}_{N^*}$ such that $b_j' - b_j \in \operatorname{im} \mathsf{L}_N$, and find u_j'' such that

$$\mathsf{L}_N u_j'' = b' - b_j. \tag{4.6.17}$$

   This is a purely nilpotent problem.

4. Then b_j and

$$u_j = u_j' + u_j'' \tag{4.6.18}$$

   satisfy the original homological equation (4.6.13).

A Termwise Description of the Normal Form When A Is in Jordan Form

The description problem for the inner product normal form can be solved explicitly when A is in Jordan canonical form. The description given here is "termwise"; that is, it characterizes the individual terms (monomials or combinations of monomials) that will appear (with arbitrary coefficients) in a vector field in normal form. The much more technical problem of giving a module description of the normal form (by giving Stanley decompositions of the ring of invariant functions and the module of equivariant vector fields) will be addressed in Section 4.7. The "termwise" approach has the disadvantage that the establishment of each term requires a separate calculation, so the normal form can be described only to a finite order in a finite amount of time. The module approach describes the normal form to all orders with a finite amount of work.

Let $A = S + N$ be in upper Jordan normal form, with $\lambda = (\lambda_1, \ldots, \lambda_n)$ being the diagonal entries of S; N, of course, will be zero except for certain ones in the superdiagonal, and will have the same block structure as S. It follows from Lemma 4.6.10 and Corollary 4.5.9 that any vector monomial $x^m e_i$ occurring in a vector field in normal form must satisfy

$$\langle m, \lambda \rangle - \lambda_i = 0. \tag{4.6.19}$$

But the set of vector fields spanned by these monomials (which is just $\ker \mathsf{L}_S$) must be intersected with $\ker \mathsf{L}_N^*$. The effect of this intersection is twofold:

1. Some monomials satisfying (4.6.19) are simply eliminated from the normal form.

2. Other monomials satisfying (4.6.19) cannot appear individually, but are grouped into combinations (vector polynomials) that appear as basis elements (each with its own coefficient) in the normal form.

(Notice that the presence of a nilpotent part N in A always leads to a *simpler* normal form than would be the case with the same S if N were zero, although the *work required* to describe the normal form can be much greater.) So our task is to give a procedure for generating the basis elements for $\mathcal{V}_j^n \cap \ker \mathsf{L}_{A^*} = \mathcal{V}_j^n \cap \ker \mathsf{L}_S \cap \ker \mathsf{L}_{N^*}$ (as a vector space over $\mathbb{R}$ or $\mathbb{C}$) in each grade j (degree $j+1$). Since the most critical step is the determination of $\ker \mathsf{L}_{N^*}$, it is convenient to consider first the case $S = 0$, or $A = N$.

The "brute-force" method for determining $\ker \mathsf{L}_{N^*}$, which can be used whether or not N is in Jordan form, is to choose a basis for $\mathcal{V}_j^n$, for instance the standard basis of monomial vector fields, express $\mathsf{L}_{N^*} : \mathcal{V}_j^n \to \mathcal{V}_j^n$ as a matrix with respect to this basis, and compute its kernel by row reduction. It has been a theme of this book, beginning with Chapter 1, that this method should be avoided if possible. The first step toward escaping from this method in the present situation is to observe that since L_{N^*} is itself nilpotent (by Lemma 4.6.9), $\mathcal{V}_j^n$ has a basis consisting of Jordan chains for L_{N^*}, each of which ends in an element of the kernel. Of course, finding these Jordan chains (by brute-force methods) would be harder than finding the kernel, but the knowledge that these chains are there is crucial. The second, equally crucial, step is to see that when N is in Jordan form, the Jordan chains of L_{N^*} can be determined from those of $\mathcal{D}_{N^*}$, which is also nilpotent and operates on a lower-dimensional space.

It is convenient to begin with an example, the single nilpotent 3×3 Jordan block. Thus,

$$A = N = N_3 = \begin{bmatrix} 0 & 1 & 0 \\ 0 & 0 & 1 \\ 0 & 0 & 0 \end{bmatrix}, \qquad (4.6.20)$$

with $S = 0$. Then

$$N^* = \begin{bmatrix} 0 & 0 & 0 \\ 1 & 0 & 0 \\ 0 & 1 & 0 \end{bmatrix}. \qquad (4.6.21)$$

Remembering that vectors written with parentheses indicate columns, we write elements of $\mathbb{R}^3$ as (x, y, z) for convenience instead of (x_1, x_2, x_3), and vector fields as $v = (f, g, h)$. Then

$$\mathsf{L}_N^* v = \begin{bmatrix} f_x & f_y & f_z \\ g_x & g_y & g_z \\ h_x & h_y & h_z \end{bmatrix} \begin{bmatrix} 0 & 0 & 0 \\ 1 & 0 & 0 \\ 0 & 1 & 0 \end{bmatrix} \begin{bmatrix} x \\ y \\ z \end{bmatrix} - \begin{bmatrix} 0 & 0 & 0 \\ 1 & 0 & 0 \\ 0 & 1 & 0 \end{bmatrix} \begin{bmatrix} f \\ g \\ h \end{bmatrix}. \qquad (4.6.22)$$

Since

$$\mathcal{D}_{N^*} = (N^*(x, y, z)) \cdot \nabla = (0, x, y) \cdot \nabla = x\frac{\partial}{\partial y} + y\frac{\partial}{\partial z}, \qquad (4.6.23)$$

it follows that

$$\mathsf{L}_{N^*} v = \begin{bmatrix} \mathcal{D}_{N^*} f \\ \mathcal{D}_{N^*} g - f \\ \mathcal{D}_{N^*} h - g \end{bmatrix}. \qquad (4.6.24)$$

This vanishes if and only if

$$\mathcal{D}_{N^*} f = 0, \qquad \mathcal{D}_{N^*} g = f, \qquad \mathcal{D}_{N^*} h = g,$$

or equivalently, if and only if h satisfies

$$\mathcal{D}_{N^*}^3 h = 0 \qquad (4.6.25)$$

and g and f are determined from h by $g = \mathcal{D}_{N^*} h$, $f = \mathcal{D}_{N^*}^2 h$, so that

$$v = \begin{bmatrix} \mathcal{D}_{N^*}^2 h \\ \mathcal{D}_{N^*} h \\ h \end{bmatrix}. \qquad (4.6.26)$$

To continue with this example, it is necessary to solve the partial differential equation (4.6.25), which is equivalent to finding a basis for the subspace of vectors in V_j^n having height ≤ 3 under $\mathcal{D}_{N^*}$, for each desired j. If the Jordan chains of $\mathcal{D}_{N^*}$ were known, this would mean choosing the bottom three elements of each chain. An efficient method for finding the Jordan chains of $\mathcal{D}_{N^*}$ will be developed in Section 4.7. For the present, it will suffice to display the results for $j = -1, 0, 1, 2$ (degrees 0, 1, 2, 3) in a table, and explain how this table can either be checked (which is simple), or else generated from scratch (which requires tedious calculations). For later convenience the table is arranged with $\mathcal{D}_{N^*}$ mapping upward, so that the bottoms of the chains appear at the tops of the columns. The vertical bars separate degrees, and each column gives the last three elements of a Jordan chain, so that the elements in the top row span ker $\mathcal{D}_{N^*}$ (the space of *invariants* for this problem), those in the first two rows span ker $\mathcal{D}_{N^*}^2$, and all three rows together span ker $\mathcal{D}_{N^*}^3$. Notice that some Jordan chains may have length shorter than three, so there can be empty spaces in the table. Here is the table:

$$
\begin{array}{c|c|cc|cc}
1 & x & x^2 & y^2 - 2xz & x^3 & xy^2 - 2x^2z \\
 & y & xy & & x^2y & y^3 - 2xyz \\
 & z & xz & & x^2z & y^2z - 2xz^2
\end{array}
\qquad (4.6.27)
$$

There is additional structure to be noticed in this table, aside from the Jordan chain structure itself. The first column in degree 2 is equal to the invariant x times the column under x in degree 1; each column in degree 3 is equal to one of the quadratic invariants (x^2 or $y^2 - 2xz$) times the linear column. This facilitates checking the action of $\mathcal{D}_{N^*}$ on these elements, since by the product rule (4.6.11), if f is an invariant and g is an arbitrary function, then

$$
\mathcal{D}_{N^*}(fg) = f\mathcal{D}_{N^*}g. \qquad (4.6.28)
$$

So it is quite easy to check that the facts recorded in the table are correct. To generate the table from scratch, without using ideas to be presented in Section 4.7, one could write out the general cubic on $\mathbb{R}^3$ as

$$
\begin{aligned}
h(x, y, z) = &\, ax^3 + bx^2y + cx^2z + dxy^2 + exyz + fxz^2 \qquad (4.6.29) \\
&+ gy^3 + hy^2z + iyz^2 + jz^3,
\end{aligned}
$$

apply (4.6.23) three times, collect like terms, and set the coefficients equal to zero. In writing out (4.6.29), the terms have been arranged in lexicographic order; to double check that no terms have been omitted, Lemma 4.5.1 implies that

$$
\dim \mathcal{F}_3^3 = \binom{3 + 3 - 1}{3} = 10.
$$

Using equation (4.6.26) and table (4.6.27), it is now possible to write down a basis (over $\mathbb{R}$) for the vector fields in normal form in each degree up to 3. For instance, in degree 2 each of the functions x^2, xy, xz, and $y^2 - 2xz$ can be used for h in (4.6.26), and then $\mathcal{D}_{N*}h$ and $\mathcal{D}_{N*}^2 h$ can be calculated from the table by moving up the column from h. Thus, the general homogeneous quadratic vector field in inner product normal form for $A = N_3$ is given by

$$
c_1 \begin{bmatrix} 0 \\ 0 \\ x^2 \end{bmatrix} + c_2 \begin{bmatrix} 0 \\ x^2 \\ xy \end{bmatrix} + c_3 \begin{bmatrix} x^2 \\ xy \\ xz \end{bmatrix} + c_4 \begin{bmatrix} 0 \\ 0 \\ y^2 - 2xz \end{bmatrix}. \tag{4.6.30}
$$

A basis for the linear terms of the normal form is

$$
\begin{bmatrix} 0 \\ 0 \\ x \end{bmatrix}, \quad \begin{bmatrix} 0 \\ x \\ y \end{bmatrix}, \quad \begin{bmatrix} x \\ y \\ z \end{bmatrix}. \tag{4.6.31}
$$

The reader is invited to write down the cubic case.

This example reveals certain structural features that continue to hold in the general case (for A in Jordan form). The notation will be as follows. The $n \times n$ matrix $A = S + N$ is in upper Jordan form with (small) Jordan block sizes $r_1, \ldots, r_\ell$, so that $r_1 + \cdots + r_\ell = n$. We set $R_1 = r_1$, $R_2 = r_1 + r_2$, ..., $R_\ell = r_1 + \cdots + r_\ell = n$, so that R_s is the index of the bottom row of the sth Jordan block of A (or of N, since these have the same Jordan structure). These rows with index R_s play an important role in the following discussion and again in Chapter 6, and will be called *main rows*:

4.6.12. Definition. A *main row* of an $n \times n$ matrix A in Jordan canonical form is the (complete) row of A containing the bottom row of a Jordan block of A. If R_s is the index of a main row of A, then the R_sth row of any $n \times n$ matrix B or $n \times 1$ column vector v will be called a *main row* of B or v (relative to the Jordan block structure of A).

If A is real (in Jordan form), all vector spaces mentioned below (such as $\mathcal{V}_j^n$, $\mathcal{F}_{j+1}^n$, or subspaces of these) will be over $\mathbb{R}$; if A is complex, they will be over $\mathbb{C}$. (If A is the complex Jordan form of a real matrix, there will be reality conditions, but we will set these aside for the present.)

The following lemma is a generalization of equation (4.6.26). The idea is that, given a particular main row R_s in N (the bottom row of the sth Jordan block, with block size r_s), and given a scalar function h that is annihilated by $\mathcal{D}_{N*}^{r_s}$, there is a unique vector field $v_{\{s,h\}}$ in $\ker \mathsf{L}_{N*}$ that has h in the same main row (the R_sth entry in the vector). The entries above h are generated by applying $\mathcal{D}_{N*}$ repeatedly until the positions corresponding to the rows of the sth Jordan block are filled. (Warning: In the notation

$v_{\{s,h\}}$, s is an integer but h is a function.) For instance, if

$$N_{2,3} = \begin{bmatrix} 0 & 1 & & & \\ 0 & 0 & & & \\ & & 0 & 1 & 0 \\ & & 0 & 0 & 1 \\ & & 0 & 0 & 0 \end{bmatrix}, \tag{4.6.32}$$

then $s \in \{1,2\}$, $r_1 = 2$, $R_1 = 2$, $r_2 = 3$, $R_2 = 5$,

$$v_{\{1,h\}} = \begin{bmatrix} \mathcal{D}_{N^*} h \\ h \\ 0 \\ 0 \\ 0 \end{bmatrix} \quad \text{for} \quad h \in \ker \mathcal{D}^2_{N^*},$$

and

$$v_{\{2,h\}} = \begin{bmatrix} 0 \\ 0 \\ \mathcal{D}^2_{N^*} h \\ \mathcal{D}_{N^*} h \\ h \end{bmatrix} \quad \text{for} \quad h \in \ker \mathcal{D}^3_{N^*}.$$

(The use of the letter h for the bottom entry is carried over from the example of (4.6.26) although the vector fields no longer have the form (f, g, h).)

4.6.13. Lemma. If $h \in \mathcal{F}^n_j$ belongs to $\ker \mathcal{D}^{r_s}_{N^*}$, then the vector polynomials $v_{\{s,h\}}$ defined by

$$v_{\{s,h\}} = \sum_{i=0}^{r_s-1} (\mathcal{D}^i_{N^*} h) e_{R_s - i}$$

belong to $\ker \mathsf{L}_{N^*}$. A basis for $\mathcal{N}_j = \mathcal{V}^n_j \cap \ker \mathsf{L}_{N^*}$ can be formed from such vector fields $v_{\{s,h\}}$ by letting s range over the blocks of N (that is, $s = 1, \ldots, \ell$) and for each fixed s, letting h range over a basis for $\mathcal{F}^n_{j+1} \cap \ker \mathcal{D}^{r_s}_{N^*}$.

Proof. It should suffice to consider the example (4.6.32). We have

$$\mathsf{L}_{N^*} \begin{bmatrix} f_1 \\ f_2 \\ f_3 \\ f_4 \\ f_5 \end{bmatrix} = \begin{bmatrix} \mathcal{D}_{N^*} f_1 \\ \mathcal{D}_{N^*} f_2 - f_1 \\ \mathcal{D}_{N^*} f_3 \\ \mathcal{D}_{N^*} f_4 - f_3 \\ \mathcal{D}_{N^*} f_5 - f_4 \end{bmatrix}.$$

For this to vanish, we must have $f_2 \in \mathcal{D}^2_{N^*}$ and $f_5 \in \mathcal{D}^3_{N^*}$, with f_1, f_3, and f_4 determined from these. Notice that the reasoning proceeds separately for the first two rows and for the last three, according to the division of N into Jordan blocks. $\qquad \square$

Although "the *reasoning* proceeds separately" for the various Jordan blocks, it is important to understand that the correct *result* cannot be obtained by treating the Jordan blocks independently. For instance, consider the quadratic terms in the normal form for $N_{2,3}$. The quadratic vector fields $v_{\{2,h\}}$ corresponding to the second block ($s = 2$, the 3×3 block) will *not* be the same as the quadratic terms for N_3 listed in (4.6.30) (even after x, y, z have been replaced by x_3, x_4, x_5). The reason is that the h that can appear in $v_{\{2,h\}}$ for the $N_{2,3}$ problem are functions of all of the variables $x_1, \ldots, x_5$ (belonging to the appropriate kernel), not just functions of x_3, x_4, x_5.

Notice that the linear vector fields in (4.6.31) correspond to the striped matrices (Definition 3.4.18)

$$\begin{bmatrix} 0 & 0 & 0 \\ 0 & 0 & 0 \\ 1 & 0 & 0 \end{bmatrix}, \quad \begin{bmatrix} 0 & 0 & 0 \\ 1 & 0 & 0 \\ 0 & 1 & 0 \end{bmatrix}, \quad \begin{bmatrix} 1 & 0 & 0 \\ 0 & 1 & 0 \\ 0 & 0 & 1 \end{bmatrix}.$$

The following definition parallels the definition of a striped matrix, and reduces to that definition if $j = 0$ (if a linear vector field Bx is identified with its matrix B). In contrast with Chapter 3, the "stripe structure" is not uniquely determined by N, but also depends on the choice of basis functions h for the various kernels:

4.6.14. Definition. Consider a fixed basis for $\mathcal{V}_j^n \cap \ker \mathsf{L}_{N^*}$ consisting of vector fields of the form $v_{\{s,h\}}$ for $s = 1, \ldots, \ell$, using a suitable finite set of functions h for each s according to Lemma 4.6.13. A linear combination

$$\sum_{s,h} c_{\{s,h\}} v_{\{s,h\}}$$

(where $c_{\{s,h\}}$ are constants, and the sum is over those pairs (s, h) appearing in the basis) may be called a *striped vector field*, each term in the sum being a *stripe*.

Lemma 4.6.13 has the following immediate corollary, which is fundamental for the study of the module structure of $\mathcal{N}$ carried out in Section 4.7:

4.6.15. Corollary. For any positive integer r, $\ker \mathcal{D}_{N^*}^r$ is a module (of scalar functions) over the ring $\ker \mathcal{D}_{N^*}$, and the normal form module $\mathcal{N} = \ker \mathsf{L}_{N^*}$ is isomorphic to the direct sum of these modules as r ranges over the block sizes occurring in N:

$$\mathcal{N} \cong \bigoplus_{s=1}^{\ell} \ker \mathcal{X}^{r_s}.$$

Proof. If $f \in \ker \mathcal{D}_{N^*}$ and $g \in \ker \mathcal{D}_{N^*}^r$, then $fg \in \ker \mathcal{D}_{N^*}^r$ by the chain rule. The module isomorphism is given by sending $(h_1, \ldots, h_\ell)$ (in the direct sum) to $v_{\{1,h_1\}} + \cdots + v_{\{\ell,h_\ell\}} \in \mathcal{N}$. $\square$

Lemma 4.6.13 gives a (termwise) description of the inner product normal form in the purely nilpotent case $S = 0$. The following theorem extends this to the general case $A = S + N$ in Jordan form. The essential point is that a vector field of the form $v_{\{s,h\}}$, as defined in Lemma 4.6.13 using the nilpotent part N of A, will automatically belong to $\ker \mathsf{L}_{A^*}$ (and not just $\ker \mathsf{L}_{N^*}$) if the vector field he_{R_s} (formed from the bottom nonzero entry of $v_{\{s,h\}}$) belongs to $\ker \mathsf{L}_S$. Equivalently, he_{R_s} must be a linear combination of vector monomials $x^m e_{R_s}$ that satisfy

$$\langle m, \lambda \rangle - \lambda_{R_s} = 0, \tag{4.6.33}$$

where $\lambda = (\lambda_1, \ldots, \lambda_n)$ are the diagonal elements of S.

4.6.16. Theorem. If $h \in \mathcal{F}_j^n$ belongs to $\ker \mathcal{D}_{N^*}^{r_s}$ and contains only monomials x^m that satisfy (4.6.33), then the vector polynomial $v_{\{s,h\}}$ defined by

$$v_{\{s,h\}} = \sum_{i=0}^{r_s-1} (\mathcal{D}_{N^*}^i h) e_{R_s-i} = \begin{bmatrix} 0 \\ \vdots \\ 0 \\ \mathcal{D}_{N^*}^{r_s-1} h \\ \vdots \\ \mathcal{D}_{N^*} h \\ h \\ 0 \\ \vdots \\ 0 \end{bmatrix}$$

belongs to $\ker \mathsf{L}_{A^*}$. A basis for $\mathcal{N}_j = \mathcal{V}_j^n \cap \ker \mathsf{L}_{A^*}$ is formed by such vector fields $v_{\{s,h\}}$, letting s range over the Jordan blocks of A (that is, $s = 1, \ldots, \ell$), and for each fixed s, letting h range over a basis for the subspace K_{js} of $\mathcal{F}_{j+1}^n$ described as follows. Let $M_{js} \subset \mathcal{F}_{j+1}^n$ be the subspace spanned by monomials x^m satisfying (4.6.33); then $\mathcal{D}_{N^*}$ maps M_{js} into itself, and K_{js} is the kernel of $\mathcal{D}_{N^*}^{r_s}$ (viewed as a map of M_{js} to itself).

Proof. Since $\ker \mathsf{L}_{A^*} = \ker \mathsf{L}_S \cap \ker \mathsf{L}_{N^*}$ when A is in Jordan form, we can begin with the description of $\ker \mathsf{L}_{N^*}$ given in Lemma 4.6.13, and add restrictions to guarantee that we stay in $\ker \mathsf{L}_S$. But $\ker \mathsf{L}_S$ is fully described by the requirement that its vector monomials $x^m e_i$ satisfy (4.6.19). Therefore, in particular, the monomials occurring in the main rows $i = R_s$ must satisfy (4.6.33). The theorem will follow if we show that this condition on the main rows implies the corresponding condition in the other rows. That is, we want to show that if x^m satisfies (4.6.33), then each monomial x^μ occurring in $\mathcal{D}_{N^*}^\ell x^m e_{R_s-\ell}$ satisfies

$$\langle \mu, \lambda \rangle - \lambda_{R_s-\ell} = 0$$

for $\ell = 1, \ldots, r_s - 1$. But if A is in Jordan form, all of the diagonal elements of S corresponding to a fixed Jordan block of N must be equal; that is,

$$\lambda_{R_s - \ell} = \lambda_{R_s} \tag{4.6.34}$$

for $\ell = 1, \ldots, r_s - 1$. Therefore, it is enough to show that

$$\langle \mu, \lambda \rangle = \langle m, \lambda \rangle$$

for all multi-indices μ that arise from m in the above manner (that is, by applying $\mathcal{D}_{N^*}$ to x^m). This can be made clear most easily by an example (which immediately generalizes). If

$$A = \begin{bmatrix} \lambda_2 & 1 & & & \\ 0 & \lambda_2 & & & \\ & & \lambda_5 & 1 & 0 \\ & & 0 & \lambda_5 & 1 \\ & & 0 & 0 & \lambda_5 \end{bmatrix}, \tag{4.6.35}$$

where the equal eigenvalues have been numbered by the main row of their block as in (4.6.34), then

$$\langle m, \lambda \rangle = (m_1 + m_2)\lambda_2 + (m_3 + m_4 + m_5)\lambda_5 \tag{4.6.36}$$

and

$$\mathcal{D}_{N^*} = x_1 \frac{\partial}{\partial x_2} + x_3 \frac{\partial}{\partial x_4} + x_4 \frac{\partial}{\partial x_5}. \tag{4.6.37}$$

Now

$$\mathcal{D}_{N^*} x^m = m_2 x_1^{m_1+1} x_2^{m_2-1} x_3^{m_3} x_4^{m_4} x_5^{m_5} + m_4 x_1^{m_1} x_2^{m_2} x_3^{m_3+1} x_4^{m_4-1} x_5^{m_5}$$
$$+ m_5 x_1^{m_1} x_2^{m_2} x_3^{m_3} x_4^{m_4+1} x_5^{m_5-1}.$$

It is clear that applying $\mathcal{D}_{N^*}$ produces monomials x^μ for which $\mu_1 + \mu_2 = m_1 + m_2$ and $\mu_3 + \mu_4 + \mu_5 = m_3 + m_4 + m_5$, so that $\langle \mu, \lambda \rangle = \langle m, \lambda \rangle$. (Before any factors with negative exponent would appear, the corresponding terms drop out by acquiring a zero coefficient.) This completes the proof of the first statement of Theorem 4.6.16 (that $v_{\{s,h\}} \in \ker \mathsf{L}_{A^*}$), and also proves that $\mathcal{D}_{N^*}$ maps M_{js} into itself. The remainder of the theorem is now obvious. $\qquad\square$

As a first illustration of this theorem, we will compute part of the description of the normal form for the (admittedly somewhat artificial) hyperbolic example

$$A = \begin{bmatrix} 1 & 1 & & & \\ 0 & 1 & & & \\ & & -1 & 1 & 0 \\ & & 0 & -1 & 1 \\ & & 0 & 0 & -1 \end{bmatrix}, \tag{4.6.38}$$

which is (4.6.35) with $\lambda_2 = 1$ and $\lambda_5 = -1$. The first step is to choose the grade j of the terms in the normal form that we wish to find, the second is to determine the spaces M_{j1} and M_{j2}, the third is to find the kernels K_{j1} and K_{j2}, and the fourth (and last) is to write down the elements $v_{\{s,h\}}$. For the quadratic terms $(j = 1)$, it is easy to see that the normal form vanishes. The reason is that

$$\langle m, \lambda \rangle = (m_1 + m_2) - (m_3 + m_4 + m_5)$$

can take only the values $2 - 0 = 2$, $1 - 1 = 0$, or $0 - 2 = -2$ for quadratic terms, and none of these values equals one of the eigenvalues 1 or -1; therefore M_{11} and M_{12} are the zero vector space. Continuing on to cubic terms, the space M_{21} is spanned by monomials x^m for which $\langle m, \lambda \rangle = 1$, that is, $m_1 + m_2 = 2$ and $m_3 + m_4 + m_5 = 1$. There are nine such monomials, each being a product of one of $\{x_1^2, x_1 x_2, x_2^2\}$ with one of $\{x_3, x_4, x_5\}$. To determine K_{21}, which is the kernel of $\mathcal{D}_{N^*}^2$ on M_{21}, we find (partial) Jordan chains of $\mathcal{D}_{N^*}$ on M_{21}. It is necessary to find these chains only up to height 2, and the result is two partial chains displayed in the following table (with $\mathcal{D}_{N^*}$ mapping up):

$$\begin{matrix} x_1^2 x_3 & x_1^2 x_4 - x_1 x_2 x_3 \\ x_1^2 x_4 & x_1 x_2 x_4 - x_2^2 x_3 \end{matrix} \qquad (4.6.39)$$

The computation of these chains can be done by "brute-force" linear algebra (for instance by writing a matrix for $\mathcal{D}_{N^*}$ using the basis for M_{21}) or else can be facilitated by methods to be developed in Section 4.7 below. Once these chains are available, we can write down the corresponding basis elements for the cubic normal form. They are

$$v_{\{1,x_1^2 x_3\}} = \begin{bmatrix} 0 \\ x_1^2 x_3 \\ 0 \\ 0 \\ 0 \end{bmatrix}, \qquad v_{\{1,x_1^2 x_4\}} = \begin{bmatrix} x_1^2 x_3 \\ x_1^2 x_4 \\ 0 \\ 0 \\ 0 \end{bmatrix},$$

$$(4.6.40)$$

$$v_{\{1,x_1^2 - x_1 x_2 x_3\}} = \begin{bmatrix} 0 \\ x_1^2 x_4 - x_1 x_2 x_3 \\ 0 \\ 0 \\ 0 \end{bmatrix}, \qquad v_{\{1,x_1 x_2 x_4 - x_2^2 x_3\}} = \begin{bmatrix} x_1^2 x_4 - x_1 x_2 x_3 \\ x_1 x_2 x_4 - x_2^2 x_3 \\ 0 \\ 0 \\ 0 \end{bmatrix}.$$

There are additional cubic normal form elements of the form $v_{\{2,h\}}$ for $h \in K_{22}$, for which the top two entries will vanish instead of the bottom three. To find these it is necessary to compute the Jordan chains of $\mathcal{D}_{N^*}$ on the twelve-dimensional space M_{22} up to height 3 (because we are dealing with a 3×3 Jordan block). The details will be left to the interested reader (preferably after studying Section 4.7).

The Projections into Inner Product Normal Form

The following theorem strengthens Theorem 4.6.16 and enables the construction of the projection into inner product normal form:

4.6.17. Theorem. There exists an *orthonormal* basis for $\ker \mathsf{L}_{A^*}$ consisting of vectors $v_{\{s,h\}}$ of the form described in Theorem 4.6.16.

Proof. By the structure of $v_{\{s,h\}}$ and the definition of the inner product, it follows that two vector fields of this form with different values of s are automatically orthogonal. Given a linearly independent set of vector fields $v_{\{s,h\}}$ with the same s and different $h \in \ker \mathcal{D}_{N^*}^{r_s}$, we show that the Gram–Schmidt process may be applied to produce an orthonormal set of vectors also having the form $v_{\{s,h\}}$.

Fix s and write $v(h) = v_{\{s,h\}}$; notice that $v(h)$ is linear in h. Let $h_1, \ldots, h_m$ be a basis for the subspace of $\ker \mathcal{D}_{N^*}^{r_s}$ spanned by monomials satisfying (4.6.33). By the Gram–Schmidt process, there exist constants c_{ij} such that the vector fields

$$u_i = \sum_j c_{ij} v(h_j) = v\left(\sum_j c_{ij} h_j\right)$$

are orthonormal. The second expression for u_i shows that these vectors have the desired form. $\qquad\square$

4.6.18. Theorem. Given a fixed *orthonormal* basis for $\mathcal{N}_j = \mathcal{V}_j^n \cap \ker \mathsf{L}_{A^*}$ consisting of vector fields $v_{\{s,h\}}$, the *projection into inner product normal form*, or *projection into striped vector fields*, is the map

$$\mathsf{P} : \mathcal{V}_j^n \to \mathcal{N}_j$$

defined by

$$\mathsf{P}w = \sum_{s,h} \langle w, v_{\{s,h\}} \rangle v_{\{s,h\}}.$$

Proof. Since $\mathcal{N}_j$ is orthogonal to $\operatorname{im} \mathcal{L}$, the orthonormal basis for $\mathcal{N}_j$ can be extended to an orthonormal basis for all of $\mathcal{V}_j^n$, with the added vectors spanning $\operatorname{im} \mathcal{L}$. When w is expressed in this basis, $\mathsf{P}w$ is the part lying in $\mathcal{N}_j$. $\qquad\square$

4.6.19. Corollary (cf. Lemma 3.4.20). A vector field w belongs to $\operatorname{im} \mathcal{L}$ if and only if

$$\langle w, v_{\{s,h\}} \rangle = 0$$

for each orthonormal basis element $v_{\{s,h\}}$.

Proof. It is clear from Theorems 4.6.16 and 4.6.17 that $w \in \ker \mathcal{L} = \ker \mathsf{L}_A$ if and only if $\mathsf{P}w = 0$. $\qquad\square$

The Simplified Normal Form

The simplified normal form for vector fields is related to the inner product normal form in the same way that the simplified normal form for matrix series is related to the inner product normal form:

1. The simplified normal form is defined only when A is in Jordan canonical form.

2. It coincides with the inner product and semisimple normal forms if A is diagonal, differing only when there are nontrivial Jordan blocks.

3. Vector fields in simplified normal form have nonzero entries only in positions corresponding to bottom rows of Jordan blocks of A.

4. A vector field in simplified normal form is in extended semisimple normal form.

5. The added simplicity is gained at the expense of some desirable mathematical features satisfied by the inner product normal form. In particular, the simplified normal form is not equivariant under either e^{At} or e^{A^*t} (except, of course, when it coincides with the other normal forms). However, by item 4 above, it is equivariant under e^{St}.

The simplified normal form style will be developed first as a normal form space $\widehat{N}_j$ in each grade j, by way of a specific choice of basis. Afterwards it will be shown that $\widehat{N}_j$ does not depend on these choices, and that the normal form spaces of each grade span a total normal form space $\widehat{N} = \oplus \widehat{N}_j$ that is a module over the ring of invariants $\ker \mathcal{D}_{A^*}$ (although the module elements are not equivariants). An example of the simplified normal form style has already been seen in (1.2.8), and others will occur later. It will play crucial roles in Sections 4.7 and 6.4.

Let N_j be the inner product normal form space of grade j for a given matrix A in Jordan canonical form, and assume that a basis for N_j is given as in Theorem 4.6.16. Rather than taking the vector polynomials $v_{\{s,h\}}$ to be orthonormal with resect to $\langle \, , \, \rangle$ (as in Theorem 4.6.17), we impose here a different requirement: The scalar polynomials h that go with a given s should be taken to be *orthogonal* with resect to $\langle \, | \, \rangle$. (This can be achieved by applying the Gram–Schmidt process to the intersection of $\ker \mathcal{D}_{N^*}^{r_s}$ with the space spanned by monomials satisfying (4.6.33).)

For each

$$v_{\{s,h\}} = \sum_{\ell=0}^{r_s-1} (\mathcal{D}_{N^*}^\ell h) e_{R_s - \ell}$$

appearing in such a basis, let

$$v_{(s,h)} = h e_{R_s}. \tag{4.6.41}$$

Thus, $v_{(s,h)}$ is obtained from $v_{\{s,h\}}$ (written out in full in Theorem 4.6.16) by setting to zero all the entries except the one in the position of the bottom row of the sth Jordan block of A. Define

$$\widehat{\mathcal{N}}_j = \text{span}\left\{v_{(s,h)}\right\} \tag{4.6.42}$$

as (s, h) ranges over the pairs occurring in the given basis for $\mathcal{N}_j$. Since the h occurring in the basis $v_{\{s,h\}}$ for a given s are linearly independent, it follows that the $v_{(s,h)}$ are linearly independent, so that

$$\dim \widehat{\mathcal{N}}_j = \dim \mathcal{N}_j. \tag{4.6.43}$$

4.6.20. Theorem. $\widehat{\mathcal{N}}_j$ is a normal form space, that is, a complement to $\text{im}\,\mathcal{L}: \mathcal{V}_j^n \to \mathcal{V}_j^n$. A vector field in simplified normal form is in extended semisimple normal form.

Proof. In view of (4.6.43), $\widehat{\mathcal{N}}_j$ has the correct dimension to be a normal form space, and it suffices to check that the intersection of $\widehat{\mathcal{N}}_j$ with $\text{im}\,\mathcal{L}$ is the zero vector. Since $\text{im}\,\mathcal{L}$ is the orthogonal complement of $\mathcal{N}_j$, it is enough to show that any nonzero element of $\widehat{\mathcal{N}}_j$ is not orthogonal to $\mathcal{N}_j$. Such an element can be written as

$$w = \sum_{s,h} c_{s,h} v_{(s,h)}.$$

Since $w \neq 0$, there is a particular coefficient $c_{s',h'}$ that is nonzero. Then

$$\langle w, v_{\{s',h'\}}\rangle = \sum c_{s,h}\langle v_{(s,h)}, v_{\{s',h'\}}\rangle.$$

First, all terms in this sum having $s \neq s'$ vanish, because $v_{(s,h)}$ and $v_{\{s',h'\}}$ have their nonzero entries in different positions. Next, for $s = s'$ it follows from the definitions of the basis vectors and the inner products that

$$\langle v_{(s',h)}, v_{\{s',h'\}}\rangle = \langle h|h'\rangle,$$

which vanishes unless $h = h'$ because of the assumption that the h going with a given s have been chosen to be orthogonal. Thus, we conclude that

$$\langle w, v_{\{s',h'\}}\rangle = c_{s',h'}\langle h'|h'\rangle \neq 0.$$

So w is not orthogonal to $\mathcal{N}_j$.

From Lemma 4.6.10, $v_{\{s,h\}} \in \ker \mathsf{L}_S$, and from Corollary 4.5.9 the conditions for membership in $\ker \mathsf{L}_S$ take the form of restrictions on the vector monomials that are allowed to appear. Since $v_{(s,h)}$ is obtained from $v_{\{s,h\}}$ by deleting vector monomials, it follows that $v_{(s,h)} \in \ker \mathsf{L}_S$. It follows that the simplified normal form is an extended semisimple normal form. $\square$

4.6.21. Remark. It is not immediately clear how to compute the projections $\widehat{\mathsf{P}}: \mathcal{V}_j^n \to \widehat{\mathcal{N}}_j$ in the general case. The argument of Theorem 4.6.18 cannot be repeated, because $\widehat{\mathcal{N}}_j$ is not orthogonal to $\text{im}\,\mathcal{L}$. In Section 3.4, the projection $\widehat{\mathbb{P}}$ was obtained by entering each stripe sum into the bottom entry of its stripe. A corresponding construction in the nonlinear case, if it is possible, would require a closer study of the "nonlinear stripe structure." There is not an immediately obvious definition of "stripe sum" in this setting.

The vector fields $v \in \widehat{\mathcal{N}}_j$ can be characterized as follows:

1. They have nonzero entries only in the positions R_s corresponding to bottom rows of Jordan blocks of A.

2. The entry in position R_s belongs to $\ker \mathcal{D}_{N*}^{r_s}$.

3. The entry in position R_s contains only monomials $x^m e_{R_s}$ satisfying $\langle m, \lambda \rangle - \lambda_{R_s} = 0$.

This characterization shows that $\widehat{\mathcal{N}}_j$ does not depend on the choices of h that were made to facilitate the proof of Theorem 4.6.20. Now suppose that $f \in \ker \mathcal{D}_{A*}$ is a polynomial of degree k. We claim that $fv \in \widehat{\mathcal{N}}_{j+k}$. First, the grade of fv is $j + k$ (that is, the degree is $j + k + 1$). Next, we check the items in the list. Clearly, fv has its nonzero entries in the same places as v. The product rule shows that $\mathcal{D}_{N*}$ "passes over" f as though it were constant, so fv belongs to the kernel of the same power as v. Finally, f contains only monomials x^μ for which $\langle \mu, \lambda \rangle = 0$, so fv satisfies the last property in this list. The following theorem is an immediate consequence of these considerations:

4.6.22. Theorem. The simplified normal form space

$$\widehat{\mathcal{N}} = \bigoplus_{j=-1}^{\infty} \widehat{\mathcal{N}}_j$$

is a module over $\ker \mathcal{D}_{A*}$.

Notes and References

The main reference for this section is Elphick et al. [41]. See also Notes and References for Section 3.4.

4.7 The Module Structure of Inner Product and Simplified Normal Forms

Section 4.6 has shown how the description problem for the inner product and simplified normal forms can be solved, up to any finite order, if A is in Jordan canonical form. There are two major steps, which must be carried out for each grade j (degree $j + 1$):

1. For each Jordan block (indexed by $s = 1, \ldots, k$) of A, determine the vector space $M_{js} \subset \mathcal{F}_{j+1}^n$ spanned by the monomials x^m for multi-indices m such that

$$\langle m, \lambda \rangle - \lambda_{R_s} = 0.$$

(Here R_s is the row index of the bottom row of the sth Jordan block, and $\lambda = (\lambda_1, \ldots, \lambda_n)$ are the eigenvalues of A.)

2. In each space M_{js}, find the kernel K_{js} of $(\mathcal{D}_{N^*})^{r_s}$, where r_s is the block size of the sth Jordan block. This can be done by finding a set of Jordan chains for $\mathcal{D}_{N^*}$ in M_{js}; K_{js} will be spanned by those elements of the Jordan chains having height $\leq r_s$. (We have written these Jordan chains with $\mathcal{D}_{N^*}$ mapping upwards, so that a basis K_{js} consists of the entries in the first r_s rows.)

When the spaces K_{js} have been determined (for a given j), we know the entries that can appear in a homogeneous vector field v of grade j in normal form (that is, $v \in \mathcal{N}_j$ or $\widehat{\mathcal{N}}_j$): The main rows (Definition 4.6.12) of v satisfy $v_{R_s} \in K_{js}$, and the entries outside the main rows are zero (in simplified normal form $\widehat{\mathcal{N}}$) or else are generated by iterating $\mathcal{D}_{N^*}$ on the main rows (in inner product normal form $\mathcal{N}$).

The purpose of this section is to extend and improve this solution of the description problem in two ways:

1. A method will be given for generating the Jordan chains of $\mathcal{D}_{N^*}$ in each M_{js} without having to solve any systems of linear equations. This facilitates the calculation of K_{js}, but does not (by itself) change the nature of the conclusion: The description problem is still solved one degree at a time.

2. Continuing further along the same line of reasoning, it will become clear how to obtain a Stanley decomposition for the module $\mathcal{N}$ over the ring $\ker \mathcal{D}_{A^*}$. This enables the description problem to be solved to all orders with a finite amount of calculation.

Since some of the work of this section is quite technical, it is best to assume at first that $A = N$ (equivalently, $S = 0$). This eliminates the step of finding M_{js}. (That is, each $M_{js} = \mathcal{F}_{j+1}^n$). The problem, then, is to describe the module $\ker \mathsf{L}_{N^*}$ over the ring $\ker \mathcal{D}_{N^*}$, for nilpotent matrices N in upper Jordan form. At the end of this section we will treat an example (the nonsemisimple $1 : 1$ resonance) for which $S \neq 0$.

This section represents a departure from our usual rule that every theorem is proved when it is stated. The crucial results that underlie this section are proved in the starred Sections 2.5 and 2.7. In order to make this section accessible to those who are not reading the starred sections, the necessary results have been restated here (in the form needed) without proof. Those who find this material interesting are invited to read those sections at such time as they wish to see the proofs.

4.7.1. Remark. Notice that the starred Section 2.6 is not required for the material in this section, and in fact is not applicable. Section 2.6 provides an algorithm for the computation of certain projection maps, but these are not the correct projection maps for the inner product and simplified normal forms. There is a separate normal form theory, the sl(2) or triad normal form, presented in starred Section 4.8 below, for which the projections calculated in Section 2.6 are correct. The present section borrows some parts of the triad theory and applies them to the normal form style based on the inner product idea, but a more powerful use of the triad notion is possible by setting aside the inner product altogether and beginning again with a normal form style based on triads.

Creating sl(2) Triads of Operators

Given the nilpotent matrix N in upper Jordan form, the first step is to create matrices M and H such that M is a nilpotent matrix with the same block structure as N but is in *modified lower Jordan form*, H is diagonal, and

$$[N, M] = H, \qquad [H, N] = 2N, \qquad [H, M] = -2M. \qquad (4.7.1)$$

"Modified lower Jordan form" means that the only nonzero entries of M lie in the subdiagonal (just as for a nilpotent matrix in lower Jordan form), but these entries are not necessarily equal to 1. Examples of such *sl(2) triads* (or simply *triads*) are

$$N_4 = \begin{bmatrix} 0 & 1 & 0 & 0 \\ 0 & 0 & 1 & 0 \\ 0 & 0 & 0 & 1 \\ 0 & 0 & 0 & 0 \end{bmatrix}, \quad M_4 = \begin{bmatrix} 0 & 0 & 0 & 0 \\ 3 & 0 & 0 & 0 \\ 0 & 4 & 0 & 0 \\ 0 & 0 & 3 & 0 \end{bmatrix}, \quad H_4 = \begin{bmatrix} 3 & & & \\ & 1 & & \\ & & -1 & \\ & & & -3 \end{bmatrix}$$
$$(4.7.2)$$

and

$$N_{3,2} = \begin{bmatrix} 0 & 1 & 0 & & \\ 0 & 0 & 1 & & \\ 0 & 0 & 0 & & \\ & & & 0 & 1 \\ & & & 0 & 0 \end{bmatrix}, \quad M_{3,2} = \begin{bmatrix} 0 & 0 & 0 & & \\ 2 & 0 & 0 & & \\ 0 & 2 & 0 & & \\ & & & 0 & 0 \\ & & & 1 & 0 \end{bmatrix}, \quad (4.7.3)$$

$$H_{3,2} = \begin{bmatrix} 2 & & & & \\ & 0 & & & \\ & & -2 & & \\ & & & 1 & \\ & & & & -1 \end{bmatrix}.$$

In order to give the procedure for obtaining M and H, it is only necessary to tell how to obtain the numbers in the diagonal of H and in the subdiagonal of M. The construction is done blockwise, and the entries in H are built first. For a Jordan block of size r in N, the diagonal entries in the corresponding block of H begin with $r - 1$ and decrease by 2 at each step until $1 - r$ is reached at the bottom of the block. The entries in the subdiagonal of the corresponding block of M are partial sums of the entries in H: The first entry in M is the first entry in H; the second is the sum of the first two entries in H; and so forth, until the block is completed.

Having obtained the triad $\{N, M, H\}$ in this way, we create two additional triads $\{X, Y, Z\}$ and $\{\mathcal{X}, \mathcal{Y}, \mathcal{Z}\}$ as follows:

$$X = M^* = M^\dagger, \qquad Y = N^* = N^\dagger, \qquad Z = H^* = H, \qquad (4.7.4)$$

and

$$\mathcal{X} = \mathcal{D}_Y, \qquad \mathcal{Y} = \mathcal{D}_X, \qquad \mathcal{Z} = \mathcal{D}_Z. \qquad (4.7.5)$$

The first of these is a triad of matrices, and it is easy to check that

$$[X, Y] = Z, \qquad [Z, X] = 2X, \qquad [Z, Y] = -2Y. \qquad (4.7.6)$$

The second is a triad of differential operators, and satisfies

$$[\mathcal{X}, \mathcal{Y}] = \mathcal{Z}, \qquad [\mathcal{Z}, \mathcal{X}] = 2\mathcal{X}, \qquad [\mathcal{Z}, \mathcal{Y}] = -2\mathcal{Y}. \qquad (4.7.7)$$

It is important to notice that (4.7.6) and (4.7.7) have the same form (except for typeface), and that this is achieved by an unexpected crossover in definition (4.7.5), where $\mathcal{X}$ is not $\mathcal{D}_X$, but rather $\mathcal{D}_Y$ (and similarly for $\mathcal{Y}$). There is a similar reversal in going from $\{N, M, H\}$ to $\{M^*, N^*, H^*\} = \{X, Y, Z\}$.

For the N_4 example (4.7.2), we have

$$X = \begin{bmatrix} 0 & 3 & 0 & 0 \\ 0 & 0 & 4 & 0 \\ 0 & 0 & 0 & 3 \\ 0 & 0 & 0 & 0 \end{bmatrix}, \qquad (4.7.8)$$

$$Y = \begin{bmatrix} 0 & 0 & 0 & 0 \\ 1 & 0 & 0 & 0 \\ 0 & 1 & 0 & 0 \\ 0 & 0 & 1 & 0 \end{bmatrix},$$

$$Z = \begin{bmatrix} 3 & & & \\ & 1 & & \\ & & -1 & \\ & & & -3 \end{bmatrix},$$

and

$$X = \mathcal{D}_Y = x_1 \frac{\partial}{\partial x_2} + x_2 \frac{\partial}{\partial x_3} + x_3 \frac{\partial}{\partial x_4}, \tag{4.7.9}$$

$$Y = \mathcal{D}_X = 3x_2 \frac{\partial}{\partial x_1} + 4x_3 \frac{\partial}{\partial x_2} + 3x_4 \frac{\partial}{\partial x_3},$$

$$Z = \mathcal{D}_Z = 3x_1 \frac{\partial}{\partial x_1} + x_2 \frac{\partial}{\partial x_2} - x_3 \frac{\partial}{\partial x_3} - 3x_4 \frac{\partial}{\partial x_4}.$$

According to Section 4.6, a vector field $v = (v_1, v_2, v_3, v_4)$ will be in inner product normal form with respect to N_4 if $Xv = 0$, where $X = L_Y = L_{N^*}$. (This X is part of another triad $\{X, Y, Z\} = \{L_Y, L_X, L_Z\}$ that can be introduced in the same manner as (4.7.6), with the same crossover of X and Y.) As in (4.6.24) and (4.6.27) for the case of N_3 (or, more generally, as in Theorem 4.6.16), the condition

$$L_{N^*}v = L_Y v = Xv = \begin{bmatrix} Xv_1 \\ Xv_2 - v_1 \\ Xv_3 - v_2 \\ Xv_4 - v_3 \end{bmatrix} = 0 \tag{4.7.10}$$

implies that

$$v = v_{\{1,h\}} = \begin{bmatrix} X^3 h \\ X^2 h \\ X h \\ h \end{bmatrix}. \tag{4.7.11}$$

with $h \in \ker X^4$. The simplified normal form consists of vector fields of the form

$$v_{(1,h)} = \begin{bmatrix} 0 \\ 0 \\ 0 \\ h \end{bmatrix} \tag{4.7.12}$$

with $h \in \ker X^4$.

For the $N_{3,2}$ example (4.7.3),

$$X = x_1 \frac{\partial}{\partial x_2} + x_2 \frac{\partial}{\partial x_3} + x_4 \frac{\partial}{\partial x_5}, \tag{4.7.13}$$

$$Y = 2x_2 \frac{\partial}{\partial x_1} + 2x_3 \frac{\partial}{\partial x_2} + x_5 \frac{\partial}{\partial x_4},$$

$$Z = 2x_1 \frac{\partial}{\partial x_1} - 2x_3 \frac{\partial}{\partial x_3} + x_4 \frac{\partial}{\partial x_4} - x_5 \frac{\partial}{\partial x_5}.$$

We will not continue to give details for the $N_{3,2}$ problem; partial results were given in Section 4.6, and the reader is invited to work out the rest using the methods presented below.

4.7.2. Remark. The sl(2) normal form, mentioned in Remark 4.7.1 and developed in Section 4.8, begins with the same triad $\{N, M, H\}$ as above, but does not apply * to these matrices. Thus, it sets $X = N$, $Y = M$, $Z = H$, and continues from there to define $\{\mathcal{X}, \mathcal{Y}, \mathcal{Z}\}$ exactly as in (4.7.6), but with the new meaning of X, Y, Z. This choice of notation may seem awkward at first, but makes it possible to use many of the same formulas in both theories: Once the matrices $\{X, Y, Z\}$ are fixed in one way or the other (either $\{N, M, H\}$ or $\{M^*, N^*, H^*\}$), most of the succeeding development is the same. The differences emerge again at the point where the specific form of X, Y, Z affects the calculations. The notation $\{N, M, H\}$ is somewhat traditional for triads. We used $\{X, Y, Z\}$ in Chapter 2 precisely to allow the usage described here. An additional triad $\{\mathsf{X}, \mathsf{Y}, \mathsf{Z}\}$ can be defined by $\mathsf{X} = \mathsf{L}_Y$, $\mathsf{Y} = \mathsf{L}_X$, $\mathsf{Z} = \mathsf{L}_Z$; this triad operates on vector fields, and $\mathcal{N} = \ker \mathsf{X}$. Part of the point of the following discussion is that when N is in Jordan form, this triad is not really necessary. The full structure of the normal form module can be obtained by studying $\{\mathcal{X}, \mathcal{Y}, \mathcal{Z}\}$.

The differential operators $\{\mathcal{X}, \mathcal{Y}, \mathcal{Z}\}$ map each vector space of homogeneous scalar polynomials $\mathcal{F}^n_{j+1}$ into itself, with $\mathcal{X}$ and $\mathcal{Y}$ being nilpotent and $\mathcal{Z}$ semisimple. The eigenvectors of $\mathcal{Z}$ (called *weight vectors*, but remember that they are actually scalar functions) are the monomials x^m, and the associated eigenvalues (called *weights*) are $\langle m, \mu \rangle$, where $\mu = (\mu_1, \ldots, \mu_n)$ are the eigenvalues (diagonal elements) of Z. That is,

$$\mathcal{Z}(x^m) = \langle m, \mu \rangle x^m. \tag{4.7.14}$$

For instance, for the example N_4 of (4.7.2), (4.7.8), and (4.7.9),

$$\mathcal{Z}(x_1^{m_1} x_2^{m_2} x_3^{m_3} x_4^{m_4}) = (3m_1 + m_2 - m_3 - 3m_4)(x_1^{m_1} x_2^{m_2} x_3^{m_3} x_4^{m_4}). \tag{4.7.15}$$

Generating Jordan Chains of $\mathcal{Y}$

We are now ready to describe the procedure for generating the Jordan chains of $\mathcal{Y}$ on any given $\mathcal{F}^n_{j+1}$. (Remember that these are the spaces M_{js} under the assumption that $A = N$, as discussed at the beginning of this section.) The procedure breaks into the following steps, which will be discussed at greater length below, using the example of N_4 (see (4.7.2)):

1. Construct a *weight table* for $\mathcal{Z}$ on $\mathcal{F}^n_{j+1}$.

2. Construct the *top weight list* derived from the weight table.

3. Determine a weight vector that *fills each position* in the top weight list. (Remember that a "vector" here is a scalar polynomial.)

4. The vectors found in the previous step will be the tops of a set of Jordan chains for the nilpotent operator $\mathcal{Y}$. Apply $\mathcal{Y}$ to these top

weight vectors to generate the Jordan chains of $\mathcal{Y}$. Make a table of these chains, in which $\mathcal{Y}$ is represented as mapping downward.

5. The vectors in the table just described will be *modified Jordan chains* for the nilpotent operator $\mathcal{X} = \mathcal{D}_Y = \mathcal{D}_{N^*}$, regarded as mapping upwards. (*Modified Jordan chain* means that $\mathcal{X}$ operating on a given vector in the table gives a nonzero constant times the vector above it, or zero if the given vector is in the top row.)

Step 1 is to construct a weight table for $\mathcal{Z}$ on $\mathcal{F}^n_{j+1}$. This is done by making a list of all the multi-indices m with $|m| = j + 1$ (so that $x^m \in \mathcal{F}^n_{j+1}$), computing $\langle m, \mu \rangle$ for each such m, and then recording the multiplicity with which each such eigenvalue of $\mathcal{Z}$ occurs. (It is easy to automate this procedure, even in Basic.) It is necessary to record only the nonnegative weights, because the set of weights is symmetrical around zero. For the case of N_4 with $j = 3$ (quartics), the nonnegative weights and their multiplicities are given by

$$
\begin{array}{lccccccc}
\text{Weight} & 12 & 10 & 8 & 6 & 4 & 2 & 0 \\
\text{Multiplicity} & 1 & 1 & 2 & 3 & 4 & 4 & 5
\end{array}
\tag{4.7.16}
$$

with symmetric multiplicities for negative weights.

Step 2 is to construct what is called a *top weight list* from the weight table. This is done by the following recipe: For each nonnegative weight w occurring in the weight table, compute the multiplicity of w minus the multiplicity of $w + 2$. (If $w + 2$ does not occur in the weight table, its multiplicity is zero.) The result is the number of times w occurs in the top weight list. (The justification for this procedure is found in Theorem 2.5.3.) In (4.7.16) the multiplicity of 12 is one greater than that of 14 (which does not occur), so 12 is entered once in the top weight list; the multiplicity of 10 is not greater than that of 12, so 10 is not entered. The complete results for N_4 with $j = 0, \dots, 3$ (that is, degrees 1 through 4, with vertical lines separating the degrees) is found to be

$$
3 \mid 6 \; 2 \mid 9 \; 5 \; 3 \mid 12 \; 8 \; 6 \; 4 \; 0. \tag{4.7.17}
$$

Step 3 is to find a weight vector (or *weight polynomial*, that is, a scalar polynomial that is an eigenvector of $\mathcal{Z}$) that *fills each position* in the top weight list, meaning that it has the required degree and eigenvalue (weight). For instance, to "fill the position" represented by the first 3 in (4.7.17), it is necessary to find a polynomial of degree 1 and weight 3. The monomial x_1 meets the requirements.

There are several techniques available to find polynomials that fill the required positions. Two of these techniques, which we call the *kernel principle* and the *multiplication principle*, are especially important, and are sufficient (in principle) to handle every problem, so we limit ourselves to these methods here. (See Remark 4.7.5 for additional methods.)

4.7.3. Lemma (Kernel principle). To find the top weight vectors of a given weight and degree, it suffices to take any basis for the kernel of the operator $\mathcal{X}$, regarded as an operator on the space spanned by the monomials of the specified weight and degree.

This lemma follows from Theorem 2.5.2. To say that the top weight polynomials are in $\ker \mathcal{X} = \ker \mathcal{D}_{N^*}$ is to say that they are *invariants* of the flow e^{N^*t}. It is helpful at this point to have kept the calculations done in step 1 (or to write your computer program so that it will give the monomials of any requested weight and degree). For instance, to fill the position of weight 2 and degree 2 in (4.7.17), we observe that x_2^2 and x_1x_3 are the monomials of this weight and degree, so the desired polynomial $ax_2^2 + bx_1x_3$ must satisfy

$$\mathcal{X}(ax_2^2 + bx_1x_3) = 0.$$

Using (4.7.9), it follows that $a = 1$, $b = -2$ will work, so that $x_2^2 - 2x_1x_3$ fills the position.

4.7.4. Lemma (Multiplication principle). Any product of weight polynomials is a weight polynomial; the degree and weight of the product is the sum of the degrees and weights of the factors.

Proof. Suppose that f and g are weight polynomials with weights w_f and w_g. Then

$$\mathcal{Z}(fg) = (\mathcal{Z}f)g + f(\mathcal{Z}g) = w_f fg + f w_g g = (w_f + w_g)fg,$$

so fg is a weight polynomial with weight $w_f + w_g$. It is clear that the degree of fg is the sum of the degrees of f and g. □

Since x_1 fills the position with degree 1 and weight 3 in (4.7.17), it follows from the multiplication principle that x_1^2, x_1^3, and x_1^4 fill the positions of weights 6, 9, and 12 (of degrees 2, 3, and 4). Having found two "basic invariants" $\alpha = x_1$ and $\beta = x_2^2 - 2x_1x_2$, the multiplication principle shows that $\alpha\beta$, $\alpha^2\beta$, and β^2 fill the positions of weights 5, 8, and 4 in (4.7.17).

The lowest position still remaining to be filled is of degree 3, weight 3; calling this γ, $\alpha\gamma$ will fill the position of degree 4, weight 6. The kernel principle is needed to determine γ, as well as the element δ of degree 4, weight 0. The results may be summarized as

$$
\begin{array}{c|cc|ccc|ccccc}
3 & 6 & 2 & 9 & 5 & 3 & 12 & 8 & 6 & 4 & 0 \\
\alpha & \alpha^2 & \beta & \alpha^3 & \alpha\beta & \gamma & \alpha^4 & \alpha^2\beta & \alpha\gamma & \beta^2 & \delta
\end{array}
\qquad (4.7.18)
$$

with

$$\alpha = x_1,$$

(4.7.19)

$$\beta = x_2^2 - 2x_1 x_3,$$

$$\gamma = x_2^3 - 3x_1 x_2 x_3 + 3x_1^2 x_4,$$

$$\delta = 9x_1^2 x_4^2 - 3x_2^2 x_3^2 - 18x_1 x_2 x_3 x_4 + 8x_1 x_3^3 + 6x_2^3 x_4.$$

4.7.5. Remark. Two additional methods for finding invariants to fill the positions 'are the *cross-section method* and the *method of transvectants*. All orbits of the flow $e^{N^* t}$ intersect the three-dimensional subspace $x_2 = 0$, which is therefore called a *cross-section* to the flow. The map sending each point $x = (x_1, x_2, x_3, x_4)$ to the point $(\alpha, 0, \beta, \gamma)$ where its orbit intersects the subspace gives three invariants, which are exactly the α, β, and γ found above. These can be found by explicitly integrating the linear vector field $N^* x$ to obtain its flow. (This is equivalent to certain calculations in the literature that are said to be done by the "method of characteristics" from partial differential equations.) The invariant δ cannot be found in this way. See Notes and References for further information about the cross-section and transvectant methods.

The polynomials found in step 3 are the chain tops for the Jordan chains of $\mathcal{Y}$ in each degree. Step 4 is to fill in the chains under each of these tops. The result will be a table exactly like (4.6.27) for the case of N_3. (We omit the first column, in degree zero, which is always the same.) As we saw in connection with (4.6.27), it is not necessary to compute the complete chains, only the entries down to a depth r equal to the size of the Jordan block in N^* that we are working on. For N_4, this means depth 4; in the general case, with blocks indexed by $s = 1, \ldots, \ell$, we must work separately with each distinct value of $r = r_s$ that appears. (The *depth* is the height with respect to $\mathcal{X}$; the depth of the top row is 1.) Furthermore, because of simplifications to be introduced later, it is not advisable to compute even these entries in detail at this point. It is sufficient to indicate in symbolic form the items to be computed, and to identify those that are zero. (Since the aim is to find a basis for ker $\mathcal{X}^r$, any zero elements must be omitted. Such elements are indicated with an asterisk in the following table.) To identify the zero entries, we use the fact (another consequence of Theorem 2.5.2) that the height of each top weight vector is one more than its weight. For instance, since the weight of β is 2, its height is 3, and its chain will be $\beta, \mathcal{Y}\beta, \mathcal{Y}^2\beta$; it follows that $\mathcal{Y}^3\beta = 0$, so the entry in the fourth row under β is $*$. At this point our result is

$$
\begin{array}{c|ccc|ccc}
\alpha & \alpha^2 & \beta & \alpha^3 & \alpha\beta & \gamma \\
\mathcal{Y}\alpha & \mathcal{Y}(\alpha^2) & \mathcal{Y}\beta & \mathcal{Y}(\alpha^3) & \mathcal{Y}(\alpha\beta) & \mathcal{Y}(\gamma) \\
\mathcal{Y}^2\alpha & \mathcal{Y}^2(\alpha^2) & \mathcal{Y}^2\beta & \mathcal{Y}^2(\alpha^3) & \mathcal{Y}^2(\alpha\beta) & \mathcal{Y}^2(\gamma) \\
\mathcal{Y}^3\alpha & \mathcal{Y}^3(\alpha^2) & * & \mathcal{Y}^3(\alpha^3) & \mathcal{Y}^3(\alpha\beta) & \mathcal{Y}^3(\gamma)
\end{array}
$$

(4.7.20)

$$
\begin{array}{ccccc|}
\alpha^4 & \alpha^2\beta & \alpha\gamma & \beta^2 & \delta \\
\mathcal{Y}(\alpha^4) & \mathcal{Y}(\alpha^2\beta) & \mathcal{Y}(\alpha\gamma) & \mathcal{Y}(\beta^2) & * \\
\mathcal{Y}^2(\alpha^4) & \mathcal{Y}^2(\alpha^2\beta) & \mathcal{Y}^2(\alpha\gamma) & \mathcal{Y}^2(\beta^2) & * \\
\mathcal{Y}^3(\alpha^4) & \mathcal{Y}^3(\alpha^2\beta) & \mathcal{Y}^3(\alpha\gamma) & \mathcal{Y}^3(\beta^2) & * \\
\end{array}
$$

Step 5 is now merely an observation, not calling for any further work. The observation is that the Jordan chains for $\mathcal{Y}$ (the top four rows of which we have just constructed) are also modified Jordan chains for $\mathcal{X}$, in the following sense: $\mathcal{X}$ applied to any entry gives a nonzero constant multiple of the entry above it. (We already know that at the top, $\mathcal{X}$ maps to zero.) More precisely, if f is an entry in (4.7.20), then $\mathcal{X}\mathcal{Y}f = pf$, where p is the sum of the weights of the entries above $\mathcal{Y}f$ in the table. This sum of weights has been called the *pressure on* $\mathcal{Y}f$ in Section 2.5. (Applying $\mathcal{Y}$ to a weight vector of weight w yields a weight vector of weight $w - 2$, so we know the weights of each entry by starting from the top weight list. All of these remarks are consequences of Theorem 2.5.2.) Since the (complete modified) Jordan chains of $\mathcal{X}$ in degree $j + 1$ span $\mathcal{F}^n_{j+1}$, it follows that the "bottom" r elements of these chains (which appear in the table as the *top* r elements of the Jordan chains of $\mathcal{Y}$) form a basis for $(\ker \mathcal{X}^r) \cap \mathcal{F}^n_{j+1}$.

This completes the procedures for finding the Jordan chains in each degree. It is necessary to carry out these steps only up to some finite degree. Later in this section it will be seen that from some point on, the "multiplication principle" suffices to determine all of the chain tops for $\mathcal{Y}$, and the "kernel principle" is no longer needed. Once this point is reached, it is possible to describe the entire module of equivariants using algebraic methods.

Simplifying the Basis for $\ker \mathcal{X}^r$ in Each Degree

It was suggested above that the computation of the entries in table (4.7.20) be postponed. The reason for this suggestion is that this table can be replaced by another that is easier to compute. The entries in this new table *do not form Jordan chains* for either $\mathcal{X}$ or $\mathcal{Y}$, but the entries in the top r rows (in each degree) still provide a basis for $\ker \mathcal{X}^r$ (in that degree). The rule for forming the new table will first be illustrated by a few examples, and then stated more precisely. The justification of the procedure (that is, the proof that it actually gives an alternative basis for $\ker \mathcal{X}^r$) will be deferred to the end of this subsection.

The entries in the new table are created by changing the position of the $\mathcal{Y}$ operators within each expression, moving them as far to the right as possible without causing the result to vanish. For instance, under α^2 in the old table is the entry $\mathcal{Y}(\alpha^2) = \mathcal{Y}(\alpha\alpha)$. According to the procedure we are describing, $\mathcal{Y}$ may be moved to the right in this product, replacing $\mathcal{Y}(\alpha^2)$ by $\alpha\mathcal{Y}\alpha$. Similarly, $\mathcal{Y}^2(\alpha^2)$ may be replaced by $\alpha\mathcal{Y}^2\alpha$, and $\mathcal{Y}^3(\alpha^2)$ by $\alpha\mathcal{Y}^3\alpha$. On the other hand, consider the column under $\alpha\beta$ in the old table. $\mathcal{Y}(\alpha\beta)$ may be replaced by $\alpha y\beta$, and $y^2(\alpha\beta)$ by $\alpha y^2\beta$, but $y^3(\alpha\beta)$ may not be replaced by $\alpha y^3\beta$ (which is equal to zero, since $y^3\beta = 0$, as can be seen from the $*$ in the column under β). Instead, $y^3(\alpha\beta)$ may be replaced by $(y\alpha)(y^2\beta)$. An alternative way to proceed is to write $\alpha\beta$ in the order $\beta\alpha$ before trying to move factors of y to the right. Since α is capable of sustaining three factors of y without vanishing, the entries under $\alpha\beta$ can now be replaced by $\beta y\alpha$, $\beta y^2\alpha$, and $\beta y^3\alpha$.

Here is the general rule: In the old table, each entry has the form of a power of y acting on a product of basic invariants. Choose an ordering of the basic invariants, for instance $\alpha < \beta < \gamma < \delta$. (This is not an ordering by magnitude, only the order in which the factors will be written.) Now expand each product by writing it without exponents, putting the factors in the chosen order; thus $y^5(\alpha\beta^2\gamma^3\delta)$ will become $y^5(\alpha\beta\beta\gamma\gamma\gamma\delta)$. Next, move each factor of y as far to the right as possible, in such a way that each y acts only on one factor of the product, without making the factor become zero. In the example, we attempt to move one factor of y to δ, obtaining $y^4(\alpha\beta\beta\gamma\gamma\gamma y\delta)$; this is rejected, because $y\delta = 0$, so instead we have $y^4(\alpha\beta\beta\gamma\gamma(y\gamma)\delta)$. Continuing to move two more factors of y to the final γ leads to $y^2(\alpha\beta\beta\gamma\gamma(y^3\gamma)\delta)$; we cannot move another factor of y to this position, because $y^4\gamma = 0$ (since γ has weight 3, hence chain length 4, from (4.7.18)). The final two factors of y are then moved to the preceding γ, yielding $\alpha\beta\beta\gamma(y^2\gamma)(y^3\gamma)\delta = \alpha\beta^2\gamma(y^2\gamma)(y^3\gamma)\delta$.

This rule is not completely defined until an ordering of the basic invariants has been selected. It is best to order the basic invariants by increasing weight (or equivalently, increasing chain length of the chains that they head); this guarantees that if there is a basic invariant of weight $\geq r$ present in a given product, then the rightmost factor will have weight $\geq r$, and (down to depth r) it will be possible to move all occurrences of y to this last factor without producing zero. (Thus, in one of the examples considered above, it is because α has greater weight than β that the $\alpha\beta$ column worked out most simply when written $\beta\alpha$.) Therefore, for the N_4 problem, we choose $\delta < \beta < \alpha < \gamma$, with weights $0 < 2 < 3 = 3$. With this choice, the new table replacing (4.7.20) becomes

$$
\begin{array}{c|cc|cccc}
\alpha & \alpha^2 & \beta & \alpha^3 & \beta\alpha & \gamma \\
y\alpha & \alpha(y\alpha) & yb & \alpha^2(y\alpha) & \beta(y\alpha) & y(\gamma) \\
y^2\alpha & \alpha(y^2\alpha) & y^2b & \alpha^2(y^2\alpha) & \beta(y^2\alpha) & y^2(\gamma) \\
y^3\alpha & \alpha(y^3\alpha) & * & \alpha^2(y^3\alpha) & \beta(y^3\alpha) & y^3(\gamma)
\end{array}
\tag{4.7.21}
$$

$$
\begin{array}{ccccc}
\alpha^4 & \beta\alpha^2 & \alpha\gamma & \beta^2 & \delta \\
\alpha^3(y\alpha) & \beta\alpha(y\alpha) & \alpha y(\gamma) & \beta(y\beta) & * \\
\alpha^3(y^2\alpha) & \beta\alpha(y^2\alpha) & \alpha y^2(\gamma) & \beta(y^2\beta) & * \\
\alpha^3(y^3\alpha) & \beta\alpha(y^3\alpha) & \alpha y^3(\gamma) & (y\beta)(y^2\beta) & *
\end{array}
$$

Observe that with this ordering there is only one place where y must be used in two factors, namely, $(y\beta)(y^2\beta)$.

Before showing that the entries in table (4.7.21) actually form a basis for ker $\mathcal{X}^4$, we should explain why this basis is considered simpler than the original basis (4.7.20). First, from a computational standpoint it is certainly easier to differentiate low-degree polynomials than ones of higher degree; in (4.7.21) the operator y need only be applied to basic invariants, whereas in (4.7.20) it is applied to products of basic invariants, which can be of arbitrarily high degree. More important still is the fact that there are only finitely many derivatives that must be computed for (4.7.21), no matter how far to the right we may wish to extend the table. These derivatives are computed in (4.7.31) below; once these are known, the rest of table (4.7.21) requires multiplication only. It is this fact that makes it possible to obtain the Stanley decomposition of the module ker $\mathcal{X}^r$ later in this section.

It is now time to justify the rules leading to table (4.7.21).

4.7.6. Lemma. The elements of a "simplified basis" for ker $\mathcal{X}^r$, constructed according to the rules described above, actually constitute a basis for that space.

Proof. It is clear that there are the right number of basis elements, since they are in one-to-one correspondence with the elements of the original Jordan chain basis. So it is enough to check that they belong to ker $\mathcal{X}^r$ and are linearly independent. The first step is to notice that applying $\mathcal{X}^{i-1}$ to any term in the table at depth i leads to a nonzero constant times the entry at the top of the column. For instance, beginning with $(y\beta)(y^2\beta)$, the first application of $\mathcal{X}$, using the product rule and the pressure formula, gives $(2\beta)(y^2\beta) + (y\beta)(2y\beta)$, the second application gives $(2\beta)(2y\beta) + (2\beta)(2y\beta) + (y\beta)(4\beta) = 12\beta y\beta$, and the third gives $24\beta^2$, which is a nonzero constant times the top of the column, β^2. In general, if $f_1 \cdots f_m$ is the top of a column, where the f_k are invariants, possibly repeated, then if a product of the form $(y^{i_1} f_1) \cdots (y^{i_m} f_m)$ occurs in the column, its depth will be $1 + i_1 + \cdots + i_m$. Applying $\mathcal{X}$ to such a product yields a sum of terms, each containing one less y, and each having a positive pressure factor. After the right number of repetitions, no factors of y remain, and the result is $f_1 \cdots f_m$ times a positive number resulting from sums and products of these pressure factors.

The first consequence of this observation is that one more application of $\mathcal{X}$ results in zero, so in fact the original entry at depth i in the table truly has depth i with respect to $\mathcal{X}$. It follows that every element down to depth r belongs to ker $\mathcal{X}^r$.

Now we confine our attention to a particular degree j, and suppose that some nontrivial linear combination of the entries of that degree (down to depth r) were zero. Let i be the depth of the deepest element in the combination having a nonzero coefficient. Applying $\mathcal{X}^{i-1}$ repeatedly to the combination would then result (using the observation again) in a nontrivial linear combination among the elements of the top row of the table. But this is impossible, since these elements are known to be linearly independent. Therefore, the simplified basis elements are linearly independent. $\qquad\square$

Finding $\mathcal{N}_j$

At this point, we pause to remind the reader that if it is required to find $\mathcal{N}_j$ or $\widehat{\mathcal{N}}_j$ only up to the degree for which a table like (4.7.20) has been made (rather than to find the entire module $\mathcal{N}$), then the result is available at once. It is only necessary to create the vector fields $v_{(1,h)}$ (for the simplified normal form) or $v_{\{1,h\}}$ (for the inner product normal form), using each entry from (4.7.21) as h; the formulas for these vector fields are given in (4.7.11) and (4.7.12). For instance, with $h = (\mathcal{y}\beta)(\mathcal{y}^2\beta)$ we have

$$
v_{(1,h)} = \begin{bmatrix} 0 \\ 0 \\ 0 \\ (\mathcal{y}\beta)(\mathcal{y}^2\beta) \end{bmatrix} \quad \text{and} \quad v_{\{1,h\}} = \begin{bmatrix} 24\beta^2 \\ 12\beta(\mathcal{y}\beta) \\ 2\beta(\mathcal{y}^2\beta) + 2(\mathcal{y}\beta)^2 \\ (\mathcal{y}\beta)(\mathcal{y}^2\beta) \end{bmatrix}.
$$

Notice that the calculations to obtain the entries above the bottom entry in $v_{\{1,h\}}$ consist of repeatedly applying $\mathcal{X} = \mathcal{D}_{N^*}$, as in Theorem 4.6.16, and these calculations are the same as those illustrated in the proof of Lemma 4.7.6 above.

If there is more than one Jordan block in N, there will be a table such as (4.7.20) or (4.7.21) for each block, having depth r_s for the sth block. Each polynomial h in the table for a given s contributes a basis vector $v_{(s,h)}$ to $\widehat{\mathcal{N}}_j$ (or $v_{\{s,h\}}$ to $\mathcal{N}_j$).

Finding $\ker \mathcal{X}$, the Full Ring of Invariants

When the steps described above have been carried out, certain *basic invariants* have been found, which generate (by the multiplication principle) a complete set of chain tops for $\mathcal{y}$ up to some chosen degree. Since these chain tops are a basis for $\ker \mathcal{X}$, we can also say that the basic invariants generate all the invariants up to the chosen degree. In the example of N_4, these basic invariants are α, β, γ, and δ, and these are known to generate all of the invariants up to degree 4. There are now two possibilities: Either the known basic invariants generate all the invariants of all orders, or there are additional basic invariants that must be calculated at some higher order. We would like to have a method for testing whether enough basic invariants have been found. If we do in fact have enough basic invariants, then the next step should be to find the relations among these invariants and use these relations to obtain a Stanley decomposition for the ring of invariants, as we have already done in Section 4.5 in several semisimple problems.

It turns out that the best way to proceed is to reverse this apparently natural order of steps, as follows:

1. Determine the relations among the known basic invariants.

2. Find a Stanley decomposition of the ring $\mathcal{R}$ of polynomials in the known basic invariants.

3. From this Stanley decomposition, develop a two-variable generating function called the *table function*, which facilitates a test to see whether $\mathcal{R}$ is in fact equal to the full ring of invariants $\ker \mathcal{X}$ or is only a subring of $\ker \mathcal{X}$.

4. In the former case we have found both the ring of invariants and a Stanley decomposition for it; in the latter case, it is necessary to go back to the previous steps and calculate more basic invariants (by the kernel principle) in higher degrees, and then repeat these steps. (The work already done is not wasted, but forms part of what is needed when more basic invariants are added.)

We will now carry out these steps for the example of N_4. It will turn out that we have already found enough basic invariants. Beginning from (4.7.19), the first step is to establish that

$$\gamma^2 = \beta^3 + \alpha^2 \delta \qquad (4.7.22)$$

and that there are no other relations. This is easy to check once it is known, but how can it be discovered? One way is that if we had continued the calculation of the top weight list up to degree 6, it would become apparent that the three products γ^2, β^3, and $\alpha^2\delta$ all have degree 6 and weight 6, but they are competing (by the multiplication principle) to fill only two positions with this degree and weight. It follows that there must be a linear combination of them equal to zero, and a little experimentation yields (4.7.22). There cannot exist another relation independent of this; the functions α, β, and γ are algebraically independent, because each of them depends upon a new variable not appearing in the previous functions in the list.

> **4.7.7. Remark.** As discussed in Section A.5 following Lemma A.5.4, there exists software, based on Gröbner basis methods, that will return equation (4.7.22) when (4.7.19) is typed in. More generally, the software will provide a Gröbner basis for the relations holding among the given basic invariants. In the present example, the ideal of relations is principal, and its single generator (4.7.22) automatically constitutes a Gröbner basis.

Now consider the ring $\mathcal{R} = \mathbb{R}[\alpha, \beta, \gamma, \delta] \subset \mathcal{F}_*^n$ of polynomials in the known basic invariants. The representation of an element of $\mathcal{R}$ as a polynomial is not unique, because of (4.7.22), but this equation itself can be used to restore the uniqueness by excluding γ^2 (or any higher power of γ). That is, whenever γ^2 appears in a polynomial in $\mathcal{R}$, it can be replaced by the right-hand side of (4.7.22), and the expression obtained in this way will be unique. Thus, a Stanley decomposition of $\mathcal{R}$ is

$$\mathcal{R} = \mathbb{R}[\alpha, \beta, \delta] \oplus \mathbb{R}[\alpha, \beta, \delta]\gamma. \tag{4.7.23}$$

Another way to say this is that any polynomial in $\mathcal{R}$ can be written uniquely as

$$f(\alpha, \beta, \delta) + g(\alpha, \beta, \delta)\gamma, \tag{4.7.24}$$

where f and g are polynomials in three variables. Still another way to say this is that each element of $\mathcal{R}$ is uniquely expressible as a linear combination of *standard monomials*, which are monomials in the basic invariants that contain at most one factor of γ (equivalently, are not divisible by γ^2). Everything stated in this paragraph remains true if "polynomial" is replaced everywhere by "formal power series," provided that (4.7.23) is rewritten as $\mathcal{R} = \mathbb{R}[[\alpha, \beta, \delta]] \oplus \mathbb{R}[[\alpha, \beta, \delta]]\gamma$ and "linear combination" is understood as "infinite linear combination."

> **4.7.8. Remark.** In the language of Section A.5, we choose a term ordering on the symbols α, β, γ, δ such that γ^2 is the leading term of $\gamma^2 - \beta^3 - \alpha^2\delta$, and then to the ideal $\langle \gamma^2 - \beta^3 - \alpha^2\delta \rangle$ we associate the monomial ideal generated by its leading term, $\langle \gamma^2 \rangle$. A standard monomial is any monomial in these symbols that does not belong to this monomial ideal.

The Stanley decomposition (4.7.24) can be abbreviated as $f \cdot 1 + g \cdot \gamma$; the functions f and g will be referred to as *coefficients*, with 1 and γ being called *Stanley basis elements*. To generate what is called the *table function* of the Stanley decomposition, replace each term in (4.7.24) by a rational function P/Q in d and w (symbols standing for "degree in x" and "weight") constructed as follows:

1. For each basic invariant appearing in a coefficient, the denominator Q is to contain a factor $1 - d^p w^q$, where p and q are the degree and weight of the invariant. In the present example, the coefficients f and g both contain α, β, and δ, so the denominator will be the same for both terms, and will be the product of factors $(1 - dw^3)$, $(1 - d^2w^2)$, and $(1 - d^4)$.

2. For each term, the numerator P will be $d^p w^q$, where p and q are the degree and weight of the Stanley basis element for that term. For the basis element 1, the degree and weight are zero, so the numerator is $d^0 w^0 = 1$. For γ, the numerator is $d^3 w^3$.

When the rational functions P/Q obtained from the terms of the Stanley decomposition are added, the result, for the example of (4.7.24), is the table function

$$T = \frac{1 + d^3 w^3}{(1 - dw^3)(1 - d^2w^2)(1 - d^4)}. \tag{4.7.25}$$

When the table function is fully expanded as a power series in d and w, it contains the term $md^r w^s$ if and only if there are exactly m standard monomials that have total degree r and weight s. Thus, the table function is an encoded version of that part of the top weight list (4.7.18), carried to all orders, that can be generated from $\alpha, \beta, \gamma, \delta$. To prove that $\mathcal{R} = \ker \mathcal{X}$, it would suffice to prove that this is the entire top weight list, and for this it is enough to prove that the number of elements in the chains having these top weights, in degree j, is equal to $\dim \mathcal{F}_j^n$. By Lemma 4.7.9 below, this is equivalent to checking that

$$\frac{\partial}{\partial w} wT \bigg|_{w=1} = \frac{1}{(1-d)^4},$$

which can easily be verified by evaluating the partial derivative.

4.7.9. Lemma. Let $\{X, Y, Z\}$ be a triad of $n \times n$ matrices satisfying (4.7.6), let $\{\mathcal{X}, \mathcal{Y}, \mathcal{Z}\}$ be defined as in (4.7.5), and suppose that $I_1, \ldots, I_k$ is a finite set of polynomials in $\ker \mathcal{X}$. Let $\mathcal{R}$ be the subring $\mathbb{R}[I_1, \ldots, I_k]$; suppose that the relations (if any) among $I_1, \ldots, I_k$ are known, and that the Stanley decomposition and its associated table function $T(d, w)$ have been determined. Then $\mathcal{R} = \ker \mathcal{X} \subset \mathcal{F}_*^n$ if and only if

$$\frac{\partial}{\partial w} wT \bigg|_{w=1} = \frac{1}{(1-d)^n}. \tag{4.7.26}$$

This is also a necessary and sufficient condition for $\mathbb{R}[[I_1, \ldots, I_k]]$ to equal $\ker \mathcal{X} \subset \mathcal{F}_{**}^n$.

Proof. First, observe that the relations, Stanley decomposition, and table function determined by the set of polynomials $I_1, \ldots, I_k$ will be the same whether we choose to work in $\mathcal{F}_*^n$ or $\mathcal{F}_{**}^n$. Regard the table function as a formal power series in d and w, and consider each term $md^r w^s$; the presence of this term indicates that the top weight list for $\ker \mathcal{X}$ contains m positions filled by monomials in $I_1, \ldots, I_k$ having degree r and weight s. (These will be standard monomials, in the sense of Remark 4.7.8, with respect to the ideal of relations among $I_1, \ldots, I_k$, but this fact is not needed for the proof of this lemma.) The only thing that is in question is whether *all* positions in the top weight list are filled in this way. (If $\mathcal{R}$ is not equal to $\ker \mathcal{X}$, then it is a proper subring of $\ker \mathcal{X}$, and there will be positions in the top weight list for $\ker \mathcal{X}$ that are not filled by elements of $\mathcal{R}$.)

To prove that $\mathcal{R} = \ker \mathcal{X}$, then, it would suffice to count the number of elements in the chains whose tops are accounted for by the table function $T(d, w)$. If the total number of elements occurring in these chains, having a given degree r, equals the dimension of $\mathcal{F}_r^n$ as given by Lemma 4.5.1, then there are no positions in the top weight list that are not accounted for. But the operations indicated by $(\partial / \partial w) wT|_{w=1}$ exactly carry out the required calculation. First, multiplying each term $md^r w^s$ by w produces $md^r w^{s+1}$, where $s+1$ is the chain length of each of the m chains of degree r and weight

s. Next, differentiating with respect to w gives $m(s+1)d^r w^{s+1}$, so that the new coefficient $m(s+1)$ equals the total number of elements in these m chains. Finally, setting $w = 1$ leaves $m(s+1)d^r$, which then combines with all other terms of degree r so that the coefficient of d^r equals the total number of elements of degree r in the chains whose tops belong to $\mathcal{R}$.

The final step of the proof is to check that $f(d) = 1/(1-d)^n$, when expanded in powers of d, has $\dim \mathcal{F}_r^n$ as the coefficient of d^r. By Taylor's theorem and Lemma 4.5.1, this requires checking that

$$\frac{f^{(r)}(0)}{r!} = \binom{n+r-1}{r},$$

which is straightforward.   □

We have shown that in fact, for the N_4 problem, $\mathcal{R}$ (as calculated above) equals $\ker \mathcal{X}$. It follows that the Stanley decomposition (4.7.23) or (4.7.24) for $\mathcal{R}$ is the correct Stanley decomposition for $\ker \mathcal{X}$. The standard monomials in $\alpha, \beta, \gamma, \delta$ can now be taken as the chain tops for the continuation of either (4.7.20) or (4.7.21) to the right to all orders.

To see an example in which the table function shows that $\mathcal{R}$ is not equal to $\ker \mathcal{X}$, suppose that we had initially done the calculations for the N_4 problem only to the third order and found α, β, and γ. These are algebraically independent (have no relations), so the Stanley decomposition of $\mathcal{R} = \mathbb{R}[\alpha, \beta, \gamma]$ is just this expression itself, and the table function is

$$T = \frac{1}{(1-dw^3)(1-d^2w^2)(1-d^3w^3)}.$$

This does not satisfy (4.7.26) with $n = 4$, so $\mathbb{R}[\alpha, \beta, \gamma] \neq \ker \mathcal{X}$, showing that it is necessary to calculate to at least the next order. When this is done, δ is found and the result is complete (as already shown).

Finding $\widehat{\mathcal{N}}$ and $\mathcal{N}$ as Modules over $\ker \mathcal{X}$

At this point, we know the structure of the ring $\ker \mathcal{X}$ (for the N_4 example) to all orders, although we have done only a finite amount of work (in contrast to the infinite amount of work that would be necessary if we continued to calculate in each degree independently). We know the structure of the top row of tables (4.7.20) and (4.7.21) continued to the right; it remains to determine the structure of the top four rows of (4.7.21) continued to the right, that is, the structure of $\ker \mathcal{X}^4$ as a module over $\ker \mathcal{X}$.

It is here that the advantage of the simplified basis shows itself. The general chain top in table (4.7.21) is a standard monomial and so can be written as $\alpha^i \beta^j \gamma^k \delta^\ell$, with k being 0 or 1. Using the ordering of variables $\delta < \beta < \alpha < \gamma$, introduced above in connection with the simplified basis, this becomes $\delta^\ell \beta^j \alpha^i \gamma^k$. To obtain the second through the fourth rows of (4.7.21) under this chain top, we introduce one, two, or three factors of $\mathcal{Y}$, and push them as far to the right as possible without making the result vanish. The chain tops can be divided into five groups as follows:

1. $k = 1$.

2. $k = 0$ but $i \neq 0$.

3. $k = i = 0$ and $j \geq 2$.

4. $k = i = 0$ and $j = 1$.

5. $k = i = j = 0$ and $\ell \geq 0$.

A monomial of the first type has the form $I\gamma$, where I is an arbitrary monomial not involving γ, that is, $I \in \mathbb{R}[\alpha, \beta, \delta]$. The entries under a chain top of this form (excluding the chain top itself) will be $I\mathcal{Y}\gamma$, $I\mathcal{Y}^2\gamma$, and $I\mathcal{Y}^3\gamma$. Such an entry will be said to have *prefix* I and *suffix* $\mathcal{Y}\gamma$, $\mathcal{Y}^2\gamma$, or $\mathcal{Y}^3\gamma$. A suffix always begins with $\mathcal{Y}$. The *prefix ring* for a given suffix is the set of admissible prefixes that can appear with that suffix; the prefix ring for each suffix $\mathcal{Y}\gamma$, $\mathcal{Y}^2\gamma$, and $\mathcal{Y}^3\gamma$ is $\mathbb{R}[\alpha, \beta, \delta]$.

A monomial of the second type has the form $I\alpha$ with $I \in \mathbb{R}[\alpha, \beta, \delta]$. The entries under it have prefix I and suffix $\mathcal{Y}\alpha$, $\mathcal{Y}^2\alpha$, or $\mathcal{Y}^3\alpha$. A monomial of type 3 has the form $I\beta^2$ with $I \in \mathbb{R}[\beta, \delta]$; notice the change in the prefix ring. The entries under it either have prefix $I\beta \in \mathbb{R}[\beta, \delta]$ with suffix $\mathcal{Y}\beta$ or $\mathcal{Y}^2\beta$, or else have prefix I with suffix $\mathcal{Y}\beta\mathcal{Y}^2\beta$. (Recall that $\mathcal{Y}^2\beta = 0$.) A monomial of type 4 has the form $\delta^\ell\beta$. Since $\mathcal{Y}\delta = 0$, there will be only two entries under this, $\delta^\ell\mathcal{Y}\beta$ and $\delta^\ell\mathcal{Y}^2\beta$. Finally, a chain top of type 5 will have no entries under it. Since every element of ker $\mathcal{X}^4$ is a linear combination of the elements of the four rows of table (4.7.21), we have obtained the following Stanley decomposition for ker $\mathcal{X}^4$:

$$\ker \mathcal{X}^4 = \mathbb{R}[\alpha, \beta, \delta] \oplus \mathbb{R}[\alpha, \beta, \delta]\gamma \qquad (4.7.27)$$
$$\oplus\, \mathbb{R}[\alpha, \beta, \delta]\mathcal{Y}\gamma \oplus \mathbb{R}[\alpha, \beta, \delta]\mathcal{Y}^2\gamma \oplus \mathbb{R}[\alpha, \beta, \delta]\mathcal{Y}^3\gamma$$
$$\oplus\, \mathbb{R}[\alpha, \beta, \delta]\mathcal{Y}\alpha \oplus \mathbb{R}[\alpha, \beta, \delta]\mathcal{Y}^2\alpha \oplus \mathbb{R}[\alpha, \beta, \delta]\mathcal{Y}^3\alpha$$
$$\oplus\, \mathbb{R}[\beta, \delta]\mathcal{Y}\beta \oplus \mathbb{R}[\beta, \delta]\mathcal{Y}^2\beta \oplus \mathbb{R}[\beta, \delta]\mathcal{Y}\beta\mathcal{Y}^2\beta.$$

(The first two summands here, $\mathbb{R}[\alpha, \beta, \delta] \oplus \mathbb{R}[\alpha, \beta, \delta]\gamma$, is just the Stanley decomposition (4.7.23) of the ring ker $\mathcal{X}$, corresponding to the top row of (4.7.21). The Stanley basis elements 1 and γ appearing here are not suffixes as defined above, since they do not contain $\mathcal{Y}$. In general, the Stanley basis elements for ker $\mathcal{X}^r$ consist of those for ker $\mathcal{X}$ together with the set of suffixes.) Another way to state this Stanley decomposition is that every element of $\ker \mathcal{X}^4$ can be written as

$$f_1(\alpha, \beta, \delta) + f_2(\alpha, \beta, \delta)\gamma \tag{4.7.28}$$
$$+ f_3(\alpha, \beta, \delta)\mathcal{y}\gamma + f_4(\alpha, \beta, \delta)\mathcal{y}^2\gamma + f_5(\alpha, \beta, \delta)\mathcal{y}^3\gamma$$
$$+ f_6(\alpha, \beta, \delta)\mathcal{y}\alpha + f_7(\alpha, \beta, \delta)\mathcal{y}^2\alpha + f_8(\alpha, \beta, \delta)\mathcal{y}^3\alpha$$
$$+ f_9(\beta, \delta)\mathcal{y}\beta + f_{10}(\beta, \delta)\mathcal{y}^2\beta + f_{11}(\beta, \delta)\mathcal{y}\beta\mathcal{y}^2\beta.$$

Since the simplified normal form $\widehat{N}$ for N_4 is obtained by placing the polynomials in $\ker \mathcal{X}^4$ into the bottom position of a vector, with zeros above, we have the following Stanley decomposition of this module:

$$\widehat{N} = (\mathbb{R}[\alpha, \beta, \delta] \oplus \mathbb{R}[\alpha, \beta, \delta]\gamma) \begin{bmatrix} 0 \\ 0 \\ 0 \\ 1 \end{bmatrix} \oplus \mathbb{R}[\alpha, \beta, \delta] \begin{bmatrix} 0 \\ 0 \\ 0 \\ \mathcal{y}\gamma \end{bmatrix} \oplus \cdots \tag{4.7.29}$$

$$\oplus \mathbb{R}[\alpha, \beta, \delta] \begin{bmatrix} 0 \\ 0 \\ 0 \\ \mathcal{y}^2\gamma \end{bmatrix} \oplus \cdots,$$

where the expression continues as in (4.7.27), with the prefixes placed before the vectors and the suffixes inside.

4.7.10. Remark. There is an important observation to make with regard to the form of (4.7.29). The first two summands could have been written

$$\widehat{N} = \mathbb{R}[\alpha, \beta, \delta] \begin{bmatrix} 0 \\ 0 \\ 0 \\ 1 \end{bmatrix} \oplus \mathbb{R}[\alpha, \beta, \delta] \begin{bmatrix} 0 \\ 0 \\ 0 \\ \gamma \end{bmatrix} \oplus \cdots,$$

but the γ (which is the Stanley basis element appearing in the second summand of (4.7.27)) can be factored out of the second vector, giving (4.7.29). On the other hand, the $\mathcal{y}\gamma$ in the third summand $\mathbb{R}[\alpha, \beta, \delta](0, 0, 0, \mathcal{y}\gamma)$ of (4.7.29) *cannot* be factored out (resulting in $\mathbb{R}[\alpha, \beta, \delta]\mathcal{y}\gamma(0, 0, 0, 1)$). The reason is that in the Stanley decomposition of a module over a ring, the coefficients of the Stanley basis elements must belong to (various subrings of) the ring. While $\mathbb{R}[\alpha, \beta, \delta]\gamma$ is a subring of $\ker \mathcal{X}$ (because γ is an invariant), $\mathbb{R}[\alpha, \beta, \delta]\mathcal{y}\gamma$ is not (because $\mathcal{y}\gamma$ is not an invariant). This is another reason for distinguishing the suffixes, containing $\mathcal{y}$, from the other Stanley basis elements in (4.7.27) that do not.

The Stanley decomposition for the inner product normal form for N_4 is obtained by applying $\mathcal{X}$ to the suffixes occurring in (4.7.29), working upwards from the bottom entry of each vector:

$$\mathcal{N} = (\mathbb{R}[\alpha, \beta, \delta] \oplus \mathbb{R}[\alpha, \beta, \delta]\gamma) \begin{bmatrix} 0 \\ 0 \\ 0 \\ 1 \end{bmatrix} \oplus \mathbb{R}[\alpha, \beta, \delta] \begin{bmatrix} 0 \\ 0 \\ xy\gamma \\ y\gamma \end{bmatrix} \tag{4.7.30}$$

$$\oplus \mathbb{R}[\alpha, \beta, \delta] \begin{bmatrix} 0 \\ x^2 y^2 \gamma \\ xy^2 \gamma \\ y^2 \gamma \end{bmatrix} \oplus \cdots .$$

It is never necessary actually to apply $\mathcal{X}$ as a differential operator in working out the entries of the column vectors in this expression. Rather, products of the form $\mathcal{X}y$ should be canceled and replaced by appropriate pressure factors, which are eigenvalues of $\mathcal{X}y$ (see Theorem 2.5.2, item 5). The following table evaluates all of the suffixes appearing in (4.7.29):

$$\frac{1}{3}y\alpha = x_2, \tag{4.7.31}$$

$$\frac{1}{12}y^2\alpha = x_3,$$

$$\frac{1}{36}y^3\alpha = x_4,$$

$$\frac{1}{2}y\beta = x_2 x_3 - 3x_1 x_4,$$

$$\frac{1}{4}y^2\beta = 2x_3^2 - 3x_2 x_4,$$

$$\frac{1}{8}y\beta y^2\beta = 2x_2 x_3^3 - 3x_2^2 x_3 x_4 - 6x_1 x_3^2 x_4 + 9x_1 x_2 x_4^2,$$

$$\frac{1}{3}y\gamma = x_2^2 x_3 - 4x_1 x_3^2 + 3x_1 x_2 x_4,$$

$$\frac{1}{12}y^2\gamma = 3x_2^2 x_4 - x_2 x_3^2 - 3x_1 x_3 x_4,$$

$$\frac{1}{12}y^3\gamma = 9x_2 x_3 x_4 - 4x_3^3 - 9x_1 x_4^2.$$

For the general case, this type of calculation gives the Stanley decomposition of each $\ker \mathcal{X}^r$. For the simplified normal form,

$$\widehat{\mathcal{N}} \cong \bigoplus_s \ker \mathcal{X}^{r_s}$$

(see Corollary 4.6.15 for the inner product case); the isomorphism is of modules over $\ker \mathcal{X}$, and is given by omitting the zero entries in the vector fields (equivalently, retaining only those entries in positions R_s). The inner product normal form is also isomorphic to the same direct sum of kernels, and is easily obtained from the simplified normal form by iterating $\mathcal{X}$ on the entries. (In other words, what is needed is to replace each $v_{(s,h)}$, defined in (4.6.41), by $v_{\{s,h\}}$, as written out in Theorem 4.6.16.)

The Nonsemisimple $1:1$ *Resonance and* $N_{2,2}$

The final examples to be studied in this section incorporate most of the ideas of Sections 4.5, 4.6, and 4.7. We present only the outlines, leaving the details to the reader.

The nonsemisimple $1:1$ resonant double center has already been briefly mentioned (4.5.82). In real coordinates (x_1, x_2, x_3, x_4) it has the linear part

$$\begin{bmatrix} 0 & -1 & 1 & 0 \\ 1 & 0 & 0 & 1 \\ 0 & 0 & 0 & -1 \\ 0 & 0 & 1 & 0 \end{bmatrix}. \tag{4.7.32}$$

In complex coordinates of the form (z_1, z_2, w_1, w_2), with $z_1 = x_1 + ix_2$, $z_2 = x_3 + ix_4$, $w_1 = x_1 - ix_2$, $w_2 = x_3 - ix_4$, this becomes

$$A = \begin{bmatrix} i & 1 & & \\ 0 & i & & \\ & & -i & 1 \\ & & 0 & -i \end{bmatrix}. \tag{4.7.33}$$

This splits into semisimple and nilpotent parts $A = S + N$ with

$$S = \begin{bmatrix} i & & & \\ & i & & \\ & & -i & \\ & & & -i \end{bmatrix}, \qquad N = N_{2,2} = \begin{bmatrix} 0 & 1 & & \\ 0 & 0 & & \\ & & 0 & 1 \\ & & 0 & 0 \end{bmatrix}. \tag{4.7.34}$$

According to Lemma 4.6.10, the normal form module for A is the intersection of those for S and N.

We begin by studying the inner product normal form of the real system of differential equations

$$\dot{x} = Nx + \cdots \tag{4.7.35}$$

with $x \in \mathbb{R}^4$ and $N = N_{2,2}$. (This problem is of interest in its own right, completely independently of the nonsemisimple $1:1$ resonance problem.) The natural triad containing N is

$$N = \begin{bmatrix} 0 & 1 & & \\ 0 & 0 & & \\ & & 0 & 1 \\ & & 0 & 0 \end{bmatrix}, \qquad M = \begin{bmatrix} 0 & 0 & & \\ 1 & 0 & & \\ & & 0 & 0 \\ & & 1 & 0 \end{bmatrix}, \qquad H = \begin{bmatrix} 1 & & & \\ & -1 & & \\ & & 1 & \\ & & & -1 \end{bmatrix}. \tag{4.7.36}$$

Notice that $M^* = N$, so the triad $\{X, Y, Z\} = \{M^*, N^*, H^*\}$ for the inner product normal form coincides with the triad $\{N, M, H\}$. The associated differential operators are

$$X = \mathcal{D}_{N^*} = x_1 \frac{\partial}{\partial x_2} + x_3 \frac{\partial}{\partial x_4}, \tag{4.7.37}$$

$$y = \mathcal{D}_{M^*} = x_2 \frac{\partial}{\partial x_1} + x_4 \frac{\partial}{\partial x_3},$$

$$Z = \mathcal{D}_Z = x_1 \frac{\partial}{\partial x_1} - x_2 \frac{\partial}{\partial x_2} + x_3 \frac{\partial}{\partial x_3} - x_4 \frac{\partial}{\partial x_4}.$$

The first step is to compute the ring of invariants $\ker X$. By creating a weight table and top weight list, one finds that there are three invariants in degrees one and two, namely,

$$\alpha = x_1, \tag{4.7.38}$$

$$\beta = x_3,$$

$$\gamma = x_1 x_4 - x_2 x_3,$$

of weights 1, 1, and 0, respectively. These are algebraically independent, and a table function argument shows that they are a complete set of basic invariants. Thus,

$$\ker X = \mathbb{R}[\alpha, \beta, \delta]. \tag{4.7.39}$$

The next step is to compute $\ker X^2$ as a module over $\ker X$; this will be used twice, because the two Jordan blocks in N both have size 2. The chains under α and β have length 2; the chain under γ has length one. Ordering the basic invariants by $\gamma < \beta < \alpha$, the standard monomials (all monomials are standard, since there are no relations) will be written $\gamma^i \beta^j \alpha^k$. There are three classes of monomials, those ending in α, β, and γ, respectively. Monomials of the first type contribute the suffix $y\alpha = x_2$, with prefix ring $\mathbb{R}[\alpha, \beta, \gamma]$; those of the second type yield suffix $y\beta = x_4$ with prefix ring $\mathbb{R}[\beta, \gamma]$; those of the third type give no suffix, since $y\gamma = 0$. Thus,

$$\ker X^2 = \mathbb{R}[\alpha, \beta, \gamma] \oplus \mathbb{R}[\alpha, \beta, \gamma] x_2 \oplus \mathbb{R}[\beta, \gamma] x_4. \tag{4.7.40}$$

Finally, according to Corollary 4.6.15, $\mathcal{N} \cong \ker X^2 \oplus \ker X^2$; explicitly,

$$\mathcal{N} = \mathbb{R}[\alpha, \beta, \gamma] v_{\{1,1\}} \oplus \mathbb{R}[\alpha, \beta, \gamma] v_{\{1, x_2\}} \oplus \mathbb{R}[\beta, \gamma] v_{\{1, x_4\}} \tag{4.7.41}$$
$$\oplus \mathbb{R}[\alpha, \beta, \gamma] v_{\{2,2\}} \oplus \mathbb{R}[\alpha, \beta, \gamma] v_{\{2, x_2\}} \oplus \mathbb{R}[\beta, \gamma] v_{\{2, x_4\}}.$$

Here

$$v_{\{1,1\}} = \begin{bmatrix} 0 \\ 1 \\ 0 \\ 0 \end{bmatrix}, \quad v_{\{1, x_2\}} = \begin{bmatrix} x_1 \\ x_2 \\ 0 \\ 0 \end{bmatrix}, \quad v_{\{1, x_4\}} = \begin{bmatrix} x_3 \\ x_4 \\ 0 \\ 0 \end{bmatrix},$$

and similarly for the others.

Turning from the "pure" $N_{2,2}$ problem to the nonsemisimple $1:1$ resonance, we must first formulate the $N_{2,2}$ results with the real system (4.7.35) replaced by a complex system with the same matrix, with the variables denoted by (z_1, z_2, w_1, w_2). In this setting the basic invariants (4.7.38) are written

$$\alpha = z_1, \tag{4.7.42}$$
$$\beta = w_1,$$
$$\gamma = z_1 w_2 - z_2 w_1,$$

and the module $\ker \mathcal{X}^2$ given by (4.7.40) becomes

$$\ker \mathcal{X}^2 = \mathbb{C}[\alpha, \beta, \gamma] \oplus \mathbb{C}[\alpha, \beta, \gamma]z_2 \oplus \mathbb{C}[\beta, \gamma]w_2. \tag{4.7.43}$$

In computing the normal form, we can ignore the equations for w, since the reality conditions (4.5.40) imply that these are the conjugates of the z equations (on the reality subspace); therefore we need only determine the vector fields of the form $v_{\{1,h\}}$. According to Theorem 4.6.16, we should use for this purpose functions h that satisfy $h \in \ker \mathcal{X}^2$ and also contain only monomials $z_1^{m_1} z_2^{m_2} w_1^{n_1} w_2^{n_2}$ satisfying $(m_1 + m_2)(i) + (n_1 + n_2)(-i) = i$, that is,

$$m_1 + m_2 = n_1 + n_2 + 1. \tag{4.7.44}$$

So let us examine the summands in (4.7.43) to determine what portion of each summand satisfies (4.7.44). Any element of $\mathbb{C}[\alpha, \beta, \gamma]$ is a sum of terms of the form $z_1^{k_1} w_2^{k_2} (z_1 w_2 - z_2 w_1)^{k_3}$. Since $z_1 w_2 - z_2 w_1$ is "balanced" with respect to powers of the z and w variables, this expression satisfies (4.7.44) if and only if $k_1 = k_2 + 1$. That is, the portion of $\mathbb{C}[\alpha, \beta, \gamma]$ that satisfies (4.7.44) is exactly $\mathbb{C}[\alpha\beta, \gamma]\alpha$, or $\mathbb{C}[\alpha\beta, \gamma]z_1$. (Notice that there is no comma between α and β here.) Similarly, the portion of $\mathbb{C}[\alpha, \beta, \gamma]z_2$ that satisfies (4.7.44) is $\mathbb{C}[\alpha\beta, \gamma]z_2$. As to the summand $\mathbb{C}[\beta, \gamma]w_2$, it is impossible for an element of this set to satisfy (4.7.44). We conclude that the set of h that should be used in building the vector fields $v_{\{1,h\}}$ is just

$$h \in \mathbb{C}[z_1 w_2, z_1 w_2 - z_2 w_1]z_1 \oplus \mathbb{C}[z_1 w_2, z_1 w_2 - z_2 w_1]z_2. \tag{4.7.45}$$

That is, writing only the equations for z_1 and z_2 restricted to the reality subspace, the normal form for the nonsemisimple 1 : 1 resonant double center is

$$\begin{bmatrix} \dot{z}_1 \\ \dot{z}_2 \end{bmatrix} = \begin{bmatrix} i & 1 \\ 0 & i \end{bmatrix} \begin{bmatrix} z_1 \\ z_2 \end{bmatrix} + f(z_1 \bar{z}_2, z_1 \bar{z}_2 - z_2 \bar{z}_1) \begin{bmatrix} 0 \\ z_1 \end{bmatrix} \tag{4.7.46}$$
$$+ g(z_1 \bar{z}_2, z_1 \bar{z}_2 - z_2 \bar{z}_1) \begin{bmatrix} z_1 \\ z_2 \end{bmatrix}.$$

From this the normal form in real or polar coordinates can be determined.

Notes and References

The use of Stanley decompositions to solve the description problem for normal forms in the nilpotent case was introduced by Richard Cushman and Jan Sanders for the sl(2) normal form. See Cushman and Sanders [36] and the other papers referenced there. Their technique was to obtain the invariants as we have done above, using table functions to verify the result, and then to repeat the same type of construction for the equivariants, using the triad $\{X, Y, Z\}$ mentioned in Remark 4.7.2; that is, they created equivariants and then used table functions (in a version for equivariants) to determine when they had enough. Alternatively, they describe a procedure using two-variable Molien integrals that predicts the correct table functions for both invariants and equivariants in advance. These procedures derive from classical invariant theory. The Molien integral is a complex line integral that (in its classical one-variable form) gives the generating function for the dimensions of the invariant ring of a group. The simplest case, in which the group is finite and the integral reduces to a sum, is Theorem 2.2.1 of Sturmfels [104].

The approach described in this section was introduced in Murdock [89] and makes it unnecessary to repeat the calculations of classical invariant theory at the level of equivariants. Instead, an algorithm is given (through the construction of suffixes and their prefix rings) that converts a Stanley decomposition of the invariant ring into a Stanley decomposition of the modules ker $\mathcal{X}^r$. This leads immediately to the structure of the normal form modules. In addition it is shown that the same method works for the inner product and simplified normal form styles (as well as for the sl(2) style). More examples and additional techniques will be found in [89], including the cross-section and transvectant methods mentioned in Remark 4.7.5.

4.8   * The sl(2) Normal Form

The sl(2) normal form style for matrix series was introduced in Section 3.5 by splitting the leading term A_0 of a matrix series $A_0 + \varepsilon A_1 + \cdots$ into its semisimple and nilpotent parts $A_0 = S + N$, and embedding N into an sl(2) triad $\{X, Y, Z\}$ (in the sense of Section 2.5), with $X = N$. In equation (3.5.5), the pseudotranspose $\widetilde{A}_0$ of A_0 was defined as $S + Y$, and in Lemma 3.5.1, ker $\mathsf{L}_{\widetilde{A}_0}$ was shown to be a normal form space. The same steps can be carried out for the nonlinear case. Given a vector field $a(x) = Ax + a_1(x) + \cdots$, A is split into $S + N$, N is embedded into a triad $\{X, Y, Z\}$ (also called $\{N, M, H\}$) with $X = N$, the pseudotranspose $\widetilde{A}$ is defined to be

$$\widetilde{A} = S + Y = S + M, \qquad (4.8.1)$$

and ker $\mathsf{L}_{\widetilde{A}}$ can then be shown to be a normal form module over ker $\mathcal{D}_{\widetilde{A}}$. We can be brief, because most of the work is nearly the same as for the inner product normal form, especially since triad methods have already been applied to the inner product normal form in Section 4.7. See Remarks 4.7.1

and 4.7.2 for an explanation of the relationship between Sections 4.7 and
4.8. (The matrices $\{N, M, H\}$ are the same in both sections, but $\{X, Y, Z\}$
and the operator triads $\{\mathcal{X}, \mathcal{Y}, \mathcal{Z}\}$ and $\{\mathsf{X}, \mathsf{Y}, \mathsf{Z}\}$ are not. However, most
formulas involving these symbols will be the same.)

The advantages and disadvantages of the sl(2) normal form style are as
follows:

1. The algorithms of Section 2.6 are available for calculating the projections. This makes the sl(2) normal form particularly suitable for
   calculations using symbolic processors.

2. Any system in sl(2) normal form is automatically in extended
   semisimple normal form. This is true for the inner product normal
   form only when A is seminormal.

3. The assumption that A is seminormal can also be removed from Algorithm 4.6.11 for splitting the computation problem into semisimple
   and nilpotent parts. This makes it feasible to compute normal forms
   when A is not in Jordan or real canonical form. This may be of
   practical use when Jordan or real canonical forms do not reflect the
   structure of the system (as in some Hamiltonian problems or problems with particular symmetries), or for problems in high dimensions
   where the Jordan form is difficult to obtain.

4. It is possible that some aspects of the sl(2) structure, in particular the
   weights of the invariants, have dynamical significance. A preliminary
   investigation of this possibility is given in Section 5.4.

5. The sl(2) normal form is slightly more complicated than the inner
   product normal form, because of the presence of pressure factors.

6. There is no natural inner product under which the projections into
   normal form are orthogonal.

7. The sl(2) normal form is less popular than the inner product normal
   form because it is harder to define, and the existing expositions depend heavily on a knowledge of representation theory and classical
   invariant theory. One goal of the starred sections of this book is to
   overcome this difficulty.

Given $A = S + N$, let $\{X, Y, Z\}$ be a triad with $X = N$ having the
additional property that X, Y, and Z commute with S. Such triads exist
according to Theorem 2.7.1. Let $\tilde{A}$ be defined by (4.8.1), and let

$$\mathcal{X} = \mathcal{D}_Y, \qquad \mathcal{Y} = \mathcal{D}_X, \qquad \mathcal{Z} = \mathcal{D}_Z,$$
$$\mathsf{X} = \mathsf{L}_Y, \qquad \mathsf{Y} = \mathsf{L}_X, \qquad \mathsf{Z} = \mathsf{L}_Z.$$

It was shown in Lemma 2.7.6 that $\{\mathfrak{X}, \mathfrak{Y}, \mathfrak{Z}\}$ and $\{X, Y, Z\}$ are sl(2) triads. Exactly as in Lemma 3.5.1, it follows that

$$\mathcal{V}_{**}^n = \operatorname{im} \mathcal{L} \oplus \ker \mathsf{L}_{\widetilde{A}} \tag{4.8.2}$$

(where $\mathcal{L} = \mathsf{L}_A$) and

$$\ker \mathsf{L}_{\widetilde{A}} = \ker \mathsf{L}_S \cap \ker \mathsf{X}. \tag{4.8.3}$$

(When it is desired to work in finite-dimensional vector spaces, these statements can be considered separately in each grade j.)

4.8.1. Definition. A formal power series vector field $a \in \mathcal{V}_{**}^n$ is in sl(2) normal form with respect to a_0 and $\{X, Y, Z\}$ if and only if the following conditions are satisfied:

1. $a_0(x) = Ax$, $A = S + N$ is the semisimple/nilpotent decomposition of A, and $\{X, Y, Z\}$ is a triad with $X = N$ such that X, Y, and Z commute with S.

2. Each a_j for $j \geq 1$ satisfies $a_j \in \ker \mathsf{L}_{\widetilde{A}}$.

4.8.2. Theorem (cf. Theorem 3.5.4). If a is in sl(2) normal form with respect to a_0 and $\{X, Y, Z\}$, and T is any invertible matrix, then $\mathsf{S}_T a$ is in sl(2) normal form with respect to $\mathsf{S}_T a_0$ and $\{\mathsf{S}_T X, \mathsf{S}_T Y, \mathsf{S}_T Z\}$.

Proof. The proof is the same as in the linear case.   □

4.8.3. Corollary. If a is in sl(2) normal form with respect to a_0 and any triad (having $X = N$), and T is any invertible matrix, then $\mathsf{S}_T a$ is in extended semisimple normal form (Definition 4.5.5).

Proof. By Theorem 4.8.2, $\mathsf{S}_T a$ is in sl(2) normal form with respect to some triad, and by (4.8.3) it follows that each a_j for $j \geq 1$ satisfies $a_j \in \ker \mathsf{L}_S$.   □

The determination of $\ker \mathsf{L}_{\widetilde{A}}$ can be carried out in two steps, first finding $\ker \mathsf{L}_S$ exactly as in Section 4.5, then finding the kernel of X restricted to $\ker \mathsf{L}_S$. The first part of this scheme is semisimple and is associated with the splitting

$$\mathcal{V}_j^n = \operatorname{im} \mathsf{L}_S \oplus \ker \mathsf{L}_S. \tag{4.8.4}$$

The second part is nilpotent, and according to (2.5.4) the splitting is

$$\ker \mathsf{L}_S = \operatorname{im} \mathsf{Y} \oplus \ker \mathsf{X} \tag{4.8.5}$$

(with X and Y restricted to $\ker \mathsf{L}_S$). The splitting (4.8.4) is computable on a symbolic processor by Lemma 4.5.18, and (4.8.5) by the pushdown algorithm (Algorithm 2.6.5). The complete computation scheme follows the same nine-step process outlined in Section 3.5. This is repeated as follows, to give the correct references for the nonlinear case.

4.8.4. Algorithm. The computation problem for the sl(2) normal form is solved by carrying out steps 1–3 (once) and steps 4–9 (repeatedly, for $j = 1, 2, \ldots$):

1. Split A into $S + N$ by either Jordan form or Algorithm E.1.2.

2. Find $\{X, Y, Z\}$ by Theorem 2.7.1 or Algorithm 2.7.2.

3. Determine projections P_λ into the eigenspaces of S, using Corollary 2.1.5 if A is not in canonical form.

4. Determine the projections P_μ into the eigenspaces of L_S. This must be done separately for each grade $j = 1, \ldots, k$, on the finite-dimensional spaces $\mathcal{V}_j^n$, up to the desired stopping point k; the values of μ will be $\langle m, \lambda \rangle - \lambda_i$ for various m and i. Lemma 4.5.18 can be used here.

5. Solve $L_S \widehat{u}_j = k_j - b'_j$ using Algorithm 4.5.19.

6. Convert $\widehat{u}_j$ into a solution u'_j of $\mathcal{L}\widehat{u}_j = k_j - b'_j$ using (4.6.17). (Note that Algorithm 3.4.23 remains valid with $\mathbb{L}$ replaced by L as long as we restrict it to a finite-dimensional space $\mathcal{V}_j^n$ and sum to the correct index of nilpotence, which will vary with j.)

7. Repeat steps 3 and 4 using Z in place of S to get the projections of each $\mathcal{V}_j^n$ into its weight spaces.

8. Solve $L_N u''_j = b'_j - b_j$ for u''_j and b_j, in the vector space $\ker L_S$, using the pushdown algorithm and $Y^\dagger$.

9. $u_j = u'_j + u''_j$.

Example: The sl(2) Normal Form for N_4

In the purely nilpotent case $(S = 0)$, the splitting (4.8.4) becomes trivial and (4.8.2) coincides with (4.8.5). In other words, $\mathcal{L} = L_A = L_N = Y$, and the normal form splitting is just

$$\mathcal{V}_j^n = \operatorname{im} Y \oplus \ker X. \qquad (4.8.6)$$

As an illustration of the sl(2) normal form, we will review the calculations for the case of N_4 given in Section 4.7 for the inner product normal form and point out where the sl(2) normal form differs.

The basic triad $\{X, Y, Z\}$ for the sl(2) version of the N_4 problem is

$$X = N = \begin{bmatrix} 0 & 1 & 0 & 0 \\ 0 & 0 & 1 & 0 \\ 0 & 0 & 0 & 1 \\ 0 & 0 & 0 & 0 \end{bmatrix},$$ (4.8.7)

$$Y = M = \begin{bmatrix} 0 & 0 & 0 & 0 \\ 3 & 0 & 0 & 0 \\ 0 & 4 & 0 & 0 \\ 0 & 0 & 3 & 0 \end{bmatrix},$$

$$Z = H = \begin{bmatrix} 3 & & & \\ & 1 & & \\ & & -1 & \\ & & & -3 \end{bmatrix}.$$

This is to be contrasted with (4.7.8) for the inner product normal form. The associated differential operators (on scalar functions) replacing (4.7.9) are then

$$\mathcal{X} = \mathcal{D}_Y = 3x_1\frac{\partial}{\partial x_2} + 4x_2\frac{\partial}{\partial x_3} + 3x_3\frac{\partial}{\partial x_4},$$ (4.8.8)

$$\mathcal{Y} = \mathcal{D}_X = x_2\frac{\partial}{\partial x_1} + x_3\frac{\partial}{\partial x_2} + x_4\frac{\partial}{\partial x_3},$$

$$\mathcal{Z} = \mathcal{D}_Z = 3x_1\frac{\partial}{\partial x_1} + x_2\frac{\partial}{\partial x_2} - x_3\frac{\partial}{\partial x_3} - 3x_4\frac{\partial}{\partial x_4}.$$

The action of $\mathsf{X} = \mathsf{L}_Y$ on a vector field $v = (v_1, v_2, v_3, v_4)$ is given in terms of $\mathcal{X}$ by

$$\mathsf{X}v(x) = \begin{bmatrix} \mathcal{X}v_1 \\ \mathcal{X}v_2 - 3v_1 \\ \mathcal{X}v_3 - 4v_2 \\ \mathcal{X}v_4 - 3v_3 \end{bmatrix}.$$ (4.8.9)

(The corresponding equation (4.7.10) in the inner product case lacks the coefficients 3, 4, 3.) If v is in normal form, $\mathsf{X}v = 0$, which implies that

$$v = \begin{bmatrix} \frac{1}{36}\mathcal{X}^3 h \\ \frac{1}{12}\mathcal{X}^2 h \\ \frac{1}{3}\mathcal{X}h \\ h \end{bmatrix},$$ (4.8.10)

with $h \in \ker \mathcal{X}^4$; this is to be compared with (4.7.11).

The computation of the ring of invariants goes almost exactly as in Section 4.7. The weight table and top weight list are the same. There are slight differences in the numerical coefficients appearing in the basic invariants, which are

$$\alpha = x_1,$$ (4.8.11)

$$\beta = 2x_2^2 - 3x_1x_3,$$

$$\gamma = 4x_2^3 - 9x_1x_2x_3 + 9x_1^2x_4,$$

$$\delta = 9x_1^2x_4^2 - 3x_2^2x_3^2 - 18x_1x_2x_3x_4 + 6x_1x_3^3 + 8x_2^3x_4.$$

The relation satisfied by these invariants is

$$\gamma^2 = 2\beta^3 + 9\alpha^2\delta.$$ (4.8.12)

Since this differs from (4.7.22) only in the coefficients, the Stanley decomposition (4.7.23) for $\ker \mathcal{X}$ and the table function (4.7.25) are unchanged. The classification of standard monomials is the same, so the suffix set and prefix ring and the resulting Stanley decomposition (4.7.27) are not changed (except, of course, for the meaning of α, β, γ, and δ). The one remaining difference concerns the Stanley decomposition of the normal form module, or module of equivariants: In constructing the analogue of (4.7.30), we must use (4.8.10) in place of (4.7.11), leading to

$$\ker \mathsf{X} = (\mathbb{R}[\alpha, \beta, \delta] \oplus \mathbb{R}[\alpha, \beta, \delta]\gamma) \begin{bmatrix} 0 \\ 0 \\ 0 \\ 1 \end{bmatrix} \oplus \mathbb{R}[\alpha, \beta, \delta] \begin{bmatrix} 0 \\ 0 \\ \frac{1}{3}\mathcal{X}\mathcal{y}\gamma \\ \mathcal{y}\gamma \end{bmatrix}$$ (4.8.13)

$$\oplus \mathbb{R}[\alpha, \beta, \delta] \begin{bmatrix} 0 \\ \frac{1}{12}\mathcal{X}^2\mathcal{y}^2\gamma \\ \frac{1}{3}\mathcal{X}\mathcal{y}^2\gamma \\ \mathcal{y}^2\gamma \end{bmatrix} \oplus \cdots .$$

Of course, the quantities occurring in these Stanley decompositions are no longer given by (4.7.31), but must be recalculated. As in (4.7.30), it is not necessary to apply the operator $\mathcal{X}$ in working out the entries of (4.8.13); it suffices to replace products $\mathcal{X}\mathcal{y}$ by the appropriate pressure factors.

Notes and References

The earliest work on the sl(2) normal form concerned the Hamiltonian case; see Notes and References to Section 4.9. The sl(2) normal form for general vector fields was developed by Richard Cushman and Jan Sanders in many papers, the most important being [34] and [36]. The N_4 example (using a slightly different notation) was worked out in [35]; see Notes and References to Section 4.7 for remarks about the method used there. The method used here (and in Section 4.7) is from Murdock [89].

4.9   The Hamiltonian Case

Space permits only a brief treatment of normal forms for Hamiltonian systems. The deeper structural features of Hamiltonian systems will not appear in this discussion. Our aim is to show that canonical transformations and Hamiltonian vector fields form a Lie group/Lie algebra pair (as in the table in Section 3.6, except that these are infinite-dimensional), a subgroup/subalgebra of the diffeomorphism/vector field pair that has been studied so far in Chapter 4. It follows that any of the generated formats (2a, 2b, and 2c) can be used to compute normal forms for Hamiltonian systems. Since the Lie algebra of Hamiltonian vector fields is isomorphic to a Lie algebra of scalar functions (the Hamiltonian functions of the vector fields), the normalization calculations can be done within this scalar algebra, making them simpler than in the general case. The examples treated are limited to the Hamiltonian version of the double center, including the nonsemisimple 1:−1 resonance (which, when perturbed, yields the "Hamiltonian Hopf bifurcation").

Definitions

Hamiltonian systems exist only on even-dimensional spaces $\mathbb{R}^n = \mathbb{R}^{2m}$, with points denoted by

$$x = (x_1, \ldots, x_n) = (p, q) = (p_1, \ldots, p_m, q_1, \ldots, q_m);$$

m is called the number of *degrees of freedom*. A *Hamiltonian function* is simply a smooth function $H : \mathbb{R}^{2m} \to \mathbb{R}$, or the germ (at the origin) of such a function, or the formal power series (about the origin) of such a germ. The term *Hamiltonian*, in connection with the word *function*, does not place any restrictions on the nature of the function, but only indicates how it will be used, namely, to create a vector field on $\mathbb{R}^{2m}$ having the form

$$\dot{p}_i = -\frac{\partial H}{\partial q_i}, \tag{4.9.1}$$

$$\dot{q}_i = +\frac{\partial H}{\partial p_i},$$

or equivalently,

$$\dot{x} = v_{\mathrm{H}}(x) = JH'(x)^{\dagger}, \tag{4.9.2}$$

where

$$J = \begin{bmatrix} 0 & -I \\ I & 0 \end{bmatrix}, \tag{4.9.3}$$

as in (3.6.6); here I is the $n \times n$ identity matrix. The transpose is needed in (4.9.2) because $H'(x)$ is a row vector. The term *Hamiltonian vector field*, or *Hamiltonian system*, does imply a restriction on the nature of the vector field, namely, that it is generated from a function H according to the formulas just given. In this case the function H is called *the Hamiltonian function*, or simply *the Hamiltonian*, of the vector field.

4.9.1. Remark. A physicist would use these terms slightly differently. Beginning with a physical system composed of mass particles, the Hamiltonian of the system is defined to be the sum of its kinetic and potential energy, expressed as functions of certain generalized position coordinates q and generalized momentum coordinates p. Then, if all of the forces acting on the system are accounted for by the potential energy, the equations of motion of the system will be (4.9.1). From this standpoint, the Hamiltonian function H is not at all arbitrary, but is dictated by the physics of the situation.

A *canonical transformation* is a diffeomorphism $\varphi : \mathbb{R}^{2m} \to \mathbb{R}^{2m}$ satisfying the condition

$$\varphi'(x)J\varphi'(x)^{\dagger} = J \tag{4.9.4}$$

for all x; that is, $\varphi'(x) \in \mathrm{Sp}(m)$ is a symplectic matrix for all x, as defined in Section 3.6. Instead of being defined on $\mathbb{R}^{2m}$, φ can be defined on an open subset, or can be a germ at the origin, or a formal power series (in which case all equalities are to be understood as holding in the class of formal power series, without being evaluated at x).

Basic Theorems

It was shown in Section 4.4 that the solution operator, of (local) flow φ^t, of a vector field is (very nearly) a one-parameter group of diffeomorphisms, and the derivative with respect to t, at $t = 0$, of a parameterized family φ_t of diffeomorphisms (not necessarily a one-parameter group) is a vector field. The same statements are true with "diffeomorphism" replaced by "canonical transformation" and "vector field" by "Hamiltonian system."

4.9.2. Lemma. Given a smooth function $H : \mathbb{R}^{2m} \to \mathbb{R}$, the flow φ^t of the vector field $JH'(x)^{\dagger}$ is a one-parameter group of canonical transformations.

The proof requires the use of second derivatives, which always create some difficulties in matrix notation. In particular, if $c(y)$ is a column vector and $y = \varphi(x)$, the chain rule says that $(c \circ \varphi)'(x) = c'(\varphi(x))\varphi'(x)$, but if $r(y)$ is a row vector, the corresponding formula is not clear, because there is no obvious convention governing the layout of the matrix $r'(y)$ (that is, of i and j in $\partial r_i / \partial y_j$, which is the row and which the column). In the following argument we write φ^t as φ_t for convenience, and we differentiate $H'(\varphi_t(x))^{\dagger}$ as a column vector rather than "differentiating under the transpose" and having to deal with a row vector. Since H'' is symmetric, there is no problem of how to arrange it.

Proof. The flow of any vector field is a one-parameter group (of diffeomorphisms), so all that needs to be proved is that φ_t is canonical for each t, that is, $\varphi'_t(x) \in \mathrm{Sp}(m)$ for all x. For any x and t, $\dot{\varphi}_t(x) = J\mathrm{H}'(\varphi_t(x))^\dagger$, and differentiating this with respect to x gives $\dot{\varphi}'_t = J\mathrm{H}''(\varphi_t(x))^\dagger \varphi'_t(x)$; in other words, for fixed x the matrix $T(t) = \varphi'_t(x)$ satisfies $\dot{T} = A(t)T$, where $A(t) = J\mathrm{H}''(\varphi_t(x))^\dagger = J\mathrm{H}''(\varphi_t(x))$. Since H'' is symmetric, $J\mathrm{H}'' \in \mathrm{hm}(m)$, so by Remark 3.6.5, $T(t) \in \mathrm{Sp}(m)$. □

4.9.3. Lemma. Let φ_t be a smooth canonical transformation depending smoothly on t, and let U be an open, connected, simply connected subset of $\mathbb{R}^{2m}$ such that for every $x \in U$, $\varphi_t(x)$ is defined for t in some interval containing 0 (which may depend on x), with $\varphi_0(x) = x$. Then there exists a smooth function $\mathrm{H} : U \to \mathbb{R}$ such that

$$\dot{\varphi}_0(x) = J\mathrm{H}'(x)^\dagger.$$

Proof. The proof is based on a technique familiar from the topic of "exact differential equations," as treated in elementary (undergraduate) differential equations texts. It is shown in such books that if $P(x,y)$ and $Q(x,y)$ are functions in the plane such that $Q_x - P_y = 0$, there exists a function $F(x,y)$ such that $P = F_x$ and $Q = F_y$. One checks, by differentiating with respect to x, that the function

$$f(y) = Q(x,y) - \int_0^x P_y(\xi,y)d\xi$$

is independent of x; then

$$F(x,y) = \int_0^x P(\xi,y)d\xi + \int_0^y f(\eta)d\eta$$

satisfies the required conditions. For these integrals to exist, the domain must be all of $\mathbb{R}^2$, or else a rectangle containing the origin, but in the case of an open, connected, simply connected set, the same argument works, replacing the integrals by line integrals over a path consisting of several vertical and horizontal segments. The same proof can be extended to show that if $v(x,y,z)$ is a vector field in $\mathbb{R}^3$ with curl $v = 0$, then $v = \mathrm{grad}\, F$ for some F, or more generally, that if $v(x)$ is a vector field on $\mathbb{R}^n$ with the matrix $v'(x)$ symmetric, then $v(x) = F'(x)^\dagger$. In our situation, since $J^{-1} = -J$, it suffices to show that $-J\dot{\varphi}'_0(x)$ is symmetric; then $-J\dot{\varphi}_0(x)$ will equal $\mathrm{H}'(x)^\dagger$ for some H. But $\varphi'_t(x)^\dagger J\varphi'_t(x) = J$; differentiating with respect to t and setting $t = 0$, using $\varphi'_0(x) = I$, yields $\dot{\varphi}'_0(x)^\dagger J + J\dot{\varphi}'_0(x) = 0$. From this and $J^\dagger = -J$ it follows that $J\dot{\varphi}'_0(x)$ is symmetric. □

Since symplectic maps and Hamiltonian systems constitute a Lie group and its associated Lie algebra, it follows that the three operations of conjugation, similarity, and Lie operator exist, as in Sections 3.6 and 4.4. These are restrictions of the corresponding operations for diffeomorphisms and vector fields, so it follows (without a new proof) that the Lie operator coincides with the Lie bracket. The crucial facts are that

1. a similarity transformation of a Hamiltonian system by a canonical map (which is just the natural coordinate change for the vector field under the canonical map) will again be Hamiltonian, and

2. the Lie bracket of two Hamiltonian vector fields will be Hamiltonian.

These facts are often proved by explicit calculations, but it is not necessary to do so, since they follow from general considerations about Lie groups and algebras. The calculations will be given below because they are useful for another purpose, namely, computing the Hamiltonians of the vector fields in each case. But before doing the calculations, it is already possible to see that the machinery exists for the normalization of Hamiltonian systems (remaining within the class of Hamiltonian systems). Indeed, the normalization could be carried out as usual for vector fields, using format 2a, 2b, or 2c, as long as the generators are restricted to be vector fields that are themselves Hamiltonian. But this is not the most efficient way to proceed, since this method would require working with the Hamiltonian vector fields (both those being normalized and the generators used for the normalization) *directly as vector fields*. It is better to recast the normalization algorithms so that we operate with the Hamiltonians themselves, which are scalar fields. It is for this purpose that we require the calculations mentioned above. The notation v_H, introduced above in (4.9.2), is useful here.

4.9.4. Lemma. If φ is a canonical transformation, the Hamiltonian vector field

$$\dot{x} = v_\mathrm{H}(x) = J\mathrm{H}'(x)^\dagger$$

is carried by the coordinate change $x = \varphi(y)$ into the Hamiltonian vector field

$$\dot{y} = v_\mathrm{K}(y) = J\mathrm{K}'(y)^\dagger,$$

where

$$\mathrm{K}(x) = \mathrm{H}(\varphi(x)). \tag{4.9.5}$$

That is,

$$S_\varphi v_\mathrm{H} = v_{\mathrm{H}\circ\varphi}. \tag{4.9.6}$$

Proof. Using $J\varphi'(y)^\dagger = \varphi'(y)^{-1}J$, which follows from $\varphi'(y) \in \mathrm{Sp}(m)$, gives

$$\dot{y} = \varphi'(y)^{-1}v_\mathrm{H}(\varphi(y)) = \varphi'(y)^{-1}J\mathrm{H}'(\varphi(y))^\dagger = J\varphi'(y)^\dagger \mathrm{H}'(\varphi(y))^\dagger$$
$$= J[\mathrm{H}'(\varphi(y))\varphi'(y)]^\dagger = J\mathrm{H}(\varphi(y))'^\dagger = v_{\mathrm{H}\circ\varphi}(y).$$

$\square$

4.9.5. Lemma. The Lie bracket of two Hamiltonian vector fields is given by

$$[v_{\mathrm{H}}, v_{\mathrm{K}}] = v_{\{\mathrm{H,K}\}}, \tag{4.9.7}$$

where $\{\cdot, \cdot\}$ is the *Poisson bracket*, defined by

$$\{\mathrm{H}, \mathrm{K}\}(x) = \mathrm{H}'(x) J \mathrm{K}'(x)^\dagger = \sum_{i=1}^{m} \left(\frac{\partial \mathrm{H}}{\partial q_i} \frac{\partial \mathrm{K}}{\partial p_i} - \frac{\partial \mathrm{H}}{\partial p_i} \frac{\partial \mathrm{K}}{\partial q_i} \right). \tag{4.9.8}$$

That is, defining the *Lie operator* $\mathcal{L}_{\mathrm{K}}$ associated with the Poisson bracket by

$$\mathcal{L}_{\mathrm{K}} = \{\cdot, \mathrm{K}\},$$

we have

$$\mathcal{L}_{\mathrm{K}} = \mathcal{D}_{v_{\mathrm{K}}}. \tag{4.9.9}$$

Proof. The one point in the following calculation that might not be obvious is that $\mathrm{H}' J \mathrm{K}'^{\dagger\prime} = (\mathrm{H}' J \mathrm{K}'^{\dagger\prime})^\dagger$; this is true because it is a scalar (or 1×1 matrix):

$$v_{\{\mathrm{H,K}\}} = J\{\mathrm{H,K}\}' = J(\mathrm{H}' J \mathrm{K}'^\dagger)' = J(\mathrm{H}'' J \mathrm{K}'^\dagger + \mathrm{H}' J \mathrm{K}'^{\dagger\prime})$$

$$= J\mathrm{H}'' \mathrm{H} \mathrm{K}'^\dagger + J(\mathrm{H}' J \mathrm{K}'^{\dagger\prime})^\dagger = J\mathrm{H}'' J \mathrm{K}'^\dagger + \mathrm{K}'^{\dagger\prime} \mathrm{H}'^\dagger$$

$$= J\mathrm{H}'' J \mathrm{K}'^\dagger - \mathrm{K}'^{\dagger\prime} J \mathrm{H}'^\dagger = (J\mathrm{H}'^\dagger)' J \mathrm{K}'^\dagger - (J\mathrm{K}'^\dagger)' J \mathrm{H}'^\dagger$$

$$= [J\mathrm{H}'^\dagger, J\mathrm{K}'^\dagger] = [v_{\mathrm{H}}, v_{\mathrm{K}}].$$

$\square$

The importance of (4.9.9) is that the scalar version of the fundamental theorem of Lie series, Theorem 4.4.9, replaces the vector version, Theorem 4.4.8, in the normal form algorithm developed below.

Normalization of Hamiltonian Systems near a Rest Point

Specializing to the case of a Hamiltonian system with a rest point at the origin, it is convenient (as usual) to work in the setting of formal power series. Hamiltonians will be written in the form

$$\mathrm{H}(x) = \mathrm{H}_0(x) + \mathrm{H}_1(x) + \mathrm{H}_2(x) + \cdots, \tag{4.9.10}$$

with H_j being a homogeneous polynomial of degree $j + 2$; the associated vector field

$$a = v_{\mathrm{H}} = a_0 + a_1 + a_2 + \cdots \tag{4.9.11}$$

has homogeneous terms $a_j = v_{\mathrm{H}_j}$ of degree $j + 1$. Both a_j and H_j are said to be of *grade* j. Thus, H_0 is a *quadratic* function giving rise to a *linear* vector field v_0, and both have grade zero. As usual, the linear part of the vector field will be written

$$a_0(x) = Ax. \qquad (4.9.12)$$

The first (or "zeroth") step in normalizing is always to bring the linear term into an acceptable canonical form. The study of linear Hamiltonian systems is interesting and important, but space does not permit developing it here, except to mention that it is not always possible to put A into Jordan form by a canonical change of coordinates. We will bypass the question of canonical forms for such systems by limiting our examples to single and (both semisimple and nonsemisimple) double centers. To describe the computations for format 2a in the general case, it is enough to assume that H_0 is already in a form that is considered acceptable; the specific form matters only for the discussion of normal form styles.

For format 2a, using the "algorithmic" version (that is, assuming that the vector field (4.9.11) is already normalized to grade $j-1$), the next step (under the procedure for general vector fields given in Section 4.4) is to find a generator u_j satisfying the homological equation

$$£u_j = a_j - b_j, \qquad (4.9.13)$$

where b_j is the projection of a_j into the chosen normal form style, and

$$£ = \mathsf{L}_{a_0} = \mathsf{L}_A. \qquad (4.9.14)$$

When u_j has been found, the new vector field will be found by expanding

$$b = e^{\varepsilon^j \mathsf{L}_{u_j}}(a_0 + \varepsilon a_1 + \cdots) \qquad (4.9.15)$$

and setting $\varepsilon = 1$. The new vector field b is then renamed as $a_0 + a_1 + \cdots$; here $a_0, \ldots, a_{j-1}$ are unchanged, a_j has become the b_j of (4.9.14), and the higher a_i are also modified.

To apply this scheme to a Hamiltonian system, we must guarantee that u_j is Hamiltonian, so that the transformation it generates will be canonical. That is, it is required that

$$u_j = v_{\mathsf{U}_j} \qquad (4.9.16)$$

for some scalar function U_j, and it is this function that should actually be found (instead of, or in advance of, u_j). We begin with a Hamiltonian $H(x)$ in the form (4.9.10), assuming it to be normalized to grade $j-1$. When normalized to grade j, the new Hamiltonian will temporarily be called

$$K = K_0 + K_1 + \cdots,$$

with $K_i = H_i$ for $i < j$ (so that $v_K = b$ above). Equation (4.9.16), together with $a_j = v_{H_j}$, $b_j = v_{K_j}$, and Lemma 4.9.5, now implies the scalar homological equation

$$£\mathsf{U}_j = H_j - K_j, \qquad (4.9.17)$$

where the new homological operator is

$$\mathcal{L} = \mathcal{L}_{H_0} = \{\cdot, H_0\}. \tag{4.9.18}$$

Given the quadratic part H_0 of the Hamiltonian, the *description problem* for the normal form is solved by determining a complement to $\mathrm{im}\,\mathcal{L}_{H_0}$ in the space of homogeneous scalar polynomials of degree $j+2$ for each j. This family of complements constitutes a *normal form style*. The *computation problem* is solved by computing the projection operator into the selected complement, applying this projection to H_j to obtain K_j, solving (4.9.17) to obtain U_j, and computing the rest of K from

$$K = e^{\varepsilon^j \mathcal{L}_{U_j}} (H_0 + \varepsilon H_1 + \cdots) \tag{4.9.19}$$

with $\varepsilon = 1$. Then K is renamed H, and the process is repeated (if desired) for the next j.

The question of choosing a normal form style breaks into two cases, one in which the matrix A of the linear vector field v_{H_0} is semisimple, and one in which it is not. In the first case, $\mathcal{L} = \mathcal{L}_{H_0}$ is semisimple and the normal form style may be taken to be $\ker \mathcal{L}_{H_0}$. This is a ring of formal power series invariant under the flow of v_{H_0}. (It is not necessary to deal with a module of equivariants.) In the second case there exists an $\mathrm{sl}(2)$ normal form style that is a ring of invariants under a modified linear flow. These points will be illustrated in the following examples. It does not appear that an inner product normal form style for the Hamiltonian case has been worked out, although it probably exists.

The Hamiltonian Single Center

A *Hamiltonian single center* is a Hamiltonian system in one degree of freedom having a Hamiltonian $H(p,q)$ of the form (4.9.10) with $x = (p,q)$ and

$$H_0 = \frac{1}{2} \left(p^2 + q^2 \right). \tag{4.9.20}$$

The corresponding system of differential equations,

$$\dot{p} = -q + \cdots, \tag{4.9.21}$$
$$\dot{q} = p + \cdots,$$

has linear part

$$A = \begin{bmatrix} 0 & -1 \\ 1 & 0 \end{bmatrix},$$

as does any single center. Since A is semisimple, the differential equations (4.9.21) will be in (semisimple) normal form if and only if the right-hand side belongs to the kernel of the (ordinary) homological operator $\mathcal{L} = \mathsf{L}_A$; this, in turn, will be the case if and only if the Hamiltonian H belongs to the kernel of the (new) homological operator $\pounds = \mathcal{L}_{H_0} = \mathcal{D}_A$. Thus, the problem of describing the normal form of the Hamiltonian is the same as determining the (scalar) invariants of the flow of A. This problem was solved in Section 1.1: All formal power series invariants are uniquely expressible as formal power series in $r^2 = p^2 + q^2$. For our purposes, it is best to introduce the *canonical polar coordinates*

$$R = \frac{1}{2}\left(p^2 + q^2\right),$$ (4.9.22)

$$\theta = \arctan\left(\frac{q}{p}\right),$$

because the transformation from (p, q) to (R, θ) is canonical, whereas the transformation to ordinary polar coordinates is not; then $H(p, q)$ is in normal form (to some finite order, or to all orders) if, when expressed in canonical polar coordinates, it is a function of R only (to finite order or to all orders):

$$H(p, q) = K(R).$$ (4.9.23)

In other words, the Hamiltonian H is constant on circles. The differential equations associated with the new Hamiltonian, truncated at the order to which the system is normalized, will take the form

$$\dot{R} = -\frac{\partial K}{\partial \theta} = 0,$$ (4.9.24)

$$\dot{\theta} = +\frac{\partial K}{\partial \theta} = \Omega(R).$$

Thus, R is an approximately conserved quantity (exactly conserved for the truncated system) called an *adiabatic invariant*; this will be discussed more fully in Section 5.2.

The Semisimple Hamiltonian Double Center

A *semisimple Hamiltonian double center* is a system in two degrees of freedom with Hamiltonian (4.9.10), where $x = (p_1, p_2, q_1, q_2)$ and

$$H_0 = \frac{1}{2}\omega_1\left(p_1^2 + q_1^2\right) + \frac{1}{2}\omega_2\left(p_2^2 + q_2^2\right).$$ (4.9.25)

The differential equations have linear part

$$A = \begin{bmatrix} & & -\omega_1 & 0 \\ & & 0 & -\omega_2 \\ \omega_1 & 0 & & \\ 0 & \omega_2 & & \end{bmatrix}.$$ (4.9.26)

We do not assume that ω_1 and ω_2 are positive, since (in contrast to the case of Section 4.5) it is not always possible to achieve this by a canonical coordinate change, which must preserve the definiteness or indefiniteness of the quadratic form H_0. The Hamiltonian H is in normal form if it is an invariant of the flow e^{At}. For the study of the ring of invariants, it is not necessary to confine ourselves to canonical transformations as long as we return to the original variables at the end of the discussion. The noncanonical transformation $(p_1, p_2, q_1, q_2) \mapsto (p_1, q_1, p_2, q_2)$ carries A into

$$
B = \begin{bmatrix} 0 & -\omega_1 & & \\ \omega_1 & 0 & & \\ & & 0 & -\omega_2 \\ & & \omega_2 & 0 \end{bmatrix},
$$

the invariants of which have already been studied in Section 4.5 in connection with the (ordinary) double center. Since the noncanonical transformation is merely a reordering of the variables, without renaming them, the step of "returning to the original variables" is actually vacuous.

In the nonresonant case (ω_1/ω_2 irrational), the ring of invariants of e^{Bt} is $\mathbb{R}[[I_1, I_2]]$, with I_1 and I_2 given by (4.5.56); in our current notation, $I_1 = p_1^2 + q_1^2$ and $I_2 = p_2^2 + q_2^2$. It is preferable to replace I_1 and I_2 by the canonical polar coordinates $R_1 = \frac{1}{2} \left(p_1^2 + q_1^2 \right)$ and $R_2 = \frac{1}{2} \left(p_2^2 + q_2^2 \right)$, because the transformation $(p_1, p_2, q_1, q_2) \mapsto (R_1, R_2, \theta_1, \theta_2)$ is canonical and can be applied to the Hamiltonians of systems. Thus, the Hamiltonian $H(p_1, p_2, q_1, q_2)$ will be in normal form if and only if it can be expressed as $K(R_1, R_2)$, that is, if and only if H is constant on tori. In this case, analogously with the single center discussed above, the differential equations will take the form

$$
\dot{R}_1 = 0, \tag{4.9.27}
$$
$$
\dot{R}_2 = 0,
$$
$$
\dot{\theta}_1 = \Omega_1(R_1, R_2),
$$
$$
\dot{\theta}_2 = \Omega_2(R_1, R_2).
$$

In the resonant case, it is convenient to write

$$
\frac{\omega_1}{\omega_2} = \frac{-b}{a} \tag{4.9.28}
$$

in place of (4.5.62), because the letters p and q are already in use; here a and b are integers, and the sign is for simplicity in later formulas. Since a and b are assumed to be relatively prime, there exist integers c and d such that $ad - bc = 1$. The transformation from canonical polar coordinates $(R_1, R_2, \theta_1, \theta_2)$ to (S_1, S_2, Φ, Ψ) defined by

$$
S_1 = dR_1 - cR_2, \tag{4.9.29}
$$
$$
S_2 = -bR_1 + aR_2,
$$
$$
\Phi = a\theta_1 + b\theta_2,
$$
$$
\Psi = c\theta_1 + d\theta_2,
$$

is canonical. It follows from (4.5.73) that the invariants can be written (nonuniquely) as functions of S_1, S_2, and Φ. (For the brief treatment we are giving, we do not require the detailed unique form given by the Stanley decomposition). Thus, if H is in normal form, it can be written as $K(S_1, S_2, \Phi)$, so that

$$\dot{S}_1 = -\frac{\partial K}{\partial \Phi}, \tag{4.9.30}$$

$$\dot{S}_2 = 0,$$

$$\dot{\Phi} = \frac{\partial K}{\partial S_1},$$

$$\dot{\Psi} = \frac{\partial K}{\partial S_2}.$$

Thus, a resonant double center has only one adiabatic invariant.

The Nonsemisimple Hamiltonian $1 : -1$ Resonance

A *nonsemisimple Hamiltonian* $1 : -1$ *resonance* is a system in two degrees of freedom with Hamiltonian H in the form (4.9.10) with $x = (p_1, p_2, q_1, q_2)$ and

$$H_0 = p_2 q_1 - p_1 q_2 - \frac{1}{2}(q_1^2 + q_2^2). \tag{4.9.31}$$

The associated differential equations have linear part

$$A = \begin{bmatrix} 0 & -1 & 1 & 0 \\ 1 & 0 & 0 & 1 \\ 0 & 0 & 0 & -1 \\ 0 & 0 & 1 & 0 \end{bmatrix}. \tag{4.9.32}$$

The semisimple and nilpotent parts of this matrix are

$$S = \begin{bmatrix} 0 & -1 & 0 & 0 \\ 1 & 0 & 0 & 0 \\ 0 & 0 & 0 & -1 \\ 0 & 0 & 1 & 0 \end{bmatrix}, \qquad N = \begin{bmatrix} 0 & 0 & 1 & 0 \\ 0 & 0 & 0 & 1 \\ 0 & 0 & 0 & 0 \\ 0 & 0 & 0 & 0 \end{bmatrix}. \tag{4.9.33}$$

Since this is exactly the same as the matrix (4.7.32) of the nonsemisimple $1 : 1$ resonance (for systems that are not required to be Hamiltonian), it may be asked why we refer to it here as a $1 : -1$ resonance. The answer is that in the theory of linear Hamiltonian systems, it makes a good deal of difference whether the (quadratic) Hamiltonian H_0 is or is not (positive or negative) definite; the terminology $1 : 1$ resonance is reserved for double centers in which the frequencies are equal and H_0 is positive definite. Since H_0 as given by (4.9.31) is indefinite, the correct terminology is $1 : -1$ resonance. This distinction does not exist in the non-Hamiltonian case.

To develop an sl(2) normal form, it is first necessary to choose an sl(2) triad $\{X, Y, Z\}$ of matrices with $X = N$; the natural choice here is

$$
X = \begin{bmatrix} 0 & 0 & 1 & 0 \\ 0 & 0 & 0 & 1 \\ 0 & 0 & 0 & 0 \\ 0 & 0 & 0 & 0 \end{bmatrix}, \quad Y = \begin{bmatrix} 0 & 0 & 0 & 0 \\ 0 & 0 & 0 & 0 \\ 1 & 0 & 0 & 0 \\ 0 & 1 & 0 & 0 \end{bmatrix}, \quad Z = \begin{bmatrix} 1 & & & \\ & 1 & & \\ & & -1 & \\ & & & -1 \end{bmatrix}.
$$

$$ (4.9.34) $$

According to Section 4.8, the vector field v_H will be in sl(2) normal form with respect to this triad, provided that it is equivariant under the flow of the pseudotranspose $\tilde{A} = S + Y$, that is, provided that it belongs to $\ker L_S \cap \ker L_Y = \ker L_S \cap \ker X$. (It is important to remember the crossover that takes place here, so that $X = L_Y$.) This will be the case, provided that the Hamiltonian H is invariant under the same flow, that is, if and only if $H \in \ker \mathcal{D}_S \cap \ker X$.

It is useful at this point to introduce the Hamiltonian functions that generate the linear vector fields Sx, Xx, Yx, and Zx, namely,

$$ S = p_2 q_1 - p_1 q_2, \qquad (4.9.35) $$

$$ X = -\frac{1}{2}\left(q_1^2 + q_2^2\right), $$

$$ Y = \frac{1}{2}\left(p_1^2 + p_2^2\right), $$

$$ Z = -(p_1 q_1 + p_2 q_2). $$

These satisfy the same bracket relations (using Poisson brackets) as do the corresponding linear vector fields or their matrices (using Lie brackets), namely,

$$ \{X, Y\} = Z, \quad \{Z, X\} = 2X, \quad \{Z, Y\} = -2Y, \qquad (4.9.36) $$
$$ \{S, X\} = \{S, Y\} = \{S, Z\} = 0. $$

In addition, we will use the trivial bracket relations $\{S, S\} = \{Y, Y\} = 0$. Since Lemma 4.9.5 implies that $\mathcal{D}_S = \mathcal{D}_{v_S} = \mathcal{L}_S$ and, similarly, $X = \mathcal{L}_Y$, it follows from these bracket relations that S, X, Y, Z are invariants of S (belong to $\ker \mathcal{D}_S$) and that S, Y are invariants of Y (belong to X).

The invariants of S have been studied already as invariants of the semisimple $1:1$ resonance; in variables (x_1, x_2, x_3, x_4) the basic invariants are $I_1 = x_1^2 + x_2^2$, $I_2 = x_3^2 + x_4^2$, $I_3 = x_1 x_3 + x_2 x_4$, and $I_4 = x_2 x_3 - x_1 x_4$. (I_1 and I_2 are given in (4.5.56); I_3 and I_4 are the real and imaginary parts of (4.5.69), which we did not compute in general but which are simple enough if $p = q = 1$.) With variables (p_1, p_2, q_1, q_2) these coincide (up to constant factors) with the functions S, X, Y, Z, so we now know that

$$ \ker \mathcal{D}_S = \mathbb{R}[[S, X, Y, Z]]. \qquad (4.9.37) $$

The noncanonical transformation $(p_1, p_2, q_1, q_1) \mapsto (p_1, q_1, p_2, q_2)$ carries N into $N_{2,2}$, the basic invariants of which were determined in (4.7.38); in the present notation they are $\alpha = p_1$, $\beta = p_2$, and $\gamma = p_1 q_2 - p_2 q_1$, which can be replaced by S. Thus,

$$\ker \mathfrak{X} = \mathbb{R}[[S, p_1, p_2]]. \tag{4.9.38}$$

From (4.9.37) and (4.9.38), it follows that

$$\ker \mathcal{D}_S \cap \ker \mathfrak{X} = \mathbb{R}[[S, Y]]. \tag{4.9.39}$$

One way to see this is to ask which functions $f(S, p_1, p_2)$ belong to $\ker \mathcal{D}_S$. Such functions satisfy

$$\mathcal{D}_S(f(S, p_1, p_2) = f_S \mathcal{D}_S S + f_{p_1} \mathcal{D}_S p_1 + f_{p_2} \mathcal{D}_S p_2 = p_2 f_{p_1} - p_1 f_{p_2} = 0,$$

which implies that f depends on (p_1, p_2) only through the combination $p_1^2 + p_2^2$. (This is the same calculation as the computation of invariants of the single center.)

It follows immediately from (4.9.39) that H is in normal form for the nonsemisimple $1 : -1$ resonance if and only if it can be expressed as

$$H(p_1, p_2, q_1 q_2) = K(S, Y). \tag{4.9.40}$$

Notes and References

The notions of Hamiltonian system and canonical transformation go back to Hamilton and Jacobi, and the first use of canonical transformations to simplify Hamiltonian systems is known as Hamilton–Jacobi theory. This theory is presented in many textbooks of classical mechanics, for instance Goldstein [48]; a more mathematical introduction is Meyer and Hall [79]. In Hamilton–Jacobi theory, Hamiltonian vector fields are obtained from their scalar Hamiltonian functions, and canonical transformations are obtained via certain scalar "generating functions," but these generate the canonical transformations in a different way than our generators; they are not used as Hamiltonians to generate a flow. This older notion of generating function can be explained nicely in terms of differential forms, but is much less convenient than the use of Hamiltonian generators.

The idea of normal forms, as that word is used in this book, was first introduced by Birkhoff in [16]; the treatment is limited to Hamiltonian multiple centers without resonance. An exposition of the normal form for Hamiltonian systems using the Hamilton–Jacobi generating function method is given in Section 30 of Siegel and Moser [103]. The use of generating functions (in this sense) can be regarded as an additional format, to be added to our formats 1ab and 2abc, but it is valid only for Hamiltonian systems and has been almost completely replaced by the use of generated formats in the Lie-theoretic sense (that is, formats 2abc). The application of these formats (like the rest of normal form theory) was worked out first

for Hamiltonian systems; a survey of the history and literature of this subject is given in Finn [45]. The sl(2) normal form style was also worked out
for Hamiltonian systems first; an important early article is Cushman, Deprit, and Mosak [38]. The literature on Hamiltonian systems is vast, and
it is impossible here to do more than mention a few additional important
books, such as Moser [80], Arnol'd [4], MacKay and Meiss [75], and Cushman [37]. Of these, [75] is a collection of important papers spanning the
years 1927 to 1987, and the others are monographs.

The definitive study of the nonsemisimple $1 : -1$ resonance and its associated unfolding, called the Hamiltonian Hopf bifurcation, is van der Meer
[108]. The results obtained there are stronger than those obtained anywhere
else; in particular, the bifurcation results are valid for values of the unfolding parameters in a full neighborhood of zero. Most other treatments, for
instance that of Meyer and Hall [79], use a scaling argument that requires
assuming that the unfolding parameters are small but not zero. Incidentally, there is a sign error in equation (2), page 219, of [79]; apparently the
authors were thinking (as I did at first) that a $1 : -1$ resonance should rotate the two coordinate planes in opposite directions. Of course, "opposite
directions" has no meaning for planes at right angles to one another, and
the true meaning of $1 : -1$ resonance relates to the indefiniteness of the
Hamiltonian. For a deeper treatment of this notion, see the article by R.S.
MacKay on pages 137–153 of [75].

4.10   Hypernormal Forms for Vector Fields

The idea of hypernormal forms has been introduced in Section 3.3; see
equation (3.3.8). In ordinary normal form theory (in format 2a), a generator
u_j of grade j is used to simplify the term a_j of the same grade; the ambiguity
in the choice of u_j does not affect the new a_j, but does affect higher orders of
the normalized series. To achieve the simplest possible form for the higher-
order terms, one "revisits" the generators of grade j to simplify these terms
without modifying the terms that are already in the desired final form.

We will indicate two ways to work with this idea, calling them "weak"
and "strong" hypernormalization. Weak hypernormalization uses format
2a; strong hypernormalization works best in format 2b. Systems will be
written in the "universal" form (4.2.1), that is,

$$\dot{x} = a(x, \varepsilon) = a_0 + \varepsilon a_1 + \varepsilon^2 a_2 + \cdots . \qquad (4.10.1)$$

For definiteness, setting 1 will be assumed, so that $a_j \in \mathcal{V}_j^n$, although the
main theorem is independent of setting.

4.10.1. Definition. A (j, k) *transformation* for the system (4.2.1), with
$1 \leq j \leq k$, is a transformation having generator $u_j \in \mathcal{V}_j^n$ such that the
terms $a_0 + \cdots + \varepsilon^{k-1} a_{k-1}$ are not affected by the transformation. Thus,

a (j, k) transformation can be used to modify the kth and higher-order terms without affecting those of lower order. The definition is for a *specific* system; a transformation that is a (j, k) transformation for one system may not be so for another.

4.10.2. Theorem. An element $u_j \in \mathcal{V}_j^n$ generates a (j, k)-transformation for (4.10.1) if and only if

$$\mathsf{L}_{a_0} u_j = \mathsf{L}_{a_1} u_j = \cdots = \mathsf{L}_{a_{k-j-1}} u_j = 0. \tag{4.10.2}$$

In this case the change Δa_k produced in the term of grade k is given by the *homological equation of type* (j, k), or (j, k)-*homological equation*,

$$\mathsf{L}_{a_{k-j}} u_j = \Delta a_k. \tag{4.10.3}$$

Proof. We will write u for u_j in the proof. The result of applying the transformation generated by u to a is $e^{\varepsilon^j \mathsf{L}_u} a$. When this series is expanded, its terms can be grouped in successive "ranges" as follows. The first range is

$$a_0 + \varepsilon a_1 + \cdots + \varepsilon^{j-1} a_{j-1};$$

the terms of this range are unchanged, simply because u is of grade j. The second range is

$$\varepsilon^j (a_j + \mathsf{L}_u a_0) + \varepsilon^{j+1}(a_{j+1} + \mathsf{L}_u a_1) + \cdots + \varepsilon^{2j-1}(a_{2j-1} + \mathsf{L}_u a_{j-1}).$$

If k falls in this range ($j \leq l \leq 2j - 1$), the theorem follows by inspection. The third range is

$$\varepsilon^{2j} \left(a_{2j} + \mathsf{L}_u a_j + \frac{1}{2} \mathsf{L}_u^2 a_0 \right) + \cdots + \varepsilon^{3j-1} \left(a_{3j-1} + \mathsf{L}_u a_{2j-1} + \frac{1}{2} \mathsf{L}_u^2 a_{j-1} \right).$$

If k falls in this range ($2j \leq k \leq 3j - 1$), the assumption that u generates a (j, k) transformation implies, first of all, that the terms of the second range are unchanged, so that $a_i \in \ker \mathsf{L}_u$ for $i = 0, \ldots, j - 1$. This implies that all terms in the third range containing squares of Lie operators drop out, and the theorem then follows as for the second-range case. The analysis of higher ranges proceeds similarly. $\qquad \square$

4.10.3. Corollary. The set of generators of (j, k)-transformations for system (4.10.1) depends only on $a_0, \ldots, a_{k-j-1}$, and is exactly the set

$$\mathcal{T}_{jk}(a_0, \ldots, a_{k-j-1}) = \mathcal{V}_j^n \cap \ker \mathsf{L}_{a_0} \cap \cdots \cap \ker \mathsf{L}_{a_{k-j-1}}.$$

Notice that $\mathcal{T}_{kk}$, the set of generators of (k, k)-transformations, is just $\mathcal{V}_k^n$ with no restrictions (even a_0 disappears from the list of terms on which $\mathcal{T}_{kk}$ depends). In fact, a (k, k) transformation is just the kind of transformation used in ordinary normal form theory.

4.10.4. Definition. The (j, k)-*removable subspace* for system (4.10.1) is the set of changes Δa_k that are producible by the use of (j, k)-transformations, or equivalently (in view of (4.10.3)), the image of $\mathcal{T}_{jk}$ under $\mathsf{L}a_{k-j}$:

$$\mathcal{R}_{jk}(a_0, \ldots, a_{k-j}) = \mathsf{L}_{a_{k-j}}(\mathcal{T}_{jk}(a_0, \ldots, a_{k-j-1})). \tag{4.10.4}$$

The *weak removable subspace* of $\mathcal{V}_k^n$ is the sum (not necessarily direct) of the (j, k)-removable subspaces for $j \leq k$:

$$\mathcal{R}_k^{\text{weak}}(a_0, \ldots, a_{k-1}) = \mathcal{R}_{1k} + \mathcal{R}_{2k} + \cdots + \mathcal{R}_{kk}. \tag{4.10.5}$$

The scheme for weak hypernormalization is now as follows. Given a system of the form (4.10.1), we first normalize a_1 as usual. Proceeding to a_2, we normalize it using both the usual $(2, 2)$-transformations and also $(1, 2)$-transformations (with respect to a_0). Notice that it does not matter in what order we apply these transformations, since the effect of either (modulo terms of grade > 2) is simply to add something to a_2, and what is added does not depend on what is already there. Therefore, if $\mathcal{N}_2$ is any complement to $\mathcal{R}_2(a_0, a_1)$, then we are able to bring a_2 into $\mathcal{N}_2$. The choice of $\mathcal{N}_2$ is the choice of a "hypernormal form style" in grade 2. Note that $\mathcal{N}_2$ will usually be a space of lower dimension than an ordinary normal form style in grade 2. Having fixed the form of a_2, it will never again change (unless we go beyond the use of (j, k) transformations in one of the ways mentioned below), and the possibilities for $(3, 1)$, $(3, 2)$, and $(3, 3)$ transformations have likewise become fixed. It is clear how to proceed.

Although weak hypernormalization is quite straightforward, it does not quite use up all of the possibilities for simplification that exist. To demonstrate this, consider the case of hypernormalizing the terms of grade 3 using a generator of the form $v_1 + \varepsilon v_2 + \varepsilon^2 v_3$ in format 2b, with $v_j \in \mathcal{V}_j^n$; as in equation (C.2.5) we write $\mathsf{L}_j = \mathsf{L}_{v_j}$. Then according to Lemma C.2.2, the transformation having this generator takes a into

$$a_0 + \varepsilon\left(a_1 + \mathsf{L}_1 a_0\right) + \varepsilon^2\left(a_2 + \mathsf{L}_1 a_1 + \mathsf{L}_2 a_0 + \frac{1}{2}\mathsf{L}_1^2 a_0\right)$$

$$+ \varepsilon^3\left(a_3 + \mathsf{L}_1 a_2 + \mathsf{L}_2 a_1 + \frac{1}{2}\mathsf{L}_1^2 a_1 + \mathsf{L}_3 a_0 + \frac{1}{2}\mathsf{L}_1 \mathsf{L}_2 a_0\right.$$

$$\left. + \frac{1}{2}\mathsf{L}_2 \mathsf{L}_1 a_0 + \frac{1}{6}\mathsf{L}_1^3 a_0\right).$$

If we require that a_1 and a_2 be unchanged, then first $\mathsf{L}_1 a_0 = 0$ (which causes the terms $\mathsf{L}_1^2 a_0$, $\mathsf{L}_2 \mathsf{L}_1 a_0$, and $\mathsf{L}_1^3 a_0$ to drop out), and then $\mathsf{L}_1 a_1 + \mathsf{L}_2 a_0 = 0$, which causes the terms $\mathsf{L}_1^2 a_1$ and $\mathsf{L}_1 \mathsf{L}_2 a_0$ to drop out. The final *general homological equation* is

$$\mathsf{L}_{a_2} v_1 + \mathsf{L}_{a_1} v_2 + \mathsf{L}_{a_0} v_3 = \Delta a_3,$$

with restrictions $\mathsf{L}_{a_0} v_1 = 0$ and $\mathsf{L}_{a_1} v_1 + \mathsf{L}_{a_0} v_2 = 0$. Since the last condition is less restrictive than requiring $\mathsf{L}_{a_1} v_1 = 0$ and $\mathsf{L}_{a_0} v_2 = 0$ separately, it is clear that this procedure of *strong hypernormalization* allows greater flexibility in eliminating terms than does weak hypernormalization, although the process remains entirely linear. The fundamental result is that given a system (4.10.1), there exists a family of *(strong) removable subspaces* $\mathcal{R}_k \subset V_k^n$ such that given any family of subspaces $\mathcal{N}_k \subset V_k^n$ complementary to $\mathcal{R}_k$, the given system can be brought into one and only one hypernormal form having $a_k \in \mathcal{N}_k$ for each k; notice that the "uniqueness" of these so-called unique normal forms is relative to the choice of a hypernormal form style. For $k = 3$, the calculations just given show that

$$\mathcal{R}_3 = \{ \mathsf{L}_{a_2} v_1 + \mathsf{L}_{a_1} v_2 + \mathsf{L}_{a_2} v_3 : \mathsf{L}_{a_0} v_1 = 0, \mathsf{L}_{a_1} v_1 + \mathsf{L}_{a_0} v_2 = 0 \}.$$

It is clear that $\mathcal{R}_3^{\mathrm{weak}} \subset \mathcal{R}_3$.

Hypernormalization of the Center

In carrying out hypernormalization, it is often helpful to carry out a complete normalization in the ordinary sense first, and then work within the Lie algebra generated by the restricted class of vector fields that remain. As an illustration of hypernormalization, and of this remark, we will give a partial proof of the following theorem:

4.10.5. Theorem. Any single center, or system of the form (4.10.1) with $a_0(x) = Ax$ where

$$A = \begin{bmatrix} 0 & -1 \\ 1 & 0 \end{bmatrix},$$

can be brought into exactly one of the following forms in polar coordinates (called *α-forms*), with uniquely determined coefficients and with $\alpha_\nu \neq 0$ and $\beta_\mu \neq 0$ (whenever these appear):

(1) $\dot{r} = 0,$
       $\dot{\theta} = 1;$

(2) $\dot{r} = 0,$
       $\dot{\theta} = 1 + \beta_\mu r^{2\mu};$

(3) $\dot{r} = \alpha_\nu r^{2\nu+1} + \alpha_{2\nu} r^{4\nu+1},$
       $\dot{\theta} = 1;$

(4) $\dot{r} = \alpha_\nu r^{2\nu+1} + \cdots + \alpha_{2\nu-\mu} r^{4\nu-2\mu+1} + \alpha_{2\nu} r^{4\nu+1},$
       $\dot{\theta} = 1 + \beta_\mu r^{2\mu}.$

Similarly, each single center may be brought into one of the following forms (called *β-forms*), in which $\alpha_\nu \neq 0$ (and there are only two classes because we have placed no restrictions on the other coefficients):

$$(1) \quad \dot{r} = 0,$$
$$\dot{\theta} = 1 + \beta_\mu r^{2\mu};$$

$$(2) \quad \dot{r} = \alpha_\nu r^{2\nu+1} + \alpha_{2\nu} r^{4\nu+1},$$
$$\dot{\theta} = 1 + \beta_1 r^2 + \cdots + \beta_\nu r^{2\nu}.$$

The α and β forms given in the theorem are examples of hypernormal form styles. It is convenient to use (x, y) rather than (x_1, x_2) as coordinates in $\mathbb{R}^2$, and to introduce the following notation for various vector fields in $\mathbb{R}^2$ that are needed in the proof of this theorem:

$$v_i = (x^2 + y^2)^i \begin{bmatrix} x \\ y \end{bmatrix}, \qquad w_i = (x^2 + y^2)^i \begin{bmatrix} -y \\ x \end{bmatrix}. \qquad (4.10.6)$$

The equations of the center in (ordinary) normal form, (1.1.24), may then be written $(\dot{x}, \dot{y}) = a(x, y)$, with

$$a = w_0 + \alpha_1 v_1 + \beta_1 w_1 + \alpha_2 v_2 + \beta_2 w_2 + \cdots. \qquad (4.10.7)$$

Since this contains no terms of odd grade (even degree), it is convenient to dilate or scale the variables by ε^2 rather than ε, to obtain

$$a = w_0 + \varepsilon(\alpha_1 v_1 + \beta_1 w_1) + \varepsilon^2(\alpha_2 v_2 + \beta_2 w_2) + \cdots; \qquad (4.10.8)$$

this is not necessary, but it helps to make clear the position of each term during the following calculations. We will refer to the power of ε as the *order* of a term; this is half the grade (in the usual grading). The following lemma shows that the set of vector fields of this form constitutes a graded Lie algebra in its own right.

4.10.6. Lemma. The vector fields v_i and w_i (for $i \geq 0$) satisfy the following bracket relations:

$$[v_i, v_j] = 2(i - j)v_{i+j},$$
$$[v_i, w_j] = -2j w_{i+j},$$
$$[w_i, w_j] = 0.$$

Proof. Since these are polynomial vector fields, the results will be true in the whole plane if they are proved in the plane with the origin removed; therefore, the calculations may be done in polar coordinates, in which

$$v_i = \begin{bmatrix} r^{2i+1} \\ 0 \end{bmatrix}, \qquad w_i = \begin{bmatrix} 0 \\ r^{2i} \end{bmatrix}.$$

The calculations in these coordinates are trivial. For instance,

$$[v_i, v_j] = \begin{bmatrix} (2i+1)r^{2i} & 0 \\ 0 & 0 \end{bmatrix} \begin{bmatrix} r^{2j+1} \\ 0 \end{bmatrix} - \begin{bmatrix} (2j+1)r^{2j} & 0 \\ 0 & 0 \end{bmatrix} \begin{bmatrix} r^{2i+1} \\ 0 \end{bmatrix}$$
$$= 2(i - j)v_{i+j}.$$

$\square$

Because of these bracket relations, we can operate within the restricted Lie algebra spanned by the v_i and w_i rather than in the full Lie algebra V_{**}^n. The first step in deriving the α-forms is to divide the (already normalized) centers into four classes. Each center belongs to exactly one of the following cases:

1. All $\alpha_i = 0$ and all $\beta_i = 0$.

2. All $\alpha_i = 0$, and β_μ is the first nonzero β_i.

3. α_ν is the first nonzero α_i, and $\beta_i = 0$ for $i \leq \nu$.

4. α_ν is the first nonzero α_i, and there is a first nonzero β_i, which is β_μ with $\mu \leq \nu$.

In case 1, there is nothing to prove: The system already has the first α-form. In case 2, we claim that β_i can be made zero for all $i > \mu$ without changing the fact that all α_i are 0. This will prove that the system can be brought into the second α-form. The proof is by induction; suppose that we have achieved

$$a = w_0 + 0 + \cdots + 0 + \varepsilon^\mu \beta_\mu w_\mu + 0 + \cdots + 0 + \varepsilon^i \beta_i w_i + \cdots .$$

Taking as generator $cv_{i-\mu}$, with c to be determined, we compute

$$e^{\varepsilon^{i-\mu} L_{cv_{i-\mu}}} (w_0 + \varepsilon^\mu \beta_\mu w_\mu + \varepsilon^i \beta_i w_i + \cdots) \tag{4.10.9}$$
$$= w_0 + \varepsilon^\mu \beta_\mu w_\mu + \varepsilon^{i-\mu}[w_0, v_{i-\mu}] + \varepsilon^i \{\beta_i w_i + c[w_\mu, v_{i-\mu}]\} + \cdots .$$

Since $[w_0, v_{i-\mu}] = 0$, and $[w_\mu, v_{i-\mu}] = 2(i-\mu)w_i$, we can take $c = -\beta_i/2(i-\mu)$ to eliminate the w_i term without disrupting any terms of lower order. (In the terminology of the beginning of this section, and remembering that the grade is twice the order, we have made a $(2\mu, 2i)$ transformation.) Before proceeding to the next step of the induction, it is essential to notice that the iterated brackets of $cv_{i-\mu}$ with higher-order w terms produce only more w terms and never any v terms, so that the inductive hypothesis remains true at the next step.

In case 3, we claim that all α_i and β_i for $i > \nu$ can be eliminated except for $\alpha_{2\nu}$. The proof is by induction on $i = \nu+1, \nu+2, \ldots$, where we observe that the argument is different for the ranges $\nu < i < 2\nu$, $i = 2\nu$, and $i > 2\nu$. For $\nu < i < 2\nu$ the induction hypothesis is

$$a = w_0 + \varepsilon^\nu \alpha_\nu v_\nu + 0 + \cdots + 0 + \varepsilon^i (\alpha_i v_i + \beta_i w_i) + \cdots .$$

We take as generator $cv_{i-\nu} + dw_{i-\nu}$, with c and d to be determined, and check that because $[w_0, cv_{i-\nu} + dw_{i-\nu}] = 0$ (notice that this is just the equivariance of the normal form terms), this does not affect the terms of order less than i. The terms of order i become

$$\alpha_i v_i + \beta_i w_i + 2(2\nu - i)c\alpha_\nu v_i - 2(i - \nu)d\alpha_\nu w_i.$$

Since $2\nu - i \neq 0$ and $i - \nu \neq 0$, it is possible to choose c and d so that this vanishes. For $i = 2\nu$ the induction hypothesis has the same form, as do the last equations, but it is no longer possible to solve for c and eliminate the $\alpha_{2\nu}v_{2\nu}$ (but d can be found to eliminate the w term). For $i > 2\nu$ the induction hypothesis changes because of the presence of the $\alpha_{2\nu}$ term; the details are left to the reader. Case 4, and the β-forms, will not be discussed here. See Notes and References.

Before leaving the topic of the center, we consider the effect of linear transformations as a source of further reduction. A linear transformation of the form

$$\begin{bmatrix} x \\ y \end{bmatrix} = \begin{bmatrix} r & -s \\ s & r \end{bmatrix} \begin{bmatrix} \xi \\ \eta \end{bmatrix}$$

leaves v_0 and w_0 unchanged, and replaces $x^2 + y^2$ by $(r^2 + s^2)(\xi^2 + \eta^2)$, so that each α_i and β_i is multiplied by $(r^2 + s^2)^i$. This leaves the structure of the α-forms and β-forms unchanged, but allows us to normalize one coefficient to ± 1. It is natural to choose the lowest nonzero coefficient, either α_ν or β_μ, to normalize in this way. These linear transformations would be classified as $(0, k)$ transformations under the general scheme for weak hypernormalization, if k is the lowest nonvanishing degree that is targeted to be normalized to ± 1, except for the fact that these transformations are not near-identity transformations and do not fit the framework of generated transformations in any format. In using these transformations we are "revisiting" the initial choice of transformation T_0 used to bring A into real canonical form prior to the first step of normalization.

Remarks About the Hypernormalization of N_2

Another system for which hypernormalization has been extensively studied is the general system with linear part

$$A = N_2 = \begin{bmatrix} 0 & 1 \\ 0 & 0 \end{bmatrix}.$$

The results for this system are quite difficult, and we will not present them here. Instead we will only set up the problem, showing the ways it is similar to and different from the case of the center, and briefly discuss one of the techniques that has played a central role in obtaining the known results.

Using the vector fields

$$v_i = x^i \begin{bmatrix} x \\ y \end{bmatrix}, \qquad w_i = x^i \begin{bmatrix} 0 \\ x \end{bmatrix}, \qquad z_0 = \begin{bmatrix} y \\ 0 \end{bmatrix},$$

the normal form (1.2.12) for this problem can be written

$$a = z_0 + \varepsilon(\alpha_1 v_1 + \beta_1 w_1) + \varepsilon^2(\alpha_2 v_2 + \beta_2 w_2) + \cdots, \tag{4.10.10}$$

using the ordinary dilation $(\varepsilon x, \varepsilon y)$ in which the power of ε equals the grade (rather than half the grade, as in the center). The bracket relations satisfied by v_i and w_i are $[v_i, v_j] = (i - j)v_{i+j}$, $[v_i, w_j] = -jw_{i+j}$, and $[w_i, w_j] = 0$; this is essentially the same as for the center (except for the factors of two, which can be removed by making a different choice of basis), and the Lie algebra spanned by these vector fields is isomorphic to that of the center. The biggest difference, which makes this problem much harder, is that the linear term z_0 does not belong to this Lie algebra, so it is not possible to carry out the calculations within this Lie algebra, but instead one must work in the whole Lie algebra $\mathcal{V}^2_{**}$.

The technique that has proved helpful in handling this problem, and should also be useful in other problems, is to introduce an alternative dilation $(\varepsilon^{\delta_1} x, \varepsilon^{\delta_2} y)$ that arranges the terms of (4.10.10) in a different order. We will show in a moment that introducing such a scaling results in a new *setting* (in the sense in which that word was used in Section 4.2); hypernormalizing in the new setting works in the same way as before, but if the scaling is chosen correctly, it can suggest the right order in which to carry out the calculations. The N_2 problem breaks down into a number of subcases, as does the center problem. The only case we will mention here is the case that α_ν is the first nonzero α_i, and all β_j are 0 for $j \leq 2\nu$. In this case, choosing $\delta_1 = 1$ and $\delta_2 = \nu + 1$ makes z_0 and $\alpha_\nu v_\nu$ have the same order (that is, the same power of ε), and these together become the new leading term. In normalizing the rest of the series with respect to this leading term, a single step in the calculation uses terms that were originally of different grades and would have to have been handled in different steps. The details are left to the original papers.

A New Grading Is a New Setting

We conclude this section with a brief indication of how the new gradings work in an n-dimensional setting. Beginning with a system

$$\dot{x} = a_0(x) + a_1(x) + a_2(x) + \cdots, \qquad (4.10.11)$$

with $a_0(x) = Ax$ and $a_j \in \mathcal{V}^n_j$, we set $x_i = \varepsilon^{\delta_i} \xi_i$ for $i = 1, \ldots, n$; here $\delta_i > 0$ are integers. Written as a system of scalar equations of the form

$$\dot{x}_i = \sum_m c_{im} x^m$$

in multi-index notation, the transformed system will be first

$$\varepsilon^{\delta_i} \xi_i = \sum_m c_{im} \varepsilon^{\langle m, \delta \rangle} \xi^m,$$

and then

$$\xi_i = \sum_m c_{im} \varepsilon^{\langle m, \delta \rangle - \delta_i}.$$

Finally, this system can be written as

$$\dot{\xi} = \sum_j \varepsilon^j b_j(\xi). \qquad (4.10.12)$$

(If desired, one can set $\varepsilon = 1$ at the end of this process and obtain a system resembling (4.10.11) except that the terms are grouped differently. Since it is not easy to "read off" the new grading from the monomials themselves, it is probably best to carry the ε as a bookkeeping device.) We have not specified the starting value of j, because this will depend on the terms that are actually generated in a given example.

From the standpoint of the ideas presented in this book, the feature of (4.10.12) that calls attention to itself is that the coefficients b_j belong to new vector spaces $\mathcal{H}_j$ spanned by all monomial vector fields $x^m e_i$ for which $\langle m, \delta \rangle - \delta_i = j$. Recall that in Section 4.2 we defined a *setting* as a family of vector spaces to which the terms of a system in "universal" form (4.2.1) are required to belong. Therefore, all of the machinery developed in this book, such as the five formats and the notions of weak and strong hypernormalization, are available for use with the modified system (4.10.12).

Notes and References

The following bibliography for the subject of hypernormal forms is nearly complete, as regards the authors (and groups of authors) that have worked on the subject, although not every paper by each author or group is listed. It seems important to collect such a bibliography, since the subject is still rather small and the authors do not always seem to be aware of each other's work.

The idea of hypernormal forms seems to have originated with G.R. Belitskii. His only paper on the subject that has been translated into English is [15]; for references to the Russian papers and book, see Robert Rousarie's review of [12] (mentioned below) in *Math Reviews*, MR 93m:58101. This work seems to have had no influence on the papers described below.

The next contribution was by Shigehiro Ushiki [106]. He provides a Lie-algebraic setting for the theory, and computes hypernormal forms up to third, fourth, or fifth degree for the 0/Hopf interaction and for the nilpotent problems N_2, N_3, and $N_{2,1}$. An interesting point made in this paper is that one sometimes has to choose between continuous dependence on parameters and degree of simplicity of the hypernormalization, and that the application one has in mind may dictate which choice is best.

Next, and by all means the most important of the papers on this subject, are the contributions of Alberto Baider, sometimes in collaboration with Richard Churchill or Jan Sanders. In [10] the β-form for the hypernormalized center is worked out in a way that is rigorous to all orders; [9] gives the general scheme for what we have called strong hypernormalization and uses it to derive the α-form for the center. The (strong) removable subspaces are defined and characterized in [11], and the theory is applied to nilpotent Hamiltonian systems in one degree of freedom (the Hamiltonian case of the N_2 problem). It is also in this paper that alternative gradings are first used. This method is further extended in [12] to treat the full (that is, non-Hamiltonian) N_2 problem. The problem is broken into three cases ($\mu < 2\nu$, $\mu = 2\nu$, $\mu > 2\nu$, where μ and ν are the first nonvanishing β_i and α_i, respectively) and solved in the first and third cases; the results for $\mu = 2\nu$ are incomplete. The authors suggest that their idea of using the Clebsch–Gordan decomposition of an sl(2) action on $\mathbb{R}^2$ as a guide to their constructions might generalize to higher dimensions; this has apparently not been followed up by any of the later workers.

The open problem concerning the $\mu = 2\nu$ case seems to have motivated the papers of Kokubu, Oka, and Wang [66] and Wang, Li, Huang, and Jiang [109]. The first of these papers introduces the class of alternative grading functions defined by an n-tuple of integers, equivalent to our construction leading to (4.10.12). The gradings already used in [11] and [12] belong to this class. These gradings are then used to study the N_2 problem in the case $\mu = 2$, $\nu = 1$ in [66], and $\mu = 2\nu$ in [109]. Although the authors claim to have solved this problem, what they have actually done is to show that the hypothesis that a certain number is nonalgebraic implies that the problem belongs to a subclass of the case $\mu = 2\nu$ for which the results of [12] were already complete. The open problem declared in [12] remains open.

The research group of Algaba, Freire, and Gamero have published a number of papers on hypernormalization, for instance [2]. They give a general formulation that seems equivalent to that of Baider, and they rederive both the α-form and β-form for the center, seemingly without awareness that Baider had done it already, although they do cite [12] without making any use of it. Since they assume that there are both a least nonvanishing α_i and a least nonvanishing β_j in the center problem, they miss the cases in which one of these does not exist. They do consider the effect of nonlinear reparameterization of time, which Baider did not, and in that way achieve some additional simplifications. In other papers they consider the 0/Hopf interaction, including the pitchfork/Hopf case. In a recent preprint with Garcia [3] they have considered alternative gradings from a Newton diagram point of view, and recomputed some cases of the N_2 or Takens–Bogdanov problem, including reparameterization of time.

Guoting Chen and Jean Della Dora have published papers [26] and [25], in which they compute hypernormal forms using Carleman linearization. This is a technique in which nonlinear systems, truncated at some jet, are replaced by very large linear systems in a large number of variables. This is exactly contrary to the philosophy guiding this book, according to which one should always work with matrices that are as small as possible; it would seem that this method would make all but the smallest problems impossible to solve. Their method has the advantage that it works without knowledge of the Jordan form or even the eigenvalues of A, but the same is true for the algorithms of Sanders presented in this book, which do not use large matrices. (We have repeatedly given algorithms that use the eigenvalues but not the Jordan form. To do without the eigenvalues, use the remark preceding Theorem 2.1.4. The characteristic equation can be found from the Frobenius canonical form, which Chen and Della Dora begin with.) Chen and Della Dora use the number of vanishing coefficients in the normal form up to a certain order as an indication that one hypernormal form is better than another; it is not clear that this is a valid test. By this standard, the α-form for the center (computed to low order) would be a "better" hypernormal form than the β-form, even though the β-form packs more information into low-order terms; actually, when both forms are considered to all orders, the number of nonvanishing coefficients is the same.

Finally, Pei Yu in [113] rediscovered the hypernormal form for the center by what he admits are "brute-force" methods using large matrices. Yuan Yuan and Pei Yu [114] do the same for the N_2 problem, and provide Maple code implementing their method.

The "weak hypernormalization" method with which this section starts began as a mistake; I was attempting to formulate Baider's theory using format 2a. When I discovered that the resulting theory was too weak, I decided to keep it anyway for pedagogical purposes, since it presents the ideas of hypernormalization in a simple way that is sufficient for some problems. In principle, strong hypernormalization can be done in format 2a; what is necessary to hypernormalize grade k is to use a product of transformations with generators of grades 1 through k, requiring that the whole product of transformations (rather than the individual transformations in the product) leave $a_0, \ldots, a_{k-1}$ unchanged. But the formulas do not seem to work out nicely this way.

As this book is receiving its final editing, Jan Sanders [98] and I are thinking, along slightly different lines, about how the hypernormal form theory fits into homological algebra. The starting point is Remark 1.1.4, which shows how ordinary normal form theory can be formulated using chain complexes. It turns out that hypernormal form theory can be formulated using a sequence of chain complexes, such that the vector spaces in each complex are the homology spaces of the preceding complex. Such a structure is known as a *spectral sequence* in homological algebra. Setting the theory up this way does not change the way the calculations are done, but it does provide a convenient way of organizing them and understanding their structure, and may lead to further discoveries such as a systematic way of defining hypernormal form styles. (At the present time there is nothing for hypernormal forms that corresponds to semisimple, inner product, or sl(2) styles.) The motivation for a spectral sequence formulation of hypernormal forms comes from [5], in which spectral sequences are used to study a rather different normal form problem (involving singularities of complex functions rather than differential equations).

5
Geometrical Structures in Normal Forms

After a system has been placed into normal form, an immediate question arises: What does the normal form tell us about the dynamics of the system? Sometimes the normalized system is simple enough to be integrable ("by quadrature," that is, its solution can be reduced to the evaluation of integrals), but this is not the usual case (except in two dimensions). We approach this question in Section 5.1 by establishing the existence of geometrical structures, such as invariant manifolds and preserved foliations, for systems in truncated normal form (that is, polynomial vector fields that are entirely in normal form, with no nonnormalized terms). These geometrical structures explain why some truncated normal forms are integrable, and give partial information about the behavior of others. In particular, the computation of a normal form up to degree k simultaneously computes the *stable, unstable, and center manifolds*, the *center manifold reduction* of the system, the *stable and unstable fibrations* over the center manifold, and various *preserved foliations*. These concepts will be defined (to the extent that they are needed) when they arise, but the reader is expected to have some familiarity with them. We do not prove the existence of these structures, except in the special case of truncated systems in normal form, where each structure mentioned above takes a simple linear form. For the full (untruncated) systems the relevant existence theorems will be stated without proof, and the normal form will be used to compute approximations.

5.0.1. Remark. Most textbooks compute the center manifold reduction of the system first, then bring the reduced system to normal form. The normal form obtained in this way represents only the behavior of the system on its center manifold, and does not compute the stable and unstable manifolds or the stable and unstable fibrations over the center manifold.

Throughout this chapter, the setting is as follows: Two smooth systems of differential equations will be considered, called the *full system*

$$\dot{x} = a(x) \tag{5.0.1}$$

and the *truncated system*

$$\dot{x} = \widehat{a}(x). \tag{5.0.2}$$

The truncated system is a polynomial vector field of degree k (grade $k-1$) of the form

$$\widehat{a}(x) = Ax + a_1(x) + \cdots + a_{k-1}(x), \tag{5.0.3}$$

with $a_j \in V_j^n$, and is assumed to be in some suitable normal form (to be discussed momentarily). The full system has the form

$$a(x) = \widehat{a}(x) + \widetilde{a}(x), \tag{5.0.4}$$

where $\widetilde{a}$ has vanishing k-jet at the origin, so that $\widehat{a} = j^k(a)$; the formal power series of a is

$$a(x) \sim Ax + a_1(x) + \cdots = \widehat{a}(x) + a_k(x) + a_{k+1}(x) + \cdots. \tag{5.0.5}$$

The terms a_j for $j \geq k$ are not assumed to be in normal form. In Section 5.3, it is more convenient to write the truncated system as

$$\dot{y} = \widehat{a}(y) = Ay + a_1(y) + \cdots + a_{k-1}(y). \tag{5.0.6}$$

Then, when solutions of the full and truncated systems are to be compared, they will be denoted by $x(t)$ and $y(t)$, with the goal being to estimate the difference $\|x(t) - y(t)\|$.

For most of this chapter, it is required that the terms $a_1, \ldots, a_{k-1}$ be only in extended semisimple normal form (Definition 4.5.5). That is, let $A = S + N$ be the semisimple/nilpotent decomposition of A; then we require that the vector field $\widehat{a}(x)$ commute with the linear vector field Sx; that is,

$$\widehat{a}(x) \in \ker \mathsf{L}_S. \tag{5.0.7}$$

(Recall from Definition 4.4.10 and Theorem 4.4.11 that two vector fields *commute* if their Lie bracket is zero, or equivalently, if their flows commute in the sense that $\varphi^t \circ \psi^s = \psi^s \circ \varphi^t$.) According to Chapter 4, $\widehat{a}$ will be in extended semisimple normal form in the following circumstances:

1. If $A = S$ is semisimple, and $\widehat{a}$ is in semisimple normal form.

2. If $A = S + N$ is in Jordan or real canonical form (or more generally, if A is seminormal), and $\widehat{a}$ is in inner product normal form (Lemma 4.6.10).

3. If $\widehat{a}$ is in simplified normal form (Theorem 4.6.20).

4. If $\widehat{a}$ is in sl(2) normal form (Corollary 4.8.3).

5. If $\widehat{a}$ has been "partially normalized" by applying Algorithm 4.6.11, stopping at step 2, or Algorithm 4.8.4, stopping at step 6. (The former requires that A be seminormal.)

The only normal form we have studied that is excluded from this list is inner product normal form in the case that A is not seminormal. In Section 5.4, which is explicitly concerned with the nilpotent part of A, the sl(2) normal form is required.

The comparison of equations (5.0.1) and (5.0.2) involves a quantitative component as well as the qualitative aspect described by "geometrical structures." The quantitative question concerns the degree to which solutions of the simplified system (5.0.2) approximate those of (5.0.1) having the same or similar initial conditions. The same question can of course be asked whenever a system is truncated, without requiring that the retained terms be in normal form, but the answer in such a case is not very satisfying; strong results are available only when the truncated system is in normal form, and these strong results (developed in Section 5.3) depend heavily on the qualitative results of Section 5.1. The strongest results (in the positive direction of time) are for the stable manifold; the next strongest involve the stable and center manifolds; and the weakest include the unstable manifold. Good estimates involving the center manifold are available (at present) only when there is no nilpotent part appearing in the linear terms on the center manifold.

5.1 Preserved Structures in Truncated Normal Forms

This section is strictly limited to the discussion of structures that exist in the truncated system (5.0.2), under the assumption (discussed above) that $\widehat{a}(x)$ commutes with Sx. The question of what these structures imply about the full system (5.0.1) is deferred to Sections 5.2 and 5.3.

Invariant Subspaces in the Truncated System

The first type of geometrical structure to be discussed is an invariant subspace. If $E \subset \mathbb{R}^n$ and $\dot{x} = a(x)$ is a system of differential equations on $\mathbb{R}^n$, E is called an *invariant subset* if every solution having its initial point in E remains in E for all time, both forward and backward. An *invariant subspace* is an invariant subset that is also a vector subspace of $\mathbb{R}^n$ (and in particular, contains the origin).

A truncated normalized system of the form (5.0.2) generally has several invariant subspaces. These include the stable, unstable, and center subspaces, as well as a variety of strong stable and unstable subspaces. All of these invariant subspaces exist as a consequence of the following theorem:

5.1.1. Theorem. Assume that the truncated system (5.0.2) is in extended semisimple normal form. Let μ be a real number, let E^+ be the direct sum of the generalized eigenspaces of A with eigenvalues λ such that $\mathrm{Re}\,\lambda \geq \mu$, and let E^- be the direct sum of the generalized eigenspaces for $\mathrm{Re}\,\lambda \leq \mu$. Then if $\mu \geq 0$, E^+ is an invariant subspace of the flow of (5.0.2), while if $\mu \leq 0$, E^- is an invariant subspace.

Proof. Assuming $\mu \leq 0$, we show that E^- is invariant; the other case is similar. Assume at first that A is in Jordan canonical form, arranged so that

$$A = \begin{bmatrix} B & \\ & C \end{bmatrix},$$

with B having eigenvalues $\lambda_1, \ldots \lambda_r$ satisfying $\mathrm{Re}\,\lambda_j \leq \mu \leq 0$ and C having eigenvalues $\lambda_{r+1}, \ldots, \lambda_n$ satisfying $\mathrm{Re}\,\lambda_j > \mu$. Writing $u = (x_1, \ldots, x_r)$, $v = (x_{r+1}, \ldots, x_n)$, it follows that $E^- = \{(u, 0) : u \in \mathbb{R}^r\}$, and system (5.0.2) may then be written

$$\dot{u} = Bu + \widehat{g}(u, v), \tag{5.1.1}$$

$$\dot{v} = Cv + \widehat{h}(u, v)$$

(where B and C are themselves in Jordan form). By hypothesis,

$$\begin{bmatrix} \widehat{g} \\ \widehat{h} \end{bmatrix} \in \ker \mathsf{L}_S, \tag{5.1.2}$$

where S is the semisimple (in this case diagonal) part of A.

To prove that E^- is invariant, it suffices to prove that $\dot{v}$ vanishes on E^-, that is, that $\widehat{h}(u, 0) = 0$; this implies that if a solution $(u(t), v(t))$ satisfies $v(0) = 0$, then $v(t) = 0$ for all t. To prove this, we show that no monomial of the form $u^m = u_1^{m_1} \cdots u_r^{m_r}$, containing only factors from u (and none from v), can occur in any component $\widehat{h}_j(u, v)$ of $\widehat{h}$. It follows from Corollary 4.5.9, which describes $\ker \mathsf{L}_S$ for S diagonal, that if $\widehat{h}_j$ were to contain such a u^m, then it would be the case that

$$m_1 \lambda_1 + \cdots + m_r \lambda_r = \lambda_j.$$

But this is impossible, because since $\mu < 0$ and $m_1 + \cdots + m_r > 0$, $\mathrm{Re}(m_1 \lambda_1 + \cdots + m_r \lambda_r) \leq (m_1 + \cdots + m_r)\mu \leq \mu$, whereas $\mathrm{Re}\,\lambda_j > \mu$.

If A is not in Jordan form, the same proof is valid, because it is possible to apply a linear transformation $x = Ty$ to the system (5.0.2) such that $T^{-1}AT$ is in Jordan canonical form and the normalization of the vector field is preserved. The argument is slightly different for each normal form style.

If A is semisimple, Theorem 4.5.6 guarantees that the similarity preserves normal form. If A is seminormal, and inner product normal form is used, Remark 3.4.11 and Theorem 4.6.8 imply that there exists a suitable T. Finally, and most generally, if A is arbitrary and (5.0.2) is in sl(2) normal form, Corollary 4.8.3 guarantees that any T taking A into Jordan form also takes (5.0.2) into a system that is in semisimple normal form (in the sense of Remark 4.5.5) with respect to the semisimple part of its linear term, and this is sufficient for the argument given above. In cases where (5.0.2) is real, but becomes complex when A is put into Jordan form, the argument above proves that the *complex* vector space E^- is invariant under the flow (in complex variables). But since the reality subspace $\mathcal{R} \subset \mathbb{C}^n$ is invariant under the flow, $E^- \cap \mathcal{R}$ is invariant, and this maps back (under the coordinate change) to the real subspace E^- for (5.0.2).   □

Using Theorem 5.1.1, all of the following invariant subspaces can be seen to exist (although some of them may be empty in any given case):

5.1.2. Definition.

1. The *stable subspace*, denoted by E^s, is the direct sum of generalized eigenspaces of A with eigenvalues in the left half-plane. Equivalently, it is the subspace E^- corresponding to a choice of $\mu < 0$ sufficiently small (in absolute value) that all λ_j with $\operatorname{Re} \lambda_j < 0$ satisfy $\operatorname{Re} \lambda_j \leq \mu$.

2. The *unstable subspace* E^u is the direct sum of generalized eigenspaces with eigenvalues in the right half-plane. This is E^+ for a sufficiently small choice of $\mu > 0$.

3. The *center subspace* E^c is the direct sum of generalized eigenspaces with eigenvalues on the imaginary axis. It is the intersection of E^+ and E^- for $\mu = 0$, and is invariant because both of these spaces are invariant ($\mu = 0$ is the only value for which both E^+ and E^- are invariant).

4. The *center-stable* and *center-unstable* spaces are defined by $E^{cs} = E^c \oplus E^s$ and $E^{cu} = E^c \oplus E^u$.

5. A *strong stable subspace*, denoted by E^{ss}, is any E^- for a value of $\mu < 0$ that separates those eigenvalues of A that lie in the left half-plane. In general, there is a "flag" (a nested set of subspaces) of strong stable subspaces corresponding to different choices of μ. Strong unstable spaces E^{su} are defined similarly, using values of $\mu > 0$ that separate the eigenvalues with positive real part.

   5.1.3. Remark. The word *center* has a slightly different meaning in the phrases "center subspace" and "center manifold" than it does in the expressions "(single) center" and "double center" in Sections 1.1 and 4.5. The "center subspace" involves any eigenvalues on the

imaginary axis, including eigenvalues equal to zero. A "center" tradi-
tionally refers to a rest point, all of whose eigenvalues occur in pure
imaginary conjugate pairs with zero excluded.

As a first example of invariant subspaces, consider the matrix A given in
(4.5.27), with normal form module described by (4.5.31) and (4.5.32). There
is a strong unstable space E^{su} spanned by $(1,0,0)$; E^u is spanned by $(1,0,0)$
and $(0,1,0)$, and E^s by $(0,0,1)$. Notice from (4.5.29) that each monomial
appearing in $(\dot{x}_2, \dot{x}_3)$ contains a factor of x_2 or x_3, as needed in the proof
of Theorem 5.1.1 to prove the invariance of E^{su}, but not every monomial
in $\dot{x}_1$ contains a factor of x_1 (as witnessed by the equivariant vector field
v_4 in (4.5.29)). Therefore, the space spanned by $(0,1,0)$ and $(0,0,1)$ is not
invariant. Many other examples, including examples with center subspaces,
will be given later in this section, after two other geometrical structures
(fibrations and foliations) have been defined.

Stable and Unstable Fibrations

One of the nice things that can happen when a system of differential equations is simplified is that a decoupling of certain subsystems may take
place. For instance, in elementary courses a linear system $\dot{x} = Ax$ is often
solved by diagonalizing A (if possible); this completely decouples the system into n separate scalar equations, solvable by exponentials. Putting a
nonlinear system into normal form does not usually produce this dramatic
a result, but a partial decoupling does take place. The next theorem can
be understood as saying that within the center-stable subspace E^{cs}, the
"center variables" decouple from the full system. The same happens within
the center-unstable subspace E^{cu}, but the center variables do not decouple from the others throughout the whole system. The best language for
formulating the result is that of *preserved fibrations*.

Roughly speaking, a *fibration*, *fiber map*, or *fiber bundle* is a mapping
$\pi : T \to B$ of a "total space" T onto a "base space" B such that all of
the preimage spaces $\pi^{-1}(b)$ for various $b \in B$ are in some sense equivalent;
$\pi^{-1}(b)$ is called the *fiber over b*, and b is the *base point* of the fiber. The
fibrations needed in this section are quite simple and require no complicated
definitions. They are simply linear projection maps of a vector space to a
subspace, with the fibers being translations of the kernel.

> **5.1.4. Remark.** Usually, when fibrations are defined, T and B are
> (at least) topological spaces, π is (at least) continuous, and the differ-
> ent fibers are (at least) homeomorphic. Various additional properties
> (such as the so-called homotopy lifting property, or a local product
> structure, or a group action, or some combination of these) are im-
> posed to define specific types of fibrations. Other related structures
> are *sheaves* (with the preimages of base points being called *stalks*
> rather than fibers) and *covering spaces*.

A fibration is *preserved* by a flow if fibers are carried into fibers. In other words, if two points x and y belong to the same fiber (so that $\pi(x) = \pi(y)$), then $\varphi^t(x)$ and $\varphi^t(y)$ belong to the same fiber for each t ($\pi(\varphi^t(x)) = \pi(\varphi^t(y))$).

5.1.5. Theorem. Assume that the truncated system (5.0.2) is in extended semisimple normal form. Then the *stable fibration* $\pi : E^c \oplus E^s \to E^c$ and *unstable fibration* $\pi : E^c \oplus E^u \to E^c$, defined as the natural projections associated with the direct sums, are preserved by the flow of the truncated system (5.0.2).

Proof. As in the proof of Theorem 5.1.1, we may, for the sake of the proof, introduce coordinates in which the linear part is in Jordan form. The theorem will be proved for the resulting system, which may need to be regarded as a system on $\mathbb{C}^n$. Write the system as

$$\dot{u} = Bu + \widehat{f}(u, v, w),$$
$$\dot{v} = Cv + \widehat{g}(u, v, w),$$
$$\dot{w} = Dw + \widehat{h}(u, v, w),$$

where B has eigenvalues in the left half-plane, C on the imaginary axis, and D in the right half-plane. To restrict to E^{cs}, set $w = 0$ and consider the system

$$\dot{u} = Bu + \widehat{f}(u, v, 0), \qquad (5.1.3)$$
$$\dot{v} = Cv + \widehat{g}(u, v, 0).$$

We claim that the v system decouples; that is, $\widehat{g}(u, v, 0)$ does not depend on u. Suppose that a monomial $u^\ell v^m$ occurs in $\widehat{g}(u, v, 0)$; then $\langle \ell, \lambda \rangle + \langle m, \mu \rangle$ must equal an eigenvalue of C and hence be pure imaginary; here λ gives the eigenvalues of B and μ those of C. This is impossible unless $\ell = 0$.

Since the v system decouples, it follows that $\pi(u, v) = v$ evolves independently of u. This means that the fibration is preserved. $\qquad \square$

Preserved Foliations

A *foliation* is another type of geometric structure similar to a fibration. Loosely speaking, a foliation of an open set in $\mathbb{R}^n$ is a decomposition of the set into a union of subsets called the *leaves* of the foliation; these leaves are usually expected to be of the same topological type, although we will not insist strictly on this requirement (that is, we allow "singularities" in the foliation). The simplest example is the foliation of the punctured plane $\mathbb{R}^2$ by level sets of the form $x^2 + y^2 = c$. Each of these leaves is topologically equivalent (they are all circles) except for the "singular leaf" $(0, 0)$ corresponding to $c = 0$ (which is a point). A "strict" foliation (in which all the leaves are circles) can be obtained by deleting the singular leaf (both from the foliation, and from the set being foliated). A foliation becomes a fibration if the set of leaves is taken as the base space, with the projection being the map assigning each point to its leaf.

The foliations used in this chapter arise from some (local) flow ψ^s defined on an open subset of $\mathbb{R}^n$. For any flow, there is a unique orbit passing through each point, and these orbits can have three topological types: intervals, circles, and points. In most examples, one of these types is "typical," and the exceptions can be regarded as singular leaves. The basic example for our purposes is the S-foliation, defined as follows:

5.1.6. Definition. Let $A = S + N$ be the semisimple/nilpotent decomposition of the matrix A appearing in the truncated system (5.0.2). The foliation of $\mathbb{R}^n$ by the orbits of the flow e^{Ss} is called the S-foliation.

As with fibrations, to say that a foliation is *preserved* by a flow φ^t does not mean that the leaves are invariant under the flow, but instead means that if two points x and y belong to the same leaf, then $\varphi^t(x)$ belongs to the same leaf as $\varphi^t(y)$ at each time t; in other words $\varphi^t(x)$ and $\varphi^t(y)$ "stay together leafwise" as they evolve across the foliation.

5.1.7. Theorem. Suppose that the truncated system (5.0.2) is in extended semisimple normal form. Then the orbits of the linear flow e^{Ss} constitute the leaves of a foliation preserved by the flow φ^t of the truncated system.

Proof. Since $[\widehat{a}, S] = 0$, the flows φ^t of $\widehat{a}$ and $\psi^s = e^{Ss}$ of Sx commute (see Theorem 4.4.11). Consider two points x and y belonging to the same leaf of the S-foliation; then there exists a value of s such that $y = \psi^s(x)$. By the commutativity of the flows, for any t,

$$\varphi^t(y) = \varphi^t(\psi^s(x)) = \psi^s(\varphi^t(x)),$$

which implies that $\varphi^t(y)$ and $\varphi^t(x)$ belong to the same leaf of the S-foliation. This is what it means for the foliation to be preserved. $\square$

> **5.1.8. Remark.** This theorem becomes vacuous if $S = 0$, because in that case each leaf of the foliation is simply a point, and such a foliation is "preserved" by any flow. So although this theorem is valid when $A = S + N$ contains a nilpotent part, it is of no help when A is *purely* nilpotent. Discussion of that case is deferred to Section 5.4.

When the orbits of e^{Ss} are intervals and circles, the leaves of the foliation determined by S are usually the level sets of $n - 1$ scalar functions, $I_1, \ldots, I_{n-1}$, which are *integrals*, or *invariants*, or *conserved quantities* of the flow of e^{Ss}; this will be illustrated in a number of examples later in this section. The invariants I of the flow e^{Ss} are just the elements of the ring $\ker \mathcal{D}_S$. The computation of this ring is a part of the computation of the normal form, so the required functions are already at hand in any application, as will be seen below. Specifically, in the semisimple case $A = S$, the ring $\ker \mathcal{D}_S$ is the same as the ring of invariants $\ker \mathcal{D}_A$ that underlies the normal form module (or module of equivariants). In the general case $A = S + N$, we either have $\ker \mathcal{D}_{A*} = \ker \mathcal{D}_S \cap \ker \mathcal{D}_{N*}$ (if A is in Jordan or real canonical form and the inner product normal form is used, Lemma 4.6.10), or $\ker \mathcal{D}_{\widetilde{A}} = \ker \mathcal{D}_S \cap \ker \mathcal{D}_M$ (if the sl(2) normal form is used, equation (4.8.3)). In either case, $\ker \mathcal{D}_S$ will have already been computed.

> **5.1.9. Remark.** The number of independent invariants needed to describe the S-foliation (when this is possible) is $n-1$. All continuous invariants can be expressed as continuous functions of a "complete" set of $n-1$ invariants. Frequently the number of invariants required to generate the ring $\ker \mathcal{D}_S$ is greater than $n-1$. This is because the ring is generated by algebraic operations, not by all possible continuous operations. The existence of "excess generators" for the ring occurs together with the existence of relations among these generators, in the sense of Remark 4.5.15.

The following corollary shows how the existence of preserved foliations leads to reduction of the order of a system, and sometimes to integrability:

5.1.10. Corollary. In any open set in which the orbits of e^{Ss} are characterized as level sets of $I = (I_1, \ldots, I_{n-1})$, the system (5.0.2) may be reduced to a system of one less dimension having the form

$$\dot{I} = F(I). \tag{5.1.4}$$

In particular, if $n = 2$, system (5.0.2) is integrable by quadrature.

First Proof. The rate of change of I measures the rate at which solutions move across the S-foliation, and by Theorem 5.1.7 this rate is the same for all points on a given leaf. Therefore, it is determined by the value of the integrals that characterize the leaf. □

Second Proof. This second proof is given with a view toward an argument in Section 5.4. In that section, we require the differential equations for I in a situation in which the foliation by level sets of I is *not* preserved; the argument will be best understood if it is first seen in the simpler case where the foliation is preserved.

The rate of change of each I_i along the flow of $\widehat{a}$ is $\dot{I}_i = \mathcal{D}_{\widehat{a}} I_i$; this function $\dot{I}_i$ is a function of x (not of t). To show that the rate of change of I_i is the same at all points of a leaf of the S-foliation, it is enough to show that $\mathcal{D}_S \dot{I}_i = 0$ everywhere, since this is the rate of change of $\dot{I}_i$ as we move along the leaf. But

$$\mathcal{D}_S \mathcal{D}_{\widehat{a}} I_i = [\mathcal{D}_S, \mathcal{D}_{\widehat{a}}] I_i - \mathcal{D}_{\widehat{a}} \mathcal{D}_S I_i = \mathcal{D}_{[\widehat{a},S]} I_i - \mathcal{D}_{\widehat{a}}(\mathcal{D}_S I_i) = 0 + 0.$$

The first term vanishes because $\widehat{a}$ and S commute, and the second because $\mathcal{D}_S I_i = 0$. □

Periodic Solutions and Invariant Tori

A final type of geometrical structure that can appear (and is often easy to detect) in truncated systems in normal form is the *invariant torus*, which includes as a special case *periodic solutions*. These occur within the center subspace, or within an even-dimensional subspace of the center subspace spanned by eigenvectors whose eigenvalues have conjugate pure imaginary eigenvalues (not equal to zero). In such a subspace of dimension $2m$ it is possible to introduce polar coordinates $(r, \theta) = (r_1, \ldots, r_m, \theta_1, \ldots, \theta_m)$, and it frequently happens that the r equations decouple from the θ equations, giving an independent subsystem of the form

$$\dot{r} = f(r). \tag{5.1.5}$$

Two situations are of particular interest, the case in which f has isolated simple zeros, and the case where f is identically zero (at least in some region). In the former case, each value of r such that $f(r) = 0$ defines an isolated invariant torus (a Cartesian product of m circles) with angular coordinates $(\theta_1, \ldots, \theta_m)$. In the latter case, there is a region that is foliated with invariant tori. (This kind of *invariant foliation* is stronger than the merely *preserved foliations* discussed above.) When $m = 1$, an invariant torus is simply a circle, and if the flow on the circle has no rest points, the circle will be a periodic orbit.

Two-Dimensional Hyperbolic Examples

For a first example, consider the system

$$\dot{x} = x(1 + f(xy)), \tag{5.1.6}$$
$$\dot{y} = y(-1 + g(xy)),$$

where f and g are polynomials with no constant term. This is the form that (5.0.2) takes when

$$A = \begin{bmatrix} 1 & 0 \\ 0 & -1 \end{bmatrix},$$

since according to Corollary 4.5.9 a monomial $x^i y^j$ can appear in $\dot{x}$ only if $i(1) + j(-1) = 1$, and in $\dot{y}$ if $i(1) + j(-1) = -1$. Notice that the x axis is E^u and the y axis is E^s, and that the normal form (5.1.6) illustrates the proof of Theorem 5.1.1; see Figure 5.1. The ring of invariants ker $\mathcal{D}_A$ consists of power series containing monomials $x^i y^j$ with $i(1) + j(-1) = 0$, which is just the ring of power series in the single generating invariant

$$I = xy. \tag{5.1.7}$$

The preserved foliation consists of the hyperbolas $I = c$, which are the orbits of e^{Ss} (with $S = A$, since A is semisimple). To find the system

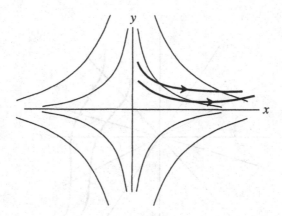

Figure 5.1. Preserved foliation of a saddle and two possible orbits.

(5.1.4), compute

$$\dot{I} = \dot{x}y + x\dot{y} = x(1 + f(xy))y + xy(-1 + g(xy)) = I(f(I) + g(I)) = F(I).$$

This system is "integrable by quadrature" (a rather antiquated phrase meaning that the solution is reducible to the evaluation of integrals, which may or may not be possible "in closed form") by writing

$$\int \frac{dI}{F(I)} = \int dt.$$

In order to see how the presence of a nilpotent part affects the two-dimensional hyperbolic case, consider a truncated system having the linear part

$$A = \begin{bmatrix} \lambda & 1 \\ 0 & \lambda \end{bmatrix}, \qquad S = \begin{bmatrix} \lambda & 0 \\ 0 & \lambda \end{bmatrix}, \qquad N = \begin{bmatrix} 0 & 1 \\ 0 & 0 \end{bmatrix} \qquad (5.1.8)$$

with real λ. Assume at first that $\lambda \neq 0$. The S-foliation then consists of half-lines directed outward from the origin, with a singular leaf at the origin. Since the origin is a limit point of each half-line, there can be no continuous invariant that distinguishes the leaves; the only invariants are constants, and (5.1.4) is trivial. Nonetheless, it is easy to understand the situation. Any system $\hat{a}$ that is in extended semisimple normal form may contain (in both $\dot{x}$ and $\dot{y}$) only monomials $x^i y^j$ satisfying $\lambda i + \lambda j = \lambda$. Since this implies $(i, j) = (1, 0)$ or $(0, 1)$, the system in normal form must in fact be linear:

$$\dot{x} = \lambda x + y,$$
$$\dot{y} = \lambda y.$$

The flow of this system does preserve the foliation by half-lines; see Figure 5.2.

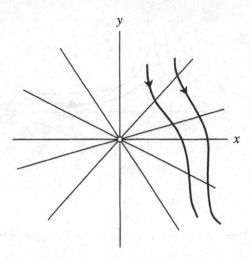

Figure 5.2. Preserved foliation by half-lines for a nonsemisimple system.

5.1.11. Remark. This is an example of the fact that a linear flow induces a flow on projective space.

The situation changes dramatically if $\lambda = 0$ (when the system is no longer hyperbolic). In that case $S = 0$ and the S-foliation is trivial (the leaves are just points). The normal form for this case is given in (1.2.8), and it does not preserve the foliation by half-lines. (It does, of course, trivially preserve the S-foliation.) To emphasize the point: the foliation by half-lines is preserved for all $\lambda \neq 0$ but not for $\lambda = 0$. The property that a specific foliation is preserved is not a property that depends continuously on parameters.

Although this two-dimensional example is much too simple to illustrate the general case, it will be seen as we proceed that the presence of a nilpotent part in A does not matter very much if the nilpotent part is confined to the stable and unstable subspaces, but matters a great deal when it appears in the center subspace. The reason is that a nilpotent part produces polynomial growth in time, which is overpowered by the exponential growth (or contraction) in the stable and unstable subspaces. In the center subspace the linear part produces no growth (only, at most, rotation), and the polynomial growth produced by the nilpotent part is significant.

Two-Dimensional Examples with a Center Subspace

For the next example, consider

$$\dot{x} = xf(x), \qquad\qquad (5.1.9)$$
$$\dot{y} = y(-1 + g(y)),$$

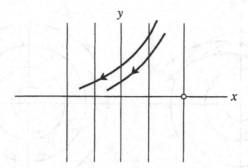

Figure 5.3. Fibers (vertical lines) and leaves (vertical half-lines and points) for a system with center subspace.

where f and g are polynomials without constant term. This is the general truncated normal form for systems with

$$A = \begin{bmatrix} 0 & 0 \\ 0 & -1 \end{bmatrix}.$$

Since this example has a center subspace (the x axis) and stable subspace (the y axis), there will be a stable fibration with projection $\pi(x, y) = x$, the fibers being the vertical lines $x = c$. As in the last example, $S = A$; the S-foliation has leaves of two types, rest points (on the x axis) and vertical half-lines; see Figure 5.3. Notice that each fiber contains three leaves. The ring of (polynomial) invariants is generated by

$$I(x, y) = x, \tag{5.1.10}$$

and the level sets of I are the fibers, not the leaves; it is not possible for a polynomial invariant, being continuous, to distinguish the singular leaves. (There exist discontinuous invariants, not belonging to the ring of polynomial invariants, that can distinguish these.) The equation (5.1.4) governing the evolution of I is just the first equation of (5.1.9). This first equation is integrable by quadrature in the form $dx/xf(x) = dt$, and then the second equation is integrable by quadrature as $dy/y = (-1 + g(x(t)))dt$.

Next, consider the (single) center having linear part

$$A = \begin{bmatrix} 0 & -1 \\ 1 & 0 \end{bmatrix}. \tag{5.1.11}$$

There are no stable or unstable manifolds or fibrations; the foliation determined by $S = A$ consists of circles, with a singular leaf at the origin. The ring of invariants is generated by $I = x^2 + y^2$. The normal form for this system was discussed in Section 1.1 and is given in (1.1.24), or in polar

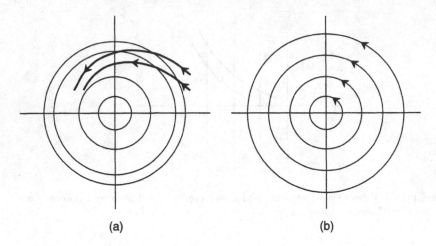

(a)   (b)

Figure 5.4. Preserved foliation for a center: (a) Circles not invariant, and
(b) circles invariant.

form in (1.1.25), namely,

$$\dot{r} = \alpha_3 r^3 + \alpha_5 r^5 + \cdots + \alpha_k r^k, \tag{5.1.12}$$
$$\dot{\theta} = 1 + \beta_3 r^2 + \beta_5 r^4 + \cdots + \beta_k r^{k-1}.$$

Except for the fact that $r = \sqrt{I}$ is not a polynomial invariant, the first
equation here is essentially (5.1.4), and illustrates the preservation of the
foliation by circles: Two orbits beginning on the same circle move across
the family of circles, but always share the same circle; see Figure 5.4. The
r equation decouples and is integrable by quadrature, after which the θ
equation can be integrated.

The first equation of (5.1.12) is an example of a subsystem of the form
(5.1.5) giving rise to invariant tori, in this case invariant circles. The right-
hand side of the equation for $\dot{r}$ has a triple root at the origin and may have
up to $k - 3$ isolated nonzero roots giving invariant circles. For sufficiently
small values of r, we have $\dot{\theta} \neq 0$, and these circles will be periodic orbits. It
is important to emphasize that these invariant circles and periodic orbits
are not automatically preserved when the nonnormalized terms (of degree
$> k$) are restored to the system, although in the present case periodic orbits
corresponding to *simple* zeros of f will be preserved if r is small enough.

The other case of interest is that in which the first equation of (5.1.12)
reduces to $\dot{r} = 0$. In this case every circle centered at the origin is invariant.
At first sight this case seems highly degenerate (since each of the numbers
$a_3, a_5, \ldots, a_k$ must separately equal zero), and hence unlikely to occur. But
in fact, it occurs whenever the system having a center at the origin is
Hamiltonian (in one degree of freedom); this is called a *Hamiltonian single
center*. To see this, recall from Section 4.9 that a Hamiltonian $H(p, q)$

defined on $\mathbb{R}^2$ having leading term

$$H_0(p, q) = \frac{1}{2}(p^2 + q^2) \tag{5.1.13}$$

(so that the associated system has linear part (5.1.11)) will be in normal form if and only if it is invariant under the flow of (5.1.11), that is, if and only if H is constant on circles. In canonical polar coordinates (4.9.22) the Hamiltonian is

$$H(p, q) = K(R);$$

the equations of motion are

$$\dot{R} = -\frac{\partial K}{\partial \theta} = 0,$$

$$\dot{\theta} = \frac{\partial K}{\partial R};$$

and we see that this system is foliated by the invariant circles $R = c$. The foliation is the same as the preserved foliation existing in the non-Hamiltonian single center, but now it is not merely preserved, but invariant.

A Three-Dimensional Example

For the next example, we take the linear term to be

$$A = \begin{bmatrix} 0 & -1 & \\ 1 & 0 & \\ & & -1 \end{bmatrix}, \tag{5.1.14}$$

writing the coordinates as (x, y, z). The plane $z = 0$ is the center subspace, and the z axis is the stable subspace. The stable fibration is $\pi(x, y, z) = (x, y, 0)$, the fibers being lines parallel to the z axis. The S-foliation $(S = A)$ is quite different. In the plane $z = 0$, the leaves are circles centered at the origin (with a singular leaf at the origin). But these circles are themselves singular leaves when looked at in the full three-dimensional space. The typical leaf is a spiral running down a cylinder centered on the z axis, converging toward the circle in which the cylinder intersects the $z = 0$ plane. These cylinders are important because they contain both the fibers and the leaves; see Figure 5.5. The linear flow e^{At} (which is explicitly solvable and which we already understand) can be described as follows: Each point (x, y, z) with $z \neq 0$ flows down one of the spiral leaves on its cylinder (these leaves are, after all, defined to be the orbits of this linear flow); two such points having the same values of (x, y) initially will continue to have equal values of (x, y) at later times. (The linear system is, trivially, in normal form with respect to its own linear part, so Theorem 5.1.5 applies and the stable fibration is preserved.) In particular, the point $(x, y, 0)$ will rotate on its circle in the $z = 0$ plane, always remaining beneath

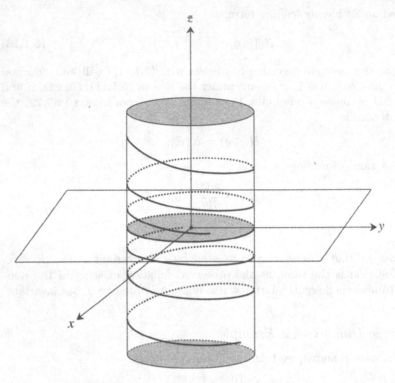

Figure 5.5. Fibers (vertical lines) and leaves (spirals and circles) for a three-dimensional example.

(x, y, z). The solutions with $z \neq 0$ are said to approach this periodic solution *asymptotically with asymptotic phase*. ("Phase" refers to an angle used to measure distance around the periodic orbit; here it is simply the polar angle θ in the (x, y) plane. "Asymptotic phase" means that the difference in this angle for the two solutions approaches zero. In the present example, the difference is identically zero.)

When we pass to the truncated nonlinear system (5.0.2) in normal form, the leaves of the S-foliation will no longer be invariant sets of the flow, but will instead be preserved (in the proper sense for a preserved foliation); the fibers will also be preserved. Understanding this geometry, we can make certain predictions about what the normal form will look like, before we even work out what the normal form is. (Alternatively, the normal form can be worked out first, and the geometry read off from the normal form.) Since the fibration $\pi(x, y, z) = (x, y, 0)$ is preserved, we can predict that the (x, y) equations will decouple from the z equation; that is, the equations for $\dot{x}$ and $\dot{y}$ will not involve z (although the $\dot{z}$ equation may still involve x and y). Since the foliation by circles in the (x, y) plane is preserved, in polar coordinates the r equation will decouple from θ (as it did in the single center, treated above). Thus, in cylindrical coordinates the system should be integrable by quadrature, beginning with r, then θ, then z.

Let us now work out the normal form and confirm these predictions. Following the pattern of (3.3.8), set

$$T = \frac{1}{2} \begin{bmatrix} 1 & 1 & 0 \\ -i & i & 0 \\ 0 & 0 & 2 \end{bmatrix},$$

so that in complex coordinates (u, v, z) with reality conditions $\bar{u} = v, \bar{z} = z$ the linear part is

$$T^{-1}AT = \begin{bmatrix} i & & \\ & -i & \\ & & -1 \end{bmatrix}.$$

Therefore, in normal form the $\dot{u}$ equation can contain $u^j v^k z^\ell$ only if $(j - k)i - \ell = 0$, implying $j = k$ and $\ell = 0$. Therefore, as predicted, z is absent from $\dot{u}$, and similarly from $\dot{v}$, and therefore also from $\dot{x}$ and $\dot{y}$; furthermore, only terms $(uv)^j = (u\bar{u})^j = (x^2 + y^2)^j = r^{2j}$ can appear, so the $\dot{r}$ equation decouples from θ. Finally, $u^j v^k z^\ell$ can appear in the $\dot{z}$ equation only if $(j - k)i - \ell = -1$, implying that $j = k$ and $\ell = 1$. So the $\dot{z}$ equation has the form $\dot{z} = zf(r)$.

The Nonresonant Double Center

Recall from Section 4.5 that a *double center* is a system with linear part

$$A = \begin{bmatrix} 0 & -\omega_1 & & \\ \omega_1 & 0 & & \\ & & 0 & -\omega_2 \\ & & \omega_2 & 0 \end{bmatrix}, \tag{5.1.15}$$

and that such a system is *nonresonant* if ω_1/ω_2 is irrational. The normal form of a nonresonant double center, written in polar coordinates, has been obtained in (4.5.61) as

$$\dot{r}_1 = \varphi_1\left(r_1^2, r_2^2\right) r_1, \tag{5.1.16}$$
$$\dot{\theta}_1 = \omega_1 + \varphi_2\left(r_1^2, r_2^2\right),$$
$$\dot{r}_2 = \varphi_3\left(r_1^2, r_2^2\right) r_2,$$
$$\dot{\theta}_2 = \omega_2 + \varphi_4\left(r_1^2, r_2^2\right).$$

(In our present context, it is the *truncated* system that takes this form in polar coordinates.)

Since the right-hand side of (5.1.16) depends only on $r = (r_1, r_2)$, the first and third equations decouple from the others, so that r evolves without regard to the values of $\theta = (\theta_1, \theta_2)$. Each fixed value of r corresponds to a 2-torus (if $r \neq (0,0)$), a circle (if $r_1 = 0$ or $r_2 = 0$ but not both), or a point (if $r = (0,0)$); these sets form a foliation by 2-tori, with singular leaves (circles or a point) in the coordinate planes. This foliation is preserved because the motion of a point across the foliation depends only on the leaf currently occupied. Any value of r that satisfies the equations

$$\varphi_1(r_1^2, r_2^2) = 0, \qquad (5.1.17)$$
$$\varphi_2(r_1^2, r_2^2) = 0,$$

defines a torus that is *invariant* under the flow (rather than merely being part of the preserved foliation).

The preserved foliation by 2-tori just described is not the same as the S-foliation e^{As} (with $S = A$) defined by (5.1.15). In fact, the S-foliation consists of curves of the form $\theta_1 = \omega_1 t + \delta_1$, $\theta_2 = \omega_2 t + \delta_2$ (with arbitrary δ_1 and δ_2) lying on each torus with fixed r. Since ω_1/ω_2 is irrational, these curves (often called "winding lines") are dense in the 2-tori. One way to obtain the foliation by tori from the S-foliation is topological: One can argue that since the winding lines give a preserved foliation, so do their closures, which are the tori. But there is another, more algebraic, way to predict the existence of the torus foliation from the matrix A (prior to, or independently of, the computation of the normal form (5.1.16)). This method, described in the next subsection, frequently gives additional preserved foliations besides the S-foliation even in cases where the topological argument (taking closures of S-orbits) fails.

Multiple Preserved Foliations

For the following discussion we confine ourselves to the semisimple case $(A = S)$. An $n \times n$ matrix B will be called *subordinate* to A if every vector field $v \in \mathcal{V}_{**}^n$ that commutes with A (that is, satisfies $\mathsf{L}_A v = 0$) also commutes with B (satisfies $\mathsf{L}_B v = 0$). It is clear (as in the proof of Theorem 5.1.7) that if B is subordinate to A, then the B-foliation (by orbits of e^{Bs}) is preserved by the flow of any vector field in semisimple normal form with respect to A.

5.1.12. Lemma. Suppose that A and B are diagonal matrices with diagonal entries $(\lambda_1, \ldots, \lambda_n)$ and $(\mu_1, \ldots, \mu_n)$, respectively. Suppose that every integer vector $k = (k_1, \ldots, k_n)$ satisfying $\langle k, \lambda \rangle = k_1\lambda_1 + \cdots + k_n\lambda_n = 0$ also satisfies $\langle k, \mu \rangle = k_1\mu_1 + \cdots + k_n\mu_n = 0$. Then B is subordinate to A.

Proof. It is enough to check that every vector monomial $x^m e_i$ that commutes with A commutes with B. But if $\mathsf{L}_A x^m e_i = 0$, then $m_1\lambda_1 + \cdots + m_n\lambda_n = \lambda_i$, or $m_1\lambda_1 + \cdots + (m_i - 1)\lambda_i + \cdots + m_n\lambda_n = 0$. By hypothesis, it follows that $m_1\mu_1 + \cdots + (m_i - 1)\mu_i + \cdots + m_n\mu_n = 0$, so that $\mathsf{L}_B x^m e_i = 0$. $\square$

5.1.13. Theorem. Let A be a (real or complex) diagonal matrix with diagonal entries $(\lambda_1, \ldots, \lambda_n)$. Suppose that every integer vector k satisfying

$\langle k, \lambda \rangle = 0$ is a linear combination of d linearly independent such vectors $k_{(1)}, \ldots, k_{(d)}$. Then any system in semisimple normal form with respect to A has $n - d$ independent preserved foliations and can be reduced to a system of dimension d.

Proof. The set of vectors μ (in $\mathbb{R}^n$ if A is real, in $\mathbb{C}^n$ is A is complex) satisfying $\langle k_{(j)}, \mu \rangle = 0$ for $j = 1, \ldots, d$ is a subspace (of $\mathbb{R}^n$ or $\mathbb{C}^n$) of dimension $n - d$ (over $\mathbb{R}$ or $\mathbb{C}$). Choose a basis $\mu_{(1)}, \ldots, m_{(n-d)}$ for this subspace, and for each $j = 1, \ldots, n - d$, create the diagonal matrix B_j having $\mu_{(j)}$ for its diagonal entries. By Lemma 5.1.12, each of these matrices B_j will be subordinate to A, and each of the B_j-foliations will be preserved.

Because the matrices B_j are diagonal, they commute with each other, and the flows $e^{B_1 s_1}, \ldots, e^{B_{n-d} s_{n-d}}$ commute with one another. An action φ of the additive group $\mathbb{R}^{n-d}$ on $\mathbb{R}^n$ (or $\mathbb{C}^n$) can be defined by $\varphi^s(x) = e^{B_1 s_1} \cdots e^{B_{n-d} s_{n-d}} x$, where $s = (s_1, \ldots, s_{n-d})$; the orbits of this action define a foliation with leaves of dimension $n - d$. (There may be singular leaves of lower dimension.) A straightforward generalization of Theorem 5.1.7 implies that this foliation is preserved. Choose d (real or complex) functions $I_1, \ldots, I_d$ such that the level sets of $I = (I_1, \ldots, I_d)$ are the leaves of the foliation; then, as in Corollary 5.1.10, there will exist a mapping F such that $\dot{I} = F(I)$.   □

5.1.14. Remark. The set of integer vectors k such that $\langle k, \lambda \rangle = 0$ is a submodule of $\mathbb{Z}^n$ of a special type (called a *pure submodule*) that always has a basis $k_{(1)}, \ldots, k_{(d)}$ (rather than merely a generating set with syzygies). Therefore, the hypothesis of Theorem 5.1.13 is always true for some d, which is commonly called the *number of resonances* satisfied by λ (but see Remark 4.5.10 and notice that in the present usage, k is a vector of integers that are not necessarily positive).

Although this theorem assumes that A is diagonal, it is applicable whenever A is semisimple; that is, a system in semisimple normal form with respect to A will have $n - d$ foliations if there are d "resonances" among the eigenvalues of A. (This follows by the use of Theorem 4.5.6.)

As an example of Theorem 5.1.13, consider

$$A = \begin{bmatrix} -1 & & \\ & 2 & \\ & & \sqrt{2} \end{bmatrix}, \tag{5.1.18}$$

for which $d = 1$ and $k_{(1)} = (2, 1, 0)$. Taking $\mu_{(1)} = (-1, 2, 0)$ and $\mu_{(2)} = (0, 0, 1)$ gives subordinate matrices

$$B_1 = \begin{bmatrix} -1 & & \\ & 2 & \\ & & 0 \end{bmatrix} \quad \text{and} \quad B_2 = \begin{bmatrix} 0 & & \\ & 0 & \\ & & 1 \end{bmatrix}.$$

Orbits of B_1 have the form $e^{B_1 s} x = (e^{-s} x_1, e^{2s} x_2, x_3)$ and are characterized by constant values of $x_1^2 x_2$ and x_3; orbits of B_2 have the form $(x_1, x_2, E^s x_3)$

and are characterized by constant values of x_1 and x_2. The combined action of $(s_1, s_2) \in \mathbb{R}^2$ on $\mathbb{R}^3$ has the form $(e^{-s_1}x_1, e^{2s_1}x_2, e^{s_2}x_3)$, with orbits being the level sets of $I = I_1 = x_1^2 x_2$; these are the leaves of the preserved foliation of dimension 2. According to Theorem 5.1.13, then, there should be a self-contained differential equation satisfied by I.

To verify this prediction, it is easy to check that the only basic invariant of A is $I = x_1^2 x_2$ (that is, $\ker \mathcal{D}_A = \mathbb{R}[[x_1^2 x_2]]$), and the basic equivariants are $(x_1, 0, 0)$, $(0, x_2, 0)$, and $(0, 0, x_3)$; there are no relations or syzygies. (The calculations for this example and the next resemble those for (4.5.27) and (4.5.24).) Thus, the normal form for a system with linear part A is

$$\dot{x}_1 = f_1(x_1^2 x_2)x_1,$$
$$\dot{x}_2 = f_2(x_1^2 x_2)x_2,$$
$$\dot{x}_3 = f_3(x_1^2 x_2)x_3.$$

(We have assimilated the linear parts $-x_1$, $2x_2$, and $\sqrt{2}x_3$ into f_1, f_2, and f_3, respectively). It then follows from $\dot{I} = 2x_1 \dot{x}_1 x_2 + x_1^2 \dot{x}_2$ that

$$\dot{I} = (2f_1(I) + f_2(I))I.$$

This is the predicted reduced equation.

Modifying the last example slightly, consider

$$A = \begin{bmatrix} 1 & & \\ & 2 & \\ & & \sqrt{2} \end{bmatrix}. \tag{5.1.19}$$

Again $d = 1$, with $k_{(1)} = (-2, 1, 0)$, $\mu_{(1)} = (1, 2, 0)$, $\kappa_{(2)} = (0, 0, 1)$, and

$$B_1 = \begin{bmatrix} 1 \\ 2 \\ 0 \end{bmatrix} \quad \text{and} \quad B_2 = \begin{bmatrix} 0 \\ 0 \\ 1 \end{bmatrix}.$$

The single scalar function $I = I_1$ defining the preserved foliation of dimension 2 is $I = x_1^{-2} x_2$, which this time is a rational function undefined when $x_1 = 0$; alternatively, however, the function $x_1^2 x_2^{-1}$ can be used as well (but fails where $x_2 = 0$). The invariants are $\ker \mathcal{D}_A = \mathbb{R}$, the equivariants are $(x_1, 0, 0)$, $(0, x_2, 0)$, $(0, x_1^2, 0)$, and $(0, 0, x_3)$, and the normal form for a system with linear part A (to any order) is simply

$$\dot{x}_1 = x_1,$$
$$\dot{x}_2 = 2x_2 + ax_1^2,$$
$$\dot{x}_3 = \sqrt{2}x_3,$$

with $a \in \mathbb{R}$ arbitrary. With $I = x_1^{-2} x_2$, the differential equation for I is $\dot{I} = a$.

As a final example we return to the nonresonant double center (5.1.15). Since A is not in diagonal form, we use Theorem 4.5.6 to diagonalize it, as discussed in detail in equations (4.5.33) to (4.5.35). With $\lambda = (i\omega_1, -i\omega_1, i\omega_2, -i\omega_2)$ we have $\langle k, \lambda \rangle = 0$ if and only if $(k_1 - k_2)\omega_1 + (k_3 - k_4)\omega_2 = 0$; since ω_1/ω_2 is irrational, this implies $k_1 = k_2$ and $k_3 = k_4$. Therefore, $d = 2$, and the generators are $k_{(1)} = (1, 1, 0, 0)$ and $k_{(2)} = (0, 0, 1, 1)$. The complex vector space of $\mu \in \mathbb{C}^4$ orthogonal to $k_{(1)}$ and $k_{(2)}$ is spanned by $\mu_{(1)} = (1, -1, 0, 0)$ and $\mu_2 = (0, 0, 1, -1)$, but in forming B_1 and B_2 we do not want to put these entries into the diagonal, because then B_1 and B_2 will not satisfy the reality conditions and cannot be carried back to the original system in real coordinates. Instead we take $\mu_{(1)} = (i, -i, 0, 0)$ and $\mu_2 = (0, 0, i, -i)$, form the matrices

$$\begin{bmatrix} i & & & \\ & -i & & \\ & & 0 & \\ & & & 0 \end{bmatrix} \quad \text{and} \quad \begin{bmatrix} 0 & & & \\ & 0 & & \\ & & i & \\ & & & -i \end{bmatrix},$$

and carry these back to real form as

$$B_1 = \begin{bmatrix} 0 & -1 & & \\ 1 & 0 & & \\ & & 0 & 0 \\ & & 0 & 0 \end{bmatrix} \quad \text{and} \quad B_2 = \begin{bmatrix} 0 & 0 & & \\ 0 & 0 & & \\ & & 0 & -1 \\ & & 1 & 0 \end{bmatrix}.$$

The flows generated by B_1 and B_2 are rotations in the (x_1, x_2) and (x_3, x_4) planes, respectively, and the joint action of (s_1, s_2) on $\mathbb{R}^4$ is expressible in polar coordinates as $(r_1, \theta_1 + s_1, r_2, \theta_2 + s_2)$; the orbits are exactly the leaves of the torus foliation in (5.1.16).

Notes and References

The geometrical structures discussed in this section (invariant manifolds, preserved fibrations, etc.) are well known in a variety of settings; see Notes and References for Section 5.2. For the theorems of this section (which show that these structures are computed by the normal form) the only reference I can give is Remarks 1 and 2 in Elphick et al. [41], where it is shown that the stable and unstable manifolds are computed by the normal form when the unstable manifold is absent. The theorems of this section originated as an attempt to see how far these remarks could be extended. On the other hand, I do not claim that these theorems are actually new. I have discussed them with a number of experts and received responses ranging from "of course" and "I would have expected that" to "I didn't know that" and even "that can't be true." But these theorems are exactly what is done in every concrete application of normal form theory. The language of fibrations may not be used, but the partial uncoupling of differential equations is noticed; foliations may not be mentioned, but a coordinate transformation expressing the foliation is employed.

The use of multiple foliations (Theorem 5.1.13) to reduce a system to a lower-dimensional system (based on the number of independent resonances that are satisfied) is closely related to a theorem of Bruno (Theorem 4 of [20], Theorem 1, page 190, in [21]; see also [22]) in which a similar reduction is achieved using his so-called power transformations. A further study of the relationship between these ideas would probably be quite useful.

After this book was completed, a paper by G. Belitskii [14] came to my attention. Although most of the content of this paper is not new, it is suggested (without complete details) that under certain circumstances it is possible to normalize a smooth vector field not only "to all orders," but "beyond all orders." We know that there exists a formal diffeomorphism that normalizes the formal power series of a smooth vector field to all orders, so that the resulting formal power series vector field commutes with an appropriate linear vector field $(A, A^*, \tilde{A}, \text{ or } S)$. By the Borel–Ritt theorem, there exists a smooth diffeomorphism that carries out this normalization; that is, it takes the original smooth vector field to a smooth vector field whose power series is normalized. (We have not emphasized this fact in this book, because the Borel–Ritt theorem is not an effective algorithm.) But the smooth vector field obtained in this way, although normalized "to all orders," is not yet normalized "beyond all orders"; that is, it does not necessarily commute with the specified linear vector field (because of the presence of a flat part that does not commute). Belitskii's suggestion is that sometimes the full smooth vector field can be made to commute. The importance of this point, in the context of Chapter 5, is that in this case the existence of certain preserved fibrations and foliations could be proved for the full system exactly as we do here for the truncated system, without the need to invoke the complicated contraction mappings that are usually constructed to prove such things. This possibility deserves further study.

5.2   Geometrical Structures in the Full System

There are two ways to attempt an extension of geometrical structures from the truncated system to the full system:

1. The geometrical structure that exists in the truncated system may be *approximately equal* to an *exactly preserved* geometrical structure that exists in the full system. This is the case for the invariant subspaces E^s, E^u, and E^c, and the stable and unstable fibrations; these are approximately equal to *stable, unstable, and center manifolds* W^s, W^u, and W^c (which are no longer linear subspaces, but which are invariant) in the full system, and to *stable and unstable fibrations* (that are no longer linear projections, but are preserved by the flow).

2. The geometrical structure that exists in the truncated system may be *exactly equal* to a geometrical structure in the full system that is *approximately preserved*. This is the case for preserved and invariant foliations. For instance, the *preserved* foliation by circles in the truncated single center is *approximately preserved* in the full system, and in the Hamiltonian case, when this foliation is *invariant* for the truncated system, it is *approximately invariant* in the full system. This is reflected in the existence of *adiabatic invariants* for the full system.

Stable, Unstable, and Center Manifolds in the Full System

To what extent do the invariant subspaces found in the truncated system correspond to invariant manifolds in the full system? As the terminology suggests, a full investigation of this question requires a grounding in the theory of manifolds, and uses techniques unrelated to those developed in this book. However, it is not difficult to state the local results (which are all that concern us), by defining what is meant by a "local manifold expressible as a graph."

5.2.1. Definition.

1. Let $E \subset \mathbb{R}^n$ be a linear subspace. A subset $M \subset \mathbb{R}^n$ is a *local manifold expressible as a graph over* E if there is an open subset $U \subset E$ containing the origin, and a subspace $F \subset \mathbb{R}^n$ with $E \oplus F = \mathbb{R}^n$, and a map $\sigma : U \to F$, such that M is the graph of σ; that is,

$$M = \{(u, v) \in E \oplus F : u \in U, v = \sigma(u)\}.$$

   (This is equivalent to choosing a coordinate system (u, v) on $\mathbb{R}^n$ such that E, F, and M are given respectively by the equations $v = 0$, $u = 0$, and $v = \sigma(u)$.) Since a point in M can be specified by giving a value of u, u may be regarded as a *system of curvilinear coordinates* on M. (It is then easy to confuse u as a point of E with u as a point of M, and it must always be made clear from the context which is intended.)

2. The local manifold is *of class C^r* if σ has continuous partial derivatives of all orders up to r, and is *smooth* if σ is smooth (i.e., of class C^∞).

3. M has *kth-order contact with E* if σ and its derivatives up to order k vanish at the origin of E.

4. A local manifold is *invariant* under a flow if the vectors of the flow are tangent to the manifold at each point. Notice that this definition of invariance is weaker than the usual notion of invariant subset (as defined at the beginning of this section); a point moving on an invariant local manifold may leave the manifold, but only (so to speak)

across its boundary. If the vector field generating the flow is expressed (using coordinates (u, v) on $\mathbb{R}^n = E \oplus F$) as

$$\dot{u} = g(u, v), \qquad\qquad (5.2.1)$$
$$\dot{v} = h(u, v),$$

then M is invariant, provided that

$$\frac{d}{dt}(u, \sigma(u)) = (h(u, \sigma(u)), g(u, \sigma(u))),$$

which is true if and only if

$$\sigma'(u)h(u, \sigma(u)) = g(u, \sigma(u)). \qquad\qquad (5.2.2)$$

5. If M is invariant under (5.2.1), then the *reduction* of (5.2.1) to M is the system

$$\dot{u} = g(u, \sigma(u)). \qquad\qquad (5.2.3)$$

This system describes the motion of a point on M under the flow of (5.2.1), using u as a coordinate system on M. It is important to understand that (5.2.3) cannot be used to describe the motion of a point $u \in E$ (which is not invariant under the flow).

The usual versions of the stable, unstable, and center manifold theorems do not assume that any terms of (5.0.1) have been brought into normal form. In this case the local manifolds have only first-order contact (that is, they are tangent to) with the stable, unstable, and center subspaces of the (fully) truncated system $\dot{y} = Ay$. This may be regarded as the case $k = 1$ of the following theorem. This theorem adds nothing to the usual versions as far as *existence* of the various manifolds is concerned. Its significance is that computing the normal form to degree k automatically computes approximations to these manifolds, in the form of linear subspaces having kth-order contact with the nonlinear manifolds. Remember that the process of obtaining the normal form involves changes of coordinates. If these coordinate changes are reversed, the linear subspaces E^s, E^c, etc., will be mapped back into nonlinear subspaces that still have kth-order contact with the true stable, etc. manifolds in the original variables.

5.2.2. Theorem (Stable, center, etc., manifold theorem). Assume that the truncated system (5.0.2) is in extended semisimple normal form. Then there exist unique smooth local stable and unstable manifolds W^s and W^u, invariant under the flow of (5.0.1), expressible as graphs over E^s and E^u respectively, having kth-order contact with E^s and E^u respectively. For each integer r with $k \le r < \infty$, there exists a (not necessarily unique) local invariant center manifold W^c of class C^r, expressible as a graph over E^c and having kth-order contact with E^c. (The "size" of the center manifold, that is, the size of the neighborhood $U \subset E^c$ over which it is defined as a graph, may decrease as $r \to \infty$.) Similarly, for any such r there exist (not necessarily unique) local invariant center-stable and center-unstable manifolds W^{cs} and W^{cu}, expressible as graphs over E^{cs} and E^{cu} with kth-order contact. Each strong stable and strong unstable subspace E^{ss} and E^{su} has an associated unique local smooth invariant manifold W^{ss} or W^{su}, expressible as a graph with kth-order contact.

Proof. See Notes and References for references to proofs of the standard versions of these theorems (without normal forms). Most discussions of the center manifold theorem contain a method for computing approximations to the center manifold having contact of any specified order; it is easy to see that if this technique is carried out to degree k in a system that has been normalized to degree k, the result is E^c. Therefore, E^c has kth-order contact with W^c. The same technique works for stable and unstable manifolds (although it is not usually presented, since there is less interest in computing these locally). □

Center Manifold Reductions

One of the goals of dynamical systems theory is to describe the behavior of a system in a neighborhood of a rest point "up to local topological conjugacy." Two flows φ^t and ψ^t, defined for all t and x, are said to be *topologically conjugate* if there is a homeomorphism H (a continuous mapping with a continuous inverse) carrying one flow to the other:

$$H(\varphi^t(x)) = \psi^t(H(x)) \qquad (5.2.4)$$

for all x. Two *local* flows, defined only for t near $0 \in \mathbb{R}$ and x near $0 \in \mathbb{R}^n$, are *locally* topologically conjugate if there is a homeomorphism H defined on a neighborhood of $0 \in \mathbb{R}^n$ such that (5.2.4) holds whenever both sides are defined. Given a dynamical system with a rest point at the origin, one seeks a simplified system such that the flows of the original and simplified systems are locally topologically conjugate. The basic example of such a theorem is *Hartman's theorem*, which states that if A is hyperbolic (has its eigenvalues off the imaginary axis), then any dynamical system $\dot{x} = Ax + \cdots$ is locally conjugate to $\dot{x} = Ax$.

> **5.2.3. Remark.** The difference between topological conjugacy and normal form theory is that in normal form theory we simplify systems using only *smooth* changes of coordinates. If H is smooth, and if a and b are the vector fields generating φ^t and ψ^t, respectively, then (5.2.4) implies $a(x) = H'(x)^{-1}b(H(x))$, so that $y = H(x)$ serves as a coordinate change between the differential equations $\dot{x} = a(x)$ and $\dot{y} = b(y)$. When H is just continuous, it can be applied only to the flows, not to the vector fields. For instance, in Hartman's theorem there need be no smooth change of coordinates that linearizes the differential equation.

The *reduction* of a system to a local invariant manifold has been defined above as part of Definition 5.1.3. The reduction to the center manifold is especially important in dynamical systems, because it is required as part of the description of the behavior of a system near a rest point up to local topological conjugacy, in cases where Hartman's theorem does not apply. The first point to notice is that computation of the normal form (under the conditions of Theorem 5.1.1) automatically computes not only the (approximate) center manifold but also the (approximate) center manifold reduction. To be precise, suppose that the full system (5.0.1) is put into the form

$$\dot{u} = Bu + g(u, v), \tag{5.2.5}$$
$$\dot{v} = Cv + h(u, v),$$

where B has eigenvalues on the imaginary axis and C is hyperbolic. (This can be done by a linear coordinate change from x to (u, v). It is possible, but not necessary, to take B and C is real canonical form.) The truncated system (which, as usual, is assumed to be in extended semisimple normal form) will be denoted by

$$\dot{u} = Bu + \widehat{g}(u, v), \tag{5.2.6}$$
$$\dot{v} = Cv + \widehat{h}(u, v).$$

Then the center subspace E^c for (5.2.6) is given by $v = 0$, and the center manifold W^c of (5.2.5) can be represented in the form $v = \sigma(u)$, where σ is a smooth function whose power series begins with terms of degree $k + 1$ (because of the kth-order contact between W^c and E^c). The center manifold reduction of (5.2.5) is the system

$$\dot{u} = Bu + g(u, \sigma(u)), \tag{5.2.7}$$

in which u is regarded as a curvilinear coordinate system on W^c. But in practice, one computes only some jet of the right-hand side of (5.2.7). Since $\widehat{g}$ is the k-jet of g, and $v = 0$ is the k-jet of $v = \sigma(u)$, the k-jet of (5.2.7) is

$$\dot{u} = Bu + \widehat{g}(u, 0). \tag{5.2.8}$$

This is precisely the same as the restriction of (5.2.6) to its invariant subspace $v = 0$, which is available without further work as soon as the normal form to degree k has been computed.

Under these circumstances, one has the following generalization of Hartman's theorem, which states that (up to local topological conjugacy) it is possible to linearize the system in the directions transverse to the center manifold, at the same time passing to the reduced system on the center manifold.

5.2.4. Theorem (Local conjugacy theorem). The full system (5.2.5) is locally topologically conjugate near the origin to the uncoupled system consisting of the full center manifold reduction (5.2.7) together with the linearized system $\dot{v} = Cv$.

The significance of this theorem is that any "local recurrent behavior" of the system (such as existence of rest points, periodic or quasiperiodic motions, or more complicated "chain-recurrent sets," which we will not define here) occurring near the origin must occur on the center manifold. This means that a knowledge of the center manifold reduction is sufficient for the determination of such behavior, and a knowledge of the k-jet of the center manifold reduction (which is given automatically by the normal form to degree k) is sufficient to establish existence of such behavior in the full system if it is sufficient to establish such behavior in the center manifold. (That is, the question of k-determinacy is reduced from the ambient system of n dimensions to the lower-dimensional reduced system on the center manifold.)

Stable and Unstable Fibrations in the Full System

The stable and unstable fibrations of E^{cs} and E^{cu} obtained for the truncated system in Theorem 5.1.5 also exist in the full system, but they are no longer linear projections. Instead, for any finite r there exist choices of the (nonunique) center-stable, center-unstable, and center manifolds W^{cs}, W^{cu}, and W^c of class C^{r+1}, and nonlinear fiber maps $\pi : W^{cs} \to W^c$ and $\pi : W^{cu} \to W^c$ of class C^r, such that these fibrations are preserved by the flow. (Since r is arbitrary, although not infinite, in our case of C^∞ vector fields, the distinction between C^{r+1} and C^r is not very important. But for vector fields of finite smoothness it is important that the fibrations lose one degree of smoothness.) Rather than define these notions more precisely, we state instead the following version for the center-stable case, using a nonlinear coordinate change of class C^r that linearizes the fibers.

5.2.5. Theorem. Assume that the truncated system (5.0.2) is in extended semisimple normal form, and that the matrix A is block-diagonalized by a real linear transformation so that the full system (5.0.1) appears as

$$\dot{u} = Bu + f(u,v,w), \tag{5.2.9}$$
$$\dot{v} = Cv + g(u,v,w),$$
$$\dot{w} = Dw + h(u,v,w),$$

where B has eigenvalues in the left half-plane, C on the imaginary axis, and D in the right half-plane. Represent the center-stable manifold as a graph $w = \sigma(u,v)$, with reduced system

$$\dot{u} = Bu + f(u,v,\sigma(u,v)), \tag{5.2.10}$$
$$\dot{v} = Cv + g(u,v,\sigma(u,v)).$$

Let $r > k + 1$ be a fixed integer. Then there exists a coordinate change $u = U + \varphi(U, V)$, $v = V + \psi(U, V)$ such that φ and ψ are of class C^r and have vanishing k-jets at the origin, and such that (5.2.10) transforms into

$$\dot{U} = BU + F(U, V), \qquad\qquad (5.2.11)$$
$$\dot{V} = CV + G(V),$$

in which G is independent of U. Thus, the V system decouples from U.

The coordinates (U, V) should be thought of as a new curvilinear coordinate system in W^{cs}, replacing the curvilinear coordinates (u, v) there. The decoupling of V is similar to the decoupling of v in the proof of Theorem 5.1.5, and indicates that the curves of constant V form a preserved foliation in W^{sc}.

Adiabatic Invariants in the Full System

Roughly speaking, an *adiabatic invariant* is a quantity that is not actually invariant, but varies so slowly that it can be treated as an invariant for a rather long period of time. A wide variety of precise definitions exist to suit various contexts. Our definition is that a quantity $I(x)$, defined near the origin of the full system (5.0.1), is an *adiabatic invariant of order k* if there exists a constant $c > 0$ such that for every integer $j = 0, \ldots, k$ and for every solution $x(t)$ satisfying $\|x(t)\| \leq \varepsilon$ for $0 \leq t \leq 1/\varepsilon^j$, it is the case that

$$|I(x(t)) - I(x(0))| \leq c\varepsilon^{k+1-j} \quad \text{for} \quad 0 \leq t \leq \frac{1}{\varepsilon^j}.$$

5.2.6. Theorem. If the k-jet

$$\dot{I}(x) = \mathcal{D}_{a(x)}I(x)$$

vanishes, I is an adiabatic invariant of order k for the system $\dot{x} = a(x)$.

Proof. Under this condition, there exists a constant $c > 0$ such that $|\dot{I}(x)| \leq c\varepsilon^{k+1}$ for $\|x\| \leq \varepsilon$. Then

$$\frac{d}{dt}|I(x(t)) - I(x(0))| \leq |\dot{I}| \leq c\varepsilon^{k+1},$$

so that $|I(x(t)) - I(x(0))| \leq c\varepsilon^{k+1}t$ as long as $\|x(t)\| \leq \varepsilon$. The result follows.   □

The simplest situation in which the condition $j^k(\dot{I}) = 0$ is satisfied is when I is one of the variables in the full system, and the corresponding differential equation has zero right-hand side in the truncation. This happens frequently in Hamiltonian systems, such as (4.9.24), (4.9.27), and (4.9.30), and accounts for the adiabatic invariants found in these systems.

Notes and References

For an introduction to invariant manifolds, see Chicone [28], Section 1.7. Many proofs of the stable and unstable manifold theorems are available, either by integral equation methods (for instance Coddington and Levinson [31], Hale [53], or Chicone [28], Chapter 4) or by graph transforms (Palis and de Melo [95] or Shub [102]).

The stable and unstable fibrations were established independently by Fenichel in [43] and [44] and Hirsch, Pugh, and Shub [59]. For reliable expositions see Haller [54] and Bronstein and Kopanskii [18]. In particular, Haller (Section 5.3.3) proves an infinite-dimensional version by an integral equation method; although the proof uses Sobelev spaces, this can be ignored, and the proof specializes easily to the finite-dimensional case. The presentation in Wiggins [112] is a good place to learn the basic ideas but not a good source for precise statements or proofs. It contains a number of errors, specifically in both the claims and proofs of smoothness and in the proposed generalization to "weakly hyperbolic" systems, which is false in general; see Chicone and Liu [27] for the last-mentioned item.

Center manifolds and center reductions are discussed in Carr [24]. The local conjugacy theorem is frequently stated (usually under the name Shoshitaishvili's theorem), but seldom proved. One source for the proof is Kirchgraber and Palmer [64]. Apparently, Shoshitaishvili proved a version that is specifically concerned with the local center manifold, at about the same time that Hirsch, Pugh, and Shub [97] proved a much stronger conjugacy result for flows in a neighborhood of a compact normally hyperbolic invariant manifold, which implies the local theorem via a compactification argument. However it is arranged, the proof is based on two additional fibrations (of a full neighborhood of the origin over the stable and unstable manifolds, respectively) besides the "Fenichel fibrations" discussed above. These additional fibrations are continuous, but not smooth in general (otherwise, the topological conjugacy would be smooth).

5.3   Error Estimates

In this section, we collect a variety of estimates for the difference between a solution of the truncated system (5.1.1) and the full system (5.2.1). To facilitate comparison of solutions, the full and truncated systems will be written as in (5.0.1) and (5.0.6), respectively; as before, it is assumed (except in a few lemmas) that $\widehat{a}(x)$ is in extended semisimple normal form. The quantity to be estimated is $\|x(t) - y(t)\|$, where $x(t)$ and $y(t)$ are specific solutions of these systems, paired according to the requirements of each specific theorem. This pairing of solutions is usually presented according to the "shadowing philosophy." According to this point of view, a solution $y(t)$ of the truncated system is relevant to the full system, provided that there exists a solution $x(t)$ of the full system that remains close to (or "shadows") $y(t)$ for t in some specified interval; it is not required that $y(t)$ and $x(t)$ satisfy the same initial conditions. According to the older "initial value" philosophy, a solution $y(t)$ should always be considered as an approximation to (and hence compared with) the solution $x(t)$ satisfying $x(0) = y(0)$. This approach arises naturally from the importance of initial value problems in applied mathematics, but unfortunately the solution $x(t)$ chosen in this manner often diverges from $y(t)$ more rapidly than some other solution $x(t)$. The shadowing point of view may be regarded as relaxing the requirement of exact equality at $t = 0$ in favor of rough equality over the longest possible interval. Whenever, without losing length of validity, the shadowing solution may be taken to satisfy $x(0) = y(0)$, this will be pointed out.

Here is a summary of the shadowing results that will be proved in this section. All results are local to the origin:

1. A solution $y(t)$ lying in E^{s} is shadowed by a solution $x(t)$ in W^{s} for all $t > 0$. The error estimate approaches zero exponentially, along with the solutions themselves. The same is true for E^{u} and W^{u} for $t < 0$.

2. A solution in E^{c} is shadowed to order ε^{k-1} by one in W^{c} for a time interval of the form $0 \le t \le T/\varepsilon$ (see below for the meaning of ε), *provided* that the restriction of A to E^{c} is semisimple.

3. The last result can be extended to E^{cs} (with the shadowing solution lying in W^{cs}).

4. In the case of a saddle ($E^{\mathrm{c}} = \{0\}$, with both E^{s} and E^{u} nontrivial), solutions $y(t)$ near the origin are shadowed by some $x(t)$ as long as they remain sufficiently near the origin. The error is uniformly bounded for all such solutions, with the time of validity varying from one solution to another.

At present, there does not appear to be a generalization of the last result to the case that E^{c} is nontrivial. Considering the known results, the best that could be hoped for is that solutions shadow for the time they remain near the origin, or for time T/ε, whichever is shorter.

Each estimate involves a small quantity ε that defines a neighborhood of the origin ($\|x\| < \varepsilon$). Alternatively, one can introduce coordinate dilations $x = \varepsilon\xi$, $y = \varepsilon\eta$ (similar to (4.2.6)) that carry (5.0.1) and (5.0.6) into

$$\dot{\xi} = A\xi + \varepsilon a_1(\xi) + \cdots + \varepsilon^{k-1} a_{k-1}(\xi) + \varepsilon^k R(\xi, \varepsilon) \qquad (5.3.1)$$

and

$$\dot{\eta} = A\eta + \varepsilon a_1(\eta) + \cdots + \varepsilon^{k-1} a_{k-1}(\eta), \qquad (5.3.2)$$

respectively, where the "remainder" R is smooth. Every family of solutions $\xi(t, \varepsilon)$ of (5.3.1) corresponds to a family $x(t, \varepsilon) = \varepsilon\xi(t, \varepsilon)$ for (5.0.1), and similarly for $\eta(t,\varepsilon)$. Then

$$\|x(t,\varepsilon) - y(t,\varepsilon)\| = \varepsilon\|\xi(t,\varepsilon) - \eta(t,\varepsilon)\|. \tag{5.3.3}$$

In this way, an error estimate for (5.3.1) and (5.3.2) implies one for (5.0.1) and (5.0.6); if the former is valid in $\|\xi\| < 1$, the latter will be valid in $\|x\| < \varepsilon$.

Estimates Involving Stable Manifolds

The main theorem of this section is based on three lemmas, at least two of which are familiar (in some form) from a basic course in differential equations; the proofs are given for completeness. These lemmas concern systems having the form of (5.0.1) and (5.0.6), but it is *not* required that $\widehat{a}(x)$ be in normal form. It is, on the other hand, assumed (for these lemmas only) that A is a stable matrix, that is, has its eigenvalues in the open left half-plane. When applied to a matrix, $\|\ \|$ denotes any matrix norm satisfying (in addition to the requirements for a vector norm)

$$\|Tx\| \le \|T\|\|x\|$$

for all $T \in \mathrm{gl}(n)$ and $x \in \mathbb{R}^n$ (or $\mathbb{C}^n$).

5.3.1. Lemma. If all eigenvalues λ of A satisfy $\mathrm{Re}\,\lambda < -\alpha < 0$, then there exists a constant $K > 0$ such that

$$\|e^{At}x\| < Ke^{-\alpha t}\|x\|$$

for all $x \in \mathbb{R}^n$ and all $t \ge 0$.

Proof. First, suppose that A is in Jordan form, with $A = S + N$ being its semisimple/nilpotent decomposition. Choose $\beta > 0$ such that all eigenvalues satisfy $\mathrm{Re}\,\lambda \le -(\alpha + \beta)$. Since S and N commute,

$$\|e^{At}x\| = \|e^{St}e^{Nt}x\| \le e^{-(\alpha+\beta)t}\|e^{Nt}x\| \le e^{-(\alpha+\beta)t}p(t)\|x\|,$$

where $p(t)$ is some polynomial (since e^{Nt} grows polynomially when N is a nilpotent matrix in Jordan form). The function $e^{-\beta t}p(t)$ is bounded for $t \ge 0$, since exponential decay dominates polynomial growth. Let K be a bound. $\qquad\square$

5.3.2. Lemma. Let A satisfy the hypothesis of Lemma 5.3.1. Then there exist constants $\varepsilon_1 > 0$ and $K > 0$ such that for $0 \le \varepsilon \le \varepsilon_1$, any solution $x(t)$ of (5.0.1) with $\|x(0)\| \le \varepsilon$ satisfies

$$\|x(t)\| \le K\varepsilon e^{-\alpha t/2}. \tag{5.3.4}$$

In particular, the origin is asymptotically stable.

Proof. Write (5.0.1) in the form $\dot{x} = Ax + f(x)$. Let $x(t)$ be a solution, and write $g(t) = f(x(t))$; then $x(t)$ satisfies the inhomogeneous linear equation

$\dot{x} = Ax + g(t)$. Writing $x(0) = x_0$, it follows that

$$x(t) = e^{At}x_0 + \int_0^t e^{A(t-s)}g(s)ds.$$

Therefore, $x(t)$ satisfies the integral equation

$$x(t) = e^{At}x_0 + \int_0^t e^{A(t-s)}f(x(s))ds. \tag{5.3.5}$$

With α and K as in Lemma 5.3.1, choose ε_0 such that for $\|x\| \le \varepsilon_0$,

$$\|f(x)\| \le \frac{\alpha}{2K}\|x\|.$$

(This is possible because $f(x)$ begins with quadratic terms.) Also choose $\varepsilon_1 = \min\{\varepsilon_0, \varepsilon_0/K\}$; as in the statement of the lemma, from this point on we assume that $0 \le \varepsilon \le \varepsilon_1$ and $\|x_0\| \le \varepsilon$. It follows that there exists $T > 0$ such that

$$\|x(t)\| \le \varepsilon_0 \quad \text{for} \quad 0 \le t \le T. \tag{5.3.6}$$

(By the end of the proof, it will be seen that this in fact holds for $T = \infty$. This technique for obtaining bounds for the solution of a differential equation is called *bootstrapping*.) Then for $0 \le t \le T$ we have

$$\|x(t)\| \le Ke^{-\alpha t}\varepsilon + \int_0^t Ke^{-\alpha(t-s)}\frac{\alpha}{2K}\|x(s)\|ds.$$

Putting

$$x(t) = e^{-\alpha t}u(t), \tag{5.3.7}$$

the last estimate becomes

$$\|u(t)\| \le K\varepsilon + \frac{\alpha}{2}\int_0^t \|u(s)\|ds.$$

The next step is to apply Gronwall's inequality, or else the following argument (which is the proof of Gronwall's inequality). Let

$$S(t) = \int_0^t \|u(s)\|ds.$$

Then

$$\frac{dS}{dt} = \|u(t)\| \le K\varepsilon + \frac{\alpha}{2}S, \tag{5.3.8}$$

or

$$\frac{dS}{dt} - \frac{\alpha}{2}S \le K\varepsilon.$$

This differential inequality can be "solved" in the same manner as the equation that would be obtained if $\le$ were $=$, namely, multiply by the integrating factor $e^{-\alpha t/2}$ and integrate. The result is

$$S(t) \le \frac{2K}{\alpha}\varepsilon(e^{\alpha t/2} - 1).$$

Using (5.3.8) and then (5.3.7) gives (5.3.4), which we now know to hold on the interval $0 \le t \le T$ on which (5.3.6) holds. But in fact it follows that both (5.3.4) and (5.3.6) hold forever. If not, let T be the first time at which $\|x(T)\| = \varepsilon_0$; then (5.3.6), and therefore (5.3.4), holds for $0 \le t \le T$. But (5.3.4) for $t = T$ implies $\|x(T)\| < \varepsilon_0$, contradicting our choice of T. $\quad\square$

The next lemma is our first estimate of the type to which this section is devoted; it is a special case of the first main theorem, stated below. For this special case, in which E^s is the whole space $\mathbb{R}^n$, the shadowing estimate reduces to an initial value estimate, and it is not required that the truncated system be in normal form:

5.3.3. Lemma. Let A satisfy the hypothesis of Lemma 5.3.1, and let $x(t)$ and $y(t)$ be solutions of (5.0.1) and (5.0.6) having the same initial condition $(x(0) = y(0))$. It is not assumed that $\hat{a}$ is in normal form. There exist constants $c > 0$ and $\varepsilon_1 > 0$ such that for every $0 \le \varepsilon \le \varepsilon_1$, if $\|x(0)\| \le \varepsilon$, then

$$\|x(t) - y(t)\| \le c\varepsilon^{k+1}e^{-\alpha t/2}.$$

Proof. Write (5.0.1) and (5.0.6) as $\dot{x} = Ax + f(x) + g(x)$ and $\dot{y} = Ay + f(y)$, respectively, where $f(x) = a_1(x) + \cdots + a_{k-1}(x) = \mathcal{O}\left(\|x\|^2\right)$ and $g(x) = \mathcal{O}\left(\|x\|^{k+1}\right)$. Let $x(t)$ and $y(t)$ be as stated in the lemma, and $z(t) = x(t) - y(t)$, so that $z(0) = 0$; then $z(t)$ satisfies the inhomogeneous linear equation

$$\dot{z} = Az + f(x(t)) - f(y(t)) + g(x(t)),$$

and

$$z(t) = \int_0^t e^{A(t-s)}\{f(x(s)) - f(y(s)) + g(x(s))\}ds,$$

implying

$$\|z(t)\| \le \int_0^t Ke^{-\alpha(t-s)}\{\|f(x(s)) - f(y(s))\| + \|g(x(s))\|\}ds.$$

Choose ε_1 sufficiently small that both $x(t)$ and $y(t)$ satisfy Lemma 5.3.2 if $\|x(0)\| \le \varepsilon_1$; in particular, if $0 < \varepsilon < \varepsilon_1$ and $\|x(0)\| \le \varepsilon$, then $\|x(t)\| \le K\varepsilon$ for all $t \ge 0$. Because f begins with quadratic terms, there exists a constant L such that for each ε with $0 < \varepsilon \le \varepsilon_1$, $L\varepsilon$ is a Lipschitz constant for f in the ball $\|x\| \le K\varepsilon$, so that $\|f(x(s)) - f(y(s))\| \le L\varepsilon\|z(t)\|$ for $s > 0$. Since g begins with terms of order $k+1$, there exists a constant $c_1 > 0$ such that

$\|g(x(s))\| \leq c_1\|x(s)\|^{k+1} \leq c_1 K^{k+1}\varepsilon^{k+1}e^{-\alpha(k+1)s/2}$. It follows that

$$\|z(t)\| \leq \int_0^t Ke^{-\alpha(t-s)}L\varepsilon\|z(s)\|ds + c_1 K^{k+1}\varepsilon^{k+1}e^{-\alpha t}\int_0^t e^{-(k-1)\alpha s/2}ds.$$

Putting $z(t) = e^{-\alpha t}u(t)$, it follows that

$$\|u(t)\| \leq KL\varepsilon\int_0^t \|u(s)\|ds + c_2\varepsilon^{k+1}$$

for $c_2 = 2c_1 K^{k+1}/(k-1)\alpha$. The rest of the proof is a Gronwall argument similar to the conclusion of the proof of Lemma 5.3.2. □

Now we are ready to state and prove the main shadowing approximation for solutions in the stable manifold. It is no longer assumed that the eigenvalues of A are in the left half-plane.

5.3.4. Theorem (Stable manifold shadowing estimate). Let system (5.0.1) be in extended semisimple normal form to degree k, and let (5.0.6) be its truncation. Suppose also that all eigenvalues λ of A that lie in the open left-half plane satisfy $\text{Re}\,\lambda < -\alpha < 0$. There exist constants $\varepsilon_1 > 0$ and $c > 0$ such that for every $0 < \varepsilon \leq \varepsilon_1$, each solution $y(t)$ of (5.0.6) lying in E^s and having $\|y(0)\| \leq \varepsilon$ is shadowed by a solution $x(t)$ of (5.0.1) that lies in W^s and satisfies

$$\|x(t) - y(t)\| \leq c\varepsilon^{k+1}e^{-\alpha t/2}$$

for all $t > 0$.

Proof. By a linear change of variables it may be assumed that A is block-diagonal such that with $x = (u, v)$, (5.0.1) takes the form

$$\dot{u} = Bu + g(u, v), \tag{5.3.9}$$
$$\dot{v} = Cu + h(u, v),$$

the eigenvalues of B lying in the open left half-plane and those of C in the closed right half-plane. The local stable manifold W^s will then be the graph of a mapping $v = \sigma(u)$ defined for u near zero, such that the k-jet of σ vanishes. The reduction of (5.3.9) to W^s is

$$\dot{u} = Bu + g(u, \sigma(u)), \tag{5.3.10}$$

with u understood as a curvilinear coordinate system in W^s. The truncated system (5.0.6) may then be written, with $y = (z, w)$, as

$$\dot{z} = Bz + \widehat{g}(z, w), \tag{5.3.11}$$
$$\dot{w} = Cw + \widehat{h}(z, w).$$

Since this system is in extended semisimple normal form, its stable manifold is a linear subspace E^s, namely, the z-space, and the reduction of the system to E^s is just

$$\dot{z} = Bz + \hat{g}(z, 0). \qquad (5.3.12)$$

Because $\hat{g}$ is the truncation of g, and because the truncation of $\sigma(u)$ is zero, (5.3.12) is the truncation of (5.3.10), and Lemma 5.3.3 applies to these systems and shows that given any solution $z(t)$ of (5.3.12) close enough to the origin, the solution $u(t)$ of (5.3.10) with $u(0) = z(0)$ satisfies

$$\|u(t) - z(t)\| \le c\varepsilon^{k+1}e^{-\alpha t/2}.$$

This estimate does not apply directly to the systems we wish to compare. However, $y(t) = (z(t), 0)$ is a solution of (5.3.11) lying in E^s, and $x(t) = (u(t), \sigma(u(t)))$ is a solution of (5.3.9) lying in W^s. We have

$$\|x(t) - y(t)\| \le \|u(t) - z(t)\| + \|\sigma(u(t))\|.$$

The first term has already been estimated, and the second term is bounded by a constant times $\|u(t)\|^{k+1}$, with $\|u(t)\|$ bounded as in Lemma 5.3.2. Each of these bounds has the form required. Notice that $x(t)$ and $y(t)$ no longer have the same initial values, so the estimate is of the shadowing type. □

Estimates Involving Center Manifolds

The argument for the center manifold estimates follows the same overall pattern as in the stable manifold case: first some lemmas assuming that all eigenvalues of A lie on the imaginary axis (so that $E^c = \mathbb{R}^n$), then a theorem for the general case. However, the details of the estimates are quite different. The fundamental difference is that on the stable manifold, the matrix A provides an exponential contraction, which is the ultimate source of all of the estimates; on the center manifold, A is neutral and is best removed (by a coordinate transformation) before the estimates are made. This argument relies on the assumption that the restriction of A to E^c is semisimple, since otherwise e^{At} is not neutral but has polynomial growth.

5.3.5. Lemma. Let A in (5.0.1) be a real matrix that is semisimple and has all of its eigenvalues on the imaginary axis. Let S be a (complex) matrix such that $S^{-1}AS = \Lambda$ is diagonal, and let $K = \|S\|\|S^{-1}\|$. Then for any $\varepsilon_0 > 0$, there exists $T > 0$ such that for $0 < \varepsilon \le \varepsilon_0$, any solution $x(t)$ having $\|x(0)\| \le \varepsilon$ satisfies

$$\|x(t)\| \le 2K\varepsilon \quad \text{for} \quad 0 \le t \le T/\varepsilon. \qquad (5.3.13)$$

(It is not required that any terms of f be in normal form.)

Proof. First observe that $\|e^{\Lambda t}\| = 1$ for all t. It follows that $\|e^{At}\| = \|Se^{\Lambda t}s^{-1}\| \le K$ for all t. Note that $K \ge 1$ (because $\|SS^{-1}\| \le \|S\|\|S^{-1}\|$). Choose $\varepsilon_0 > 0$; then (because f begins with quadratic terms) there exists

$c > 0$ such that $\|f(x)\| \leq c\|x\|^2$ for all x such that $\|x\| \leq \varepsilon_0$. Then the integral equation (5.3.5) implies that

$$\|x(t)\| \leq K\|x_0\| + \int_0^t Kc\|x(s)\|^2 ds,$$

provided that $\|x(s)\| \leq \varepsilon_0$ for $0 \leq s \leq t$.

Consider any ε with $0 < \varepsilon \leq \varepsilon_0$, and any x_0 with $\|x_0\| \leq \varepsilon$. Since $K \geq 1$, there exists an interval of time (beginning at $t = 0$) during which $\|x(t)\| \leq 2K\varepsilon$; during this interval, we have

$$\|x(t)\| \leq K\varepsilon + \int_0^t Kc\varepsilon^2 ds = K\varepsilon(1 + c\varepsilon t).$$

Therefore (as in the bootstrapping argument of Lemma 5.3.2), the condition $\|x(t)\| \leq 2K\varepsilon$ cannot fail before $K\varepsilon(1 + c\varepsilon t)$ reaches $2K\varepsilon$, which occurs at $t = 1/c$. Since this does not depend on ε (as long as $0 < \varepsilon \leq \varepsilon_0$), setting $T = 1/c$ proves the lemma. $\qquad\square$

5.3.6. Lemma. Suppose that A satisfies the hypotheses of Lemma 5.3.5 and that (5.0.6) is in semisimple normal form. There exist constants $\varepsilon_0 > 0$, $c > 0$, and $T > 0$ such that for any ε with $0 < \varepsilon \leq \varepsilon_0$, if $x(t)$ and $y(t)$ are solutions of (5.0.1) and (5.0.6) with $x(0) = y(0)$ and $\|x(0)\| \leq \varepsilon$, then

$$\|x(t) - y(t)\| \leq c\varepsilon^k \quad \text{for} \quad 0 \leq t \leq T/\varepsilon.$$

Proof. This proof will be done using the dilated forms (5.3.1) and (5.3.2) of the full and truncated equations. By the previous lemma, it is possible to choose ε_0 and T so that all solutions $\xi(t, \varepsilon)$ and $\eta(t, \varepsilon)$ beginning in the unit ball remain in a compact set (specifically, the ball of radius $2K$) for $0 \leq t \leq T/\varepsilon$, provided that $0 < \varepsilon \leq \varepsilon_0$. Applying the bounded coordinate changes $\xi = e^{At}u$ and $\eta = e^{At}v$ results in

$$\dot{u} = \varepsilon a_1(u) + \cdots + \varepsilon^{k-1}a_{k-1}(u) + \varepsilon^k e^{-At}R(e^{At}u, \varepsilon)$$

and

$$\dot{v} = \varepsilon a_1(v) + \cdots + \varepsilon^{k-1}a_{k-1}(v)$$

(since $e^{-At}a_j(e^{At}u) = a_j(u)$ for $j = 1, \ldots, k - 1$); this is a statement of the equivariance of each a_j under the flow of the linear part, which is characteristic of the semisimple normal form. (See Theorem 4.4.11 and the discussion prior to Lemma 4.5.7.) Since the coordinate change is the identity at $t = 0$ and is bounded for all time, it remains true that solutions starting in the unit ball remain in a fixed compact set for $0 \leq t \leq T/\varepsilon$. Let L be a Lipschitz constant for $a_1(u) + \varepsilon a_2(u) + \cdots + \varepsilon^{k-2}a_{k-1}(u)$ and let M be a bound for $\|R(u, \varepsilon)\|$, both valid in this compact set; then KM (with K as in Lemma 5.3.5) is a bound for $\|e^{-At}R(e^{At}u, \varepsilon)\|$ as well. Then

$$\frac{d}{dt}\|u(t) - v(t)\| \leq L\varepsilon\|u(t) - v(t)\| + KM\varepsilon^k.$$

It follows by a Gronwall argument, using $u(0) = v(0)$, that

$$\|u(t) - v(t)\| \leq \frac{KM}{L}\varepsilon^{k-1}(e^{L\varepsilon t} - 1).$$

The quantity $e^{L\varepsilon t} - 1$ is bounded for $0 \leq t \leq T/\varepsilon$, so for a suitable c we have

$$\|\xi(t) - \eta(t)\| \leq c\varepsilon^{k-1}$$

on this interval. In view of (5.3.3), the estimate stated in the lemma follows.

$\square$

5.3.7. Remark. This estimate is sharp; in particular, the lemma cannot be strengthened to "$\|x(t)-y(t)\| \leq c\varepsilon^{k}$ for as long as $\|x(t)\| \leq 2\varepsilon$." To see this, consider the example $\dot{x} = x^{3}$ (for $x \in \mathbb{R}$), with truncation $\dot{y} = 0$. These equations are explicitly solvable, and with $x(0) = y(0) = \varepsilon$, $|x(t) - y(t)|$ is strictly $\mathcal{O}(\varepsilon^{2})$ (not smaller) for time $\mathcal{O}(1/\varepsilon)$, and the error becomes greater after that even though $x(t)$ is still within the ball of radius 2ε. This implies that the estimate of Lemma 5.3.6 breaks down before that of Lemma 5.3.5, so that Lemma 5.3.5 is not the sole reason for the restriction $0 \leq t \leq T/\varepsilon$ in Lemma 5.3.6.

5.3.8. Theorem (Center manifold shadowing estimate). Let system (5.0.1) be in extended semisimple normal form to degree k, and let (5.0.6) be its truncation. Suppose also that the restriction of A to E^{c} is semisimple, or equivalently, that each eigenvalue of A lying on the imaginary axis has a full set of eigenvectors (i.e., the generalized eigenspace equals the eigenspace). There exist constants $\varepsilon_0 > 0$, $c > 0$, and $T > 0$ such that for $0 < \varepsilon \leq \varepsilon_0$, every solution $y(t)$ of (5.0.6) lying in E^{c} and having $\|y(0)\| \leq \varepsilon$ is shadowed for time T/ε by a solution $x(t)$ of 5.0.1 that lies in W^{c}; the shadowing estimate is

$$\|x(t) - y(t)\| \leq c\varepsilon^{k} \quad \text{for} \quad 0 \leq t \leq T/\varepsilon.$$

Proof. We use the notation of (5.2.5), so that u is the center part and v the hyperbolic part; B has eigenvalues on the imaginary axis and is assumed to be semisimple, while C has eigenvalues off the imaginary axis. Exactly as in the proof of Theorem 5.3.4 (but with W^{s} replaced by W^{c}), the center manifold reduction of the full system may be written $\dot{u} = Bu + g(u, \sigma(u))$, and its truncation will coincide with the center manifold reduction $\dot{z} = Bz + \hat{g}(z, 0)$ of the truncated system. The difference $\|u(t) - z(t)\|$ between solutions with $u(0) = v(0)$ can be estimated by Lemma 5.3.6, and will be $\mathcal{O}(\varepsilon^{k})$ for time $\mathcal{O}(1/\varepsilon)$. Again as in the proof of Theorem 5.3.4, $\|x - y\| \leq \|u - z\| + \|\sigma(u)\|$. By Lemma 5.3.5, $\|u(t)\| = \mathcal{O}(\varepsilon)$ for time $\mathcal{O}(1/\varepsilon)$, so $\|\sigma(u(t))\| = \mathcal{O}(\varepsilon^{k+1})$, which is smaller than the main term. $\square$

Estimates on the Center-Stable Manifold

The next theorem extends the last result from the center to the center-stable manifold:

5.3.9. Theorem. Under the same hypotheses as Theorem 5.3.8, any solution $y(t)$ in E^{cs} with $\|y(0)\| \leq \varepsilon$ is shadowed by a solution $x(t)$ in W^{cs} with error ε^k for time $\mathcal{O}(1/\varepsilon)$.

Proof. The idea of the proof is simple: The center manifolds and stable fibers of the full and truncated systems are $\mathcal{O}(\varepsilon^{k+1})$-close; the motion of the base points of the fibers in their center manifolds are $\mathcal{O}(\varepsilon^k)$-close for time $\mathcal{O}(1/\varepsilon)$; and the contraction in the fibers is exponential. For the details, we use the notations of Theorem 5.2.5, so that W^{cs} can be represented as

$$(u, v, w) = (U + \varphi(U, V), V + \psi(U, V), \sigma(U + \varphi(U, V), V + \psi(U, V))),$$

with the flow (of the full system) on W^{cs} governed by (5.2.11):

$$\dot{U} = BU + F(U, V), \tag{5.3.14}$$
$$\dot{V} = CV + G(V).$$

The flow of the truncated system on its center-stable subspace E^{cs} satisfies (5.1.3), which we write as

$$\dot{u} = Bu + \widehat{f}(u, v, 0), \tag{5.3.15}$$
$$\dot{v} = Cv + \widehat{g}(-, v, 0),$$

to emphasize that $\widehat{g}$ is independent of u when $w = 0$. Since φ, ψ, and σ are $\mathcal{O}(\varepsilon^{k+1})$ in a $\mathcal{O}(\varepsilon)$ neighborhood of the origin, they may be ignored, and the problem reduces to estimating the difference between solutions of (5.3.14) and (5.3.15) having the same initial conditions. The important point here is that (5.3.15) is the truncation of (5.3.14). To see this, observe that

$$\begin{bmatrix} BU + F(U, V) \\ CV + G(V) \end{bmatrix} = \left(I + \begin{bmatrix} \varphi_U & \varphi_V \\ \psi_U & \psi_V \end{bmatrix} \right)^{-1} \begin{bmatrix} Bu + f(u, v, \sigma(u, v)) \\ Cv + g(u, v, \sigma(u, v)) \end{bmatrix},$$

where $u = U + \varphi(U, V)$ and $v = V + \psi(U, V)$. The matrix here is the identity plus terms of degree k and higher (since φ and ψ begin with terms of degree $k + 1$), and this matrix is multiplied by a vector field that begins with linear terms; therefore the effect of φ, ψ, and their derivatives on the right-hand side affects only terms of degree $k + 1$ and higher, so these can be ignored when taking the k-jet. So (5.3.15) can be written as

$$\dot{u} = Bu + \widehat{F}(u, v), \tag{5.3.16}$$
$$\dot{v} = Cv + \widehat{G}(v).$$

The v and V portions of (5.3.14) and (5.3.16) are uncoupled from the u and U parts, and $\|v - V\|$ can be estimated by Lemma 5.3.6. Then $u(t)$ and $U(t)$ satisfy the nonautonomous equations $\dot{u} = Bu + \widehat{F}(u, v(t))$ and

$\dot{U} = BU + F(U, V(t))$. Since B is contracting, $\|u - U\|$ can be estimated by a slight modification of Lemma 5.3.3 (to allow for the time dependence).   □

Estimates for Saddles

None of the previous estimates in this section has allowed the unstable manifold to be actively involved. (An unstable manifold can be present, but the solutions studied lie entirely in W^{sc}.) As soon as the unstable manifold becomes active, the length of validity of any error estimate valid near the origin is, in general, reduced to a finite time interval. For instance, the solution of $\dot{x} = x$ for $x \in \mathbb{R}$ having $x(0) = \varepsilon$ is $x(t) = \varepsilon e^t$, which reaches 2ε at time $t = \ln 2$. On the other hand, there exist solutions that take an arbitrarily long time to escape from a neighborhood of the origin. In the special case where $E^c = \{0\}$, it is possible to give uniform shadowing estimates for all solutions close to the origin, valid for as long as they remain close.

5.3.10. Theorem (Saddle shadowing estimate). Suppose that A in (5.0.1) is hyperbolic and that $a_1, \ldots, a_{k-1}$ are in extended semisimple normal form. Then there exist constants $\varepsilon_0 > 0$ and $c > 0$ such that for $0 < \varepsilon \leq \varepsilon_0$, every solution $y(t)$ of (5.0.6) that satisfies $\|y(t)\| \leq \varepsilon$ for any interval of time $t_1 \leq t \leq t_2$ (where t_1 may equal $-\infty$ and t_2 may equal $+\infty$) is shadowed by a solution $x(t)$ of (5.0.1) satisfying

$$\|x(t) - y(t)\| \leq c\varepsilon^{k+1} \quad \text{for} \quad t_1 \leq t \leq t_2.$$

Since the techniques of proof for this theorem are unrelated to the rest of this book, two proofs will be sketched without complete details. For more information see Notes and References.

First Proof. Since (by Hartman's theorem) both (5.0.1) and (5.0.6) are topologically conjugate to their linear parts, they are conjugate to each other. If the techniques of the proof of Hartman's theorem are applied directly to obtain a conjugacy between (5.0.1) and (5.0.6), it can be shown that within a neighborhood of size ε, the conjugacy moves points by an amount $\mathcal{O}(\varepsilon^{k+1})$. Since the conjugacy maps solutions to solutions, we may define $x(t)$ to be the solution conjugate to $y(t)$, and the error estimate follows.   □

Second Proof. Write the full system in the form

$$\begin{aligned} \dot{u} &= Bu + f(u, v) + F(u, v), \\ \dot{v} &= Cv + g(u, v) + G(u, v), \end{aligned} \qquad (5.3.17)$$

where B has eigenvalues in the left half-plane and C in the right, and F and G are the terms to be deleted for the truncated system. The truncated system has the u-space for its stable space E^s and the v-space for E^u; the full system has a stable manifold W^s given by $v = \sigma(u)$ and an unstable manifold W^{u} given by $u = \tau(v)$. The change of coordinates $\tilde{u} = u - \tau(v)$, $\tilde{v} = v - \sigma(u)$ is well-defined in the *box neighborhood* $\{(u, v) : \|u\| \leq \varepsilon_0, \|v\| \leq \varepsilon_0\}$ for some ε_0, and produces a system having the same form as (5.3.17) after the tildes are dropped; only F and G are changed (because σ and τ have vanishing k-jets), and the new system has $W^s = E^s$ and $W^{\mathrm{u}} = E^{\mathrm{u}}$ in the box neighborhood. Furthermore, it suffices to prove the theorem for the new system and its truncation, because the coordinate change moves points at a distance ε from the origin by an amount $\mathcal{O}(\varepsilon^{k+1})$ (for $\varepsilon \leq \varepsilon_0$). Now (for the new system (5.3.17), with flattened stable and unstable manifolds) we define the *box data problem*, as follows: Let α belong to the sphere of radius ε_0 (centered at the origin) in the u-space, let β belong to the sphere of radius ε_0 in the v-space, and let $T > 0$. Then the solution of (5.3.17) with box data (α, β, T) is the solution satisfying $u(0) = \alpha$ and $v(T) = \beta$. It can be shown by an integral equation argument that the box data problem is well-posed; that is, there exists a unique solution, which remains in the box neighborhood on the interval $0 \leq t \leq T$ and depends smoothly on the right-hand side of (5.3.17) even at $T = \infty$. (This point is a bit delicate. For $T = \infty$ the "solution" of the box data problem should be understood as the union of two solutions, one passing through α and approaching the origin, the other leaving the origin and passing through β. To define this correctly it is necessary to scale the time along each solution by defining $\tau = t/T$, so that all solutions, including the "limit solutions" as $T \to \infty$, are defined for $0 \leq \tau \leq 1$.) Finally, there exists a uniform bound of the form $c\varepsilon^{k+1}$ for the distance between any two solutions of the full and truncated systems having the same box data.   □

The General Case

Reviewing the results that have been proved in this section, we see that shadowing estimates have been proved *for all solutions near the origin* only in cases where at least one of the subspaces E^s, E^c, and E^{u} is trivial. (If E^s is trivial and E^{u} is not, the required estimates follow by reversing time so that E^s and E^{u} exchange places.) In the general case, when all three subspaces are nontrivial, we have results only in E^{cs} and E^{cu}. It is natural to ask whether there is an estimate valid in a full neighborhood of the origin, and our experience teaches us not to ask for more than the following conjectured theorem:

5.3.11. Conjecture (Conjectured shadowing estimate). We shall assume that the restriction of A to E^c is semisimple. There exist constants $\varepsilon_0 > 0$, $c > 0$, and $T > 0$ such that for $0 < \varepsilon \leq \varepsilon_0$, every solution $y(t)$ of (5.0.6) having $\|y(0)\| \leq \varepsilon$ is shadowed by a solution $x(t)$ of (5.0.1) in the sense that

$$\|x(t) - y(t)\| \leq c\varepsilon^k$$

for as long as both of the following conditions are met:

1. $\|y(t)\| \leq 2\varepsilon$, and

2. $0 \leq t \leq T/\varepsilon$.

Such a theorem has apparently never been proved. It seems that a proof similar to the first proof of Theorem 5.3.10 is not possible, for the following reason. While it does follow from the local conjugacy theorem, Theorem 5.2.4, that (5.0.1) and (5.0.6) are locally conjugate, it seems impossible that this conjugacy could move points at a distance ε from the origin by an amount $\mathcal{O}(\varepsilon^k)$. If this were true, Lemma 5.3.6 could be strengthened so that the estimate held as long as $\|x(t)\| \leq 2\varepsilon$, and we have seen in Remark 5.3.7 that this is false. This leaves the possibility of a proof along the lines of the second proof of Theorem 5.3.10. Such a proof might use box data for the stable and unstable parts of the motion and initial data for the center part, but the argument has not been attempted.

Notes and References

The error estimates for stable manifolds are modeled on standard arguments for ordinary differential equations whose linear part has its spectrum in the left half-plane. There are many ways to do these arguments; our proofs follow Lefschetz [70], page 89.

The estimates for center manifolds are based on those commonly given for the method of averaging, for instance in Perko [96]. The estimates for both stable and center manifolds ultimately rely on Gronwall's inequality, but in the stable case one retains the linear part and exploits its contracting properties to obtain results valid for all (future) time, whereas in the center case one removes the linear part (using its boundedness and the equivariance of the normal form terms). Once the linear part is gone, the problem resembles the case of averaging.

The estimate for saddles, presented here without complete details, is similar to research I have published under the heading of "elbow orbits." The argument using Hartman's theorem occurs in [83], and the one using box neighborhoods is in [85] and [86].

5.4 The Nilpotent Center Manifold Case

In the previous sections of this chapter, very little use was made of the nilpotent part of A; sometimes it was necessary to assume the absence of a nilpotent part, while in other cases the nilpotent part was harmless but was not used. This may seem surprising considering that in Chapters 3 and 4 the existence of a nontrivial N always permits additional simplifications of the normal form beyond those possible due to S alone, but there is a reason why it is difficult to exploit these simplifications. In fact, it is only the quadratic and higher-order terms that become simpler when $A = S$ is replaced by $A = S + N$; the linear term itself has become more complicated. As a result the nonlinear terms, although they still commute with S, no longer commute with A. This makes many of the arguments that are useful in the semisimple case impossible.

The best-understood example of a system with nontrivial nilpotent part is the (two-dimensional) Takens–Bogdanov system, introduced in Section 1.2. This system will be studied in further detail in Section 6.6. The arguments that are successful in this case have not yet been generalized to cover nonsemisimple systems in general. The present section does not pretend to provide a solution to these questions, but only to indicate a few possibilities for further research.

A preliminary observation is that if $A = S + N$, it is always possible to remove S from the truncated normal form. Thus, the change of variables $x = e^{St}z$ carries (5.0.2) into

$$\dot{z} = Nz + a_1(z) + \cdots + a_k(z), \tag{5.4.1}$$

as long as (5.0.2) is in extended semisimple normal form; this is because the right-hand side of (5.0.2) is equivariant under e^{St}. Making this change of variables is not necessarily desirable. In particular, the stable and unstable subspaces (if any), and the S-foliation, are lost after the coordinate change. But any results that follow from (5.4.1) can be carried back to the x coordinates and used there together with the other facts that are known. For this reason, and for simplicity, we assume now that $S = 0$, so that $A = N$ is entirely nilpotent. This implies also that $E^c = \mathbb{R}^n$. Thus, the truncated system then has the form

$$\dot{x} = \hat{a}(x) = Nx + a_1(x) + \cdots + a_k(x) = Nx + f(x), \tag{5.4.2}$$

and the full system is

$$\dot{x} = a(x) = Nx + f(x) + F(x), \tag{5.4.3}$$

where the Taylor expansion of F begins with degree $k + 1$. It will be assumed that $a_1, \ldots, a_k$ are in sl(2) normal form with respect to a triad $\{N, M, H\} = \{X, Y, Z\}$ (see Remark 4.7.2 for the notation); this means that f is equivariant under the one-parameter group e^{Ms}, or

$$e^{-Ms}f(e^{Ms}x) = f(x). \tag{5.4.4}$$

5.4.1. Definition. If (5.4.2) is in sl(2) normal form, the M-*foliation* is the foliation of $\mathbb{R}^n$ by orbits of e^{Ms}.

In the simplest case, when N is in upper Jordan form and consists of a single block (see (4.7.2) for an example), the M-foliation has a line $(x_1 = \cdots = x_{n-1} = 0)$ of singular leaves (rest points); the other leaves are (topologically) intervals, which are transverse to the hyperplane $x_2 = 0$

except where $x_1 = 0$. We exclude the hyperplane $x_1 = 0$ from the following discussion; the leaves of the M-foliation are then given by the level sets of $n - 1$ integrals or invariants $I_1, \ldots, I_{n-1}$ of the M-flow (or N^*-flow). The invariants required are a subset of those calculated when the normal form module is determined according to the methods of Section 4.8.

In Theorem 5.1.9, it was seen that the S-foliation is preserved under the flow of the truncated normalized system. In the present situation, the M-foliation is *not* preserved by the flow of (5.4.2). However, there is a sense in which we can "measure the failure of the foliation to be preserved," and find that the failure is not great; in fact, in a precise sense, the failure is linear.

5.4.2. Lemma. Assume that (5.4.2) is in sl(2) normal form, and that I is a weight invariant, that is, $\mathcal{D}_M I = 0$ and $\mathcal{D}_H I = \text{wt}(I)I$. Let $\dot{I} = \mathcal{D}_{\hat{a}} I$. Then

$$\mathcal{D}_M \dot{I} = \text{wt}(I)I. \qquad (5.4.5)$$

Proof. The calculation parallels that of the second proof of Corollary 5.1.10, using $[\hat{a}, M] = [N + f, M] = [N, M] + [f, M] = [N, M] = H$:

$$\mathcal{D}_M \mathcal{D}_{\hat{a}} I = [\mathcal{D}_M, \mathcal{D}_{\hat{a}}]I - \mathcal{D}_{\hat{a}}(\mathcal{D}_M I) = \mathcal{D}_{[\hat{a}, M]} I = \mathcal{D}_H I = \text{wt}(I)I.$$

$\square$

It is clear that weight invariants of weight zero have a special importance, since they satisfy $\mathcal{D}_M \dot{I} = 0$. If there were $n - 1$ invariants of weight zero, then the second proof of Corollary 5.1.10 could be repeated, and the M-foliation would be preserved. But there do not exist $n - 1$ invariants of weight zero. A gain in the number of such invariants can be obtained by noting that Lemma 5.4.2 applies to *rational* invariants as well as *polynomial* invariants, but this is still not sufficient. For instance, in the case of N_4, it may seem that we have the required $n - 1 = 3$ invariants of weight zero, because not only δ but also β^3/α^2 and γ/α (where $\alpha, \beta, \gamma, \delta$ are as in (4.8.11)) will serve. But only two of these invariants are functionally independent, because according to (4.8.12), $(\gamma/\alpha)^2 = (2\beta^3/\alpha^2) + 9\delta$. So we are not able to reduce the order of the system to $n - 1$, but nevertheless, the invariants that are available do contain some geometrical information about the flow. It remains an open problem to find a way of making use of this information.

Invariants of nonzero weight also contain some dynamical information. In particular, it follows from Lemma 5.4.2 that for such an I, there is a hypersurface through the origin on which $\dot{I} = 0$, with $\dot{I} > 0$ on one side and $\dot{I} < 0$ on the other. Thus, I is somewhat like a weak Lyapunov function, giving some indication of the direction of the flow.

Notes and References

Bruno's power transformations have already been mentioned in connection with Theorem 5.1.13 (see Notes and References to Section 5.1). He has also used power transformations to study nilpotent systems; see [21], Chapter II, Section 3, and Chapter IV, Section 1.5 (devoted to what he calls nonelementary singular points). The other methods commonly used for such problems are the method of blowing up expounded in the work of Dumortier [39], and the rescaling method discussed in Section 6.6 below, but these are of value particularly for the unfolded system (and often require that at least one of the unfolding parameters be nonzero).

6

Selected Topics in
Local Bifurcation Theory

As the title indicates, this short final chapter is not intended to be a complete treatment of bifurcation theory, or even a complete overview. Instead, we focus on certain specific topics, chosen because they are closely related to the main themes of this book, or because they are not developed extensively in other monographs. Among these topics are the questions of jet sufficiency, the computation of unfoldings, and the rescaling of unfolded systems.

The first three sections are devoted to bifurcations from a single zero eigenvalue in a single scalar differential equation. For such problems, the method of normal forms provides no simplification, since a scalar equation with a rest point at the origin is automatically in (semisimple) normal form with respect to its linear part. Therefore, Sections 6.1 to 6.3 can be read independently of the rest of the book. (On the other hand, the one-dimensional problem can result from a center manifold reduction of a higher-dimensional problem with a single zero eigenvalue on the imaginary axis; this center manifold reduction is provided by the normal form, according to Section 5.1.) For these easiest-possible bifurcation problems, which include the fold, pitchfork, and transcritical bifurcations, our first goal (in Sections 6.1 and 6.2) is to compare two approaches (called "neoclassical" and "modern") to the question of jet sufficiency. The "classical" approach to this problem assumes that the vector field is analytic, and relies upon the Weierstrass preparation theorem at one crucial point (except in the simplest cases). The "neoclassical" method (Section 6.1) is the same, except that the Weierstrass preparation theorem is replaced by the Malgrange preparation theorem, so that the results now apply to the smooth (and not just the analytic) case. The preparation theorem is discussed in some detail (although not proved) in Appendix A, Sections A.6 and A.7. The "modern" approach (Section 6.2) is based on ideas borrowed from the theory of singularities of smooth mappings. It is interesting that although the second method may seem harder than the first, the difficult Malgrange preparation theorem is not needed for this approach; the much easier Nakayama's lemma is sufficient.

The problems treated in Sections 6.1 and 6.2 contain a bifurcation parameter η that is specified when the problem is posed. A more general way to approach bifurcation theory is to begin with an "unperturbed problem" and ask what bifurcations take place when the problem is given a sufficiently small, but otherwise arbitrary, perturbation; the unperturbed problem is often called the "organizing center" for the resulting bifurcations. The first step is to introduce a finite number of parameters that will (in some sense, that is, up to some equivalence relation) account for all possible perturbations; this is called finding an *unfolding* of the organizing center. Our goal here is to show that such unfoldings can be computed by the method of normal forms, provided that the equivalence relation is defined by a preassigned choice of the asymptotic order to which the unfolding will be computed. Such an unfolding is called an *asymptotic unfolding*. These are developed in Section 6.3 for the case of a single zero eigenvalue (continuing the problems studied in Sections 6.1 and 6.2), and then in Section 6.4 for the general case.

Sections 6.5 and 6.6 give brief treatments of two important bifurcation problems in two dimensions, the Hopf and Takens–Bogdanov bifurcations. The Hopf bifurcation is so important that it cannot be ignored, even though it is well treated in many books; our approach is to show that it reduces to a one-dimensional problem similar to those of Sections 6.1 and 6.2 and can be treated by those methods, even in degenerate cases. A truly complete treatment of the Takens–Bogdanov bifurcation does not exist in any one textbook, but it poses a difficulty in that it requires a variety of methods that have not been developed in this book. On the other hand, this book deals at length with nonsemisimple normal forms, of which the Takens–Bogdanov bifurcation is the best-understood application. Our way of coping with this situation is to present those parts of the theory that are closely related to the rest of this book, and provide guidance (in the Notes and References for Section 6.6) to other widely available expositions that cover what is omitted here.

Finally, Section 6.7 addresses several more complicated bifurcations commonly known as mode-interaction problems. The treatment given to these is limited to computing the required unfoldings by the theory of Section 6.4. At this point we are close to the boundary of what is known. Our understanding of complicated bifurcations is rapidly changing, and the literature is a mixture of experimental and numerical results, quasi-mathematical results by nonrigorous methods, and, occasionally, solid proofs.

6.1 Bifurcations from a Single-Zero Eigenvalue: A "Neoclassical" Approach

The root meaning of the word "bifurcation" is "splitting in two." Its original use in mathematics and physics was for the splitting of a stable equilibrium point into two as a parameter is varied. For instance, under light pressure at the ends a yardstick remains straight, but under stronger pressure it begins to bend to one side or the other; the original (straight) stable equilibrium state is replaced by two symmetrical bent stable equilibrium states. In fact, the straight state remains as a third, unstable, equilibrium state that is not physically realizable, and the splitting into two is actually a splitting into three, known as a *pitchfork bifurcation*. Today the word "bifurcation" has come to mean not only the splitting of rest points (regardless of the number of rest points involved), but any change in the qualitative behavior of a system of differential equations as a parameter is varied.

The simplest context in which to study bifurcation is a scalar differential equation depending on one scalar parameter,

$$\dot{x} = a(\eta, x), \tag{6.1.1}$$

with $x \in \mathbb{R}$, $\eta \in \mathbb{R}$, and $a : \mathbb{R}^2 \to \mathbb{R}$. (The variable η is written first in $a(\eta, x)$ in this section because we will often want to draw graphs in which η is the first, or horizontal, axis.) It is easy to exhibit examples of the possible behavior of such systems in a neighborhood of $x = 0$ for values of η near zero. First observe that if $a(0,0) > 0$, then $a(\eta, x) > 0$ for all small x and η, so all solutions simply move to the right on the x axis and there is no change in the qualitative behavior as η crosses zero, although of course the speed of the motion to the right will vary. (It is important to understand that when we speak of changing η, this does not mean changing η *with time*. This is simply a manner of speaking, and refers to a comparison of systems with different values of η. See Remark 1.1.11.) If $a(0,0) < 0$, the same thing happens except that the motion is to the left. So it is clear that nothing interesting happens unless $a(0,0) = 0$; that is, the origin is a rest point of the unperturbed ($\eta = 0$) system. A simple example that meets this condition is

$$\dot{x} = \eta - x, \tag{6.1.2}$$

which for each η has a rest point at

$$x = \hat{x}(\eta) = \eta. \tag{6.1.3}$$

(The "hat" symbol $\hat{\ }$ will be used to denote a function whose values are denoted by the same letter without the hat. Those who do not wish to be fastidious may simply write $x = x(\eta)$.) This example illustrates the *continuation* of a rest point, without bifurcation; see Figure 6.1(a). The

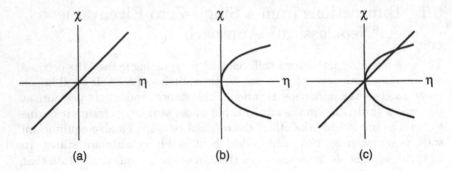

Figure 6.1. Examples of bifurcation diagrams: (a) continuation, (b) fold, and (c) pitchfork with transcritical bifurcation.

next example,

$$\dot{x} = x^2 - \eta, \tag{6.1.4}$$

has two rest points

$$\widehat{x}_+(\eta) = \sqrt{\eta} \quad \text{and} \quad \widehat{x}_-(\eta) = -\sqrt{\eta} \tag{6.1.5}$$

for $\eta > 0$, none for $\eta < 0$ (and of course only one for $\eta = 0$). See Figure 6.1(b). This is called a *fold*, or *pair*, or *saddle-node* bifurcation. (The name "saddle-node" is actually appropriate only in higher-dimensional versions of this situation, where one of the bifurcating rest points is a saddle and the other is a node.) The "product" of these last two examples,

$$\dot{x} = (\eta - x)(x^2 - \eta), \tag{6.1.6}$$

has all of the rest points given in (6.1.3) and (6.1.5), and is an example of the already mentioned *pitchfork* bifurcation; see Figure 6.1(c). Notice that the pitchfork structure applies only in a neighborhood of $x = 0$ and $\eta = 0$; away from the origin, two of the rest points cross in what is called a *transcritical* bifurcation. (See Theorem 6.1.11.)

The goal of this section is to develop methods to analyze an arbitrary equation of the form (6.1.1) in a neighborhood of $(\eta, x) = (0, 0)$. Thus, we expect to find criteria on the function a (assumed to satisfy $a(0, 0) = 0$), under which the unperturbed rest point either continues for $\eta \neq 0$, or undergoes a fold bifurcation, pitchfork bifurcation, transcritical bifurcation, or something else. Note that these results will be purely local in nature. We do not expect to give conditions that will reveal both of the bifurcations in Figure 6.1(c) at the same time, but only the pitchfork bifurcation at the origin. (The transcritical bifurcation in this picture would be studied using coordinates centered at the point where this bifurcation occurs.)

Our attention will be devoted to the existence of these rest points, not their stability. For these one-dimensional problems, the stability is easily determined by the sign of $a(\eta, x)$ between the rest points. For instance, the reader should check that in (6.1.6), the single rest point for $\eta < 0$ is stable, while of the three rest points that exist for η just a little larger than zero, the middle one is unstable and the outer two are stable. (This is shown in Figure 6.4 below.) Stability considerations are part of almost all treatments of bifurcation theory listed in Notes and References.

It is naturally to be expected that for x and η small, the behavior of (6.1.1) will generally be decided by the lower-order terms in the Taylor expansion of a in these variables. It is convenient to write this Taylor expansion as

$$a(\eta, x) \sim \sum_{ij} a_{ij} \eta^i x^j, \tag{6.1.7}$$

with

$$a_{ij} = \frac{1}{i!j!} \frac{\partial^{i+j} a}{\partial \eta^i \partial x^j}(0,0), \tag{6.1.8}$$

and to picture the coefficients a_{ij} as forming an infinite matrix. This matrix is usually arranged to the right and upward from a_{00} at the lower left-hand corner, thus:

$$
\begin{array}{cccccc}
\vdots & \vdots & \vdots & \vdots & \vdots & \\
a_{40} & a_{41} & a_{42} & a_{43} & a_{44} & \cdots \\
a_{30} & a_{31} & a_{32} & a_{33} & a_{34} & \cdots \\
a_{20} & a_{21} & a_{22} & a_{23} & a_{24} & \cdots \\
a_{10} & a_{11} & a_{12} & a_{13} & a_{14} & \cdots \\
a_{00} & a_{01} & a_{02} & a_{03} & a_{04} & \cdots
\end{array}
\tag{6.1.9}
$$

These coefficients a_{ij} should be pictured as attached to the integer lattice points (i, j) in the first quadrant of a system of (η, x) axes; this set of lattice points (with various additional markings) is then called a *Newton diagram*. Our use of such diagrams will differ slightly from the classical version.

6.1.1. Definition.

1. A *Newton diagram* is a partition of the terms in (6.1.9) into three classes, designated as *low terms*, *intermediate terms*, and *high terms*. This partition is described by marking each lattice point in the first quadrant with a zero (or open dot), to indicate a low term; a solid dot, to indicate an intermediate term; or no mark, for a high term. As the names suggest, the low terms (for those Newton diagrams that are actually useful) will generally be clustered near the lower left-hand corner of the diagram, the intermediate terms will be next, and the high terms will be the remaining terms running off to infinity; but the definition places no restriction on which terms may be classified as low, high, or intermediate.

2. A function $a(\eta, x)$ is said to *belong to* a Newton diagram if its low terms vanish and its middle terms are known. More precisely, a func-

tion $a(\eta, x)$ with Taylor series (6.1.7) belongs to a Newton diagram if $a_{ij} = 0$ whenever the lattice point (i, j) is marked with a zero in the Newton diagram, and if the values of a_{ij} corresponding to lattice points marked with a solid dot are known. (Of course, being "known" is not a mathematical property, and this last condition has no mathematical meaning. Its significance lies in the fact that the intermediate terms will appear in various formulas to follow, while the high terms will not. Therefore, to apply these formulas it is necessary to know the intermediate terms.)

3. A *bifurcation theorem* associated with a Newton diagram is any theorem describing (in qualitative form) the zero set of all functions $a(\eta, x)$ belonging to the Newton diagram that satisfy certain conditions imposed on the intermediate terms. These conditions vary from one bifurcation theorem to another, and are part of the statement of the theorem; they are of two types, *degeneracy conditions* and *nondegeneracy conditions*, as described below.

4. A *degeneracy condition* on a function $a(\eta, x)$ is an equation stating that a certain algebraic expression in the coefficients a_{ij} equals zero. The most common type of degeneracy condition is simply that a particular coefficient vanishes. (Thus, to say that a function a belongs to a Newton diagram already imposes one degeneracy condition for each lattice point marked with a zero.) Another type of degeneracy condition is that a particular combination of intermediate terms vanishes. (An example is the condition $\Delta = 0$ in Theorem 6.1.12 below.)

5. A *nondegeneracy condition* is an inequality stating that a particular combination of intermediate terms is nonzero, or positive, or negative.

These definitions may be summarized as follows: If a bifurcation theorem associated with a particular Newton diagram is known, then the behavior of the zero set of any function associated with that Newton diagram is known, provided that the intermediate terms satisfy the degeneracy and nondegeneracy conditions specified in the theorem. Specifically, it is not necessary to know anything about the high-order terms.

6.1.2. Remark. Classically, a Newton diagram is defined as follows. It is assumed that $a(\eta, x)$ is a polynomial. The lattice points corresponding to terms with nonzero coefficients are marked with a solid dot. Then certain of these dots are connected with a polygonal line, such that the resulting polygon is concave upwards and all solid dots lie on or above the polygon. Roughly speaking, the unmarked terms below the polygon correspond to our low terms, but the other marked and unmarked terms do not correspond to our intermediate and high terms, because some marked terms may actually be high (that is, they may not need to be known) and some unmarked terms may actually be intermediate (that is, the result of the classical pro-

cedure may depend on the fact that some of these terms are zero).
The classical version is designed to treat only one particular problem
at a time, whereas ours is designed to treat classes of problems. As
a consequence, any randomly chosen (finite) number of solid dots
can constitute a legitimate classical Newton diagram, but for us a
given assignment of zeros and solid dots constitutes a "valid" Newton diagram only if there is an associated theorem stating that the
high terms are ignorable if the low terms are zero and the intermediate terms are nondegenerate. Although we use the diagrams in this
somewhat unorthodox way, the methods of proof for our theorems
correspond closely to the classical arguments. A still different use
of Newton diagrams (to indicate monomial ideals) is described and
used in Appendix A.

There is a considerable amount of "philosophy" that can be associated
with these Newton diagrams, connected with the words "genericity" and
"jet sufficiency" or "determinacy." (Recall that a k-jet is a kth-degree Taylor polynomial.) Genericity is an attempt to describe how likely it is that a
given bifurcation diagram will be encountered in an application. The idea
is that any given condition imposed on the a_{ij} in the form of a strict equality is unlikely to be satisfied if the a_{ij} are chosen "arbitrarily," whereas a
condition in the form of an inequality is likely to be satisfied. Thus, it is
most likely that none of the a_{ij} are equal to zero. But if we impose the
condition that a_{00} is equal to zero (which we have seen is necessary if there
is to be a bifurcation problem), then we still expect that a_{10} and a_{01} are
nonzero; furthermore, we hope that these terms suffice to decide the bifurcation picture (i.e., that "the 1-jet of a is sufficient"). This is the situation
described in Newton diagram (6.1.10) below. The terms a_{10} and a_{01} are
designated as "intermediate," and are subject to the "generic" (or likely to
be true) nondegeneracy condition that a_{01}, at least, is not zero. But what
happens if this condition fails? In this "degenerate" case, a_{01} joins a_{00} as a
"low term," and we move to the Newton diagram (6.1.11). The new nondegeneracy conditions (which are generic in this new context, i.e., likely to
be true given that $a_{01} = 0$ holds) are that a_{02} and a_{10} are nonzero, and
the hope is that the "lopsided jet" defined by these terms is "sufficient"
(to decide the bifurcation). As we go to higher levels of degeneracy, new
nondegeneracy conditions arise. These are always inequalities (as opposed
to the degeneracy conditions, which are equalities), but as the degeneracies
get more complicated, so do the nondegeneracy conditions, and eventually
it is necessary to classify more terms as intermediate (that is, as terms that
need to be known) than might be expected. (See (6.1.18).)

6.1.3. Remark. There is a technical definition of genericity that is
sometimes needed in dynamical systems theory, but it is not really
required here. A subset A of a topological space X is *generic*, or
residual, if it is a countable intersection of open dense sets. In our
setting it suffices to take for X the Euclidean space whose points

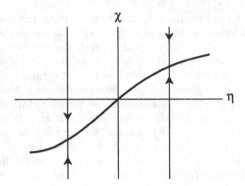

Figure 6.2. Continuation of a stable rest point, $a_x(0,0) < 0$.

are vectors made from the intermediate terms of a given Newton diagram. There are only a finite number of nondegeneracy conditions imposed on these quantities, so the set of nondegenerate points is open and dense, hence (trivially) generic.

6.1.4. Remark. This discussion of genericity has not included the important effect of symmetries. For a simple illustration, if $a(\eta, x)$ is known to be odd in η, then the even terms along the vertical axis of Newton's diagram are forced to be zero, in spite of the fact that this is otherwise unlikely. This changes the Newton diagrams that one expects to encounter for this problem and its possible degenerate cases.

We now turn to an examination of some of the basic bifurcations.

Continuation

The first bifurcation theorem (actually a nonbifurcation theorem) is associated with the Newton diagram

$$
\begin{matrix}
\bullet & \\
0 & \bullet
\end{matrix}
\tag{6.1.10}
$$

6.1.5. Theorem (Continuation). If $a(0,0) = 0$ and $a_x(0,0) \neq 0$, then equation (6.1.1) has a unique rest point $\widehat{x}(\eta)$ near $x = 0$ for each η near zero; the rest point continues with no bifurcation, as in Figure 6.2. If in addition $a_\eta(0,0) \neq 0$, then $\widehat{x}(\eta) = c_1\eta + \mathcal{O}(\eta^2)$ for some $c_1 \neq 0$.

Proof. We apply the implicit function theorem to the equation $a(\eta, x) = 0$ for $x = \widehat{x}(\eta)$ near the starting solution $a(0,0) = 0$. The condition $a_x(0,0) \neq 0$ implies the existence of a unique smooth function $\widehat{x}(\eta)$ defined for η near zero, satisfying the conditions $\widehat{x}(0) = 0$ and $a(\eta, \widehat{x}(\eta)) = 0$. This shows the existence of the desired rest point; differentiation of the last formula

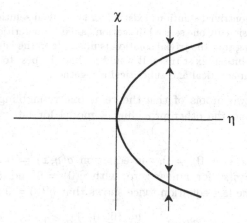

Figure 6.3. Supercritical fold bifurcation.

implies $\hat{x}'(0) = -a_\eta(0,0)/a_x(0,0)$, so if $a_\eta(0,0) \neq 0$, then $c_1 = \hat{x}'(0) \neq 0$. Furthermore, the graph of $\hat{x}$ coincides with the level set $a(\eta, x) = 0$ in some neighborhood of the origin in $\mathbb{R}^2$. This shows that there are no other rest points in this neighborhood. $\qquad\qquad\square$

The Fold Bifurcation

The Newton diagram for the next bifurcation theorem is

$$
\begin{array}{l}
\bullet \\
0 \\
0 \ \bullet
\end{array}
\qquad (6.1.11)
$$

6.1.6. Theorem (Fold). If $a(0,0) = a_x(0,0) = 0$ but $a_{02} = \frac{1}{2}a_{xx}(0,0) \neq 0$ and $a_{10} = a_\eta(0,0) \neq 0$, then (6.1.1) exhibits a fold bifurcation, supercritical (as in Figure 6.3) if a_{02} and a_{10} have opposite signs, subcritical if the signs are the same. In the supercritical case there are exactly two rest points near $x = 0$ for small $\eta > 0$, none for small $\eta < 0$; in the subcritical case this is reversed. In either case one rest point is stable; the other is unstable. The two rest points satisfy $\hat{x}_\pm(\eta) = \pm c_{1/2}|\eta|^{1/2} + \mathcal{O}(\eta)$.

6.1.7. Remark. We adopt the simple point of view that a bifurcation is *supercritical* if the more complicated picture occurs for positive η, *subcritical* if the more complicated picture occurs for negative η. *Transcritical* is reserved for the specific situation described in Theorem 6.1.11 below. An alternative point of view defines supercritical and subcritical in terms of stability. This definition is tenable only in the (at one time standard) situation where there is a continuing "trivial" solution that loses stability at the bifurcation point. It was generally assumed that the trivial solution was stable below the bifurcation point and unstable above. Thus, "supercritical" came to mean that nontrivial solutions exist when the trivial solution is unstable. Occasionally one sees a bifurcation called "supercritical to the left," meaning that the trivial solution is unstable to the left and the nontrivial solutions exist there. It would perhaps be best to abandon the terms supercritical and subcritical altogether.

We will give two proofs of this theorem, one resembling the proof of Theorem 6.1.5 and the other providing a model for later proofs in this section:

First Proof. Since $a_\eta(0,0) \neq 0$, the equation $a(\eta, x) = 0$ can be solved uniquely for $\eta = \widehat{\eta}(x)$ for x near zero, with $\widehat{\eta}(0) = 0$ and $a(\widehat{\eta}(x), x) = 0$. Differentiating the last equation twice shows that $\widehat{\eta}'(0) = 0$ and

$$\widehat{\eta}''(0) = -\frac{a_{xx}(0,0)}{a_\eta(0,0)} \neq 0.$$

The sign of this quantity determines the direction of bending of the curve near the origin, and hence whether the bifurcation is supercritical or subcritical.   □

Second Proof. Assume that a_{02} and a_{10} have opposite signs. (This yields the supercritical case; the other case can be treated similarly.) The hypotheses of Theorem 6.1.6 imply that there exist smooth functions f_1, f_2, f_3 such that

$$a(\eta, x) = a_{02}x^2 + a_{10}\eta + x^3 f_1(\eta, x) + \eta x f_2(\eta, x) + \eta^2 f_3(\eta, x). \quad (6.1.12)$$

(This is an example of either of the monomial divisions theorems, Theorems A.2.10 and A.2.16 in Appendix A; see Remarks A.2.11 and A.2.17.) Consider $\eta > 0$, let $\mu = \sqrt{\eta} > 0$, and define y by $x = \mu y$. Then

$$a(\eta, x) = a(\mu^2, \mu y) = \mu^2 g(\mu, y),$$

where

$$g(\mu, y) = a_{02}y^2 + a_{10} + \mu[y^3 f_1(\mu^2, \mu y) + y f_2(\mu^2, \mu y) + \mu f_3(\mu^2, \mu y)].$$

Notice that for $\eta > 0$, we have $a(\eta, x) = 0$ if and only if $g(\mu, y) = 0$. We will apply the implicit function theorem to the latter equation to solve for y as a function of μ (even though the implicit function theorem cannot be used to solve the former equation for x as a function of η). Let y_+ and y_- be defined by

$$y_\pm = \pm\sqrt{-\frac{a_{10}}{a_{02}}}.$$

Then $g(y_\pm, 0) = 0$ and $g_y(y_\pm, 0) = 2a_{02}y_\pm \neq 0$. So there exist two smooth functions $\widehat{y}_\pm(\mu)$, defined for μ near zero, such that $\widehat{y}_\pm(0) = y_\pm$ and $g(\widehat{y}_\pm(\mu), \mu) = 0$. (These functions are defined even for negative μ near

zero, although such values have no relation to the original problem.) Then
the functions

$$\widehat{x}_\pm(\eta) = \eta^{1/2}\widehat{y}_\pm(\eta^{1/2}), \qquad (6.1.13)$$

defined for small $\eta > 0$, provide the two supercritical branches of rest
points. Notice that the functions $\widehat{x}_\pm(\eta)$ given by (6.1.13) are not differentiable at $\eta = 0$. Since the $\widehat{y}_\pm(\mu)$ are smooth, they can be expanded in
formal power series in μ; this allows (6.1.13) to be expanded in a "fractional
power series" (or Puiseux series) in η, that is, a formal power series in $\eta^{1/2}$.

This establishes the "existence part" of the theorem; it remains to prove
that there are no other rest points near $x = 0$ for small η. (In the first
proof, this did not require a separate argument, but followed directly from
the implicit function theorem applied "to the wrong variable.")

The Malgrange preparation theorem (see Section A.7), together with
$a(0,0) = a_x(0,0) = 0$ and $a_{xx}(0,0) \neq 0$, implies that there exists a smooth
function $u(\eta, x)$ that is nonzero in a neighborhood of $(0,0)$ and satisfies

$$u(\eta, x)a(\eta, x) = p(\eta, x) = x^2 + q_1(\eta)x + q_0(\eta), \qquad (6.1.14)$$

where q_1 and q_2 are smooth functions of η. Notice that $p(\eta, x)$ is a polynomial in x, but is smooth only in η. Since u is nonzero near $(0,0)$, the level
sets $a = 0$ and $p = 0$ coincide near the origin. For each η, there are at most
two (real) roots of the quadratic polynomial equation $p(\eta, x) = 0$, so there
are at most two branches to the solution of $a(\eta, x) = 0$ locally. For $\eta > 0$,
we have exhibited these two branches in (6.1.13). Therefore, there do not
exist any other rest points near $x = 0$ for small $\eta > 0$.

For small $\eta < 0$, we wish to show that $a(\eta, x)$ (which, we recall, is defined
only for real x) has no zeros near $x = 0$. If it did, then $p(\eta, x) = 0$ would
have real roots for $\eta < 0$. So it suffices to prove that the two roots of the
quadratic polynomial $p(\eta, x)$ are nonreal when $\eta < 0$. Let $c = u(0,0) \neq 0$;
then $u(\eta, x) = c + \mathcal{O}(x) + \mathcal{O}(\eta)$ and

$$u(\eta, x)a(\eta, x) = ca_{02}x^2 + ca_{10}\eta + \mathcal{O}(x^3) + \mathcal{O}(\eta x^2) + \mathcal{O}(\eta^2).$$

It follows from (6.1.14) that $q_0(0) = q_1(0) = 0$ and that $q_0'(0) = a_{10}/a_{02} <$
0. Then the discriminant of the quadratic equation $p(\eta, x) = 0$ is $q_1(\eta)^2 -
4q_0(\eta) = 14q_0'(0)\eta + \mathcal{O}(\eta^2)$, which is negative for small $\eta < 0$, and the proof
is complete.   □

> **6.1.8. Remark.** The ad hoc argument at the end of this proof, us-
> ing the quadratic discriminant, can be replaced by a more abstract
> argument that works in many similar situations. This argument is
> given at the end of Section A.7 in Appendix A. With this argument
> in hand, we could have completed the second proof of Theorem 6.1.6
> simply by observing that the formal expansion of (6.1.13) has a non-
> real leading term when $\eta < 0$. But since this argument is roundabout
> and involves ideas that may not be familiar, we will continue to offer
> ad hoc arguments for the nonexistence of branches in the theorems

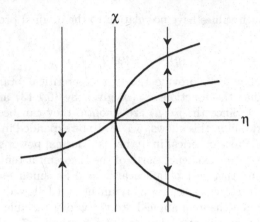

Figure 6.4. Supercritical pitchfork bifurcation.

of this section. The next section uses a different approach in which the preparation theorems are not necessary.

The Pitchfork Bifurcation

The next bifurcation theorem goes with the Newton diagram

$$
\begin{array}{lll}
\bullet & & \\
0 & & \\
0 & \bullet & \\
0 & 0 & \bullet
\end{array}
\tag{6.1.15}
$$

6.1.9. Theorem (Pitchfork). If $a_{00} = a_{01} = a_{02} = a_{10} = 0$ and a_{03} and a_{11} are nonzero, then (6.1.1) undergoes a pitchfork bifurcation, supercritical if a_{03} and a_{11} have opposite signs, subcritical otherwise. (See Figure 6.4 for the supercritical case.) The two outer rest points (on the side with three branches) are at a distance of strict order $\mathcal{O}(|\eta|^{1/2})$ from the origin. If also $a_{20} \neq 0$, the central rest point is at a distance strictly $\mathcal{O}(\eta)$ from the origin; otherwise, it is closer.

Proof. Assume that a_{03} and a_{11} have opposite signs; we will establish a supercritical bifurcation. (The subcritical case can be handled similarly.) By the monomial division theorem,

$$
\begin{aligned}
a(\eta, x) = {} & a_{03}x^3 + a_{11}\eta x + a_{20}\eta^2 + x^4 f_1(\eta, x) + \eta x^2 f_2(\eta, x) \\
& + \eta^2 x f_3(\eta, x) + \eta^3 f_4(\eta, x).
\end{aligned}
$$

For $\eta > 0$, introduce $\mu = \sqrt{\eta} > 0$ and $x = \mu y$; then

$$
a(\eta, x) = \mu^3 g(\mu, y)
$$

with

$$g(\mu, y) = a_{03}y^3 + a_{11}y + \mathcal{O}(\mu).$$

(Notice that a_{20} does not appear in the leading order, so we need not assume $a_{20} \neq 0$ for this part of the argument. Also observe that all of the degeneracy conditions stated in the theorem are used here.) Then $g(y, 0) = 0$ has three simple roots ("starting solutions"), which we write using superscript $+$ to denote $\eta > 0$ and subscripts $+, -, 0$ to distinguish the roots:

$$y_0^+ = 0, \qquad y_+^+ = \sqrt{\frac{-a_{11}}{a_{03}}}, \qquad y_-^+ = \sqrt{\frac{-a_{11}}{a_{03}}},$$

with $g_y(y_0^+, 0) = a_{11} \neq 0$ and $g_y(y_\pm^+, 0) = -2a_{11} \neq 0$. The implicit function theorem implies the existence of smooth functions $\widehat{y}_0$ and $\widehat{y}_\pm$ of μ, defined for all small μ, taking these starting values at $\mu = 0$ and with graphs lying in the level set $g = 0$. Notice that $\widehat{y}_\pm^+(\mu) = y_\pm + \mathcal{O}(\mu)$ has a nonzero constant term, whereas for $\widehat{y}_0^+(\mu) = \mathcal{O}(\mu)$ we do not know the exact leading order. Remembering $x = \mu y$, we have established rest points for $\eta > 0$ given by

$$\widehat{x}_\pm^+(\eta) = \eta^{1/2} y_\pm + \mathcal{O}(\eta),$$
$$\widehat{x}_0^+(\eta) = \mathcal{O}(\eta).$$

Notice that the strict order (the order of the leading term) of $\widehat{x}_\pm^+$ is $\eta^{1/2}$, but that of $\widehat{x}_0^+$ is not determined. Although the functions $\widehat{y}^+$ are defined for all small μ, these $\widehat{x}^+$ are defined only for $\eta > 0$. To study $\eta < 0$ we set $\mu = \sqrt{-\eta} > 0$, or $\eta = -\mu^2$, and $x = \mu y$, obtaining $a(\eta, x) = \mu^3(a_{03}y^3 - a_{11}y) + \mathcal{O}(\mu^4)$; the argument runs as before, except that now the starting solutions $y_\pm^-$ are rejected because they are imaginary, and only $y_0^- = 0$ is used. In this way we obtain the existence of one rest point for $\eta < 0$. It remains to prove the nonexistence of other rest points (on either side).

Before doing this, we observe that the treatment of the "continuing" rest point just given is somewhat unsatisfactory, both because it does not reveal the leading term and because the cases $\eta > 0$ and $\eta < 0$ are treated separately. If there is truly a "continuing" rest point, it should be expressible as a function of η in a full neighborhood of zero. It is possible to remedy both defects, but only by making an assumption about the order of the first nonzero term on the η axis of the Newton diagram. Therefore, we now add the assumption that $a_{20} \neq 0$, and define y by $x = \eta y$. Then $a(\eta, x) = \eta^2 g(y, \eta)$ with $g(y, \eta) = a_{11}y + a_{20} + \mathcal{O}(\eta)$, with starting solution $y_0 = -a_{21}/a_{11} \neq 0$. The implicit function theorem now yields $\widehat{y}_0(\eta) = y_0 + \mathcal{O}(\eta)$, so

$$\widehat{x}_0(\eta) = y_0\eta + \mathcal{O}(\eta^2),$$

which is of strict order η and is defined for all η near zero.

Turning to the nonexistence of other rest points, the Malgrange preparation theorem implies the existence of a smooth function $u(\eta, x)$, nonzero near $(0,0)$, such that

$$u(\eta, x)a(\eta, x) = p(\eta, x) = x^3 + q_2(\eta)x^2 + q_1(\eta)x + q_0(\eta)$$

is a cubic polynomial in x with coefficients that are smooth functions of η. Therefore, there is a maximum of three rest points for each η in the neighborhood where $u \neq 0$. For $\eta > 0$, we have found all three, and the proof that there are no others is complete. For $\eta < 0$ we have found one, and it remains to prove that the other roots of $p = 0$ are nonreal. The necessary calculation was essentially done when we investigated the case $\eta < 0$ and found that $y_{\pm}^-$ were imaginary; the only difficulty is that we did this calculation for a rather than p, and since a is defined only for real values of x, the fact that the $y_{\pm}^-$ are imaginary meant only that we *failed* to establish real roots for a, whereas instead we must now *succeed* at establishing that the roots of p are nonreal. We can either appeal to the general argument mentioned in Remark 6.1.8, which says that the asymptotic series solutions of $p = 0$ and $a = 0$ are the same, or we can compute the necessary terms of p and repeat the construction of the solution. To take the latter course, write $u(\eta, x) = c + \mathcal{O}(x) + \mathcal{O}(\eta)$ with $c \neq 0$, and multiply this by $a(\eta, x)$, observing that the effect on the matrix of coefficients (the Newton diagram) is to leave the zero entries as they are; then multiply the entries a_{03}, a_{11}, and a_{20} by c, making $ca_{03} = 1$, while eliminating all entries above the fourth row and all entries in the fourth row after the first, and modifying the unmarked entries in the first three rows in an undetermined manner. The new problem $p = 0$ is now well-defined for complex x, and when we repeat the analysis for $\eta < 0$ (using $\eta = -\mu^2$ and $x = \mu y$), the fact that $y_{\pm}^-$ is imaginary now proves that these roots of $p = 0$ are nonreal. (The argument requires an implicit function theorem for maps $g : \mathbb{C} \times \mathbb{R} \to \mathbb{C}$ that are polynomials, or at least analytic, in the complex variable and smooth in the real variable. The derivative $g_y(y, \mu)$ is the complex derivative with respect to y. This can be reformulated using the real implicit function theorem if $\mathbb{C}$ is replaced by $\mathbb{R}^2$.) $\square$

6.1.10. Remark. Part of the utility of Newton diagrams lies in the fact that they reveal the proper "scalings" for the application of the implicit function theorem. In the Newton diagram for the last theorem, the line joining the points a_{03} and a_{11} on the Newton diagram has slope $-\frac{1}{2}$; this indicates that if we set $x = \eta^{1/2}y$, the terms a_{03} and a_{11} will come into balance (at the same fractional power of η) and will dominate the term a_{20} (which lies above this line in the diagram). Similarly, the line joining a_{11} and a_{20} has slope -1, and the scaling $x = \eta y$ makes the terms a_{11} and a_{20} balance and dominate a_{03}, which now lies above the line. The rule is that a slope of $-p/q$ suggests the scaling $x = \eta^{p/q}y$. A "Newton polygon" is a broken line connecting some or all of the solid dots in the Newton diagram, which joins the two axes and is concave upwards; the lattice points below the Newton polygon must be low terms (equal to zero). Each

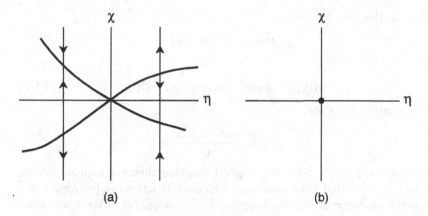

Figure 6.5. Transcritical bifurcation (with $a_{02} > 0$) and isola.

line segment gives a scaling that coincides with the strict order (in η) of some of the roots, and the use of this scaling will reveal the leading-order term of the root and also enable the use of the implicit function theorem. It is unusual that in the proof of the last theorem, the $\eta^{1/2}$ scaling was successful (with regard to the implicit function theorem) for all three roots; notice that this scaling was not successful for obtaining the leading term of the middle root.

The Transcritical Bifurcation and the Isola

The next Newton diagram to be considered is

$$\begin{matrix} \bullet & & \\ 0 & \bullet & \\ 0 & 0 & \bullet \end{matrix} \qquad (6.1.16)$$

6.1.11. Theorem (Transcritical/isola). Suppose that $a(\eta, x)$ satisfies the degeneracy conditions $a_{00} = a_{10} = a_{01} = 0$ and the nondegeneracy conditions $a_{02} \neq 0$ and $\Delta = a_{11}^2 - 4a_{02}a_{20} \neq 0$. (Note that this nondegeneracy condition involves the three intermediate terms on the Newton diagram, but is no longer simply the condition that these terms be nonzero.) Then the rest point at the origin for $\eta = 0$ either undergoes a *transcritical bifurcation* (if $\Delta > 0$) or is an *isola* (if $\Delta < 0$); that is, in the former case there exist two rest points $\hat{x}_{\pm}(\eta)$, defined for small η and crossing at $\eta = 0$, having strict order $\mathcal{O}(\eta)$. In the latter case the rest point disappears for small $\eta \neq 0$. See Figure 6.5.

Proof. Since a line joining the three intermediate terms on the Newton diagram has slope -1, there is only one scaling needed, namely $x = \eta y$, valid on both sides of $\eta = 0$. After the usual appeal to the monomial division theorem, we have

$$a(\eta, x) = \eta^2 g(\eta, y)$$

with

$$g(\eta, y) = a_{02}y^2 + a_{11}y + a_{20} + \mathcal{O}(\eta). \tag{6.1.17}$$

The starting roots are

$$y_\pm = \frac{-a_{11} \pm \sqrt{\Delta}}{2a_{02}}.$$

If these are real ($\Delta > 0$), the implicit function theorem implies (because $g_y(\eta_\pm, 0) \neq 0$) that they continue as smooth functions $\widehat{y}_\pm(\eta)$ for small η, proving existence of the two branches $\widehat{x}_\pm(\eta) = \eta\widehat{y}_\pm(\eta)$ of the transcritical bifurcation, and showing that they are strictly $\mathcal{O}(\eta)$. It remains to prove nonexistence of other branches. The Malgrange preparation theorem says that there is a function $u(\eta, x)$, nonzero near the origin, such that $ua = p$ is a quadratic polynomial in x. Thus, there are at most two roots, locally, and if $\Delta > 0$, we have found them. If $\Delta < 0$, we claim that the roots of p are nonreal. Because $u(\eta, x) = c + \mathcal{O}(x) + \mathcal{O}(\eta)$ with $c \neq 0$, the dominant terms in the Newton diagram of p are the same (up to a constant factor) as those of a, although p is now meaningful for complex x (because its upper rows have been removed). But the starting values for the roots of $p = 0$ are the same $y_\pm$ calculated above, and these are nonreal if $\Delta < 0$.   □

Secondary Newton Diagrams

Theorem 6.1.11 contains a nondegeneracy condition $\Delta \neq 0$ that is not expressible simply as the existence of a specific nonzero term in the Newton diagram. If we ask what happens when this nondegeneracy condition is violated, we are imposing a new kind of degeneracy condition, one that cannot be expressed as the vanishing of a term in the Newton diagram. The next theorem illustrates the classical approach to such problems, which entails the creation of secondary (and perhaps higher-order) Newton diagrams.

The (primary) Newton diagram associated with the next theorem is

$$
\begin{array}{cccc}
\bullet & & & \\
\bullet & \bullet & & \\
0 & \bullet & \bullet & \\
0 & 0 & \bullet & \bullet
\end{array}
\tag{6.1.18}
$$

Recall that for us, a (primary) Newton diagram is a partition of the terms into three classes, low, intermediate, and high. There are three low terms in this diagram, which are assumed to be zero. There are seven intermediate terms, which must be known and must satisfy the hypotheses of Theorem 6.1.12, which include both degeneracy and nondegeneracy conditions on the intermediate terms. The high terms are not mentioned in the theorem,

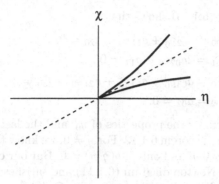

Figure 6.6. Cusp bifurcation. (The dotted line has slope y_0.)

and need not be known. In the course of proving the theorem a function $b(\eta, u)$ is introduced. The (primary) Newton diagram of this function is called the *secondary Newton diagram* of $a(\eta, x)$.

6.1.12. Theorem (Cusp). Suppose that $a(\eta, x)$ satisfies the degeneracy conditions

$$a_{00} = a_{10} = a_{01} = 0 \quad \text{and} \quad \Delta = a_{11}^2 - 4a_{02}a_{20} = 0,$$

and the nondegeneracy conditions

$$a_{02} \neq 0 \quad \text{and} \quad a_{03}y_0^3 + a_{12}y_0^2 + a_{21}y_0 + a_{30} \neq 0,$$

where

$$y_0 = -\frac{a_{11}}{2a_{02}}.$$

Then the bifurcation is a *cusp*, with two rest points bifurcating from the origin for $\eta > 0$ and none for $\eta < 0$. See Figure 6.6.

Proof. The argument begins as in the proof of Theorem 6.1.11, but since $\Delta = 0$, the two roots $y_\pm = -a_{11}/2a_{02}$ are not distinct, and we designate this value as y_0. Since $g(0, y)$ now has a double root at y_0, $g_y(0, y_0) = 0$, the implicit function theorem no longer applies. To analyze this situation requires a more detailed calculation of g, replacing (6.1.17) by

$$g(\eta, y) = a_{02}y^2 + a_{11}y + a_{20} \tag{6.1.19}$$
$$+ \eta \left[a_{03}y^3 + a_{12}y^2 + a_{21}y + a_{30} \right] + \mathcal{O}(\eta^2).$$

To center the root y_0 at zero (in a new variable), introduce u by

$$y = y_0 + u,$$

and define

$$b(\eta, u) = g(\eta, y_0 + u) = \sum b_{ij}\eta^i u^j.$$

A calculation using (6.1.19) shows that

$$b_{00} = a_{02}y_0^2 + a_{11}y_0 + a_{20} = 0,$$
$$b_{01} = 2a_{02}y_0 + a_{11} = 0,$$
$$b_{10} = a_{03}y_0^3 + a_{12}y_0^2 + a_{21}y_0 + a_{30} \neq 0,$$
$$b_{02} = a_{02} \neq 0.$$

The first two are zero by the properties of y_0, and the last two are nonzero by the hypotheses of Theorem 6.1.12. For $\eta \neq 0$, we know that $a(\eta, x) = 0$ if and only if $g(\eta, y) = 0$, if and only if $b(\eta, u) = 0$. But our calculations show that $b(\eta, u)$ has the Newton diagram (6.1.11) and satisfies the conditions of Theorem 6.1.6. It follows that b has two roots $\widehat{u}_\pm(\eta)$ for $\eta > 0$ and one for $\eta < 0$, and that these roots are of strict order $\eta^{1/2}$. Since $x = \eta y = \eta(y_0+u)$, the roots $\widehat{x}_\pm(\eta)$ are of strict order η, but the two roots are equal to that order; the next order in their asymptotic expansion is $\eta^{3/2}$, and this term distinguishes the roots. $\square$

Notes and References

This section is based on the presentation in Vainberg and Trenogin [107], but with the Weierstrass preparation theorem replaced by the Malgrange version. The definition of Newton diagram used in this section is influenced by the definition of lower-order, intermediate-order, and higher-order terms in Golubitsky and Schaeffer [50]. A good reference for classical Newton diagrams in the complex case is Hille [58], volume 2, chapter 12.

6.2   Bifurcations from a Single-Zero Eigenvalue: A "Modern" Approach

In this section, we continue the study of the bifurcation of rest points of a scalar differential equation

$$\dot{x} = a(\eta, x) \tag{6.2.1}$$

containing a small parameter. The final results obtained will be similar to those in the last section, but the method is entirely different. Our experience in Section 6.1 suggests that the function $a(\eta, x)$ can often be broken into two parts, which we will write as

$$a(\eta, x) = g(\eta, x) + h(\eta, x), \tag{6.2.2}$$

such that g consists of low-order and intermediate-order terms that must be known in order to understand the bifurcation, while h consists of higher-order terms that, while they modify the results in small numerical ways, do not affect the qualitative behavior of the system, and therefore do not need to be known if only the qualitative behavior is sought. (In fact, g contains only what we have called intermediate terms, since the low terms are those that are required to be zero in the situation being considered. An example is (6.1.12), with $g(\eta, x) = a_{20}x^2 + a_{01}\eta$ and $h(\eta, x) = x^3 f_1 + \eta x f_2 + \eta^2 f_3$.) In Section 6.1, we proved the unimportance of h in each case by studying the zero set of a (with h included), and showing that the qualitative results did not depend on p. In the present section, we show directly that h can be removed, before even beginning to study the behavior of the bifurcation. Once h is gone, the bifurcation of the roots of g is usually obvious.

6.2.1. Definition. Given smooth functions g and h of two real variables x and η, we say that h is *removable* from $a = g + h$ if there exist smooth functions Y and S such that the following conditions are satisfied for all (η, x) in some neighborhood of the origin:

$$S(\eta, x)a(\eta, Y(\eta, x)) = g(\eta, x),$$
$$Y(0,0) = 0,$$
$$Y_x(\eta, x) > 0,$$
$$S(\eta, x) > 0.$$

To understand the significance of the definition, notice that since $Y_x(\eta, x) > 0$ near the origin, $y = Y(\eta, x)$ is a valid coordinate change. Since $Y(0,0) = 0$, this coordinate change does not move the origin. Finally, since $S(\eta, x) > 0$ near the origin, $g(\eta, x) = 0$ if and only if $a(\eta, y) = 0$. Therefore, the zero sets of g and a are qualitatively the same near the origin, and the removal of h does not change the nature of the bifurcation. It is to be stressed that "removable" means removable only with regard to determining the structure of the zero set, *not with regard to determining the dynamics*. In particular, the behavior of the solutions of the differential equation $\dot{x} = g(\eta, x) + h(\eta, x)$ is not expected to be the same as that of $\dot{x} = g(\eta, x)$.

The problem, then, is to find circumstances under which h is removable, and to find a way to prove this fact. From our experience in Section 6.1, we expect theorems that take the following form: If certain specified low terms are zero, and certain specified intermediate terms satisfy appropriate nondegeneracy conditions, then the remaining terms will be removable, which justifies counting them as high terms.

Motivating the Method to Be Used

To prove theorems of this type, we will use what is sometimes called an *invariant embedding method*. That is, the problem of removing h from $a = g + h$ will be embedded into a one-parameter family of problems, namely, removing th from

$$G(\eta, x, t) = g(\eta, x) + th(\eta, x) \tag{6.2.3}$$

for $0 \leq t \leq 1$. For $t = 0$, this problem is trivial: There is nothing to remove. For $t = 1$, the problem is equivalent to the original problem. For values of t between 0 and 1, the problem is (we imagine) intermediate in difficulty, and we attempt to solve the problem for $t = 1$ by working our way up starting from $t = 0$. Thus, we pose the problem of finding $Y(\eta, x, t)$ and $S(\eta, x, t)$ such that

$$S(\eta, x, t)G(\eta, Y(\eta, x, t), t) = g(\eta, x). \tag{6.2.4}$$

Since nothing needs to be done when $t = 0$, the natural choices for Y and S at $t = 0$ are

$$Y(\eta, x, 0) = x, \tag{6.2.5}$$
$$S(\eta, x, 0) = 1.$$

Since the origin is never to be moved, we also require

$$Y(0, 0, t) = 0. \tag{6.2.6}$$

Embedding methods are commonly used in applied mathematics as numerical methods. For instance, a boundary value problem may be embedded in a family of boundary value problems that can then be reformulated as an initial value problem in a different variable, and solved numerically as such. In the present case we are not interested in a numerical solution, but the idea is the same: Equations (6.2.4), (6.2.5), and (6.2.6) will be converted into an initial value problem for a differential equation in t (which is not to be confused with the original time variable in (6.2.1). The existence theorem for solutions of initial value problems will then be invoked to demonstrate the existence of Y and S on the interval $0 \leq t \leq 1$; this existence at $t = 1$ will then imply the removability of h. Of course, the whole procedure will work only when g satisfies certain hypotheses.

To obtain the initial value problem, notice that the right-hand side of (6.2.4) is independent of t. Supposing, then, that $S > 0$ and Y satisfying (6.2.4) exist, it follows that

$$\frac{d}{dt} S(\eta, x, t)G(Y(\eta, x, t), \eta, t) = 0,$$

or

$$S_t G + S G_x Y_t + S G_t = 0.$$

Define functions α and δ by

$$\alpha(\eta, x, t) = -\frac{S_t(\eta, x, t)}{S(\eta, x, t)}, \tag{6.2.7}$$
$$\delta(\eta, x, t) = -Y_t(x, \sigma, t).$$

Then, using $G_t = p\eta$, we have

$$S[-\alpha G - \delta G_x + h] = 0,$$

which, since $S > 0$, implies

$$h = \alpha G + \delta G_x.$$

According to (6.2.6), $Y(0,0,t) = 0$, and it follows that $\delta(0,0,t) = -Y_t(0,0,t) = 0$. Therefore, $\delta(\eta, x, t)$ is small for small η and x, and according to a variant of Lemma A.2.5 it can be written

$$\delta(\eta, x, t) = \beta(\eta, x, t)\eta + \gamma(\eta, x, t)x.$$

Therefore, we finally have

$$h = \alpha G + \beta \eta G_x + \gamma x G_x. \tag{6.2.8}$$

The argument just given can be considered (in classical Greek style) as the "analysis phase" of our problem. That is, we have supposed that the solution exists, and derived some consequences. The actual solution of the problem takes place in reverse, in the "synthesis phase." From what has just been written, we see that the starting point will be to write h in the form (6.2.8), for certain functions α, β, γ. It is here that the specific hypotheses of each bifurcation type will play a role: The degeneracy and nondegeneracy conditions will determine what the higher-order terms are that can be written in such a form, and so then be removed. To establish the removability of the terms, we must then show the solvability of the differential equations

$$S_t(\eta, x, t) + \alpha(\eta, x, t)S(\eta, x, t) = 0,$$
$$Y_t(\eta, x, t) + \beta(\eta, x, t)\eta + \gamma(\eta, x, t)x = 0, \tag{6.2.9}$$

and prove that these solutions have the required properties $S > 0$ and $Y_x > 0$.

The Basic Theorems

We now start over, motivated by the discussion just given. Given a function $g(\eta, x)$, our goal is to determine the class of functions $h(\eta, x)$ such that h is removable from $g + h$. From this point of view there is something unsatisfactory about condition (6.2.8), because we would like the removable terms h to be characterized in terms of g, but (6.2.8) characterizes them in terms of $G = g + th$. In other words, h actually appears on both sides of (6.2.8), although it is not visible on the right-hand side. Therefore, we begin by looking for conditions on h, expressed in terms of g, that imply (6.2.8). Two such conditions are developed in Lemma 6.2.3 and Corollary 6.2.5; another such condition is given later in Corollary 6.2.9. These results convince us that (6.2.8) is at least sometimes verifiable. Knowing this, we will then show in Theorem 6.2.6 that (6.2.8) implies that h is removable.

Before proceeding, it is worthwhile to develop some notation. All functions that we consider will either be smooth functions of (η, x) defined in a neighborhood of $(0,0)$, or will belong to the class of smooth functions $f(\eta, x, t)$ defined on a neighborhood of the line segment $\{(0,0,t) : 0 \leq t \leq 1\}$ in $\mathbb{R}^3$. (Functions that do not depend upon t can be considered to depend trivially on t, when it is useful to consider them as members of the latter class.) Given functions $f_1, \ldots, f_r$ in one of these classes, the set of linear combinations $u_1 f_1 + \cdots + u_r f_r$ (where $u_1, \ldots, u_r$ are also functions in the same class) is called the *ideal* generated by $f_1, \ldots, f_r$, and will be denoted by $\langle f_1, \ldots, f_r \rangle$. Of particular importance is the ideal

$$\mathcal{M} = \langle \eta, x \rangle.$$

The *product* of $\mathcal{M}$ and $\langle f_1, \ldots, f_r \rangle$ is the ideal

$$\mathcal{M}\langle f_1, \ldots, f_r \rangle = \langle \eta f_1, x f_1, \eta f_2, x f_2, \ldots, \eta f_r, x f_r \rangle.$$

In particular, $\mathcal{M}^2 = \langle \eta^2, \eta x, x^2 \rangle$, and more generally, $\mathcal{M}^k$ consists of functions that are homogeneous polynomials of total degree k in η and x (for any fixed t, when the function depends on t). In this notation, (6.2.8) can be written

$$h \in \langle G, \eta G_x, x G_x \rangle, \tag{6.2.10}$$

where $G = g + th$.

6.2.2. Remark. More technically, if we identify two functions when they are equal in a neighborhood of the line segment $(0,0,t)$, the resulting set is a ring (of what might be called "semigerms" of functions) and $\langle f_1, \ldots, f_r \rangle$ is an ideal, as discussed in Appendix A. Section A.2 is especially relevant to the issues that will be discussed here. $\mathcal{M}$ is not a maximal ideal in this ring, but is the maximal ideal in the ring $\mathcal{E}^2$ of germs (at the origin) of functions of η and x; our functions belong to this ring when t is fixed. The definition of $\mathcal{M}\langle f_1, \ldots, f_r \rangle$ is a special case of (A.2.8).

The next lemma shows that (6.2.16) holds whenever the terms of h are "high enough." Specifically, it tells us that if h and two other functions (ηh_x and $x \eta_x$) derived from h have their terms far enough out from the origin in the Newton diagram that they are expressible as certain linear combinations of other functions (spelled out in (6.2.12) below), then (6.2.10) holds. Define the *restricted tangent space* of g to be the ideal

$$RT(g) = \langle g, \eta g_x, x g_x \rangle. \tag{6.2.11}$$

6.2.3. Lemma. If h satisfies the condition that h, ηh_x, and $x h_x$ belong to $\mathcal{M}RT(g) = \mathcal{M}\langle g, \eta g_x, x g_x \rangle$, then for each fixed t in $0 \leq t \leq 1$, h belongs to $\langle G, \eta G_x, x G_x \rangle$, where $G = g + th$.

Proof. The hypothesis is equivalent to the existence of functions $a_{ij} \in \mathcal{M}$ such that

$$h = a_{11}g + a_{12}\eta g_x + a_{13}xg_x, \tag{6.2.12}$$
$$\eta h_x = a_{21}g + a_{22}\eta g_x + a_{23}xg_x,$$
$$x h_x = a_{31}g + a_{32}\eta g_x + a_{33}xg_x,$$

which can be written in matrix form as

$$\begin{bmatrix} h \\ \eta h_x \\ x h_x \end{bmatrix} = \begin{bmatrix} a_{11} & a_{12} & a_{13} \\ a_{21} & a_{22} & a_{23} \\ a_{31} & a_{32} & a_{33} \end{bmatrix} \begin{bmatrix} g \\ \eta g_x \\ x g_x \end{bmatrix}.$$

It follows from $G = g + th$ that

$$\begin{bmatrix} G \\ \eta G_x \\ x G_x \end{bmatrix} = \begin{bmatrix} 1 + ta_{11} & ta_{12} & ta_{13} \\ ta_{21} & 1 + ta_{22} & ta_{23} \\ ta_{31} & ta_{32} & 1 + ta_{33} \end{bmatrix} \begin{bmatrix} g \\ \eta g_x \\ x g_x \end{bmatrix}.$$

Since the entries a_{ij} of the matrix A are in $\mathcal{M}$, they are small for small η and x; the matrix $I + tA$ occurring in the last equation is invertible (for (η, x, t) in a neighborhood of the line segment $(0, 0, t)$), giving

$$\begin{bmatrix} g \\ \eta g_x \\ x g_x \end{bmatrix} = (I + tA)^{-1} \begin{bmatrix} G \\ \eta G_x \\ x G_x \end{bmatrix}.$$

Substituting the resulting expressions for g, ηg_x, and xg_x into the first line of (6.2.12) expresses h as an element of $\langle G, \eta G_x, x G_x \rangle$.   □

6.2.4. Remark. This lemma is a special case of what is called Nakayama's lemma, which applies to the ring of germs $\mathcal{E}^n$ discussed in Appendix A. One version of Nakayama's lemma states that if $q_1, \ldots, q_r \in \mathcal{M}\langle p_1, \ldots, p_r \rangle$, then

$$\langle p_1 + q_1, \ldots, p_r + q_r \rangle = \langle p_1, \ldots, p_r \rangle. \tag{6.2.13}$$

Inclusion in one direction is trivial, and in the other direction the proof is the same as the proof of our lemma.

The following corollary is a sufficient determination of high terms for some easy applications:

6.2.5. Corollary. If $\mathcal{M}^k \subset RT(g)$, then any $h \in \mathcal{M}^{k+1}$ satisfies $h \in \langle G, \eta G_x, x G_x \rangle$, where $G = g + th$.

Proof. Suppose $h \in \mathcal{M}^{k+1}$. Then there exist functions u_i, $i = 0, \ldots, k+1$, such that

$$h = \sum_{i=0}^{k+1} u_i \eta^{k+1-i} x^i.$$

It follows that

$$h_x = \left(\sum (u_i)_x \eta^{k+1-i} x^i \right) + \left(\sum i u_i \eta^{k+1-i} x^{i-1} \right).$$

The first sum in parentheses belongs to $\mathcal{M}^{k+1}$ as it is, and the second becomes a member of $\mathcal{M}^{k+1}$ when it is multiplied by either η or x. Thus, h, ηh_x, and $x h_x$ belong to $\mathcal{M}^{k+1} \subset \mathcal{M}\langle g, \eta g_x, x g_x \rangle$. By Lemma 6.2.3, it follows that $h \in \langle G, \eta G_x, x G_x \rangle$. □

Finally, we turn to the proof that (6.2.8) implies removability.

6.2.6. Theorem. Suppose that (6.2.8) or (6.2.10) holds, that is, that

$$h \in \langle g + th, \eta(g_x + th_x), x(g_x + th_x) \rangle$$

for each t with $0 \le t \le 1$. Then h is removable from $g + h$ in the sense of Definition 6.2.1.

Proof. By the hypothesis of the theorem, there exist smooth functions $\alpha(\eta, x, t)$, $\beta(\eta, x, t)$, and $\gamma(\eta, x, t)$ such that

$$h = \alpha G + \beta \eta G_x + \gamma x G_x,$$

where $G = g + th$; these functions are defined on a neighborhood of the line segment $(0, 0, t)$, $0 \le t \le 1$, and this neighborhood contains some tube $|\eta| < \eta_0$, $|x| < c_0$, $t_1 \le t \le t_2$ (where $t_1 < 0$ and $t_2 > 1$). Consider the following differential equation containing the parameter η:

$$\frac{dy}{dt} + \beta(\eta, y, t)\eta + \gamma(\eta, y, t)y = 0, \tag{6.2.14}$$

with initial condition $y = x$ at $t = 0$. By the fundamental existence theorem for differential equations, this has a unique solution $y = Y(\eta, x, t)$ satisfying $Y(\eta, x, 0) = x$, defined for $|\eta| < \eta_0$, $|x| < c_0$, and for t in some interval $0 \le t \le t_0(\eta, x)$, where t_0 is a continuous function of η and x. For $\eta = 0$ and $x = 0$, this solution is $Y(0, 0, t) = 0$, and it exists on $t_1 \le t \le t_2$. It follows that for sufficiently small η and x, $Y(\eta, x, t)$ is defined for $0 \le t \le 1$. (It may be necessary to reduce the domain to $|\eta| \le \eta_1$, $|x| < c_1$, with $0 < \eta_1 < \eta_0$ and $0 \le c_1 < c_0$.)

Now consider the differential equation

$$\frac{ds}{dt} + \alpha(\eta, Y(\eta, x, t), t)s = 0 \tag{6.2.15}$$

with initial condition $s = 1$ when $t = 0$. In this equation both η and x are parameters. Since the equation is linear and is defined for all t, the solution $s = S(\eta, x, t)$ is defined for all t.

It remains to check that these solutions satisfy $Y_x > 0$ and $S > 0$. For S this is simple: Equation (6.2.15) admits the trivial solution $s = 0$, and no other solution can cross this. So a solution having $s = 1$ for $t = 0$ remains positive. For Y_x the argument is similar, except that we differentiate (6.2.14) with respect to x to obtain

$$\frac{dY_x}{dt} + (\beta_x \eta + \gamma + \gamma_y Y)Y_x = 0.$$

The sum in parentheses may be regarded as a known function of t, given the solution for Y, so Y_x satisfies a linear equation of the form $du/dt + f(t)u = 0$ and cannot cross zero. Since $Y = x$ at $t = 0$, $Y_x = 1$ at $t = 0$ and it remains positive. $\qquad\square$

Easy Examples

To see how these theorems are used, we will re-prove parts of Theorems 6.1.5 and 6.1.11 concerning continuation and the transcritical/isola. Under the present approach, our concern is not to determine the behavior of the bifurcation, only the removability of the high terms in the Newton diagrams (6.1.10) and (6.1.16). Once the high terms have been removed, the behavior of the bifurcation is obvious in these simple examples, because the zero set is explicitly computable.

6.2.7. Theorem (Continuation). If $g(\eta, x) = a_{01}x + a_{10}\eta$ with $a_{01} \neq 0$, then all functions $h \in \mathcal{M}^2 = \langle \eta^2, \eta x, x^2 \rangle$ are removable from $g + h$. These are exactly the unmarked (high) terms of the Newton diagram (6.1.10).

Proof. By Corollary 6.2.5, it is sufficient to prove that $\mathcal{M} \subset RT(g)$. We will show that these two ideals are actually equal. Since $g_x = a_{01}$,

$$RT(g) = \langle g, \eta g_x, x g_x \rangle = \langle a_{10}\eta + a_{01}x, a_{01}\eta, a_{01}x \rangle$$
$$= \langle a_{10}\eta + a_{01}x, \eta, x \rangle = \langle \eta, x \rangle = \mathcal{M}.$$

To check this calculation, it is clear that any function of the form $u(a_{10}\eta + a_{10}x) + v(a_{01}\eta) + w(a_{01}x)$, where u, v, and w are functions of (η, x), can be rewritten as a linear combination of η and x. The converse is also true, because we have assumed $a_{01} \neq 0$. $\qquad\square$

6.2.8. Theorem (Transcritical/isola). If $g(\eta, x) = a_{02}x^2 + a_{11}\eta x + a_{20}\eta^2$ with $a_{02} \neq 0$ and $a_{11}^2 - 4a_{02}a_{20} \neq 0$, then all functions $h \in \mathcal{M}^3$ are removable from $g + h$. These are exactly the unmarked (high) terms of the Newton diagram (6.1.16).

Proof. By Corollary 6.2.5, it is sufficient to prove that $\mathcal{M}^2 \subset \langle g, \eta g_x, x g_x \rangle$, and again we will see that these are actually equal. For convenience we rewrite g as

$$g = ax^2 + b\eta x + c\eta^2.$$

It is clear that $\langle ax^2 + b\eta x + c\eta^2, 2a\eta x + b\eta^2, 2ax^2 + b\eta x \rangle \subset \langle x^2, \eta x, \eta^2 \rangle$; what is needed is to show the converse, that is, that given functions A, B, C there exist functions D, E, F such that

$$Ax^2 + B\eta x + C\eta^2 = D(ax^2 + b\eta x + c\eta^2)$$
$$+ E(2a\eta x + b\eta^2) + F(2ax^2 + b\eta x).$$

This will be the case if the system

$$A = aD + 2aF,$$
$$B = bD + 2aE + bF,$$
$$C = cD + bE,$$

is solvable for D, E, F, which will be the case if

$$\det \begin{bmatrix} a & 0 & 2a \\ b & 2a & b \\ c & b & 0 \end{bmatrix} = a(b^2 - 4ac) \neq 0.$$

This is exactly the nondegeneracy condition assumed in the theorem. $\square$

Harder Examples

Corollary 6.2.5 does not suffice to reprove Theorem 6.1.6 concerning the fold bifurcation or Theorem 6.1.9 concerning the pitchfork, because the high terms of their Newton diagrams (6.1.11) and (6.1.15) do not have the form $\mathcal{M}^{k+1}$ for some k. But Lemma 6.2.3 does suffice. It frequently happens, as in this example, that $\mathcal{M}RT(g)$ consists of removable terms, and it is perhaps worthwhile to isolate this as another corollary of Lemma 6.2.3, although the proof is entirely trivial.

6.2.9. Corollary. If the ideal $\mathcal{I} = \mathcal{M}RT(g)$ satisfies the condition

$$\text{if} \quad h \in \mathcal{I}, \quad \text{then} \quad \eta h_x \in \mathcal{I} \quad \text{and} \quad x h_x \in \mathcal{I}, \tag{6.2.16}$$

then any $h \in \mathcal{M}RT(g)$ is removable from $g + h$.

Proof. If $h \in \mathcal{I}$, then (6.2.16) implies that h, ηh_x, and $x h_x$ belong to $\mathcal{M}\langle g, \eta g_x, x g_x \rangle$. Lemma 6.2.3 then states that h is removable. $\square$

6.2.10. Theorem (The fold). If $g(\eta, x) = a_{02}x^2 + a_{10}\eta$ with $a_{02} \neq 0$ and $a_{10} \neq 0$, then any function h whose Taylor series contains only unmarked (high) terms of the Newton diagram (6.1.11) is removable from $g + h$.

Proof. For convenience we will write $a_{02} = a$ and $a_{10} = b$, so that $g(\eta, x) = ax^2 + b\eta$ with $a \neq 0$ and $b \neq 0$. Then $RT(g) = \langle ax^2 + b\eta, 2a\eta x, 2ax^2 \rangle = \langle \eta, x^2 \rangle$, so that $\mathcal{I} = \mathcal{M}RT(g) = \langle x^3, \eta x, \eta^2 \rangle$, which are exactly the unmarked terms of the Newton diagram. It remains to check that $\mathcal{I}$ satisfies (6.2.16). But if $h \in \mathcal{I}$, then $h = px^3 + q\eta x + r\eta^2$, where p, q, and r are functions. Then $h_x = (p_x x^3 + q_x \eta x + r_x \eta^2) + (3px^2 + q\eta)$. The terms belonging to the first set of parentheses already belong to $\mathcal{I}$, and the terms in the second set of parentheses belong to $\mathcal{I}$ after they are multiplied by either η or x. $\square$

6.2.11. Theorem (The pitchfork). If $g(\eta, x) = a_{03}x^3 + a_{11}\eta x + a_{20}\eta^2$ with $a_{03} \neq 0$ and $a_{11} \neq 0$, then any function h whose Taylor series contains only unmarked (high) terms of the Newton diagram (6.1.15) is removable from $g + h$.

Proof. The first step is to show that

$$RT(g) = \langle x^3, \eta x, \eta^2 \rangle. \tag{6.2.17}$$

It follows that $\mathfrak{J} = \mathcal{M}RT(g) = \langle x^4, \eta x^2, \eta^2 x, \eta^3 \rangle$, which are exactly the unmarked terms of the Newton diagram. The proof is completed by showing that $\mathfrak{J}$ satisfies condition (6.2.16).

For convenience, we write

$$g(\eta, x) = ax^3 + b\eta x + c\eta^2,$$

where a, b, and c are constants with $a \neq 0$ and $b \neq 0$ but with no restriction on c. Then

$$\langle g, \eta g_x, x g_x \rangle = \langle ax^3 + b\eta x + c\eta^2, 3a\eta x^2 + b\eta^2, 3ax^2 + b\eta x \rangle. \tag{6.2.18}$$

To prove (6.2.17), it is simplest to begin with $\langle x^3, \eta x, \eta^2 \rangle$ and build up to $\langle g, \eta g_x, x g_x \rangle$. First,

$$\langle x^3, \eta x, \eta^2 \rangle = \langle ax^3 + b\eta x + c\eta^2, b\eta^2, 3ax^3 + b\eta x \rangle; \tag{6.2.19}$$

this holds because

$$\det \begin{bmatrix} a & b & c \\ 0 & 0 & b \\ 3a & b & 0 \end{bmatrix} = 2ab^2 \neq 0.$$

Comparing (6.2.19) with (6.2.18), we see that only the term $3a\eta x^2$ is missing. But this term belongs to $\mathcal{M}\langle x^3, \eta x, \eta^2 \rangle$, and according to Nakayama's lemma in the form (6.2.13), elements of $\mathcal{M}\langle x^3, \eta x, \eta^2 \rangle$ can be added to any set of generators for $\langle x^3, \eta x, \eta^2 \rangle$ without changing the ideal.

To prove that $\mathfrak{J}$ satisfies (6.2.16), observe that any element of $\mathfrak{J}$ can be written

$$h = px^4 + q\eta x^2 + r\eta x + s\eta^2,$$

where p, q, r, s are functions. Then

$$h_x = (p_x x^4 + a_x x^2 + r_x \eta x + s_x \eta^2) + (4px^3 + 2q\eta x + r\eta + s\eta^2).$$

The first sum in parentheses already belongs to $\mathfrak{J}$, and the second becomes a member of $\mathfrak{J}$ when it is multiplied by either η or x. $\qquad\square$

The last example to be considered corresponds to Theorem 6.1.12 and Newton diagram (6.1.18). Our aim is to show that under the degeneracy and nondegeneracy conditions stated in that theorem, the terms lying above the solid dots in the Newton diagram are removable; these are the terms $\eta^i x^j$ with $i + j \geq 4$, that is, $\mathcal{M}^4$. But even though this set of terms has the form specified in Lemma 6.2.3, it is difficult or impossible to apply Lemma 6.2.3 to this end. It is best here to combine the methods of Sections 6.1 and 6.2. First, the transformations $x = \eta y$ and $y = y_0 + u$ introduced in the proof of Theorem 6.1.12 change the problem $a(\eta, x) = 0$ to $b(\eta, u) = 0$ with Newton diagram (6.1.11). Theorem 6.2.10 implies that $\eta^\ell u^k$ is removable from $b(\eta, u)$ if $2\ell + k > 0$. If $i + j \geq 4$, then substituting $x = \eta y_0 + \eta u$ and dividing by η^2 carries the term $\eta^i x^j$ in $a(\eta, x)$ into a sum of terms in $b(\eta, y)$ containing η^{i+j-2}. These terms are removable from $b(\eta, y)$, and we are finished.

Notes and References

This section is entirely based on parts of Golubitsky and Schaeffer [50], although the exposition is rather different. The book [50] begins with a notion of a "normal form" for a bifurcation, which (in this paragraph) we will call a *standard form* to distinguish it from the idea of normal form treated in this book. A standard form is a simple example of a function exhibiting a particular bifurcation. Given a standard form, the *recognition problem* for that standard form is to determine conditions under which an arbitrary function exhibits the same bifurcation. These conditions specify a set of low-order terms that must be zero, a set of intermediate-order terms that must be known, and a set of specific conditions that the intermediate-order terms must satisfy. An important part of the proof of these conditions is to show that the remaining (higher-order) terms are ignorable. Our exposition sets aside the idea of a standard form. Instead of asking that a function have the same bifurcation as a given standard form, we ask only that it have the same bifurcation (whatever that may be) as its own intermediate terms would have (if the high terms were deleted). This results in fewer conditions on the intermediate terms.

Certain terminology from [50] has been eliminated or simplified in our treatment. The reason for these choices is to simplify the presentation, not to claim that our method is better. Our Corollary 6.2.9 is definitely weaker than Theorem 8.7 in Chapter II of [50], but it is sufficient for our examples. It should be pointed out that still stronger theorems than those of [50] are given in Melbourne [77]. Another reference for this approach to bifurcation theory is Govaerts [51].

6.3   Unfolding the Single-Zero Eigenvalue

The topic to be studied in this section and the next was originally known as *imperfect bifurcation*. Afterward, it was recognized as an instance of the more general concept of *universal unfolding*. The version presented here differs in certain ways from both of these conceptions, but seems to be the most suitable way to organize the topic, both from the viewpoint of dynamics (in which truly universal unfoldings are usually impossible) and of algorithmic computability (which requires that all calculations be completed in finite time).

Imperfect Pitchfork Bifurcations

A pitchfork bifurcation is characterized by the Newton diagram (6.1.15), having $a_{00} = a_{01} = a_{02} = a_{10} = 0$. There are two reasons why a quantity may be zero: It may be *exactly* zero, which is usually the case for some theoretical reason; or it may be *approximately* zero, which often means merely that it is too small to measure. Which is the case with regard to the zeros in this Newton diagram? Suppose that we start with $\dot{x} = a(\eta, x)$, a vector field with one parameter η on a line ($x \in \mathbb{R}$). Nontrivial local dynamics occurs only near a rest point, so we assume that there is a rest point and that the origin is placed there. This justifies the assumption that $a_{00} = 0$ exactly. Next, nontrivial bifurcation occurs only if $a_{01} = 0$ exactly. (This is the content of Theorem 6.1.5.) There is no apparent theoretical justification for the assumption that a_{02} and a_{10} are exactly zero. (If it were known, in a specific application, that the vector field was odd in x, this would constitute a reason to assume $a_{02} = 0$.) This being the case, it is worthwhile to ask what happens if a_{02} and a_{10} are merely small, rather than exactly zero. It is in fact the case that experimentalists studying situations that were expected to produce pitchfork bifurcations as in Figure 6.4 observed behavior of a different kind instead; the explanation was that these terms were not exactly zero.

It might be objected that if a_{02} and a_{10} are not zero, then in fact the case falls under Theorem 6.1.6: There is a fold bifurcation, and nothing more to say. It is indeed true that Theorem 6.1.6 says that there will be a fold bifurcation, taking place in some neighborhood of the origin in (η, x) space. However, this theorem is for *fixed* a_{02} and a_{10}. If these quantities are varied, the size of this neighborhood will change, and if they approach zero, it will shrink. Somehow the fold bifurcation has to change into a pitchfork bifurcation at the moment that a_{02} and a_{10} become zero. Theorem 6.1.6 by itself is inadequate to explain this situation. This suggests that there is something more to be learned by studying the system

$$\dot{x} = (a_{03}x^3 + a_{11}\eta x + a_{20}\eta^2) + (\nu_1 x^2 + \nu_2 \eta). \tag{6.3.1}$$

The first sum in parentheses here represents the solid dots in the pitchfork Newton diagram (6.1.15). The second sum in parentheses contains two new small parameters ν_1 and ν_2, representing what might be called "grey dots" replacing two of the zeros in the diagram. Higher-order terms have been ignored; it must, of course, be shown, eventually, that this does not change the conclusions.

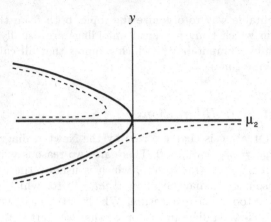

Figure 6.7. Graph of $y^3 + \mu_1 + \mu_2 y = 0$ with $\mu_1 = 0$ (perfect pitchfork, solid lines) and with $\mu_1 > 0$ (imperfect pitchfork, dotted lines).

The following quick analysis will be replaced by a more careful study below. By a change of time units, it may be assumed that $a_{03} = \pm 1$; we assume the case $+1$. It will be shown below (in the proof of Theorem 6.3.2) that the term $\nu_1 x^2$ can be removed by a small η-dependent shift of the origin, introducing a new variable y. The terms $a_{20}\eta^2$ and $\nu_2\eta$ are both small constants (independent of x) and can be replaced by a single small parameter μ_1. The quantity $a_{11}\eta$ can be replaced by a small parameter μ_2. This leads us to consider the system·

$$\dot{y} = y^3 + \mu_1 + \mu_2 y. \tag{6.3.2}$$

The cubic equation $y^3 + \mu_1 + \mu_2 y = 0$ has either one or three real roots for each choice of μ_1 and μ_2, and if $\mu_1 = 0$, it clearly exhibits a (subcritical) pitchfork bifurcation as μ_2 is varied. On the other hand, if μ_1 is slightly positive, then for $\mu_2 = 0$ there is a single (negative) root, which continues as μ_2 is varied, and in addition there is a (subcritical) fold bifurcation (as μ_2 is varied), occurring at a point (μ_2^*, x^*) close to the origin in the second quadrant; see Figure 6.7. Notice that what was the middle branch of the pitchfork has become the bottom branch of the fold, together with the right-hand portion of the continuing solution. These figures can be verified by considering the graph of $f(y) = y^3 + \mu_1 + \mu_2 y$ as μ_2 varies, for fixed μ_1. (If μ_1 is fixed at a small positive value, and μ_2 is decreased from zero, the stationary point of the graph of f splits into a maximum and a minimum, and the minimum soon crosses the x axis to product two new roots, the fold bifurcation.) The behavior shown in Figure 6.7 is typical of what is seen experimentally in many purported pitchfork bifurcations.

6.3.1. Remark. In passing from (6.3.1) to (6.3.2) we eliminated the parameter η in favor of two new small parameters μ_1 and μ_2. The original system can be recovered by allowing μ_1 and μ_2 to become functions of η (and also undoing the coordinate shift that eliminates the x^2 term). In some approaches to imperfect bifurcation and unfolding, the parameter η is regarded as having a physical meaning that is essential and should be preserved at all costs. In this case, η is referred to as a "distinguished parameter," and the unfolding process consists in adding additional parameters without eliminating η. We will not follow that procedure here, but the reader should be aware that calculations of "codimension" (roughly, the number of parameters that must be added) will differ from one author to another according to which approach is taken.

Computing the Asymptotic Unfolding of x^k

Motivated by the example of the imperfect pitchfork bifurcation, we ask the following question: Given the vector field

$$\dot{x} = x^k, \tag{6.3.3}$$

how can we describe all nearby vector fields? Behind this question lies the idea that any vector field having a rest point at the origin has the form $\dot{x} = a_k x^k + \cdots$, with $a_k \neq 0$, for some positive integer k, and that the leading term should dominate near the origin; furthermore, the a_k can be removed by changing the time units.

Unfolding theorems are often considered to be rather deep. That is because as usually formulated, they contain information about the ignorability of higher-order terms. The version contained in the following theorem is not at all deep; in fact, it is almost trivial, precisely because the question of dropping higher-order terms is not addressed, but rather postponed. The important point in the theorem is the absence of the term y^{k-1}.

6.3.2. Theorem. Every system of the form

$$\dot{x} = x^k + \eta f(x)$$

can be reduced, modulo the ideal $\langle y^{k+1}, \eta y^k, \eta^2 \rangle$, to the form

$$\dot{y} = y^k + \mu_1 + \mu_2 y + \cdots + \mu_{k-1} y^{k-2}, \tag{6.3.4}$$

where $\mu_i = \mathcal{O}(\eta)$ are linear functions of η.

Proof. We treat the case $k = 3$; the general case goes the same way. The arbitrary perturbation of $\dot{x} = x^3$ can be written

$$\dot{x} \equiv x^3 + \eta(a + bx + cx^2),$$

where $\equiv$ denotes equality in the ring of formal power series modulo the ideal $\langle x^4, \eta x^3, \eta^2 \rangle$. Making the small shift of coordinates

$$x = y + \eta h, \tag{6.3.5}$$

where h is to be specified later, leads to

$$\dot{y} \equiv (y + \eta h)^3 + \eta[a + b(y + \eta h) + c(y + \eta h)^2$$
$$\equiv y^3 + 3y^2\eta h + \eta a + \eta by + \eta cy^2.$$

If we choose

$$h = -\frac{1}{3}c,$$

this reduces to

$$\dot{y} \equiv y^3 + \eta(a + by).$$

Setting $\mu_1 = \eta a$ and $\mu_2 = \eta b$, this becomes

$$\dot{y} \equiv y^3 + \mu_1 + \mu_2 y,$$

which is the same as (6.3.2), and also the same as (6.3.4) with $k = 3$.   □

We refer to (6.3.4) as the *asymptotic unfolding* of $\dot{x} = x^k$ to degree k.

6.3.3. Remark. Theorem 6.3.2 can be extended as follows: Every system of the form

$$\dot{x} = x^k + \eta f(\eta, x)$$

can be reduced to the form (6.3.4) modulo $\langle x^{k+1}, \eta^{\ell+1}\rangle$ for any ℓ. In this case the functions μ_i, which are still $\mathcal{O}(\eta)$, are polynomials in η of degree ℓ. The reduction is achieved by a shift of coordinates of the form

$$y = x + \eta h_1 + \eta^2 h_2 + \cdots + \eta^\ell h_\ell,$$

where each h_i is determined so as to eliminate the term in $\eta^i y^{k-1}$. This represents a partial computation of the power series for the function $h(\eta)$ appearing in the next theorem.

Justifying the Unfolding of x^k

As explained in the preface to this book, our general approach to unfoldings is "pragmatic" rather than "ideological." That is, we rely on easy results such as Theorem 6.3.2 to provide unfoldings up to a specified degree, and then try to determine what aspects of the behavior of the system are successfully unfolded, rather than require ahead of time that an unfolding be universal with respect to a specified equivalence relation. In the present case, however, it is not hard to show that the unfolding just obtained is in fact a universal unfolding with respect to the "static" equivalence relation that has already been used in Definition 6.2.1.

6.3.4. Theorem. Given any smooth function $f : \mathbb{R}^2 \to \mathbb{R}$, there exists a smooth function $u : \mathbb{R}^2 \to \mathbb{R}$ with $u(0,0) \neq 0$ and a change of variables $x = y + \eta h(\eta)$ such that

$$u(\eta, x)(x^k + \eta f(\eta, x)) = y^k + \mu_1(\eta) + \mu_2(\eta)y + \cdots + \mu_{k-1}(\eta)y^{k-2}$$

for certain smooth functions $\mu_i(\eta)$. That is, for small η and x, and "up to" the change of coordinates from x to y, the set of rest points of any equation of the form

$$\dot{x} = x^k + \eta f(\eta, x)$$

can be found among the rest points of the $(k-1)$-parameter differential equation (6.3.4).

Proof. By the Malgrange preparation theorem (Theorem A.7.2 in Appendix A), there exists a smooth function $u(\eta, x)$ with $u(0,0) \neq 0$ such that

$$u(\eta, x)(x^k + \eta f(\eta, x)) = x^k + p_{k-1}(\eta)x^{k-1} + \cdots + p_1(\eta)x + p_0(\eta),$$

where $p_i(\eta)$ are smooth. Introduce $x = y + \eta h(\eta)$ into this expression; the coefficient of y^{k-1} will be

$$k\eta h(\eta) + p_{k-1}(\eta).$$

This coefficient can be made to vanish by taking $h(\eta) = -p_{k-1}(\eta)/k\eta$, which is permissible (by Theorem A.2.2), since $p_{k-1}(0) = 0$. $\square$

Notes and References

The notion of universal unfolding for bifurcation problems is developed in Golubitsky and Schaeffer [50], in the context of problems with a distinguished parameter (see Remark 6.3.1 above). Here we have eliminated the distinguished parameter, which makes our treatment closer in some ways to that of Wiggins in Section 3.1D of [111], except that our unfoldings are asymptotic in character. The notion of asymptotic unfolding was introduced in Murdock [87], in the context described in Section 6.4 below. The application to the unfolding of the scalar differential equation $\dot{x} = x^k$ has not been published elsewhere. It is motivated by the elementary treatment of the unfolding of the mapping t^k (regarded as a mapping of $\mathbb{R}$ to itself) in Chapter 6 of Bruce and Giblin [19].

6.4   Unfolding in the Presence of Generic Quadratic Terms

It was seen in the last section that an unfolding of the scalar system $\dot{x} = x^k$ can be computed by adding an arbitrary perturbation and then considering the effect of a shift of the origin. This differential equation is highly degenerate, since all terms of degree less than k vanish. In this section an unfolding will be computed for systems of arbitrary dimension, but with strict limits on the amount of degeneracy allowed. The linear term is allowed to have zero as an eigenvalue, with arbitrary multiplicity and

Jordan structure; shifts of the origin (confined to shifts within this zero eigenspace) will again play an important role in computing the unfolding. But the quadratic terms will (when A contains nontrivial nilpotent Jordan blocks) be required to satisfy certain nondegeneracy conditions with respect to this same eigenspace. These shifts are not used to eliminate high-order terms (as the shift in Section 6.3 eliminated the term of degree $k - 1$). Instead, the effect of the shifts will be confined to the constant and linear terms.

In addition to shifts, the derivation of the unfolding relies on normal form theory for both matrix series (Chapter 3) and for vector fields (Chapter 4). The argument, as presented here, is valid only when the simplified normal form style is used (Sections 3.4 and 4.6). See Notes and References for further discussion of the situation with other normal form styles.

> **6.4.1. Remark.** Two easy examples of the calculations explained in this section have already appeared in Chapter 1, one leading to the unfolding (1.1.31) of the (single) center, the other to the unfolding (1.2.17) of the nonsemisimple double-zero eigenvalue. The latter is the more typical of these two examples, since it contains a nontrivial nilpotent block. The symbols k and h, used below to denote two kinds of shifts (called "primary" and "secondary"), were chosen to correspond (roughly) to the notation in Section 1.2; actually, k and h here are vectors, and correspond to $(0, k)$ and $(h, 0)$ in Section 1.2.

The First-Order Unfolding

The goal is to find a simple form representing all systems close to a given system $\dot{x} = Ax + Q(x) + \cdots$, where A is a given matrix and Q a given quadratic part. All perturbations of this system can be obtained by adding $\eta\{p + Bx + \cdots\} + \cdots$, where $p \in \mathbb{R}^n$ is an arbitrary constant vector, B is an arbitrary $n \times n$ matrix, η is a small parameter, and the second set of dots represents terms of higher order in η. Since the most important part of the calculation concerns the linear and quadratic terms of the unperturbed system, and the constant and linear terms of the perturbation, only these terms have been written explicitly. Defining the equivalence relation $\equiv$ to be congruence modulo cubic terms in x, quadratic terms in η, and terms that are jointly linear in η and quadratic in x, the system to be simplified appears as

$$\dot{x} \equiv Ax + Q(x) + \eta\{p + Bx\}. \qquad (6.4.1)$$

We assume that A is in Jordan form (which may entail the use of complex variables with associated reality conditions), and that Q is in simplified normal form with respect to A. Our aim is to simplify p and B in (6.4.1), reducing these $n + n^2$ quantities to a much smaller number of unfolding parameters $\mu_1, \ldots, \mu_c$. The result will be called the *first-order unfolding* of $\dot{x} = Ax + Q(x) + \cdots$.

6.4.2. Remark. The reason for using η (rather than ε) as the small parameter is to allow for using the dilation $x = \varepsilon \xi$ if it is desired; if that is done, then $\equiv$ becomes congruence modulo $\langle \varepsilon^3, \eta^2, \varepsilon^2 \eta \rangle$. The reader will recall that the formulas in Chapter 4 were arranged to match those of Chapter 3 when the dilated form is used; that is, ε in Chapter 4 plays a role that corresponds to ε in Chapter 3. In the present section the two independent small parameters ε and η both play a similar role. The normalization of quadratic (and, in a later subsection, higher-order) terms of order η^0 proceeds as in Chapter 4 (using ε or not, as desired). The normalization of the linear terms $Ax + \eta Bx + \mathcal{O}(\eta^2)$ proceeds as in Chapter 2 for the matrix series $A + \eta B + \cdots$, with η in place of ε. In the general case (when the relation $\equiv$ is relaxed) there will be mixed terms of order $\varepsilon^i \eta^j$.

The simplification of (6.4.1) will be carried out in three stages:

1. A coordinate shift

$$x = y + \eta k \tag{6.4.2}$$

will be performed, using a vector k having $k_i = 0$ whenever the ith row of A is the top row of a Jordan block having eigenvalue zero. Such a shift is called a *primary shift*; by choosing k correctly, it will be possible to simplify p while making uncontrolled changes to B.

2. A linear coordinate transformation

$$y = z + \eta T z \tag{6.4.3}$$

will be used to simplify the (modified) B by bringing it into simplified normal form for matrices, as in Section 3.4.

3. A coordinate shift

$$z = w + \eta h \tag{6.4.4}$$

will then be performed, this time using h having $h_i = 0$ *except* when the ith row of A is the top row of a Jordan block with eigenvalue zero. This is called a *secondary shift*. It has no effect on p, and modifies the (already simplified) matrix B without taking it out of simplified normal form. It is at this point that the nondegeneracy conditions on Q, mentioned above, come into play: They make it possible for a careful choice of h to simplify the matrix B still further.

At this point, all of the simplifications that have been achieved in p and B take the form that certain entries must equal zero, while the others remain arbitrary. These arbitrary terms will define the unfolding parameters. Now we turn to the details.

Applying the transformation (6.4.2) to (6.4.1) results in

$$\dot{y} \equiv Ay + Q(y) + \eta\{(p + Ak) + (By + Q'(y)k)\}; \tag{6.4.5}$$

the unperturbed part is unchanged (except for replacing x by y), the constant term in the perturbation becomes $p+Ak$, and the linear term becomes $By+Q'(y)k$. Note that since $Q(y)$ is quadratic in y, $Q'(y)$ is a matrix whose entries are linear in y, and $Q'(y)k$ is a vector whose entries are bilinear in y and k. In step 1, the aim is to simplify p by making a good choice of k. Clearly, if $k \in \ker A$, the transformation will not change p, so we should choose a complement $\mathcal{C}$ to $\ker A$ in $\mathbb{R}^n$ and restrict k to lie in $\mathcal{C}$; shifts with $k \in \mathcal{C}$ will be primary shifts used to simplify p, whereas shifts using $h \in \ker A$ will be saved for use as secondary shifts targeted at the simplification of B. Since a vector belongs to $\ker A$ if and only if its only nonvanishing entries occur in positions corresponding to top rows of nilpotent Jordan blocks in A, a convenient complement $\mathcal{C}$ consists of vectors that vanish in those positions (and are arbitrary elsewhere). Calculations of the form

$$\begin{bmatrix} 0 & 1 & 0 \\ 0 & 0 & 1 \\ 0 & 0 & 0 \end{bmatrix} \begin{bmatrix} 0 \\ a \\ b \end{bmatrix} = \begin{bmatrix} a \\ b \\ 0 \end{bmatrix}$$

show that A maps $\mathcal{C}$ one-to-one onto the set of vectors that vanish in the positions corresponding to *bottom* rows of nilpotent blocks of A. (Jordan blocks that are not nilpotent are invertible, and since vectors $k \in \mathcal{C}$ have arbitrary entries in positions corresponding to these blocks, so do their images Ak.) It follows that $k \in \mathcal{C}$ can be chosen uniquely so that

$$q = p - Ak$$

has nonzero entries only in positions corresponding to bottom rows of nilpotent blocks. With this choice of k, the linear vector field $By + Q'(y)k$ can be written as Cy for some matrix C, and our system takes the form

$$\dot{y} \equiv Ay + Q(y) + \eta\{q + Cy\}. \tag{6.4.6}$$

The difference between this system and (6.4.1) is that q has only one nonzero entry for each nilpotent Jordan block in A. Using notation from Sections 4.5 and 4.6, it also follows from this discussion that

$$q \in \widehat{N}_{-1}. \tag{6.4.7}$$

(It was pointed out in connection with (4.5.8) that the portions of the normal form module having grades -1 and 0 would appear in connection with unfoldings. A constant vector belongs to $\widehat{N}_{-1}$, grade -1 of the simplified normal form module, if its nonzero entries are confined to *main rows* (in the sense of Definition 4.6.12), that is, the positions R_s corresponding to bottom rows of Jordan blocks of A. Here q is still more restricted, since its nonzero entries are confined to correspond to bottom rows of *nilpotent* Jordan blocks of A. These rows may be called *principal rows*.)

Applying (6.4.3) to (6.4.6) causes C to be replaced with $D = C - TA + AT = C - [T, A] = C - \mathbb{L}_A T$. That is, there is a homological equation

$$\mathbb{L}_A T = C - D, \tag{6.4.8}$$

which has the same form as those discussed in Chapter 3. It follows that we can take $D = \widehat{\mathbb{P}}C$ to be the projection of C into simplified normal form, defined as in (3.4.19); thus D will have nonzero entries only in main rows, and even within these rows there will be nonzero entries only in positions that are at the bottoms of stripes of the stripe structure associated with A. The system is now

$$\dot{z} \equiv Az + Q(z) + \eta\{q + Dz\}. \tag{6.4.9}$$

Again, this does not look different from the previous form, but the difference lies in the number of nonzero entries in D. Since D is in simplified normal form for matrices if and only if Dz is in simplified normal form for linear vector fields, we have

$$Dz \in \widehat{\mathcal{N}}_0. \tag{6.4.10}$$

When (6.4.4) is applied to (6.4.9), the result is structurally the same as (6.4.5), except that $Ah = 0$. Therefore, q is not changed, and Dz is replaced by $Dw + Q'(w)h$. As pointed out in step 1, this is a linear vector field; we may introduce a matrix $E(h)$, with entries depending linearly on h, such that

$$Q'(w)h = E(h)w, \tag{6.4.11}$$

and in this notation the transformed system is

$$\dot{w} \equiv Aw + Q(w) + \eta\{q + (D + E(h))w\}. \tag{6.4.12}$$

The entries of the matrix $E(h)$ can be thought of as "injected" into D from the quadratic part Q as a result of the secondary shift h. The aim is to choose h so as to simplify D. Before discussing this, it is essential to verify that adding $E(h)$ does not destroy the simplification already achieved in D. This is established in the following lemma:

6.4.3. Lemma. $E(h)$ is in simplified normal form for matrices, with respect to A.

Proof. It suffices, instead, to show that $Q'(w)h$ is in simplified normal form for linear vector fields. According to Section 4.6, vector fields in simplified normal form are characterized as follows:

1. They have nonzero entries only in main rows.

2. The entry in position R_s belongs to $\ker \mathcal{D}_{N^*}^{r_s}$.

3. The entry in position R_s contains only monomials $x^m e_{R_s}$ satisfying $\langle m, \lambda \rangle - \lambda_{R_s} = 0$.

Given that Q is quadratic and satisfies these conditions, and that $h \in \ker A$, we need to prove that $Q'(w)h$ satisfies these same conditions. Since the vector Q vanishes outside of main rows, so does the matrix $Q'(w)$ and the vector $Q'(w)h$, establishing property 1. Since each operator $\partial/\partial x_i$ commutes with $'$, $\mathcal{D}_{N*}$ commutes with $'$ (assuming that $\mathcal{D}_{N*}$ acts componentwise on both vectors and matrices). Therefore,

$$\mathcal{D}_{N*}^{r_s} Q'(w)h = (\mathcal{D}^{r_s} Q)'(w)h,$$

and since $\mathcal{D}^{r_s}$ annihilates the entry in the R_sth position of Q, the R_sth row of $(\mathcal{D}^{r_s}Q)'(w)$ vanishes. It follows that the R_sth entry of $Q'(w)h$ belongs to $\ker \mathcal{D}_{N*}^{r_s}$, proving property 2. Finally, since Q is quadratic, property 3 for $Q(w)$ says that $Q_{R_s}(w)$ contains only monomials $w_i w_j$ for which $\lambda_i + \lambda_j = \lambda_{R_s}$ (with $i = j$ permitted). Then $Q_{R_s}(w)$ may be written

$$Q_{R_s}(w) = \sum q_{R_s ij} w_i w_j,$$

where the sum is over $i \leq j$ such that $\lambda_i + \lambda_j = \lambda_{R_s}$. It follows that the R_sth entry of $Q'(w)k$ is

$$Q'_{R_s}(w)k = \sum q_{R_s ij}(k_i w_j + k_j w_i).$$

Since $h \in \ker A$, h_i is zero unless the ith row of A is the top row of a nilpotent Jordan block, in which case $\lambda_i = 0$, implying that $\lambda_j = \lambda_{R_s}$. That is, w_j actually appears in $Q'_{R_s}(w)k$ (with a nonzero coefficient) only if $\lambda_j = \lambda_{R_s}$ (and the same is true with j replaced by i). This proves property 3 for $Q'(w)h$.   □

Knowing that $E(h)$ is already in the same simplified normal form as D, and that it depends linearly on $h \in \ker A$, we now wish to choose h so as to eliminate some terms from $D + E(h)$, that is, to make some entries vanish that do not already vanish as a consequence of being in normal form. The details here can be seen most clearly in the examples, but the idea is as follows. Suppose that d_{ij} is a nonzero entry in D. The corresponding entry of $D + E(h)$ will be

$$d_{ij} + e_{ij}(h) = d_{ij} + e_{ij1}h_{(1)} + \cdots + e_{ij\ell}h_{(\ell)}, \tag{6.4.13}$$

where ℓ is the number of nilpotent Jordan blocks in A and $h_{(1)}, \ldots, h_{(\ell)}$ are the independent variables in h, which occur in the positions corresponding to the top rows of these blocks (not the position indicated by the subscript on h). The numbers $e_{ij1}, \ldots, e_{ij\ell}$ are the coefficients showing how $e_{ij}(h)$ depends linearly on the free variables in h; these coefficients are determined by Q. Since there are ℓ free variables in h, we can choose ℓ nonzero components d_{ij} that we will try to eliminate. Once the selection of d_{ij} is made, the elimination can be done, provided that the ℓ linear functionals e_{ij} defined on $\ker A$ are linearly independent. This can be expressed as the nonvanishing of an $\ell \times \ell$ determinant of the appropriate coefficients e_{ijt}, which is a nondegeneracy condition on Q that is generically satisfied. The simplest example has already been seen in the condition $\alpha \neq 0$ associated with equation (1.2.16). At this stage our system takes the form

$$\dot{w} \equiv Aw + Q(w) + \eta\{q + Fw\}. \tag{6.4.14}$$

Notice that the number of entries that have been eliminated from F is exactly the same as the number of nonzero entries occurring in q. Thus, the total number c of parameters present in q and F is exactly the same as the number of nonzero terms in the simplified normal form with respect to A. This number is the *codimension* of A (see Definition 3.4.18). The final step in constructing the first-order unfolding is to replace the nonzero entries in ηq and ηF by independent small parameters $\mu_1, \ldots, \mu_c$ called the *unfolding parameters*. The vector and matrix resulting from this replacement will be called $q(\mu)$ and $F(\mu)$, respectively, so that the first-order unfolding can be written as

$$\dot{w} \equiv Aw + Q(w) + \{q(\mu) + F(\mu)w\}. \tag{6.4.15}$$

This will be most easily understood through an example.

The following example is somewhat artificial, but illustrates all the possibilities that can arise in computing the first-order unfolding of a system, except for those that concern complex systems with reality conditions. The handling of reality conditions will be explained in detail by way of examples in Sections 6.5, 6.6, and 6.7.

Let A be the 8×8 matrix

$$A = \begin{bmatrix} 0 & 1 & & & & & & \\ 0 & 0 & & & & & & \\ & & 0 & 1 & 0 & & & \\ & & 0 & 0 & 1 & & & \\ & & 0 & 0 & 0 & & & \\ & & & & & 2 & & \\ & & & & & & 2 & 1 \\ & & & & & & 0 & 2 \end{bmatrix}, \tag{6.4.16}$$

having two nilpotent Jordan blocks and two nonnilpotent Jordan blocks with the same eigenvalue. (It is important that A have Jordan blocks of different sizes with the same eigenvalue, both in the nilpotent and nonnilpotent parts, if it is to be truly "typical." That is why we use such a large matrix for our example.) For this matrix $\ell = 4$, $r_1 = 2$, $r_2 = 3$, $r_3 = 1$, $r_4 = 2$, $R_1 = 2$, $R_2 = 5$, $R_3 = 6$, $R_4 = 8$. Primary shifts have the form $k = (0, *, 0, *, *, *, *, *)$, and the secondary shifts are $h = (*, 0, *, 0, 0, 0, 0, 0) = (h_{(1)}, 0, h_{(2)}, 0, 0, 0, 0, 0)$, in the notation used in (6.4.13). The stripe structure defined by A (see Definition 3.4.18 and compare (3.4.10)) is

$$
B = \begin{bmatrix}
b & & d & & & & \\
a & b & c & d & 0 & & \\
 & & i & & & & \\
f & & h & i & & & \\
e & f & g & h & i & & \\
 & & & & j & k & 0 \\
 & & & & & n & \\
 & & & & l & m & n
\end{bmatrix},
$$

which also gives the inner product normal form for matrix series beginning with A. The entries that are not marked are zero; the two entries that are explicitly indicated as 0 are entries that occur in main rows R_s and are within one of the large blocks (of the block structure of A), but are zero because they do not belong to a stripe (in the stripe structure of A). Such entries arise whenever there are Jordan blocks of unequal size with the same eigenvalue. The simplified normal form is

$$
D = \begin{bmatrix}
0 & & & & & & \\
a & b & c & d & 0 & & \\
0 & & & & & & \\
0 & & & & & & \\
e & f & g & h & i & & \\
 & & & & j & k & 0 \\
 & & & & & 0 & \\
 & & & & l & m & n
\end{bmatrix}. \tag{6.4.17}
$$

(The additional entries marked 0 are simply to aid in counting rows.) The codimension of A is $c = 14$. To construct the first-order unfolding (6.4.15), it is necessary only to exhibit $\{q(\mu) + F(\mu)w\}$. Since q has nonzero entries only in positions corresponding to bottom rows of nilpotent blocks of A, it is clear that $q(\mu) = (0, \mu_1, 0, 0, \mu_2, 0, 0, 0)$. For $F(\mu)$, it is necessary to choose two entries from (6.4.17) to be removed by the secondary shift. If we choose to eliminate a and f, the resulting $\{q(\mu) + F(\mu)w\}$ is

$$
\left\{ \begin{bmatrix} 0 \\ \mu_1 \\ 0 \\ 0 \\ \mu_2 \\ 0 \\ 0 \\ 0 \end{bmatrix} + \begin{bmatrix} 0 & & & & & & \\ 0 & \mu_3 & \mu_4 & \mu_5 & 0 & & \\ 0 & & & & & & \\ 0 & & & & & & \\ \mu_6 & 0 & \mu_7 & \mu_8 & \mu_9 & & \\ & & & & \mu_{10} & \mu_{11} & 0 \\ & & & & & 0 & \\ & & & & \mu_{12} & \mu_{13} & \mu_{14} \end{bmatrix} x \right\}. \tag{6.4.18}
$$

To determine the genericity (or nondegeneracy) condition on Q under which the unfolding (6.4.18) is valid, we write down the expressions (6.4.13) corresponding to the terms $a = d_{21}$ and $f = d_{52}$ that we want to eliminate, and set these equal to zero. Since $\ell = 2$, these equations have the form

$$a + e_{211}h_{(1)} + e_{212}h_{(2)} = 0,$$
$$f + e_{521}h_{(1)} + e_{522}h_{(2)} = 0.$$

The genericity condition is then

$$\begin{vmatrix} e_{211} & e_{212} \\ e_{521} & e_{522} \end{vmatrix} \neq 0.$$

To understand this condition in greater detail requires a computation of the quadratic terms Q in simplified normal form, in order to see which terms of Q contribute the quantities e_{ijk} appearing in this determinant. In this example the quadratic terms in normal form are spanned by 37 vector fields, only four of which contribute to the determinant, so the genericity condition affects only these four terms. For details see Notes and References.

Unfoldings of Higher Order

The first-order unfolding can be extended to higher-order terms, both in x and in η, without any additional work. This is seen by changing the equivalence relation $\equiv$ to whatever (double) jet equivalence relation we wish, and observing that all of the additional normal form calculations that are required merely repeat those that have already been done, either in putting the unperturbed system $\dot{x} = Ax + Q(x) + \cdots$ into normal form to degree k, or in computing q and F as above.

The simplest case to consider is the one in which the quadratic terms of the perturbation (6.4.1) are restored. Taking $\equiv$ to be congruence modulo cubic terms in x and quadratic terms in η, (6.4.1) is replaced by

$$\dot{x} \equiv Ax + Q(x) + \eta\{p + Bx + R(x)\}, \tag{6.4.19}$$

where $R(x)$ is homogeneous quadratic. The three normalization steps described above bring this system into the following form corresponding to (6.4.14):

$$\dot{w} \equiv Aw + Q(w) + \eta\{q + Fw + S(w)\}, \tag{6.4.20}$$

where S is quadratic. One final transformation

$$w = x + \eta q(x),$$

in which q is quadratic and we reuse the letter x for the new vector variable, brings us to

$$\dot{x} \equiv Ax + Q(x) + \eta\{q + Fx + T(x)\},$$

where T is quadratic and

$$L_A q = S - T.$$

This is exactly the same homological equation as the one that would have been used when Q itself was brought into simplified normal form (which we assumed had been done before we began). Therefore, T can be taken to be in the same simplified normal form as Q, so that the complete quadratic part $Q(x) + \eta T(x)$ can be obtained simply by perturbing the normal form coefficients already present in Q. For instance, the first-order unfolding (1.2.17) can be extended to second order by introducing two new unfolding parameters μ_3 and μ_4 to perturb the quadratic normal form coefficients α and β:

$$\begin{bmatrix} \dot{x} \\ \dot{y} \end{bmatrix} \equiv \begin{bmatrix} 0 \\ \mu_1 \end{bmatrix} + \begin{bmatrix} 0 & 1 \\ 0 & \mu_2 \end{bmatrix} \begin{bmatrix} x \\ y \end{bmatrix} + \begin{bmatrix} 0 \\ (\alpha + \mu_3)x^2 + (\beta + \mu_4)y^2 \end{bmatrix}.$$

But it is scarcely necessary to do this; it is enough to regard α and β themselves as variables rather than constants (remembering that the non-degeneracy condition $\alpha \neq 0$, required for the validity of the unfolding, cannot be violated).

The same result holds at higher degrees as well; if we define $\equiv$ so as to retain the cubic terms of order η^0 and η^1, both of these terms can be brought into simplified cubic normal form, and the order-η contribution simply varies the normal form coefficients of the unperturbed cubic part. We can either introduce further unfolding parameters to do this, or regard the normal form coefficients themselves as variables.

Finally, suppose that we wish to include terms of higher order in η as well (but with a finite bound to the order), let us say terms of order η^2. Then, through a primary shift $\eta^2 k$, followed by a near-identity linear transformation of the form $I + \eta^2 T$, followed by a secondary shift $\eta^2 h$, the constant and linear terms that are quadratic in η can be brought into the same form as the constant and linear terms of order η. That is, they can be brought into the same unfolded form as (6.4.15) by allowing each μ_i to contain a linear term and a quadratic term in η.

Putting all of this together, we have the following theorem:

6.4.4. Theorem. Given a vector field

$$\dot{x} = a(x) = Ax + a_1(x) + \cdots + a_{k-1}(x) \tag{6.4.21}$$

in simplified normal form (which entails that A is in Jordan form) truncated at degree k, an asymptotic unfolding (to degree k) of this vector field may be constructed as follows. Form the general matrix $G(\mu)$ in simplified normal form with respect to A, denoting its nonzero entries by $\mu_1, \ldots, \mu_c$, where c is the codimension of A. Letting d be the number of nilpotent Jordan blocks in A, move d of the entries μ_i from $G(\mu)$ to a vector $q(\mu)$, placing them in principal rows (that is, in the positions corresponding to the bottom rows of the nilpotent blocks). (The other entries of $q(\mu)$ are zero.) Let $F(\mu)$ be $G(\mu)$ with the "moved" entries set equal to zero. Then an unfolding of

(6.4.21) is

$$\dot{x} = \tilde{a}(x, \mu) = q(\mu) + (A + F(\mu))x + \tilde{a}_1(x) + \cdots + \tilde{a}_{k-1}(x), \qquad (6.4.22)$$

where $\tilde{a}_i$ means that the normal form coefficients appearing in a_i are allowed to vary in a neighborhood of their original values (by adding additional unfolding parameters to these coefficients, if desired). Given any integer $j \geq 1$ and any perturbation of (6.4.21) with perturbation parameter η, there exist generically valid nondegeneracy conditions on Q under which it is possible to bring the perturbed system into the form (6.4.22) modulo terms of degree $k + 1$ in x and degree $j + 1$ in η by a smooth coordinate change, with the unfolding parameters becoming smooth functions of η vanishing at $\eta = 0$.

The total number of unfolding parameters, including those that merely modify normal form coefficients, is the *codimension* of the unfolding (not to be confused with the codimension of A). Notice that the codimension approaches infinity as k is increased. It is frequently possible to reduce the number of parameters by one through a change of the time scale, which is not one of the coordinate changes that has been used in this section; this has already been illustrated in Remark 1.1.10. Further reductions of codimension may be obtained by changing the units of length on the individual x_i axes; see Remark 6.6.1. The unfolding constructed in Theorem 6.4.4 is called an *asymptotic unfolding to grade k* (or to degree $k + 1$). It "captures" all behavior of systems near (6.4.21) that can be detected using asymptotic expansions to grade k in x and to arbitrary finite degree in η, whether this behavior is topological in character or not, but "misses" behavior that cannot be detected by finite asymptotic expansions. Thus, the unfolding may contain parameters that are unnecessary from a purely topological standpoint, and yet may miss parameters that are necessary from that standpoint. This is not to be considered a drawback; in fact, there frequently cannot exist a topological unfolding of finite codimension.

Notes and References

The asymptotic unfolding (as defined here) first appeared in Murdock [87]. The development here is considerably simpler, since in [87] it was necessary to work out the simplified normal form (to quadratic order) at the same time, and the two issues were not always clearly separated. Here the simplified normal form has already been developed separately in Sections 3.4 and 4.6, and can simply be cited when it is needed. The main parts of [87] that have not been incorporated into the present treatment are the detailed discussion in Section 1 of the equivalence relation with respect to which the asymptotic unfolding is universal, and the remarks in Section 7 giving further details about the normalization of the higher-order terms. The details of the genericity condition for (6.4.18), including the identification of the 37 terms in the quadratic normal form, is contained in Section 6.

It is still an unsolved problem how to carry out an asymptotic unfolding using an equivariant normal form style (the inner product or sl(2) normal forms). The difficulty is that secondary shifts do not preserve the normal form already achieved for the linear part, unless the simplified normal form is used. This failure of what was there called the "injection condition" is illustrated in [87] for the case of the nonsemisimple double zero, for which the inner product and sl(2) normal forms coincide. The problem might be resolvable by carrying out the normalization of the perturbed linear term and the secondary shift simultaneously, rather than successively. It would also be helpful to find a coordinate-free formulation of the injection condition, rather than one that depends on having A in Jordan form. I have attempted this in several ways, but so far without success.

6.5 Bifurcations from a Single Center (Hopf and Degenerate Hopf Bifurcations)

In this section and the next, we return to the two examples studied in Chapter 1, the (single) nonlinear center and the nonsemisimple double-zero eigenvalue. The normal form and unfolding for each of these problems were derived by elementary methods in Chapter 1; we will quickly rederive these results by applying the methods of Sections 4.5 and 6.4 (for the center) and Sections 4.6, 4.7, and 6.4 (for the double zero). Then we will address the bifurcations resulting from each of these unfoldings, giving complete results for the center and referring to the literature for additional techniques needed to complete the treatment of the double zero.

The Normal Form and Unfolding of the Center

The linear part of the single center is given by

$$A = \begin{bmatrix} 0 & -1 \\ 1 & 0 \end{bmatrix}. \tag{6.5.1}$$

Writing (x, y) in place of (x_1, x_2), the linear map

$$z = x + iy, \tag{6.5.2}$$
$$w = x - iy,$$

transforms the system into

$$\dot{z} = iz + f(z, w), \tag{6.5.3}$$
$$\dot{w} = -iz + g(z, w),$$

with reality subspace $\mathcal{R}$ defined by $w = \bar{z}$. (See the discussion beginning with (4.5.35) for additional details.) The system (6.5.3) will be in complex normal form (with reality conditions), provided that f contains only monomials $z^j w^k$ satisfying $j(i) + k(-i) = i$ (that is, $j = k + 1$) and g satisfies $g(z, \bar{z}) = \bar{f}(z, \bar{z})$. For such a system the full information is contained in the $\dot{z}$ equation restricted to $\mathcal{R}$, which can be written as the formal power series

$$\dot{z} = iz + \sum_{\ell=1}^{\infty} (\alpha_{2\ell+1} + i\beta_{2\ell+1}) z(z\bar{z})^{\ell}. \qquad (6.5.4)$$

Returning to real form via (6.5.2) this becomes

$$\begin{bmatrix} \dot{x} \\ \dot{y} \end{bmatrix} = \begin{bmatrix} 0 & -1 \\ 1 & 0 \end{bmatrix} \begin{bmatrix} x \\ y \end{bmatrix} \qquad (6.5.5)$$
$$+ \sum_{\ell=1}^{\infty} \left(\alpha_{2\ell+1} \left(x^2 + y^2 \right)^{\ell} \begin{bmatrix} x \\ y \end{bmatrix} + \beta_{2\ell+1} \left(x^2 + y^2 \right)^{\ell} \begin{bmatrix} -y \\ x \end{bmatrix} \right),$$

which is the same as (1.1.24). For the polar form (1.1.25) it is simplest to apply $z = re^{i\theta}$ directly to (6.5.4) to obtain

$$\dot{r} = \sum_{\ell=1}^{\infty} \alpha_{2\ell+1} r^{2\ell+1}, \qquad (6.5.6)$$

$$\dot{\theta} = 1 + \sum_{\ell=1}^{\infty} \beta_{2\ell+1} r^{2\ell}.$$

Since the linear part

$$\Lambda = \begin{bmatrix} i & \\ & -i \end{bmatrix} \qquad (6.5.7)$$

of (6.5.3) is diagonal, once the system is in normal form it is already in simplified normal form and is ready to be unfolded by the methods of Section 6.4. According to Theorem 6.4.4, this is done by first forming the general matrix

$$G(\mu_1, \mu_2) = \begin{bmatrix} \mu_1 & 0 \\ 0 & \mu_2 \end{bmatrix} \qquad (6.5.8)$$

that is in simplified normal form with respect to Λ, and then adding this matrix to A. (In the general case, the theorem calls for moving certain entries from G to the constant term, but since A has no zero eigenvalues, and therefore no nilpotent Jordan blocks, there is nothing to move. Notice also that because no entries are moved, there are no genericity conditions to be imposed on the quadratic terms.) In Section 6.4, the issue of reality conditions was postponed to the examples, and this is the first such example: If the unfolded version of (6.5.3) is to satisfy the reality conditions, we must have $\mu_2 = \bar{\mu}_1$. Thus, we set

$$\mu_1 = \mu + i\nu \qquad (6.5.9)$$

with μ and ν real, and $\mu_2 = \mu - i\nu$. In real coordinates the resulting unfolded system is

$$\begin{bmatrix} \dot{x} \\ \dot{y} \end{bmatrix} = \begin{bmatrix} \mu & -1-\nu \\ 1+\nu & \mu \end{bmatrix} \begin{bmatrix} x \\ y \end{bmatrix} \qquad (6.5.10)$$

$$+ \sum_{\ell=1}^{\infty} \left(\alpha_{2\ell+1} \left(x^2 + y^2 \right)^{\ell} \begin{bmatrix} x \\ y \end{bmatrix} + \beta_{2\ell+1} \left(x^2 + y^2 \right)^{\ell} \begin{bmatrix} -y \\ x \end{bmatrix} \right),$$

and in polar coordinates it is

$$\dot{r} = \mu r + \sum_{\ell=1}^{\infty} \alpha_{2\ell+1} r^{2\ell+1}, \qquad (6.5.11)$$

$$\dot{\theta} = 1 + \nu + \sum_{\ell=1}^{\infty} \beta_{2\ell+1} r^{2\ell}.$$

(This agrees with (1.1.32) if we put $\mu = -\varepsilon\delta$ and truncate at $\ell = 1$.) As explained in Remark 1.1.10, we can (and henceforth will) assume $\nu = 0$. Also, according to Theorem 6.4.4, we should allow the normal form coefficients $\alpha_{2\ell+1}$ and $\beta_{2\ell+1}$ to vary in a neighborhood of their original values. The precise meaning of these results is stated in the following lemma:

6.5.1. Lemma. Any system of the form

$$\begin{bmatrix} \dot{x} \\ \dot{y} \end{bmatrix} = A(\eta) \begin{bmatrix} x \\ y \end{bmatrix} + f(x, y, \eta), \qquad (6.5.12)$$

where η is a small parameter, f is a smooth vector field whose Taylor series begins with quadratic terms, and

$$A(0) = \begin{bmatrix} 0 & -1 \\ 1 & 0 \end{bmatrix},$$

can be reduced (by a change of the variables x, y, and t) to the form

$$\dot{r} = \mu(\eta)r + \sum_{\ell=1}^{k} \alpha_{2\ell+1}(\eta) r^{2\ell+1} + r^{2k+2} R(r, \theta, \eta), \qquad (6.5.13)$$

$$\dot{\theta} = 1 + \sum_{\ell=1}^{k} \beta_{2\ell+1} r^{2\ell} + r^{2k+1} \Theta(r, \theta, \eta).$$

Behavior of the Truncated System

There are two forms of the truncated system that can be studied. On the one hand, we can truncate (6.5.11) to obtain

$$\dot{r} = f(\mu, r, \alpha) = \mu r + \alpha_3 r^3 + \cdots + \alpha_{2k+1} r^{2k+1}, \tag{6.5.14}$$

$$\dot{\theta} = u(r, \beta) = 1 + \beta_3 r^2 + \cdots + \beta_{2k+1} r^{2k},$$

where $\alpha = (\alpha_1, \ldots, \alpha_{2k+1})$ and similarly for β, and the notation u is intended to indicate that this function is a unit in the ring of germs of smooth functions; that is, it is nonzero in a neighborhood of the origin. Alternatively, we can expand (6.5.13) in powers of η and r, truncate this at degree $2k+1$ in r and at some degree in η, and write the result as

$$\dot{r} = F(\eta, r) = f(\mu(\eta), r, \alpha(\eta)), \tag{6.5.15}$$

$$\dot{\theta} = U(\eta, r) = u(r, \beta(\eta)).$$

(This, of course, assumes that the functions $\mu(\eta)$, $\alpha(\eta)$, and $\beta(\eta)$ are given.) There is something to be learned from each approach. For the remainder of this section, we will confine ourselves to a disk $0 \leq r \leq r_0$ in which $u \neq 0$ and $U \neq 0$. (This entails fixing a compact set of β in the case of u, and a bounded interval $0 \leq \eta \leq \eta_0$ for U.) It follows immediately that all orbits of (6.5.14) and (6.5.15) rotate around the origin, and that the only periodic solutions are circles characterized by $f = 0$ or $F = 0$. Thus, the study of the periodic solutions in $0 \leq r \leq r_0$ reduces to the study of the zeros of the polynomial function $f(\mu, r)$ or $F(\eta, r)$, which can be carried out by the methods of Section 6.1 or 6.2. Here we will use those of Section 6.1. All zeros of f and F are symmetrical around the origin, and we need only consider positive roots.

It is to be noticed that the Newton diagrams of f and F are not arbitrary. The case of f is the simplest; here we think of α as fixed and draw the Newton diagram of f as a function of μ and r. There is only one term containing μ, namely the term μr, so the Newton diagram has a solid dot in position $(1, 1)$. If $\alpha_3 \neq 0$, there will be a solid dot in position $(0, 3)$; if $\alpha_3 = 0$ but $\alpha_5 \neq 0$, there will be a solid dot in position $(0, 5)$; and so forth:

$$
\begin{array}{c}
\bullet \\
0 \\
0 \\
\bullet \quad\quad 0 \\
0 \quad\quad\quad 0 \\
0 \;\bullet \quad\quad 0 \;\bullet \\
0 \; 0 \; 0 \;, \quad\quad 0 \; 0 \; 0 \;, \quad\quad \cdots
\end{array}
\tag{6.5.16}
$$

Each of these Newton diagrams is determinate, in the sense that the bifurcation it predicts can be determined from the solid dots (intermediate terms) alone as long as the entries in these positions are not zero. (It is to be stressed that this means only that the higher-order terms in f are irrelevant to the bifurcation behavior of the truncated system (6.5.14). The permissibility of taking this truncation at all is a separate issue, to be addressed later.) In the case that α_3 is nonzero, the dominant terms of f are $\mu r + \alpha_3 r^3$, and the positive root will have the form

$$r = \sqrt{-\mu/\alpha_3} + \mathcal{O}(\mu). \tag{6.5.17}$$

Thus, there will be a single periodic solution of amplitude $\mathcal{O}(|\mu|^{1/2})$ bifurcating from the origin for $\mu > 0$ if $\alpha_3 < 0$, and for $\mu < 0$ if $\alpha_3 < 0$. Similarly, if $\alpha_3 = 0$ but $\alpha_5 \neq 0$, the bifurcation will again take place on one side of $\mu = 0$ only, and will have amplitude $\mathcal{O}(|\mu|^{1/4})$. The case of (6.5.17) is the classical Hopf bifurcation, and the other cases are degenerate Hopf bifurcations. From this point of view there can never be more than a single bifurcating periodic orbit, and it can exist on only one side of the bifurcation point; the only difference between ordinary and degenerate Hopf bifurcations is the order of the amplitude. But this point of view does not do justice to the fact that not only μ, but the components of α, can vary. By treating α as fixed we are limiting the kind of degeneracies that can appear.

A truly complete treatment of degenerate Hopf bifurcations would treat μ and the components of α as independent small parameters. Instead of this, we will treat a few examples of (6.5.15), which is the general one-parameter subfamily of the full multiparameter family. The possible Newton diagrams for F (with axes η and r) are obtained from those for f (with axes μ and r) by substituting $\mu = \mu(\eta)$ and $\alpha = \alpha(\eta)$. For instance, suppose $\mu = \eta^2$ and $\alpha_3 = a + b\eta$ with $a \neq 0$. Since $\alpha_3 \neq 0$ for small η, the governing Newton diagram for f will be the first diagram in (6.5.16), with dominant terms $\mu r + \alpha_3 r^3$; after the substitution, the dominant terms are $\eta^2 r + a r^3$, and the Newton diagram (with axes η and r) is

$$
\begin{array}{l}
\bullet \\
0 \\
0 \quad 0 \quad \bullet \\
0 \quad 0 \quad 0
\end{array}
\tag{6.5.18}
$$

The amplitude of the bifurcating periodic solution is

$$r = \sqrt{-\eta^2/a} + \mathcal{O}\left(\eta^2\right), \tag{6.5.19}$$

if this quantity is real. The behavior of this bifurcation is quite different from what was found from (6.5.14): If $a > 0$, then no periodic solutions exist at all, while if $a < 0$, a periodic solution bifurcates as η moves in either direction from zero.

A still more interesting possibility arises if, for instance, $\alpha_5 = 1$, $\alpha_3 = \eta$, and $\mu = \eta^3$. In this case both the α_3 and α_5 terms of f contribute significantly to the Newton diagram of F, which is

$$
\begin{array}{cccc}
\bullet & & & \\
0 & & & \\
0 & \bullet & & \\
0 & 0 & 0 & \\
0 & 0 & 0 & \bullet \\
0 & 0 & 0 & 0
\end{array}
\qquad (6.5.20)
$$

The two slopes of $-\frac{1}{2}$ and -1 occurring in this Newton diagram indicate that there will be two bifurcating periodic solutions, one with amplitude $\mathcal{O}\left(\eta^{1/2}\right)$ and the other with amplitude $\mathcal{O}(\eta)$.

Justification of the Truncation

It remains to be shown that the results obtained above for the truncated differential equations accurately reflect the behavior of the full system. For this we will treat the case of (6.5.13) in complete generality. Stated briefly, Theorem 6.5.3 below says that periodic solutions of (6.5.13) correspond to positive roots of F (defined in (6.5.15)), provided that F is taken to be a high enough jet that it is determinate (or jet-sufficient), in the ordinary sense of sufficient to determine zeros. That is, whenever the Newton method arguments given above are adequate to determine the periodic solutions of the truncated differential equations (6.5.15), they are also adequate to determine the periodic solutions of the full equations (6.5.13). The proof uses the *Caesari–Hale method*, which is based on the following lemma (stated only in the form that we need).

6.5.2. Lemma. Let $f(\eta, r, \theta)$ be a smooth function with period 2π in θ such that $f(\eta, 0, \theta) = 0$ for all η and θ. Then there exists a unique solution $r(\theta, \eta, a)$, smooth in all its arguments, of the functional differential equation

$$
\frac{d}{d\theta} r(\theta, \eta, a) = f(\eta, r(\theta, \eta, a), \theta) - \frac{1}{2\pi} \int_0^{2\pi} f(\eta, r(\varphi, \eta, a), \varphi) d\varphi
$$

with the side condition

$$
\frac{1}{2\pi} \int_0^{2\pi} r(\varphi, \eta, a) d\varphi = a.
$$

The function $r(\theta, \eta, a)$ is of period 2π in θ.

Proof. The proof is by a contraction mapping argument, similar to the standard existence/uniqueness theorem for ordinary differential equations. See Notes and References.   □

The idea of the Caesari–Hale method is that the differential equation

$$
\frac{dr}{d\theta} = f(\eta, r, \theta)
$$

has $r(\theta, \eta, a)$ (given by the lemma) as a periodic solution if and only if a satisfies $\Psi(\eta, a) = 0$, where

$$\Psi(\eta, a) = \frac{1}{2\pi} \int_0^{2\pi} f(\eta, r(\varphi, \eta, a), \varphi) d\varphi,$$

because under this condition the differential and functional differential equations coincide (along this particular solution).

The following theorem is stated in slightly greater generality than we need. Namely, F and U are not required to have the specific form that they have in (6.5.15); for instance, even powers of r are permissible in F. The number k in the theorem would be $2k + 2$ in (6.5.15):

6.5.3. Theorem. Let $\widetilde{F}(\eta, r, \theta)$ and $\widetilde{U}(\eta, r, \theta)$ be smooth functions, with $U \neq 0$ for $0 \leq r \leq r_0$ and $0 \leq \eta \leq \eta_0$. Suppose that the jets of $\widetilde{F}$ and $\widehat{U}$ to degree $k-1$ in r, and to some specified degree in η, are independent of θ, and write these jets as $F(\eta, r)$ and $U(\eta, r)$. Assume also that $F(\eta, 0) = 0$, so that the differential equations (6.5.21) below make sense in a neighborhood of the origin in the polar coordinate plane, with the origin as a rest point. Then there exists a smooth function $\Phi(\eta, a)$, called the *determining function*, having the following properties:

1. The solution $r(t, \eta, a)$, $\theta(t, \eta, a)$ of the initial value problem

$$\dot{r} = \widetilde{F}(\eta, r, \theta), \qquad (6.5.21)$$
$$\dot{\theta} = \widetilde{U}(\eta, r, \theta),$$
$$r(0, \eta, a) = a,$$
$$\theta(0, \eta, a) = 0,$$

is periodic if and only if $\Phi(\eta, a) = 0$.

2. The jet of $\Phi(\eta, a)$ is $F(\eta, a)$.

Therefore, if the polynomial $F(\eta, a)$ is sufficient to determine the root structure (near the origin) of any smooth function having F as its jet, then it is sufficient to determine the bifurcation structure of periodic solutions of (6.5.21).

Proof. Since $U \neq 0$ in the indicated region, F/U is defined and smooth; let G be its jet. The system (6.5.21) is then equivalent to a single scalar equation

$$\frac{dr}{d\theta} = G(\eta, r) + H(\eta, r, \theta), \qquad (6.5.22)$$

where the jet of H is zero. Since G is equal to its own average over θ, the Caesari–Hale functional differential equation associated with this differential equation is

$$\frac{dr}{d\theta} = H(\eta, r, \theta) - \frac{1}{2\varphi} \int_0^{2\pi} H(\eta, r(\varphi), \varphi) d\varphi.$$

Let the periodic solution with mean value a (which exists by Lemma 6.5.2) be denoted by $r(\theta, \eta, a)$; this will satisfy (6.5.22) if $\Psi(\eta, a) = 0$, where

$$\Psi(\eta, a) = \frac{1}{2\pi} \int_0^{2\pi} \left(G(\eta, r(\varphi, \eta, a)) + H(\eta, r(\varphi, \eta, a), \varphi) \right) d\varphi.$$

Since $r(\theta, \eta, a)$ is smooth, it can be expanded in a power series in a to degree $k - 1$ with smooth remainder, in the form

$$r(\theta, \eta, a) = r_0(\theta, \eta) + a r_1(\theta, \eta) + a^2 r_2(\theta, \eta) + \cdots$$
$$+ a^{k-1} r_{k-1}(\theta, \eta) + a^k \rho(\theta, \eta, a).$$

Substituting this into the functional differential equation and its side condition, and using the fact that the jet of H is zero, it follows that $r_0(\theta, \eta) = 0$, $r_1(\theta, \eta) = 1$, and $r_2 = \cdots = r_{k-1} = 0$, so that

$$r(\theta, \eta, a) = a + a^k \rho(\theta, \eta, a).$$

It follows at once that $\Psi(\eta, a) = G(\eta, a) + \mathcal{O}(a^k)$, so that the jet of Ψ is G. This is not quite what we wanted, but since $U \neq 0$, $\Phi = U\Psi$ is as good a determining function as Ψ, and its jet is F.   □

Notes and References

Every book on dynamical systems (for instance [111], [52], and [67]) contains a treatment of the Hopf bifurcation, but most do not treat the degenerate cases. Unfortunately, one of the best-known sources, Marsden and McCracken [76], is said to contain an error in its most general stability condition (Theorem 3B.4, page 93). The most complete treatment including degenerate cases is probably Chapter VIII of Golubitsky and Schaeffer [50].

The Caesari–Hale method is a modification of classical perturbation methods for periodic solutions. It is set forth in Chapters VIII and IX of Hale [53]. Under the hypotheses of Lemma 6.5.2, the differential equation $dr/d\theta = f$ can be reduced to the form of (1.1) on page 253 of [53] by dilating r by ε.

6.6   Bifurcations from the Nonsemisimple Double-Zero Eigenvalue (Takens–Bogdanov Bifurcations)

As in the last section, we recalculate the normal form and unfolding for the nonsemisimple double zero (obtained first in Section 1.2) using the methods of Chapter 4 (in this case especially Section 4.7) and Section 6.4. Then we discuss the scaling of the unfolded equations, and compare this with the

transplanting theory of Section 3.7. Then we briefly discuss the behavior
of the unfolded system, referring to the literature for the details.

The Normal Form and Unfolding

The system of differential equations familiarly called the "nonsemisimple
double zero" is the system $\dot{x} = Nx + \cdots$ with linear part

$$N = N_2 = \begin{bmatrix} 0 & 1 \\ 0 & 0 \end{bmatrix}. \tag{6.6.1}$$

To determine the simplified normal form, we first find the ring of invariants
$\ker \mathcal{X}$, where

$$\mathcal{X} = \mathcal{D}_{N^*} = x_1 \frac{\partial}{\partial x_2} = x \frac{\partial}{\partial y},$$

writing $(x, y) = (x_1, x_2)$. (See equations (4.7.4) and (4.7.5).) By inspection,
$\alpha = x$ is one invariant, and we claim that this generates the entire ring;
that is, we propose that

$$\ker \mathcal{X} = \mathbb{R}[[x]]. \tag{6.6.2}$$

To check this, notice that the weight of x (that is, its eigenvalue under
$\mathcal{Z} = x\partial/\partial x - y\partial/\partial y$) is one, so the table function of $\mathbb{R}[[x]]$ is

$$T = \frac{1}{1 - dw}.$$

Since

$$\frac{\partial}{\partial w} wT \bigg|_{w=1} = \frac{1}{(1 - d)^2},$$

Lemma 4.7.9 implies (6.6.2).

The next step, since N contains one Jordan block of size 2, is to compute
$\ker \mathcal{X}^2$ as a module over $\ker \mathcal{X}$. Since this entails going to "depth 2" in the
Jordan chain for $\mathcal{Y} = y\partial/\partial x$ with chain top x, we compute

$$\mathcal{Y}x = y$$

and conclude (since no "simplification of the basis" is needed) that the
required Stanley decomposition is

$$\ker \mathcal{X}^2 = \mathbb{R}[[x]] \oplus \mathbb{R}[[x]]y. \tag{6.6.3}$$

That is, every formal power series $h \in \ker \mathcal{X}^2$ can be written

$$h(x, y) = f(x) + g(x)y.$$

The Stanley basis of the simplified normal form is then $v_{(1,1)} = (0, 1)$ and
$v_{(1,y)} = (0, y)$, and the Stanley decomposition is

$$\widehat{N} = \mathbb{R}[[x]] \begin{bmatrix} 0 \\ 1 \end{bmatrix} \oplus \mathbb{R}[[x]] \begin{bmatrix} 0 \\ y \end{bmatrix}.$$

That is, the differential equations in simplified normal form are

$$\dot{x} = y,$$ (6.6.4)

$$\dot{y} = f(x) + g(x)y = (\alpha_1 x + \beta_1 y)x + (\alpha_2 x + \beta_2 y)x^2 + \cdots.$$

Notice that since only $\widehat{N}_+$, the quadratic-and-higher part of the normal form module, is used in the normal form (see (4.5.7), $f(x) = \alpha_1 x^2 + \alpha_2 x^3 + \cdots$ begins with the quadratic term, while $g(x) = \beta_1 x + \beta_2 x^2 + \cdots$ begins with the linear term. The normal form (6.6.4) agrees with (1.2.8) as far as that was calculated.

To unfold the system, by Theorem 6.4.4 we first write the general matrix in simplified normal form with respect to N,

$$G(\mu) = \begin{bmatrix} 0 & 0 \\ \mu_1 & \mu_2 \end{bmatrix},$$

and then move one unfolding parameter from G to the constant term. If we choose to move μ_1, the unfolded normal form will be

$$\dot{x} = y,$$ (6.6.5)

$$\dot{y} = \mu_1 + \mu_2 y + (\alpha_1 x + \beta_1 y)x + (\alpha_2 x + \beta_2 y)x^2 + \cdots.$$

The genericity condition on the quadratic terms under which this is valid is determined by examining the effect of a secondary shift on the quadratic terms of the normal form. The necessary calculation (according to the general theory of Section 6.4) is exactly the one that was done in equation (1.2.15), with c and d for μ_1 and μ_2; the genericity condition was found to be (in our present notation) $\alpha_1 \neq 0$. That is, any system of the form

$$\begin{bmatrix} \dot{x} \\ \dot{y} \end{bmatrix} = A(\eta) \begin{bmatrix} x \\ y \end{bmatrix} + F(\eta, x, y),$$

where $A(0) = N_2$ and F begins with quadratic terms, can be brought into the form (6.6.5) with $\mu_1, \mu_2, \alpha_1, \ldots, \beta_1, \ldots$ becoming functions of η, provided that the normal form of the system with $\eta = 0$ has x^2 appearing with a nonzero coefficient in the $\dot{y}$ equation. As we saw in looking at degenerate Hopf bifurcations, the way in which the coefficients depend upon η will have a strong effect on the bifurcations of the system, but nonetheless, the most important part of the work is to determine how (6.6.5) itself behaves under variation of its coefficients.

Scaling the Truncated Equation

For the next step, we truncate (6.6.5) at the quadratic terms and rewrite it in a form closer to that of Chapter 1 as

$$\dot{x} = y,$$ (6.6.6)

$$\dot{y} = \mu + \nu y + \alpha x^2 + \beta xy.$$

By changing the units of measurement in x, y, and t by constant factors, it can be arranged that $\alpha = 1$ (since we have already assumed $\alpha \neq 0$) and $\beta = 1, -1$, or 0. We take the case $\beta = 1$, and study

$$\dot{x} = y, \tag{6.6.7}$$
$$\dot{y} = \mu + \nu y + x^2 + xy.$$

6.6.1. Remark. This fact has a bearing on the meaning of the unfolding parameters and the codimension of the system. According to the definition of asymptotic unfolding given in Section 6.4, (6.6.6) is a codimension-4 unfolding; that is, it contains four unfolding parameters: μ and ν vary in a neighborhood of zero, and α and β vary in a neighborhood of their original values (or we add to α and β new unfolding parameters that vary near zero). Now suppose that the original values of α and β are nonzero, and (for convenience) are both positive; then any nearby system (even with α and β slightly varied) can be brought into the form (6.6.7). (The required change of units will vary with α and β.) Thus, under a slightly wider equivalence relation, the system has codimension two: The only remaining unfolding parameters are μ and ν. Suppose, on the other hand, that the original value of β is zero; then slight variations of β will have opposite signs, and it is no longer possible to reduce the system to (6.6.7). It is still possible to eliminate α, and the system has codimension 3. The changes of units for x and y correspond to a linear coordinate change using a diagonal matrix. This is not a near-identity transformation; it fits into the framework of hypernormalization, since it is a linear transformation that does not affect the linear term of the system but instead is used to simplify the higher-order terms.

Preliminary investigation of (6.6.7) shows that it has two rest points at $(x, y) = (\pm\sqrt{-\mu}, 0)$. It is easy to check as an exercise (the details are given in sources listed in Notes and References below) that one of these rest points is always a saddle, and the other has two complex conjugate eigenvalues that cross the imaginary axis when (μ, ν) crosses the curve $\mu = -\nu^2$; this suggests that this rest point undergoes a Hopf bifurcation and sheds a limit cycle as this curve is crossed. The interesting and difficult question is to follow the limit cycle after it bifurcates from the rest point and see how it disappears as μ and ν continue to be varied. Although we will not carry out the details of this study, since it requires several techniques unrelated to those developed in this book, we will focus momentarily on one step that is usually simply introduced as a given, without motivation. This step has a close resemblance to the transplanting or shearing method developed for linear systems in Section 3.7. We are going to perform an additional scaling of the variables x and y (and eventually t), and of the parameter ν, but this time not by a constant factor (as we just did to eliminate α and β) but by a fractional power of μ. Thus, the resulting transformation will not be defined at $\mu = 0$; if we think of (6.6.7) as a

two-parameter family (a "plant") having its "root" at the special case of
(6.6.7) with $(\mu, \nu) = (0, 0)$ (the organizing center of the Takens–Bogdanov
family), the system obtained after the scaling will have been "transplanted"
to a different root or organizing center. In Chapter 3, the original and
transplanted systems were linear, and the effect of the transplanting was
to break the nonsemisimplicity of the leading term. Here the effect will be
more complicated, because we are not simply shuffling around linear terms.

Although in the last paragraph we said that we would use fractional
powers of μ, it is more convenient (and equivalent) to introduce a new
parameter ε and write $\mu = -\varepsilon^r$, where r is an integer to be determined.
(The negative sign is necessary because the rest points exist for $\mu < 0$, and
the correct choice of r turns out to be even.) Thus, in (6.6.7) we set

$$x = \varepsilon^p u, \tag{6.6.8}$$
$$y = \varepsilon^q v,$$
$$\mu = -\varepsilon^r,$$
$$\nu = \varepsilon^s \delta.$$

The new variables are u, v, and the new parameters are ε, δ. The result of
this substitution is

$$\dot{u} = \varepsilon^{q-p} v, \tag{6.6.9}$$
$$\dot{v} = -\varepsilon^{r-q} + \varepsilon^s \delta v + \varepsilon^{2p-q} u^2 + \varepsilon^p uv.$$

The rest points of this system have $\varepsilon^{2p-q} u^2 - \varepsilon^{r-q} = 0$ and $v = 0$; it seems
natural to impose $2p - q = r - q$, or $r = 2p$, so that these rest points are
fixed at $(\pm 1, 0)$. It is also natural to make the leading terms of the two
equations equal. Since the leading term of the second equation is not clear
until the exponents are fixed, we must experiment and see what works.
Since $r - q$ is now equal to $2p - q$, setting $q - p = 2p - q$ (or $2q = 3p$)
makes three terms equal in order. (In the language of perturbation theory,
an assignment of exponents that makes two terms equal in importance
and dominant over the others is a "significant degeneration." One that
makes three terms equal is a "richer" significant degeneration because it
packs more information into the leading term.) The conditions $r = 2p$ and
$2q = 3p$, with p, q, r integers, have smallest solution $p = 2, q = 3, r = 4$. At
this stage (6.6.9) takes the form

$$\dot{u} = \varepsilon v,$$
$$\dot{v} = \varepsilon(u^2 - 1) + \varepsilon^s \delta v + \varepsilon^2 uv.$$

It is now natural to take $s = 2$ and to scale time by $\tau = \varepsilon t$, so that (with
$' = d/d\tau$) the system becomes

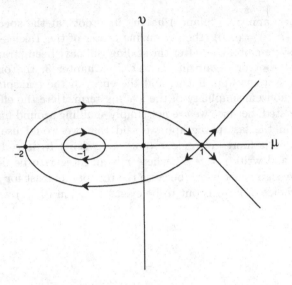

Figure 6.8. The dynamics of the root of the transplanted system.

$$u' = v,$$
$$v' = u^2 - 1 + \varepsilon(\delta v + uv).$$
(6.6.10)

The transplanted system (6.6.10) has for its unperturbed system (or "root," or organizing center) a system having no rest point at the origin, but instead having two rest points at a finite distance from the origin. This is a sign that this problem no longer belongs to the subject of local dynamical systems (and that we are very near to the end of this book.) From this point on, it becomes necessary to use global techniques, such as the study of homoclinic bifurcations, even though the original problem that we are concerned with was a local one; the reason for this change is the operation of transplanting to a new organizing center. We now sketch very briefly the treatment of the behavior of (6.6.10); see Notes and References for further information. The unperturbed system $\varepsilon = 0$ (the new organizing center) is Hamiltonian with $H(u, v) = v^2/2 + u - u^3/3$; its orbit structure is shown in Figure 6.8. All orbits within the homoclinic loop (from the saddle point to itself) are periodic. Since this is a transplanted root, it does not correspond to any actual behavior of the original system; our interest is actually in the periodic orbits that exist for small $\varepsilon > 0$. Let $(u(t, \varepsilon), v(t, \varepsilon))$ denote the solution with $(u(0, \varepsilon), v(0, \varepsilon)) = (u_0, 0)$ with $-2 < u_0 < -1$, and let $T(\varepsilon)$ be the first time at which this orbit again crosses the segment $-2 < u < -1$;

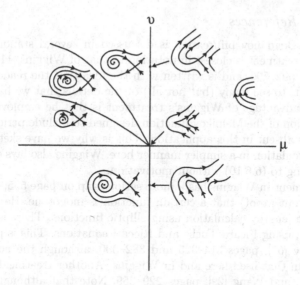

Figure 6.9. The Takens–Bogdanov bifurcation.

for $\varepsilon = 0$ this will be the period. Let $u_1(\varepsilon) = u(T(\varepsilon), \varepsilon)$. Then

$$H(u_1(\varepsilon), 0) - H(u_0, 0) = \int_0^{T(\varepsilon)} \frac{d}{dt} H(u(t, \varepsilon), v(t, \varepsilon)) dt$$

$$= \varepsilon \int_0^{T(0)} [\delta + u(t, 0)] v(t, 0)^2 dt + \mathcal{O}(\varepsilon^2)$$

$$= \varepsilon M(u_0, \delta) + \mathcal{O}(\varepsilon^2).$$

If this quantity is zero, the solution through u_0 is periodic. The implicit function theorem implies that any simple root u_0 of the *Melnikov function* M gives rise to a family $u_0(\varepsilon)$ of initial conditions of periodic orbits. The Melnikov function is an integral, along the unperturbed periodic orbit, of an expression derived from the perturbation terms. The treatment of this integral for $u_0 = -2$ is slightly different; it becomes an integral from $-\infty$ to ∞ around the unperturbed homoclinic loop. The remainder of the analysis consists in showing that the Melnikov function always has either exactly one simple zero, or no zeros, depending on δ, and that in the original coordinates this translates into the behavior indicated in Figure 6.9: The periodic solution originates in a Hopf bifurcation along the curve $\mu = -\nu^2$ and ends in a homoclinic bifurcation along the curve $\mu = -49\nu^2/25 + \mathcal{O}(\nu^{5/2})$.

Notes and References

The Takens–Bogdanov bifurcation is discussed in several standard text-books. Our treatment is closest in notation to that of Wiggins [111], pages 321–330 and 592–596, and is written with the idea that the reader should go there next, to see many (but not all) of the details that we have omitted. One disadvantage of Wiggins's treatment is that he employs a more difficult version of the Melnikov function designed to include periodic forcing, which is absent in this application. This is why we have sketched the Melnikov calculation in a simpler manner here. Wiggins also lays down the scaling leading to (6.6.10) without motivation.

The treatment in Wiggins has a more serious gap on page 595, where he claims (without proof) that a certain function is monotone. He indicates that this is a lengthy calculation using elliptic functions. There is another way to do it, using Picard–Fuchs and Riccati equations. This is presented in Kuznetsov [67], pages 314–325 and 382–390, although the notation is different from that used here and in Wiggins. Another treatment is given in Chow, Li, and Wang [29], pages 229–259. Note that although the homoclinic bifurcation in the Takens–Bogdanov system seems similar to the Andronov–Leontovich bifurcation treated in [67], it is different, because an unperturbed system that is Hamiltonian cannot satisfy the nonzero trace condition in the Andronov–Leontovich theorem.

The scaling that leads to (6.6.10), or to its equivalent in other notations, is sometimes referred to as "blowing up" the system, but this is an unfortunate terminology, since blowing up has another, distinctly different, meaning. In fact, this other kind of blowing up can also be applied to the Takens–Bogdanov problem. This technique is sketched in Guckenheimer and Holmes [52], where the Takens–Bogdanov bifurcation occupies pages 361–371, and is developed in more detail in the many writings of Dumortier, such as [39].

All of these sources cite the original papers by Takens and Bogdanov, as well as other contributions that should no doubt be consulted by anyone wishing to go more deeply into this subject.

6.7   Unfoldings of Mode Interactions

The phrase "mode interactions" is an oxymoron. The original context for the concept of modes was a diagonalizable linear system with pure imaginary eigenvalues. In diagonalized form, such a system reduces to a collection of uncoupled oscillators. Therefore, in the original coordinates the solutions are linear combinations of independent oscillations, or modes, each with its own frequency. The very notion of mode is dependent on the two facts that the system is linear and that the modes do not interact.

When nonlinear terms are added, the modes immediately cease to exist, except as approximations having some validity very close to the origin. At the same time these (nonexistent) modes are said to interact with each other through the nonlinear coupling terms, which do not disappear when the linear part is diagonalized. Although these terms cannot be completely eliminated, they can be simplified, through the use of normal form methods. Today, the term *mode interaction* has become synonymous with "system whose center manifold has dimension greater than two." It is generally assumed that a center manifold reduction has been performed before beginning the study, so that the linear part is an $n \times n$ matrix with $n \geq 3$ having all its eigenvalues on the imaginary axis; the value zero is not excluded (another contrast to the original idea of mode). It is the bifurcations of the system that are of greatest interest, so the unfolding must be computed (with the original linear system as organizing center). Thus, the simplest mode interaction problems are:

1. the "0/Hopf" interaction, with $n = 3$ and eigenvalues of 0 and $\pm i$ (or $\pm i\omega$);

2. the "double Hopf" interaction, which is the unfolding of the double center and comes in nonresonant and resonant flavors, with the $1 : 1$ resonance allowing a nonsemisimple variant;

3. the "00/Hopf" (or at greater length, the "nonsemisimple double-zero/Hopf" or "Takens–Bogdanov/Hopf") interaction, a four-dimensional system in which the linear part is block diagonal with a two-by-two nilpotent block and a single center.

Our (very limited) task in this final section of the book is to determine the asymptotic unfoldings of each of these systems, and to give references for their further study. In particular, we want to determine the number of parameters required in the constant and linear terms of the unfolding, because this gives a good idea of the codimension of the system. Of course, perturbations of the higher-order normal form coefficients must be allowed as further asymptotic unfolding parameters up to the desired order, and then this number of parameters is subject to further reduction through scaling of the variables, as in Remark 6.6.1. Working out these unfoldings and normal forms constitutes a review of many of the ideas in this book.

The 0/Hopf Interaction

The 0/Hopf interaction has for its linear part

$$A = \begin{bmatrix} 0 & & \\ & 0 & -1 \\ & 1 & 0 \end{bmatrix}$$

in real variables $x = (x_1, x_2, x_3)$, or in diagonal form,

$$\Lambda = \begin{bmatrix} 0 & & \\ & i & \\ & & -i \end{bmatrix}.$$

The diagonalized system uses variables (u, z, w), where $u = x_1$ is real, $z = x_2 + ix_3$, and $w = x_2 - ix_3$; the reality condition is that the third equation (for $\dot{w}$) must be the conjugate of the second on the reality subspace $\mathcal{R} \subset \mathbb{R} \times \mathbb{C}^2$ defined by $w = \bar{z}$. It will also be convenient to think in the cylindrical coordinates (u, r, θ), with $r^2 = x_2^2 + x_3^2 = z\bar{z}$.

In (u, z, w) coordinates, the $\dot{u}$ equation in normal form will contain monomials $u^j z^k w^\ell$ with $j(0) + k(i) + \ell(-i) = 0$, so that j is arbitrary and $k = \ell$. The $\dot{z}$ equation will contain monomials with $j(0) + k(i) + \ell(-i) = i$, so that j is arbitrary and $k = \ell + 1$. Thus, the quadratic terms in the complex normal form (with reality condition) will be

$$Q(u, z, w) = \begin{bmatrix} \alpha u^2 + \beta zw \\ (\gamma + i\delta)uz \\ (\gamma - i\delta)uw \end{bmatrix},$$

with $\alpha, \beta, \gamma, \delta \in \mathbb{R}$.

The invariants of e^{At} will be formal scalar fields that are unchanged by rotations around the u axis, so they will be power series of the form $f(u, r^2)$. The basic equivariants will be

$$v_1 = \begin{bmatrix} 1 \\ 0 \\ 0 \end{bmatrix}, \qquad v_2 = \begin{bmatrix} 0 \\ x_2 \\ x_3 \end{bmatrix}, \qquad v_3 = \begin{bmatrix} 0 \\ -x_3 \\ x_2 \end{bmatrix}$$

in Cartesian coordinates, or

$$v_1 = \begin{bmatrix} 1 \\ 0 \\ 0 \end{bmatrix}, \qquad v_2 = \begin{bmatrix} 0 \\ r \\ 0 \end{bmatrix}, \qquad v_3 = \begin{bmatrix} 0 \\ 0 \\ 1 \end{bmatrix}$$

in polar. Since the linear part in polar coordinates is $\dot{u} = 0$, $\dot{r} = 0$, $\dot{\theta} = 1$, the normal form will be

$$\dot{u} = f(u, r^2),$$
$$\dot{r} = rg(u, r^2),$$
$$\dot{\theta} = 1 + h(u, r^2),$$

where $f(u, r^2) = \alpha u^2 + \beta r^2 + \cdots$, $g(u, r^2) = \gamma u + \cdots$, and $h(u, r^2) = \delta u + \cdots$. That is, the normal form equals the linear part plus the most general vector field, beginning with quadratic terms (in Cartesian coordinates), that lies in the module over the ring of invariants generated by the basic equivariants. The S-foliation consists of the circles $u = c_1$, $r = c_2$; the preservation of the S-foliation is reflected in the decoupling of the (u, r) subsystem.

We compute the unfolding using the complex normal form. The general matrix in (semisimple, and therefore simplified) normal form with respect to A, and also satisfying the reality conditions, is

$$G(\mu) = \begin{bmatrix} \mu_1 & & \\ & \mu_2 + i\mu_3 & \\ & & \mu_2 - i\mu_3 \end{bmatrix}$$

for μ_1, μ_2, μ_3 real. Since the top row is principal (since it is the bottom row of a 1×1 nilpotent block), we move μ_1 to the constant term according to Theorem 6.4.4, so that the constant and linear terms of the unfolded system will be

$$\dot{u} = \mu_1 + \cdots,$$
$$\dot{z} = (\mu_2 + i(\mu_3 + 1))z + \cdots.$$

In polar coordinates the unfolded system (through terms that are quadratic in Cartesian coordinates) is (generically)

$$\dot{u} = \mu_1 + \alpha u^2 + \beta r^2 + \cdots,$$
$$\dot{r} = \mu_2 r + \gamma r u + \cdots,$$
$$\dot{\theta} = 1 + \delta u + \cdots.$$

To determine the genericity condition on the quadratic terms under which this unfolding is valid, we determine the secondary shifts, compute the matrix $E(h)$ defined by (6.4.11), and find the expression (6.4.14) corresponding to the entry μ_1 in $G(\mu)$ that we desire to shift to the constant term. Since the top row of A is not only the bottom row of its (nilpotent) Jordan block, but also the top row, (6.4.4) implies that the secondary shift will have the form $h = (h_{(1)}, 0, 0)$. Then

$$Q'h = \begin{bmatrix} 2\alpha u & \beta w & \beta z \\ (\gamma + i\delta)z & (\gamma + i\delta)u & 0 \\ (\gamma - i\delta)w & (\gamma - i\delta)u & 0 \end{bmatrix} \begin{bmatrix} h_{(1)} \\ 0 \\ 0 \end{bmatrix} = \begin{bmatrix} 2\alpha h_{(1)}u \\ (\gamma + i\delta)h_{(1)}z \\ (\gamma - i\delta)h_{(1)}w \end{bmatrix}$$

$$= \begin{bmatrix} 2\alpha h_{(1)} & & \\ & (\gamma + i\delta)h_{(1)} & \\ & & (\gamma - i - d)h_{(1)} \end{bmatrix} \begin{bmatrix} u \\ z \\ w \end{bmatrix} = E(h) \begin{bmatrix} u \\ z \\ w \end{bmatrix}.$$

Thus, the upper left entry in $D + E(h)$, as given by (6.4.14), will be $d_{11} + 2\alpha h_{(1)}$, and the condition for eliminating this entry (and thus "moving" μ_1 from this position to the constant term) will be

$$\alpha \neq 0.$$

The Double Hopf Interaction

Since the normal form of the double center has been discussed thoroughly in Section 4.5 and (for the nonsemisimple $1 : 1$ resonance) in Section 4.7, it is only necessary to determine the unfolding. The unfolding has three different forms according to the following division of cases:

1. The nonresonant case and all resonant cases other than $1 : 1$.

2. The nonsemisimple $1 : 1$ resonance.

3. The semisimple $1 : 1$ resonance.

These cases are listed in order of increasing difficulty.

In cases of type 1 in the above list, the linear part in complex diagonal form is

$$A = \begin{bmatrix} i\omega_1 & & & \\ & -i\omega_1 & & \\ & & i\omega_2 & \\ & & & -i\omega_2 \end{bmatrix}.$$

All eigenvalues are distinct, and the general matrix in normal form with respect to A, and also satisfying the reality conditions, is

$$G(\mu) = \begin{bmatrix} \mu_1 + i\mu_2 & & & \\ & \mu_1 - i\mu_2 & & \\ & & \mu_3 + i\mu_4 & \\ & & & \mu_3 - i\mu_4 \end{bmatrix}.$$

There are four real unfolding parameters in the linear part, and none of them should be moved to the constant term, since there are no zero eigenvalues (nilpotent blocks). Since no parameters are to be moved, there is no genericity condition imposed on the quadratic terms.

In cases of type 2, the linear part takes the form (4.5.82), namely,

$$A = \begin{bmatrix} i & 1 & & \\ 0 & i & & \\ & & -i & 1 \\ & & 0 & -i \end{bmatrix}.$$

The general matrix in simplified normal form with respect to A will have a single entry at the bottom of each stripe in the stripe structure associated with A; when the reality condition is included, this will be

$$G(\mu) = \begin{bmatrix} 0 & 0 & & \\ \mu_1 + i\mu_2 & \mu_3 + i\mu_4 & & \\ & & 0 & 0 \\ & & \mu_1 - i\mu_2 & \mu_3 - i\mu_4 \end{bmatrix}.$$

Again there are four real parameters, none to be moved, and no genericity condition.

For type 3, the linear part is (4.5.81):

$$A = \begin{bmatrix} i & & & \\ & i & & \\ & & -i & \\ & & & -i \end{bmatrix}.$$

This time there are two 1×1 Jordan blocks with the same eigenvalue i, so each of the four entries in the upper left corner of A is a stripe of the stripe structure; the same is true for the four entries in the lower right. Taking account of reality conditions,

$$G(\mu) = \begin{bmatrix} \mu_1 + i\mu_2 & \mu_3 + i\mu_4 & & \\ \mu_5 + i\mu_6 & \mu_7 + i\mu_8 & & \\ & & \mu_1 - i\mu_2 & \mu_3 - i\mu_4 \\ & & \mu_5 - i\mu_6 & \mu_7 - i\mu_8 \end{bmatrix}.$$

There are eight real parameters, none to be moved, and no genericity condition. This example illustrates the fact that nonsemisimple problems have a much lower codimension than semisimple problems with the same eigenvalues. If there are repeated eigenvalues of multiplicity greater than two, the case with the lowest codimension will be the one in which all of these eigenvalues belong to a single Jordan block. It is also the case that nonsemisimple problems (or problems with Jordan blocks as large as possible) are much more likely to be encountered in applications (given that there are repeated eigenvalues). Thus, in studying the 1 : 1 resonance, preference should be given to the nonsemisimple case, since it is both simpler and more prevalent. It has even been argued that one *should not* study problems of high codimension, since they are unlikely to arise. This is perhaps the advice of laziness. What is much more nearly true is that one should not study a problem of high codimension, such as the semisimple 1 : 1 resonance, unless one is willing to study the full problem with all of its unfolding parameters; the study of special cases is unlikely to be useful, since the "actual" differential equations (if there are such things) governing a physical system are likely to be small perturbations of the model equations that one is studying. But there are exceptions even to this, namely when the actual system satisfies some symmetry condition that forces some of the unfolding parameters to vanish. This is the reason that so many studies of complicated mode interactions take place under the assumption of various symmetry conditions.

The 00/Hopf Interaction

The nonsemisimple double-zero/Hopf interaction is the system with organizing center

$$\begin{bmatrix} 0 & 1 & & \\ 0 & 0 & & \\ & & 0 & -1 \\ & & 1 & 0 \end{bmatrix}$$

in real coordinates $x = (x_1, x_2, x_3, x_4)$. With $u = x_1$, $v = x_2$, $z = x_3 + ix_4$, $w = x_3 - ix_4$ it becomes

$$A = \begin{bmatrix} 0 & 1 & & \\ 0 & 0 & & \\ & & i & \\ & & & -i \end{bmatrix}.$$

The reality condition is that u and v are real and $w = \bar{z}$. The semisimple and nilpotent parts of A are

$$S = \begin{bmatrix} 0 & & & \\ & 0 & & \\ & & i & \\ & & & -i \end{bmatrix}, \qquad N = \begin{bmatrix} 0 & 1 & & \\ 0 & 0 & & \\ & & 0 & \\ & & & 0 \end{bmatrix}.$$

The general matrix in simplified normal form with respect to A, with reality conditions, is

$$G(\mu) = \begin{bmatrix} 0 & 0 & & \\ \mu_1 & \mu_2 & & \\ & & \mu_3 + i\mu_4 & \\ & & & \mu_3 - i\mu_4 \end{bmatrix}.$$

Thus, there are four parameters, one of which should be shifted to the second row of the constant term. If μ_1 is shifted, for instance, the constant and linear terms of the unfolding will be

$$\dot{u} = v + \cdots,$$
$$\dot{v} = \mu_1 + \mu_2 v + \cdots,$$
$$\dot{z} = (\mu_3 + (1 + \mu_4)i)z + \cdots,$$
$$\dot{w} = (\mu_3 - (1 + \mu_4)i)w + \cdots.$$

There will be a genericity condition on the quadratic terms in order for the shifting of μ_1 to be possible; to work this out requires that we work out the quadratic terms of the normal form. For this we use Theorem 4.6.16 together with (4.6.41). Since A has three Jordan blocks, s can take the values $1, 2, 3$. Each s contributes basis elements $v_{(s,h)} = he_{R_s}$ to the quadratic normal form, where $h \in K_{2s}$. (There is a slight notational conflict here: $v_{(s,h)}$ is not to be confused with the coordinate v.)

For $s = 1$, the space M_{21} is spanned by those quadratic monomials $u^j v^k z^\ell w^m$ for which $j(0) + k(0) + \ell(i) + m(-i) = 0$, so that j and k are unrestricted and $\ell = m$; these monomials are u^2, uv, v^2, and zw. Finally, K_{21} is the subspace of M_{21} in the kernel of $\mathcal{D}_{N^*}^2$, where

$$\mathcal{D}_{N^*} = u\frac{\partial}{\partial v}.$$

Since $\mathcal{D}_{N^*}^2(au^2 + buv + cv^2 + dzw) = 2cu^2$, we must have $c = 0$, so K_{21} is spanned by u^2, uv, and zw. Thus, $s = 1$ contributes the generators $u^2 e_2$, uve_2, and zwe_2. For $s = 2$, M_{22} is spanned by uz and vz, and $K_{22} = \ker \mathcal{D}_{N^*}$ is spanned by uz, contributing uze_3. It is not necessary to compute the $s = 3$ contribution, since the reality conditions dictate that the fourth entry in the vector field must be the conjugate of the third. So the quadratic term of the simplified normal form is

$$Q = \begin{bmatrix} 0 \\ \alpha u^2 + \beta uv + \gamma zw \\ (\delta + i\varepsilon)uz \\ (\delta - i\varepsilon)uw \end{bmatrix}.$$

The secondary shift is $h = (h_{(1)}, 0, 0, 0)$, and

$$Q'h = \begin{bmatrix} 0 \\ (2\alpha u + \beta v)h_{(1)} \\ (\delta + i\varepsilon)zh_{(1)} \\ (\delta - i\varepsilon)wh_{(1)} \end{bmatrix},$$

so

$$E(h) = \begin{bmatrix} 0 & 0 & 0 & 0 \\ 2\alpha h_{(1)} & \beta h_{(1)} & 0 & 0 \\ 0 & 0 & (\delta + i\varepsilon)h_{(1)} & 0 \\ 0 & 0 & 0 & (\delta - i\varepsilon)h_{(1)} \end{bmatrix}.$$

Thus, the genericity condition for moving μ_1 to the constant term is $\alpha \neq 0$; the condition for moving μ_2 instead would be $\beta \neq 0$.

The computation of the remaining terms of the normal form is left as an exercise for the reader, with the following hints to get started. The triad $\{X, Y, Z\}$ will be the same for the inner product, simplified, and sl(2) normal forms, and will have $Y = N^*$, so that $X = \mathcal{D}_{N^*} = u\partial/\partial v$. Then the ring of invariants of the nilpotent part of the problem is

$$\ker X = \mathbb{R}[[u, z, w]].$$

The table function that proves this to be the whole ring is

$$T = \frac{1}{(1 - dw)(1 - d)^2},$$

since z and w have weight zero. The rest of the calculation follows the same lines as the example of the nonsemisimple 1 : 1 resonance worked out in

Section 4.7: From the ring of invariants of the nilpotent part, one works out the module of its equivariants and intersects this with the equivariants of the semisimple part.

Notes and References

For the 0/Hopf interaction, see Langford and Zhan [68] and [69]. Apart from this, I am not sufficiently familiar with the literature on mode interactions to make recommendations beyond the basic texts by Wiggins [111], Guckenheimer and Holmes [52], and Kuznetsov [67]. With regard to [52], be sure to read only the "second printing, revised and corrected." I am told that the first printing, while not actually wrong, proposed some erroneous conjectures concerning mode interactions. W.F. Langford is in the early stages of preparing a book on mode interactions, which is likely to be important, but it will not appear for several years.

Appendix A
Rings

A.1 Rings, Ideals, and Division

By a *ring*, we will always mean a *commutative ring with identity*, that is, a collection of objects that can be added, subtracted, and multiplied, but not necessarily divided, and that satisfy all the rules of high school algebra except those involving division. In particular, every ring contains elements denoted by 0 and 1 that function respectively as additive and multiplicative identities. The cancellation law (that if $fh = gh$ and $h \neq 0$, then $f = g$) holds in a field because we can divide by h, but it is not automatically true in a ring; it holds in some rings but not in others. Since division is not assumed, it is natural that questions about the possibility of division in special cases will form a large part of the theory of rings. Most of Appendix A is devoted to exactly this question.

The easiest example of a ring is the set $\mathbb{Z}$ of integers, and issues of divisibility among integers forms a substantial part of number theory. We say that $n \in \mathbb{Z}$ *is divisible by* $m \in \mathbb{Z}$ if there exists $k \in \mathbb{Z}$ with $n = km$. The set of all integers divisible by m (equivalently, all *multiples* of m) is called the *ideal* $\langle m \rangle$ generated by m. Although not every integer is divisible by a given integer m, every integer *can be divided by m with remainder*, as long as $m \neq 0$. That is, every integer n can be written uniquely as

$$n = km + r, \tag{A.1.1}$$

for some integers k and r with $r \in \{0, 1, \ldots, |m|-1\}$. Notice that $km \in \langle m \rangle$; we have decomposed an arbitrary integer n into a part km that belongs to the ideal $\langle m \rangle$ and a part r that, when it is nonzero, is in some loose sense orthogonal to the ideal. Our goal in this appendix is to generalize these ideas to certain other rings.

> **A.1.1. Remark.** Warning: In spite of the ordinary meaning of English words, "divisible" is a technical word and is not equivalent to "can be divided." For instance, 6 is divisible by 3, and 5 is not, although 5 can certainly be divided by 3 with quotient 1 and remainder 2. "Divisible" means *exactly* divisible, with remainder zero.

The following definitions are fundamental:

1. If $\mathcal{R}$ is a ring, a nonempty subset $\mathcal{I} \subset \mathcal{R}$ is called an *ideal* if $\mathcal{I}$ is closed under addition (so that if $f, g \in \mathcal{I}$ then $f + g \in \mathcal{I}$) and "absorptive" under multiplication by elements of $\mathcal{R}$ (so that if $f \in \mathcal{R}$ and $g \in \mathcal{I}$ then $fg \in \mathcal{I}$). Notice that the closure rule for addition requires two elements of $\mathcal{I}$, but the "absorptive" rule for multiplication allows one factor to come from outside of $\mathcal{I}$. Observe that $\langle m \rangle \subset \mathbb{Z}$, as defined above, is an ideal in this sense.

2. A *finitely generated ideal* $\mathcal{I} = \langle f_1, \ldots, f_s \rangle$ in a ring $\mathcal{R}$ is the set of all "linear combinations" $g_1 f_1 + \cdots + g_s f_s$ with $g_1, \ldots, g_s \in \mathcal{R}$. (It is assumed that $f_1, \ldots, f_s \in \mathcal{R}$.) It is easy to check that $\mathcal{I}$ is an ideal, and that every ideal containing $f_1, \ldots, f_s$ contains $\mathcal{I}$; therefore $\mathcal{I}$ can be characterized as the smallest ideal containing the elements $f_1, \ldots, f_r$, which are known as *generators* of the ideal or as a *basis* for it. (We will avoid the word basis in this sense, except in certain phrases such as "Gröbner basis" and "standard basis" where it has become accepted. The reason is that our ideals are usually also vector spaces, and a vector space basis is quite different from a set of generators for an ideal, since it allows only scalar coefficients in forming linear combinations.)

3. A finitely generated ideal with a single generator is called a *principal ideal*, and a ring that has only principal ideals is called a *principal ideal ring*. The ring $\mathbb{Z}$ of integers is an example.

4. A *subring* of a ring is a nonempty subset that is itself a ring, with the operations of addition and multiplication being the same as those of the larger ring. This is equivalent to requiring the nonempty subset to be closed under addition, additive inverse, and multiplication, and to contain the multiplicative identity. Under this definition, an ideal is (usually) not a subring, because the only ideal that contains the identity is the full ring itself. (An ideal is a subring in all other respects, and is called a subring by authors who do not require rings to contain an identity.) A subring is (usually) not an ideal, because it is not (usually) absorptive.

5. A *ring homomorphism* is a map of rings that preserves addition and multiplication. (In symbols, $h : \mathcal{R}_1 \to \mathcal{R}_2$ is a ring homomorphism if $h(f + g) = h(f) + h(g)$ and $h(fg) = h(f)h(g)$ for all $f, g \in \mathcal{R}_1$.) The kernel $h^{-1}(0)$ of any ring homomorphism $h : \mathcal{R}_1 \to \mathcal{R}_2$ is an ideal in $\mathcal{R}_1$. An invertible homomorphism is called an isomorphism, and two rings are isomorphic (essentially the same) if there exists an isomorphism from one to the other.

If $\mathcal{R}$ is a ring, $\mathcal{I}$ is an ideal in $\mathcal{R}$, and $f, g \in \mathcal{R}$, we define

$$f \equiv g \quad \text{if and only if} \quad f - g \in \mathcal{I}, \tag{A.1.2}$$

and in this case, we say f is *congruent* to g *modulo* $\mathcal{I}$. (The word *modulo* is the ablative case of the Latin word *modulus* and means "with respect to the modulus.") The equivalence classes are called *cosets*. It is easy to check that if $f_1 \equiv g_1$ and $f_2 \equiv g_2$, then $f_1 + f_2 \equiv g_1 + g_2$ and $f_1 f_2 \equiv g_1 g_2$; therefore, the cosets form a ring, denoted by $\mathcal{R}/\mathcal{I}$. Although the term *quotient ring* is used for $\mathcal{R}/\mathcal{I}$, and the process of passing from $\mathcal{R}$ to $\mathcal{R}/\mathcal{I}$ is often called *dividing the ring by the ideal*, this is not division in the sense that we use the term in this appendix. It is common among mathematicians speaking informally to call the process "modding out" by the ideal (from the word *modulo*), or even "squashing" or "collapsing" the ideal, because in the quotient ring all elements of the ideal are squashed or collapsed into a single element (namely the zero element of the quotient ring).

The following definitions are not standard:

A.1.2. Definition. If $\mathcal{I}$ is an ideal in a ring $\mathcal{R}$, a *strong division process* for $\mathcal{R}$ with respect to $\mathcal{I}$ is a map $\psi : \mathcal{R} \to \mathcal{R}$ (not assumed to be a ring homomorphism) with the following properties:

1. $g \equiv \psi(g)$ for all $g \in \mathcal{R}$;

2. if $g \equiv h$, then $\psi(g) = \psi(h)$; and

3. $\psi(0) = 0$.

In other words, ψ maps each coset to a distinguished representative of that coset, and for the coset $\mathcal{I}$, the distinguished representative is zero. If ψ is a strong division process, $\psi(g) = r$ is called the *remainder* upon dividing g by $\mathcal{I}$ using process ψ.

Notice that property 1 of a strong division process implies the converse of property 2: If $\psi(g) = \psi(h)$, then $g - h = (g - \psi(g)) - (h - \psi(h)) \in \mathcal{I}$ and $g \equiv h$. The reader should check that properties 1, 2, and 3 imply:

4. $\psi(\psi(g)) = \psi(g)$; and

5. $g \in \mathcal{I}$ if and only if $\psi(g) = 0$.

A *weak division process* is a map $\psi : \mathcal{R} \to \mathcal{R}$ satisfying property 1 and a weak form of property 5:

5'. If $\psi(g) = 0$, then $g \in \mathfrak{I}$.

An example of a weak division process will appear in Section A.5.

An example of a strong division process is the ordinary division in the ring $\mathbb{Z}$ of integers: Let $\mathfrak{I} = \langle m \rangle$, and let $\psi(n) = r$ be the remainder upon dividing n by m as in (A.1.1). In this example the set of possible remainders is $\psi(\mathbb{Z}) = \{0, 1, \ldots, |m| - 1\}$; these numbers are just those that form the familiar "ring of integers modulo m," which is isomorphic to the quotient ring $\mathbb{Z}/\langle m \rangle$. The ring structure on $\{0, 1, \ldots, |m| - 1\}$ is defined by first applying the ordinary ring operations on these numbers as elements of $\mathbb{Z}$, followed by taking their remainder upon dividing by m. We will now show that the same construction leads to a "concrete realization" of $\mathcal{R}/\mathfrak{I}$, given a strong division process for $\mathcal{R}$ with respect to $\mathfrak{I}$.

A.1.3. Lemma. Let $\mathcal{R}$ be a ring, $\mathfrak{I}$ an ideal in $\mathcal{R}$, and ψ a strong division process for $\mathcal{R}$ with respect to $\mathfrak{I}$. Then the set $\psi(\mathcal{R})$, with the operations

$$a \oplus b = \psi(a + b),$$
$$a \odot b = \psi(ab),$$

forms a ring isomorphic to $\mathcal{R}/\mathfrak{I}$, and the zero element in this ring structure on $\psi(\mathcal{R})$ is the zero element of $\mathcal{R}$. (In the displayed equations, $a + b$ and ab on the right-hand side denote the sum and product in $\mathcal{R}$.)

> **A.1.4. Remark.** In the sequel we will use $a + b$ and ab for $a \oplus b$ and $a \odot b$, relying on the context to make clear which ring operations are intended.

Proof. By the first two properties of a strong division process, ψ establishes a one-to-one correspondence between cosets in $\mathcal{R}$ and elements of $\psi(\mathcal{R})$. Since $\psi(a + b) \equiv a + b$, $a \oplus b$ is the distinguished representative of the coset containing $a + b$, which is the sum (in the ring structure on $\mathcal{R}/\mathfrak{I}$) of the cosets determined by a and b. Similarly, $\odot$ corresponds to multiplication of cosets. Finally, the third property of strong division implies that the zero elements of $\psi(\mathcal{R})$ and $\mathcal{R}$ coincide. $\qquad\square$

Although $\psi : \mathcal{R} \to \mathcal{R}$ is not a homomorphism, $\psi : \mathcal{R} \to \psi(\mathcal{R})$ is a homomorphism using the ring structure on $\psi(\mathcal{R})$. In fact, it is the concrete version of the natural projection homomorphism $\mathcal{R} \to \mathcal{R}/\mathfrak{I}$.

One example of Lemma A.1.3 has already been mentioned, namely, the integers modulo m. For another example, let $\mathcal{R}$ be the ring of formal power series $g(x) = \sum a_i x^i$ in a real variable x, with real coefficients, and let $\mathfrak{I} = \langle x^{k+1} \rangle$ be the subset of formal power series that are multiples of x^{k+1} (so that their first nonzero coefficient is a_j for some $j \geq k + 1$). A strong division process may be specified by the truncation map

$$\psi(g)(x) = a_0 + a_1 x + \cdots + a_k x^k. \tag{A.1.3}$$

Then $\psi(\mathcal{R})$ is the ring whose elements are polynomials of degree $\leq k$, with ordinary polynomial addition, but with multiplication defined as ordinary multiplication of polynomials followed by truncation at degree k. (The addition is ordinary because it does not generate terms that must be truncated.)

The following is a list of the rings that are used in this book and that will be studied in the remainder of this appendix. *Unless otherwise stated, the word "ring" from now on refers to one of the rings in this list.* The first column gives the symbol we use for the ring; see also Definition 4.1.1. The second column gives the symbol (if any) commonly used for the same ring in the algebraic literature. The third column names the ring in words. For each of these rings, the addition and multiplication are defined in the usual manner for the objects in question; the cases that are less familiar are discussed below. Loosely speaking, we will sometimes refer to elements of any of these rings as "functions." Strictly speaking, only the first two rings are rings of functions, because only elements of these have definite values at points other than the origin. Recall from Definition 4.1.1 that $\mathcal{F}_j^n$ denotes homogeneous polynomials in n variables of degree j. Replacing j by a star indicates finite sums over j, while a double star indicates infinite (formal) sums. Do not confuse $\mathcal{F}_j^n$ with $\mathcal{F}_{[k]}^n$, a symbol that was not defined in Chapter 4:

$\mathcal{F}_*^n$	$\mathbb{R}[x_1,\ldots,x_n]$	real polynomials in n variables
$\mathcal{F}^n$	$\mathbb{C}^\infty(\mathbb{R}^n,\mathbb{R})$	smooth functions $f : \mathbb{R}^n \to \mathbb{R}$
$\mathcal{E}^n$		smooth germs at $0 \in \mathbb{R}^n$
$\mathcal{F}_{**}^n$	$\mathbb{R}[[x_1,\ldots,x_n]]$	formal power series
$\mathcal{F}_{[k]}^n$	$\mathbb{R}[[x_1,\ldots,x_n]]/\mathcal{M}^{k+1}$	truncated power series (modulo $\mathcal{M}^{k+1}$)

The symbol $\mathcal{M}^{k+1}$ will be defined in Section A.2 below, but $\mathcal{F}_{[k]}^n$ is simply the ring of polynomials of degree $\leq k$, with the usual addition and with multiplication defined as ordinary multiplication truncated at degree k; this is a generalization of (A.1.3). The only other ring in this list that may require explanation is $\mathcal{E}^n$. Let $\mathcal{I} \subset \mathcal{F}^n$ be the set of smooth real-valued functions on $\mathbb{R}^n$ that vanish in a neighborhood of the origin. (The neighborhood is not fixed, but may depend on the function.) Then $\mathcal{I}$ is an ideal in $\mathcal{F}^n$, and $\mathcal{E}^n = \mathcal{F}^n/\mathcal{I}$. That is, a germ is an equivalence class of functions under the relation $f \equiv g$ if and only if $f - g$ vanishes near the origin. A member of such an equivalence class is called a *representative* of the germ, and we will commonly ignore the distinction between a germ and any one of its representatives. Since a smooth function defined near the origin can always be extended to a smooth function on all of $\mathbb{R}^n$, such a "local function" suffices to define a germ, and it is common to view a germ as an equivalence class of local functions.

A germ has a value at the origin, because all functions belonging to (or representing) the germ must have the same value at the origin, but a germ cannot properly be said to have a value at any other point, because at any point other than the origin there will be representatives of the germ that take different values. However, the partial derivatives (to all orders) of a germ are well-defined at the origin: Any two representatives of the germ must agree on some neighborhood of the origin, and hence must have the same derivatives at the origin. The *Taylor series* of a germ f, also known as the *infinity-jet*, is the formal power series

$$(j^\infty f)(x) = \sum_m \frac{(D^m f)(0)}{m!} x^m; \qquad (A.1.4)$$

the *kth-degree Taylor polynomial*, or *k-jet*, is the truncation

$$(j^k f)(x) = \sum_{|m| \leq k} \frac{(D^m f)(0)}{m!} x^m. \qquad (A.1.5)$$

These equations use multi-indices $m = (m_1, \ldots, m_n)$ of nonnegative integers, with the following conventions:

$$x^m = x_1^{m_1} \cdots x_n^{m_n}, \qquad (A.1.6)$$
$$m! = m_1! \cdots m_n!,$$
$$D^m = \frac{\partial^{m_1}}{\partial x_1^{m_1}} \cdots \frac{\partial^{m_n}}{\partial x_n^{m_n}},$$
$$|m| = m_1 + \cdots + m_n.$$

The motivation for the definition of the Taylor series is that on polynomials, j^∞ is the identity ($j^\infty f = f$). One version of Taylor's theorem, which asserts the asymptotic validity of the Taylor series of smooth functions, will be proved below; see Corollary A.2.15.

There are natural maps from each ring listed above to the one following it. Every polynomial is a smooth function, so the map $\mathcal{F}^n_* \to \mathcal{F}^n$ is simply the inclusion. The map $\mathcal{F}^n \to \mathcal{E}^n$ takes a function to its equivalence class as just defined. The map $j^\infty : \mathcal{E}^n \to \mathcal{F}^n_{**}$ assigns to each germ its formal Taylor series, or infinity-jet. The map $\mathcal{F}^n_{**} \to \mathcal{F}^n_{[k]}$ is the truncation at degree k, so that the composite of the last two maps is j^k, which assigns to each germ its kth Taylor polynomial, or k-jet. Each of these maps is a ring homomorphism. (The fact that j^∞ is a ring homomorphism involves the higher-order product rules for partial derivatives.)

Each of the rings $\mathcal{R}$ in our list is also an *algebra* over $\mathbb{R}$. This means that in addition to being a ring, $\mathcal{R}$ is a vector space over $\mathbb{R}$, and the ring and vector space structures are related as follows. The ring structure and vector space structure each contain an addition operation; these are the same operation (namely, the ordinary addition in each of our rings). The multiplication by real numbers c (part of the vector space structure) relates to the multiplication in the ring structure by the rule $c(fg) = (cf)g$. (All of our rings $\mathcal{R}$ contain $\mathbb{R}$ as a subring in a natural way, so this last condition is automatic.) All ring homomorphisms that we encounter are also algebra homomorphisms (that is, they are simultaneously ring homomorphisms and linear maps over $\mathbb{R}$). The only ring ever mentioned in this book that is not an algebra is the ring of integers $\mathbb{Z}$. We will seldom use the word "algebra" in the sense described in this paragraph; instead we will simply say "ring." (We do make considerable use of Lie algebras, but these are not algebras in the present sense because they are not rings in our sense. Recall that by ring we mean commutative ring with identity. Lie algebras are not commutative, or even associative, nor do they have an identity.)

One's first thought with regard to division in $\mathcal{F}^n$ might be that f/g is defined if and only if g is nowhere zero. This is correct if we think of division as done in the real number system, *after* f and g are evaluated, but that is not what we mean by division in a ring. Rather, just as in the case of integers, f is said to be divisible by g if there exists a "quotient" function $q \in \mathcal{F}^n$ such that $f = gq$, where the product of g and q is taken in the ring (which in turn means $f(x) = g(x)q(x)$ for all x). Thus, g may be zero at some points and still divide f, if f is zero at the same points. (This is a necessary, not a sufficient, condition.) There is no requirement here that q be unique. For instance, if f and g vanish on an open set, q is arbitrary there (except for the smoothness requirement). Although it is sometimes possible to "divide," in this sense, by a function that has zeros, the case in which g is never zero is still of importance, because such a g can divide any function. In particular, g divides the function that is identically one, so there exists a smooth function q with $g(x)q(x) = 1$; it follows that $q(x) = 1/g(x)$ (with the latter division performed in $\mathbb{R}$), and q is therefore unique. The relationship between g and q is then symmetrical (q is nonvanishing, and $g(x) = 1/q(x)$). An element u of a ring $\mathcal{R}$ that divides the multiplicative identity $1 \in \mathcal{R}$, and hence has a multiplicative inverse, is called a *unit*. We have just shown that the units in $\mathcal{F}^n$ are the nonvanishing functions. (Of course, the multiplicative inverse of a function with respect to a ring structure has nothing to with the inverse of the function in the mapping sense.)

From the standpoint of this appendix, the most important reason for passing from $\mathcal{F}^n$ to $\mathcal{E}^n$ is that it makes many more units available. A function is a unit if and only if it is nowhere zero; a germ is a unit if and only if it (or more precisely, one of its representatives) is nonzero at the origin. (It will then be nonzero in a neighborhood of the origin, and its reciprocal in the neighborhood defines the germ that is the multiplicative inverse.)

Notes and References

Excellent general references for ring theory are Zariski and Samuel [115] and Eisenbud [40]. The books by Cox, Little, and O'Shea, [32] and [33], and Adams and Loustaunau [1], emphasize the ideas surrounding division processes and Gröbner bases, and are thus most closely related to this appendix. The notions of strong and weak division processes have not previously been defined as such, although there is nothing new involved. These definitions owe something to the (more technical) notion of *normal form* as defined on page 121 of Lunter [74]; this is not a normal form in the sense of this book, but rather, "heuristically ... performs a division."

A.2   Monomials and Monomial Ideals

A *monomial* in n variables is a product

$$x^m = x_1^{m_1} x_2^{m_2} \cdots x_n^{m_n}, \tag{A.2.1}$$

where $m = (m_1, \ldots, m_n)$ is a multi-index as in (A.1.6). The *degree* of the monomial is

$$|m| = m_1 + \cdots + m_n. \tag{A.2.2}$$

Each ring in the list in the last section contains monomials, and the maps between these rings preserve the monomials. (In the case of $\mathcal{E}^n$, a monomial is, technically speaking, replaced by its equivalence class, but the monomial in the usual sense reappears when we pass to a jet.) In this section we consider the following question, which has different answers in different rings: When is a function f divisible by a monomial?

The Case $n = 1$

For functions of one variable ($n = 1$), the answers are trivial. Writing x for x_1 and m for m_1, the following statements are clear:

A.2.1. Lemma. A polynomial $a_0 + a_1 x + \cdots + a_k x^k$ is divisible by x^m if and only if $a_j = 0$ for $j < m$. The same is true for formal power series.

The polynomials in Lemma A.2.1 can be replaced by analytic functions (that is, *convergent* power series) without changing the proof: If some leading terms of a convergent power series are zero, it is possible to factor out a power x^m, and the series that remains will still converge. The corresponding fact for $\mathcal{F}^1$ and $\mathcal{E}^1$ is not so trivial, and cannot be done by power series arguments: The same x^m can be factored out of the Taylor series, but since it need not converge, the series that remains does not define a function. (If the Borel–Ritt theorem, given in Section A.3, is invoked here, one obtains a function, but it is not unique and its product with x^m need not equal the original function.) Instead, an argument using integration is successful. Notice that the hypothesis of the following lemma reduces to that of Lemma A.2.1 for polynomials and formal power series (using formal differentiation):

A.2.2. Lemma. A smooth function f of one real variable is divisible by x^m if and only if $f(0) = f'(0) = \cdots = f^{(m-1)}(0) = 0$, that is, if and only if $j^{m-1}(f) = 0$.

Proof. If $f(0) = 0$, then by the fundamental theorem of calculus and the substitution $s = tx$,

$$f(x) - 0 = \int_0^x f'(s)ds = xg(x),$$

where

$$g(x) = \int_0^1 f'(tx)dt,$$

which is smooth (by the theorem permitting "differentiation under the integral sign"). Thus, f is divisible by x. Since $f'(x) = g(x) + xg'(x)$, if also $f'(0) = 0$, then $g(0) = 0$, so the argument can be repeated: $g(x) = xh(x)$, so $f(x) = x^2 h(x)$. The pattern continues.   □

The Case $n > 1$

It is somewhat surprising that while the analogue of Lemma A.2.1 is true in any number of variables, the natural candidate for a generalization of Lemma A.2.2 is false. (Valid generalizations will be developed later in this section.) Define

$$\mu \prec m \quad \text{if and only if for some } i, \quad \mu_i < m_i. \tag{A.2.3}$$

A.2.3. Lemma. In the rings $\mathcal{F}_*^n$ and $\mathcal{F}_{**}^n$, an element

$$f(x) = \sum a_\mu x^\mu$$

(summed over finitely many or infinitely many multi-indices μ) is divisible by x^m (that is, belongs to the ideal $\langle x^m \rangle$) if and only if $a_\mu = 0$ for all $\mu \prec m$. It is false that a function $f \in \mathcal{F}^n$ is divisible by x^m if $(D^\mu f)(0) = 0$ for all $\mu \prec m$.

Proof. The condition on a_μ implies that each term occurring (with a nonzero coefficient) in f is divisible by x^m, and therefore f is divisible. The falsity of the theorem for $\mathcal{F}^n$ follows from the existence of a function $f(x) = \varphi(x_1)$, depending on the first variable only, for which all derivatives vanish at the origin but f is not identically zero. (Such "flat" functions will be discussed in the next section.) Although all derivatives of f are zero at the origin, it is not possible to write $f(x) = x_2 g(x)$, because it would follow (by setting $x_2 = 0$) that f is identically zero, which is not true.   □

The condition on a_μ in the last lemma is easily visualized using a Newton diagram (see Section 6.1) in the case $n = 2$. Placing the numbers a_μ at the lattice points in the first quadrant of the (μ_1, μ_2) plane, a polynomial or formal power series is divisible by $x_1^2 x_2$, for instance, if it belongs to the following Newton diagram, that is, if it has zero entries at the places marked zero (which extend to infinity in both directions) and arbitrary entries at the other points:

$$
\begin{array}{cccccc}
\uparrow & \uparrow & \cdot & \cdot & \cdot & \\
0 & 0 & \cdot & \cdot & \cdot & \\
0 & 0 & * & \cdot & \cdot & \\
0 & 0 & 0 & 0 & \rightarrow &
\end{array}
\tag{A.2.4}
$$

The position marked $*$ is the lattice point associated with the generator $x_1^2 x_2$ of the ideal $\langle x_1^2 x_2 \rangle$; Lemma A.2.3 states that a power series belongs to the Newton diagram if and only if it belongs to the ideal. Notice that every monomial that belongs to the ideal (i.e., is a multiple of $x_1^2 x_2$) corresponds to a lattice point lying in the infinite rectangle whose lower left corner is the asterisk, and a power series belongs to the ideal if and only if each of its terms belongs to the ideal. The counterexample to the lemma in the smooth case can be pictured (roughly) as having a nonzero term "at infinity on the μ_1 axis" of the Newton diagram that does not appear at any finite order but still prevents the function from belonging to the ideal.

The statement (in the last paragraph) that "a series belongs to an ideal if and only if its terms belong to the ideal" is false in general. It is true for ideals of the form $\langle x^m \rangle$, and more generally for *monomial ideals*, defined as follows:

A.2.4. Definition. $\mathfrak{I} \subset \mathcal{R}$ is a *monomial ideal* if and only if there exists a finite set $\{f_1, \ldots, f_r\}$ such that each f_i is a monomial and $\mathfrak{I} = \langle f_1, \ldots, f_r \rangle$.

Notice that a monomial ideal need not be *presented as* generated by monomials. For instance, in $R[x, y]$, $\langle x-y, x+y \rangle$ is a monomial ideal because it is equal to $\langle x, y \rangle$. The use of Newton diagrams to represent monomial ideals (for $n = 2$) is very useful, but loses much of its effectiveness in later sections of this appendix, where we consider ideals that are not generated by monomials.

There are two ways to overcome the difficulty presented by flat functions in the smooth case. The first (Theorem A.2.7 and its corollaries) is to weaken both the hypothesis and the conclusion; we assume only that finitely many partial derivatives vanish, and conclude that f belongs to an ideal generated by finitely many monomials (rather than just one). The second (Theorem A.2.16) is to strengthen the hypothesis on f by imposing an asymptotic order condition, which is stronger than the vanishing of the infinitely many derivatives $D^\mu f(0)$ mentioned in Lemma A.2.3. This stronger hypothesis implies divisibility by a single monomial. We begin with a special case of the weak version, which is motivated by following the proof technique of Lemma A.2.2 rather than its statement.

A.2.5. Lemma. Let $f \in \mathcal{F}^n$ or $\mathcal{E}^n$, with $f(0) = 0$. Then there exist smooth functions (or germs) $g_1, \ldots, g_n$ such that

$$f(x) = g_1(x)x_1 + \cdots + g_n(x)x_n. \qquad (A.2.5)$$

Proof. Let $f_i = \partial f / \partial x_i$ and

$$g_i(x) = \int_0^1 f_i(tx)dt.$$

Then by the fundamental theorem of calculus and the chain rule, $f(0) = 0$ implies

$$f(x) = \int_0^1 \frac{d}{dt} f(tx)dt = \int_0^1 \sum_{i=1}^n f_i(tx)x_i dt = \sum_{i=1}^n x_i g_i(x).$$

$\square$

Let

$$\mathcal{M} = \langle x_1, \ldots, x_n \rangle \qquad (A.2.6)$$

be the ideal generated by the coordinate functions $x_1, \ldots, x_n$. (Each x_i is to be viewed as an element of $\mathcal{F}^n$, namely, the projection $(x_1, \ldots, x_n) \mapsto x_i$.) Then the conclusion (A.2.5) of Lemma A.2.5 may be stated as $f \in \mathcal{M}$. There are actually several different ideals $\langle x_1, \ldots, x_m \rangle$, one in each of the rings listed in the last section, but we will refer to all of these as $\mathcal{M}$, relying on the context to distinguish the meaning. In each of these rings it is the case that $f(0) = 0$ if and only if $f \in \mathcal{M}$; for the polynomial and power series rings this is obvious, and for smooth functions and germs it is proved in Lemma A.2.5. For $\mathcal{F}_*^2$ and $\mathcal{F}_{**}^2$, the ideal $\mathcal{M}$ may be represented by the following Newton diagram:

$$\begin{array}{ccc} \cdot & \cdot & \cdot \\ * & \cdot & \cdot \\ 0 & * & \cdot \end{array} \qquad (A.2.7)$$

That is, a polynomial or formal power series in two variables belongs to $\mathcal{M}$ if and only if its constant term (corresponding to the lower left-hand lattice point in the Newton diagram) is zero, and $\mathcal{M}$ is generated by the monomials corresponding to the lattice points with asterisks, $\mathcal{M} = \langle x_1, x_2 \rangle$. The monomials belonging to the ideal correspond to lattice points in the union of the infinite rectangles with corners at the asterisks, and once again a series belongs to the ideal if and only if each term does. However, unlike (A.2.4), this Newton diagram is valid also for $\mathcal{F}^2$ and $\mathcal{E}^2$; that is, the ideal $\mathcal{M}$ in each of these rings is generated by the same starred monomials, and a function or germ f belongs to $\mathcal{M}$ if and only if $f(0) = 0$. (This is the content of Lemma A.2.5.) It will shortly become apparent that the reason for this difference between the cases represented by (A.2.4) and (A.2.7) is that in the latter, there are only finitely many monomials not belonging to the ideal. This is visible on the Newton diagrams in the fact that in (A.2.7) the stars separate off a finite number of lattice points marked zero, whereas in (A.2.4) the zeros extend to infinity.

A.2.6. Remark. In each of these rings $\mathcal{M}$ is a *maximal ideal*, that is, one that is not contained in any larger ideal other than the ring itself; equivalently, $\mathcal{I} \subset \mathcal{R}$ is maximal if $\mathcal{R}/\mathcal{I}$ is a field. (In our cases, this field is $\mathbb{R}$.) In each ring in the list *other than* $\mathcal{F}_*^n$ and $\mathcal{F}^n$, $\mathcal{M}$ is the only maximal ideal. Rings with a unique maximal ideal are called *local rings*. To see that $\mathcal{F}_*^n$ and $\mathcal{F}^n$ are not local rings, consider the case $n = 1$. In $\mathcal{F}_*^1$, $\mathcal{M} = \langle x \rangle$ is a maximal ideal, and so is $\langle x + 1 \rangle$, because the quotient ring by either ideal is $\mathbb{R}$; since there is more than one maximal ideal, $\mathcal{F}_*^1$ is not a local ring. This does not happen in $\mathcal{F}_{**}^1$ and $\mathcal{E}^1$, where $1+x$ is a unit and therefore $\langle 1+x \rangle$ is the entire ring, not a maximal ideal.

The *product* of two finitely generated ideals is the ideal generated by the products of the generators:

$$\langle f_1, \ldots, f_r \rangle \langle g_1, \ldots, g_s \rangle = \langle f_1 g_1, \ldots, f_1 g_s, f_2 g_1, \ldots, f_2 g_s, \ldots, f_r g_s \rangle.$$
$$(A.2.8)$$

(We will not need the more general definition of the product of two arbitrary ideals.) In particular, the $(k+1)$st power $\mathcal{M}^{k+1}$ of the maximal ideal $\mathcal{M}$ (in any of our rings) is the ideal generated by all monomials x^m with $|m| = k+1$. For instance, in $\mathcal{F}_{**}^2$ the ideal $\mathcal{M}^3$ is represented by the Newton diagram

$$
\begin{array}{ccccc}
 \cdot & \cdot & \cdot & \cdot & \cdot \\
* & \cdot & \cdot & \cdot & \\
0 & * & \cdot & \cdot & \\
0 & 0 & * & \cdot & \\
0 & 0 & 0 & * & \cdot
\end{array}
\qquad (A.2.9)
$$

The generators of $\mathcal{M}^3 = \langle x_1^3, x_1^2 x_2, x_1 x_2^2, x_2^3 \rangle$ are marked by asterisks; a monomial belongs to the ideal if and only if it lies in the union of the rectangles determined by the asterisks; a power series belongs to the ideal if and only if its terms belong, or equivalently, if and only if the coefficients corresponding to the lattice points below the stars are zero. The content of the next theorem is that these statements remain true for smooth functions and germs (with the lattice points understood as representing the coefficients of the Taylor series).

A.2.7. Theorem. Let $f \in \mathcal{F}^n$ or $\mathcal{E}^n$, and suppose that f and all of its partial derivatives of order $\leq k$ vanish at the origin. Then $f \in \mathcal{M}^{k+1}$; that is, there exist smooth functions (or germs) g_m, defined for each multi-index m such that $|m| = k + 1$, such that

$$f(x) = \sum_{|m|=k+1} g_m(x) x^m. \qquad (A.2.10)$$

Proof. The proof is by induction on k, with $k = 0$ being Lemma A.2.5. Suppose, as the induction hypothesis, that the lemma is known for $\mathcal{M}^k$, and suppose that f satisfies the hypotheses of the lemma as stated (for $\mathcal{M}^{k+1}$). Then by the $\mathcal{M}^k$ case, we have $f \in \mathcal{M}^k$, so that f can be written

$$f(x) = \sum_{|m|=k} h_m(x) x^m. \tag{A.2.11}$$

We claim that each h_m satisfies $h_m(0) = 0$. When this is proved, each h_m may be written as in (A.2.5), and when these expressions are substituted into (A.2.11), the result is (A.2.10).

Fix μ with $|\mu| = k$, and apply D^μ to (A.2.11). For $m \neq \mu$, we have $D^\mu x^m = 0$ at $x = 0$, either because some entry of μ is greater than the corresponding entry of m (and hence $D^\mu x^m = 0$ identically) or because some entry of μ is less than the corresponding entry of m (and then $D^\mu x^m$ contains a factor of one of the x_i). Therefore,

$$D^\mu f(0) = \mu! h_\mu(0).$$

But $D^\mu f(0) = 0$ by hypothesis, and therefore $h_\mu(0) = 0$. $\square$

Theorem A.2.7 can be restated in the following form, using Definition (A.1.2):

A.2.8. Corollary. The map $\psi = j^k$, defined in (A.1.5), is a strong division process for $\mathcal{F}^n$ (or $\mathcal{E}^n$) with respect to the ideal $\mathcal{M}^{k+1}$.

Proof. Since $f - j^k(f)$ satisfies the hypotheses of Theorem A.2.7, it belongs to $\mathcal{M}^{k+1}$, and therefore $f \equiv j^k(f)$, which is property 1 of a strong division process according to Definition A.1.2. The other properties are obvious. $\square$

For applications it is necessary to extend this to "asymmetrical" or "lopsided" jets. (See, for instance, the second proof of Theorem 6.1.6.) These jets are associated with *monomial ideals of finite codimension* in $\mathcal{E}^n$ or $\mathcal{F}^n$. The following definition is given in greater generality than we need at the moment:

A.2.9. Definition. An ideal $\mathcal{I}$ (whether it is a monomial ideal or not) in one of the rings $\mathcal{R} = \mathcal{F}^n_*$, $\mathcal{F}^n$, $\mathcal{E}^n$, $\mathcal{F}^n_{**}$, or $\mathcal{F}^n_{[k]}$ has *finite codimension* if there are only finitely many monomials in $\mathcal{R}$ that do not belong to $\mathcal{I}$. The *codimension* is the number of monomials not belonging to $\mathcal{I}$.

It is not hard to see that an ideal has finite codimension if and only if $\mathcal{M}^{k+1} \subset \mathcal{I}$ for some $k \geq 0$. (It is only necessary to choose k large enough that the generating monomials of $\mathcal{M}^{k+1}$ belong to $\mathcal{I}$. This is possible because there are only finitely many monomials that do not belong to $\mathcal{I}$.) In the present section, we are concerned only with *monomial* ideals. For $n = 2$, a monomial ideal $\mathcal{I}$ has finite codimension when its generating monomials (the starred monomials in the Newton diagram) form a "staircase path" descending from the m_2 axis to the m_1 axis, with finitely many monomials "under the stairs," marked with zeros and not belonging to the ideal.

Let f be an element of $\mathcal{F}^n$ or $\mathcal{E}^n$, and let $\mathcal{J} \subset \mathcal{R}$ be a monomial ideal of finite codimension. Let a_m denote the mth Taylor coefficient of f, so that $j^\infty f(x) = \sum a_m x^m$. Define

$$\psi(f)(x) = \sum_{\{m : x^m \notin \mathcal{J}\}} a_m x^m. \tag{A.2.12}$$

This is a polynomial and can be regarded as an element of $\mathcal{F}^n$ or $\mathcal{E}^n$, so that $\psi : \mathcal{R} \to \mathcal{R}$. For $n = 2$ we may think of $\psi(f)$ as the sum of the terms "under the staircase of stars" in the Newton diagram. In general, $\psi(f)$ is a "lopsided jet" of f. The following corollary can be thought of as the smooth division theorem for monomial ideals of finite codimension.

A.2.10. Corollary. Let $\mathcal{J}$ be a monomial ideal of finite codimension in $\mathcal{R} = \mathcal{F}^n$ or $\mathcal{E}^n$, and let ψ be defined as above. Then ψ is a strong division process.

Proof. Let

$$\varphi(f) = f - \psi(f). \tag{A.2.13}$$

Note that we may not write $\varphi(f) = \sum_{\{m : x^m \in \mathcal{J}\}} a_m x^m$, since the series need not converge. To prove the first property of a strong division process, that $\psi(f) \equiv f$, is to show that $\varphi(f) \in \mathcal{J}$. This follows from Theorem A.2.7 and the fact that for large enough k, $\mathcal{M}^{k+1} \subset \mathcal{J}$. Namely, $\varphi(f) - j^k(\varphi(f)) \in \mathcal{M}^{k+1} \subset \mathcal{J}$, so it suffices to prove that $j^k(\varphi(f)) \in \mathcal{J}$. But $j^k(\varphi(f)) = \varphi(j^k(f))$ is a polynomial containing only terms that belong to $\mathcal{J}$. Properties 2 and 3 of a strong division process are obvious.   □

> **A.2.11. Remark.** As an example of Corollary A.2.10, see equation (6.1.12) in the second proof of Theorem 6.1.6. This equation can be obtained by considering the monomial ideal of finite codimension $\langle x^3, \varepsilon x, x^2 \rangle$. The hypotheses of the theorem imply that $\psi(a) = a_{20} x^2 + a_{01} \varepsilon$, because the other terms "under the staircase" are zero. Then $a - \psi(a) \in \mathcal{J}$, meaning that $\varphi(a) = f_1 x^3 + f_2 \varepsilon x + f_3 x^2$ for some f_1, f_2, f_3. For another way to prove (6.1.12) and similar equations, see Remark A.2.17 below.

Although our motivation for developing these results for $\mathcal{F}^n$ and $\mathcal{E}^n$ was the failure of Lemma A.2.3 to hold in these rings, it is actually quite useful to state the following result corresponding to Corollary A.2.10 for the rings $\mathcal{F}^n_*$ and $\mathcal{F}^n_{**}$, in which Lemma A.2.3 does hold. As usual in these rings, the proof is trivial. The other important difference is that we do not need to assume finite codimension.

A.2.12. Lemma. Let $\mathcal{J}$ be a monomial ideal in $\mathcal{F}^n_*$ or $\mathcal{F}^n_{**}$, and define ψ as in (A.2.12). Then ψ is a strong division process.

Each of the rings $\mathcal{R}$ that we are discussing is a vector space over $\mathbb{R}$, and the maps φ and ψ defined by (A.2.12) and (A.2.13) are linear maps of vector spaces. It is easy to see that

$$\mathfrak{I} = \operatorname{im}\varphi = \ker\psi$$

and

$$\mathcal{R} = \mathfrak{I} \oplus \operatorname{im}\psi; \qquad (A.2.14)$$

the maps ψ and φ are the projections associated with the splitting. If a vector space V satisfies $V = E \oplus F$, and F is finite-dimensional, the dimension of F is called the *codimension* of E (with respect to V). If $\mathfrak{I}$ is a monomial ideal of finite codimension, a basis for $\operatorname{im}\psi$ is given by the (finite) set of monomials not in $\mathfrak{I}$. This justifies referring to the number of such monomials as the codimension of the ideal.

A.2.13. Definition. If $\mathfrak{I}$ is a monomial ideal, the monomials belonging to $\mathfrak{I}$ are called *nonstandard monomials*. The *standard monomials* with respect to this ideal are the monomials that *do not* belong to it. The phrase "standard monomials" is used only in regard to monomial ideals.

> **A.2.14. Remark.** In Section A.5, we will associate to each ideal $\mathfrak{I}$ a monomial ideal $\tilde{\mathfrak{I}}$, and use standard monomials with respect to $\tilde{\mathfrak{I}}$ to study both $\mathcal{R}/\tilde{\mathfrak{I}}$ and $\mathcal{R}/\mathfrak{I}$. But we will avoid calling these standard monomials with respect to $\mathfrak{I}$, because $\tilde{\mathfrak{I}}$ is not unique: It depends on a choice of term ordering.

Suppose that $\{f_1, \ldots, f_r\}$ is a finite set of monomials. There is an ideal $\mathfrak{I} = \langle f_1, \ldots, f_r \rangle$ in each of our rings $\mathcal{R}$ in our list; the same symbol $\mathfrak{I}$ is used for each of these ideals, although they are different. Nevertheless, *the set of standard monomials with respect to $\mathfrak{I}$ is the same regardless of the ring*. Since the standard monomials don't depend upon the ring, one might suspect that $\operatorname{im}\psi$ does not depend upon the ring, but it is necessary to be careful here. Here is how the standard monomials determine $\operatorname{im}\psi$ in each ring, in the general case where $\mathfrak{I}$ is a monomial ideal that is not necessarily of finite codimension:

1. In $\mathcal{F}_*^n$, $\operatorname{im}\psi$ consists of all finite linear combinations of the standard monomials.

2. In $\mathcal{F}_{**}^n$, $\operatorname{im}\psi$ consists of all formal infinite linear combinations of the standard monomials.

3. In $\mathcal{F}^n$ or $\mathcal{E}^n$, $\operatorname{im}\psi$ consists of all smooth functions or germs having Taylor series that are formal infinite linear combinations of the standard monomials.

The one case in which these coincide is when $\mathfrak{I}$ has finite codimension, so that there are only finitely many standard monomials. In this case, $\operatorname{im}\psi$ is independent of the ring $\mathcal{R}$, and since $\operatorname{im} \psi$ gives a concrete representation of $\mathcal{R}/\mathcal{I}$, we see that

$$\mathcal{F}_*^n/\mathcal{I} = \mathcal{F}^n/\mathcal{I} = \mathcal{E}^n/\mathcal{I} = \mathcal{F}_{**}^n/\mathcal{I}. \tag{A.2.15}$$

The significance of this result is that whenever we work modulo an ideal of finite codimension, all distinction between polynomials, power series, and smooth functions disappears.

Another corollary of Theorem A.2.7 is the following version of Taylor's theorem:

A.2.15. Corollary (Taylor's theorem). Let $f \in \mathcal{F}^n$. Then

$$f = j^k f + \mathcal{O}(\|x\|^{k+1}).$$

Proof. By Theorem A.2.7,

$$f(x) - (j^k f)(x) = \sum_{|m|=k+1} g_m(x) x^m.$$

But each term satisfies $g_m(x) x^m = \mathcal{O}\left(\|x\|^{k+1}\right)$: Since $\|g_m(x)\|$ is bounded near $x = 0$, it suffices to check that $\|x^m\|/\|x\|^{k+1}$ is bounded in a punctured neighborhood of the origin; but it is bounded on the unit sphere (by compactness), and constant on rays from the origin (because numerator and denominator are homogeneous of degree $k + 1$). For instance, in $\mathbb{R}^2$, $xy^3/(x^2 + y^2)^{3/2}$ is bounded. $\square$

The second way to generalize Lemma A.2.2 to dimensions $n > 1$ is to impose an asymptotic order condition strong enough to imply that f is divisible by a single monomial. The ideal generated by such a monomial has infinite codimension if $n > 1$.

A.2.16. Theorem (Monomial division theorem). A smooth function $f \in \mathcal{F}^n$ (or its germ in $\mathcal{E}^n$) is divisible by x^m if and only if

$$f(x) = \mathcal{O}(x^m) \quad \text{as} \quad x \to 0,$$

that is, if and only if there exist constants $c_1 > 0$ and $c_2 > 0$ such that

$$|f(x)| < c_1 |x^m| \quad \text{for} \quad |x| < c_2.$$

Proof. Suppose that $f(x) = \mathcal{O}(x^m)$ with $m_1 > 0$. Then $f(x) = \mathcal{O}(x_1)$, and it follows that $f(0, x_2, \ldots, x_n) = 0$ for all $x_2, \ldots, x_n$. Then by the argument of Lemma A.2.2, we have

$$f(x) = x_1 g(x)$$

with

$$g(x) = \int_0^1 f_1(tx_1, x_2, \ldots, x_n) dx.$$

(Here $f_1 = \partial f/\partial x_1$.) It is easy to check that $g(x) = \mathcal{O}(x_1^{m_1-1} x_2^{m_2} \cdots x_n^{m_n})$. Thus, if $m_1 - 1 > 0$, we can factor out another x_1. Continuing in this way we can factor out $x_1^{m_1}$, then $x_2^{m_2}$, until we have factored out x^m. □

A.2.17. Remark (Continuation of Remark A.2.11). Theorem A.2.16 can be used to prove results such as (6.1.12), provided that an appropriate version of Taylor's theorem is also used. Thus, in the situation of (6.1.12) we can first assert, by a form of Taylor's theorem, that

$$a(x, \varepsilon) = a_{20}x^2 + a_{01}\varepsilon + \mathcal{O}(x^3) + \mathcal{O}(\varepsilon x) + \mathcal{O}(\varepsilon^2).$$

Then Theorem A.2.16 gives the existence of f_1, f_2, f_3.

Notes and References

Monomial ideals in the ring of polynomials are thoroughly discussed in Cox, Little, and O'Shea [32]. The case of rings of germs of smooth functions is treated in Golubitsky and Schaeffer [50].

A.3 Flat Functions and Formal Power Series

In the last section, we saw that the existence of *flat functions* makes the study of $\mathcal{F}^n$ and $\mathcal{E}^n$ much harder than the study of $\mathcal{F}_*^n$, $\mathcal{F}_{**}^n$, and $\mathcal{F}_{[k]}^n$. In this section, we will clarify the role of flat functions and show that in some sense it is *only* flat functions that create the difficulty: The flat functions form an ideal (denoted by $\mathcal{M}^\infty$) in either $\mathcal{F}^n$ or $\mathcal{E}^n$, and when this ideal is "modded out," $\mathcal{F}_{**}^n$ is what remains.

A *flat function* is a function whose partial derivatives, to all orders, are zero at the origin. Equivalently, f is flat if and only if $j^\infty(f) = 0$. (Often the function $f(x) = 0$ is excluded from the definition, but it should be included if the flat functions are to form an ideal or even just a vector subspace.) One can show directly that e^{-1/x^2} is a flat function in $\mathcal{F}^1$, thus proving that there are flat functions other than zero. Since $\mathcal{M}^{k+1}$ is the set of functions whose derivatives up to order k vanish at the origin, it is natural to denote the set of flat functions by $\mathcal{M}^\infty$. The product rule shows that this is an ideal. The ring homomorphism

$$j^\infty : \mathcal{F}^n \to \mathcal{F}_{**}^n$$

has $\mathcal{M}^\infty$ as its kernel, and it follows from ring theory that

$$\operatorname{im} j^\infty \cong \mathcal{F}^n/\mathcal{M}^\infty,$$

where $\cong$ denotes ring isomorphism. Thus, to prove that

$$\mathcal{F}_{**}^n \cong \mathcal{F}^n/\mathcal{M}^\infty \cong \mathcal{E}^n/\mathcal{M}^\infty \tag{A.3.1}$$

requires showing only that j^∞ is onto. This is the content of the Borel–Ritt theorem, which will be proved at the end of this section.

Let $\mathcal{I} = \langle f_1, \ldots, f_s \rangle$ be any ideal with polynomial generators. (As before, there will be an ideal $\mathcal{I}$ with these generators in each ring, and the same symbol $\mathcal{I}$ will designate all of them.) If $\mathcal{M}^\infty \subset \mathcal{I}$, it follows from (A.3.1) that

$$\mathcal{F}_{**}^n / \mathcal{I} \cong \mathcal{F}^n / \mathcal{I} \cong \mathcal{E}^n / \mathcal{I}. \tag{A.3.2}$$

In other words, if we "mod out" by an ideal that contains $\mathcal{M}^\infty$, we might as well mod out by $\mathcal{M}^\infty$ first, and then by the rest of the ideal; therefore it is not necessary to be concerned with functions or germs, but only with formal power series. In particular, any ideal of finite codimension (see Definition A.2.9) contains $\mathcal{M}^{k+1}$ for some k and therefore contains $\mathcal{M}^\infty$; therefore ideals of finite codimension satisfy (A.3.2), and it is even possible to include $\mathcal{F}_*^n / \mathcal{I}$ in the list of isomorphic rings (which is not the case for ideals of infinite codimension, such as $\mathcal{M}^\infty$ itself). This is a generalization of (A.2.15).

The ideal $\mathcal{M}^\infty$ is an example of an ideal that does not have a division process: There is no natural, unique way to decompose a function $f \in \mathcal{F}^n$ into two parts $f = \varphi(f) + \psi(f)$ such that $\varphi(f)$ is flat (and can be thought of as "the flat part of f"). There is a special case in which this is possible. If the formal power series of f is convergent, it defines an analytic function that could be taken as $\psi(f)$. Then $\varphi(f) = f - \psi(f)$ is flat, and f decomposes into an analytic part and a flat part. When the power series of f is divergent, the Borel–Ritt theorem shows only that there exist smooth functions with this power series (which, since f is given, we already knew). It does not give a method of selecting a "privileged" representative $\psi(f)$ of the equivalence class of f modulo the flat functions.

A.3.1. Remark. There is one sense in which a division process for $\mathcal{M}^\infty$ does exist. The flat functions form a vector subspace of $\mathcal{F}^n$, and every vector subspace has a complementary subspace, which exists by an argument using Hamel bases. Hamel bases exist by virtue of the axiom of choice, and complementary subspaces obtained in this way are not algorithmically computable, either in finite time or by a sequence of approximations. Also, the projections into the complementary subspaces are not continuous. Although Definition A.1.2 does not explicitly require continuity (it does not even mention a topology) or constructibility in any specific sense, each division process that is actually useful has these properties in some way. For instance, the division processes in Section A.5 below terminate in finite time, and those in Section A.6 converge in a "filtration topology" on the space of formal power series. When we say that there is no division process for $\mathcal{M}^\infty$, we mean that there is no useful one.

A.3.2. Theorem (Borel–Ritt). Every formal power series is the Taylor series of a smooth function (which is not unique; any two smooth functions with the same Taylor series differ by a flat function).

Proof. To obtain a general idea of the proof before filling in the details, let $\sum a_m x^m$ be a formal power series (with $x \in \mathbb{R}^n$, where m is a multi-index), let $h_k(x) = \sum_{|m|=k} a_m x^m$ be the terms of total order k, and consider the function f defined as follows: $f(x) = h_0(x) = a_0$ for $\frac{1}{2} < \|x\| \leq 1$; $f(x) = h_0(x) + h_1(x)$ for $\frac{1}{4} < \|x\| \leq \frac{1}{2}$; $f(x) = h_0(x) + h_1(x) + h_2(x)$ for $\frac{1}{8} < \|x\| \leq \frac{1}{4}$; and so on. For every $x \neq 0$, $f(x)$ is a finite sum and is well defined, and for $x = 0$ we have $f(0) = a_0$. On every open "annulus" $1/2^{s+1} < \|x\| < 1/2^s$ this function is smooth. Consider now some derivative $D^m f$, with $|m| = r$. On any annulus with $s > r$, $D^m f(x) = m! a_m + \cdots$, and the omitted terms (taken individually) approach zero at the origin, and hence as $s \to \infty$. With luck, then, it would be at least plausible that $D^m f(0) = m! a_m$, were it not for the fact that f is discontinuous at each boundary between the annuli and so cannot be differentiable at all at the origin (since differentiability implies continuity). In order to make f continuous, it is necessary to add the successively higher-order terms to f gradually, rather than suddenly, as we get closer to the origin. To make sure that the resulting function is smooth at the origin, the easiest way is to control the C^r norm of the terms for each r. Therefore, we proceed as follows.

There exists a smooth (C^∞) function $\eta(x)$ with $0 \leq \eta(x) \leq 1$ such that $h(x) = 1$ for $\|x\| < \frac{1}{2}$ and $h(x) = 0$ for $\|x\| > 1$. We choose a sequence $\delta_k \to 0$ as $k \to \infty$, and set $\eta_k(x) = \eta(x/\delta_k)$. Some further conditions governing the selection of the δ_k will be specified later. Once δ_k and η_k are fixed, we define

$$f(x) = \sum_k \eta_k(x) h_k(x) = \sum_k \eta_k(x) \left(\sum_{|m|=k} a_m x^m \right). \qquad \text{(A.3.3)}$$

This series converges everywhere, since at any point there are only finitely many nonzero terms. (This is true for $x \neq 0$ because $\delta_k \to 0$, and at the origin because the only nonzero term there is a_0.) Clearly f is smooth at all points $x \neq 0$. To make it smooth at $x = 0$, it is enough to make (A.3.3) be a Cauchy sequence in the C^r norm for each r. For this it is sufficient to make $\eta_k h_k$ sufficiently small, say less than 2^{-k}, in the C^{k-1} norm, for all $k \geq 1$. (Notice the "diagonalization" here: We estimate the successively higher-order terms in successively higher-order norms. For each r, the terms having $k \geq r$ will be Cauchy in the C^r norm.) To make $\eta_1(x) h_1(x)$ smaller than 1 in the C^0 norm, take δ_1 small enough that $|h_1(x)| < 1$ for $\|x\| \leq \delta_1$; it follows that $|\eta_1(x) \eta_1(x)| < 1$ for all x. To make $\eta_2(x) h_2(x)$ less than $\frac{1}{2}$ in the $\mathbb{C}^1$ norm requires that we make both this function and its first partial derivatives small in the C^0 norm. The function itself will be small if δ_2 is small. To see what happens to the first derivatives, consider the case $n = 1$, in which there is only one such derivative, namely,

$$\frac{d}{dx}\eta(|x|/\delta_2)a_2x^2 = \frac{\operatorname{sgn} x}{\delta_2}\eta'(|x|/\delta_2)a_2x^2 + 2\eta(|x|/\delta_2)a_2x.$$

The second term is small for small ξ, hence for all x if δ_2 is small. The first term is zero except for $\delta_2/2 < \|x\| < \delta_2$, and in that annulus is small with δ_2 because $x^2/\delta_2 < \delta_2$. Now the general case is clear: Each term of $D^m\eta_k(x)h_k(x)$ with $|m| \le k$ will contain more factors of the variables x_i in the numerator than factors of δ_k in the denominator, and hence can be made small by taking δ_k sufficiently small.

Once it is arranged that (A.3.3) is Cauchy in each C^r norm, it is not only smooth at the origin but can be differentiated termwise, showing that $D^m f(x) = m!a_m$.   □

Notes and References

Treatments of the Borel–Ritt theorem are given in Erdélyi [42] and Narasimhan [92].

A.4   Orderings of Monomials

When a polynomial or power series is written down, its terms must be arranged in some order. For many purposes the order does not matter. But there are purposes for which the order does matter. These include situations in which the monomials are arranged in their order of importance asymptotically, and algorithms in which the terms are processed in the order in which they are written. The next two sections involve such algorithms, and for this reason we inject a short section on orderings of monomials.

For $n = 1$, there are only two useful orderings, the "upward" ordering $a_0 + a_1x + a_2x^2 + \cdots$ and the "downward" ordering $\cdots + a_2x^2 + a_1x + a_0$. The upward ordering reflects behavior for small $|x|$ (that is, asymptotic order as $x \to 0$), and the downward order reflects behavior for large $|x|$. Polynomials are commonly written in either order; power series are usually written only in the upward order. (But there is a familiar exception in complex analysis: A power series is written in downward order when it represents the principal part of a meromorphic function at infinity on the Riemann sphere.)

For polynomials and power series in several variables, there are three approaches to the question of ordering the monomials. Each of the three approaches comes in various versions, many of which can be classified as "upward" or "downward" in character. We speak only of power series, regarding polynomials as a subset of these:

1. Partial ordering by division: Given two monomials x^m and x^μ, we may ask whether one divides the other. A power series is upwardly ordered by division if each term precedes any terms that it divides, downwardly ordered by division if the opposite is true. Since some pairs of monomials are "not comparable" (because neither divides the other), this criterion does not completely dictate the ordering of the terms.

2. Partial ordering by degree or weighted degree: A power series is upwardly ordered by degree if the terms are arranged as constant term; linear terms; quadratic terms; and so on. Terms of the same degree are "tied" and can be arranged arbitrarily. The definition of the downward ordering is obvious. Instead of working by degree, one can use *weighted degree*

$$L(m) = \ell_1 m_1 + \cdots + \ell_n m_n, \tag{A.4.1}$$

where the ℓ_i are positive numbers, usually integers. Weighted degrees are related to asymptotic scalings. For instance, the Newton diagram (6.1.15) for the pitchfork bifurcation involves two different weighted degrees, $m_1 + 2m_2$ and $m_1 + m_2$, associated with the two scalings discussed in Remark 6.1.10.

3. Total orderings involving lexicographic considerations: Sometimes it is necessary (for precise definitions of algorithms) to have a total, or linear, ordering of the monomials. Usually this is done by ordering the *exponent multi-indices* in some variant of a lexicographic, or "alphabetic," manner. We can say that x^m comes before x^μ if the first (or last) component of m that differs from the corresponding component of μ is smaller (for "upward" versions, or larger, for "downward" versions) than that component of μ; taking all combinations, this gives four possible versions of lexicographic ordering. In addition to these four, it is possible to combine the lexicographic and degree orderings: First the series is arranged in upward or downward order by degree, and then within each degree, one of the lexicographic orderings is used to "break the ties." These orderings have confusing code names in the literature, such as "grevlex" and "grlex."

For our purposes, the most natural ordering is almost always an upward ordering by total degree, with the simplest lexicographic ordering to break ties. That is, x^m comes before x^μ if $|m| < |\mu|$; if $|m| = |\mu|$, then x^m comes first if $m_1 < \mu_1$, or if $m_1 = \mu_1$ but $m_2 < \mu_2$, or if $m_1 = \mu_1$ and $m_2 = \mu_2$ but $m_3 < \mu_3$, or if $\ldots$.

Upward orderings are natural for local dynamical systems because the interest is always in behavior for small $\|x\|$, and terms of lower degree are dominant in this situation. In spite of this, there are occasions that call for a different ordering. Specifically, in Sections 4.6 to 4.8 we construct Stanley decompositions of certain subrings of $\mathcal{F}_{**}^n$ called "rings of invariants." It turns out that in each degree, these rings are the same as subrings of $\mathcal{F}_*^n$, so it is sufficient to work with polynomials instead of power series. Once this transition is made, it seems best to write the polynomials with a downward ordering, because this guarantees that the algorithms needed in the calculations will terminate. For this reason, Section A.5 below is devoted to a study of polynomial rings with downward orderings. Then in Chapter 6 we have occasion to use the Malgrange preparation theorem. This is a theorem involving smooth germs, and is not algorithmic in character. But modulo flat functions, germs become power series, and at this level the Malgrange preparation theorem has an algorithmic version. The ordering required for this version is upward in character (and therefore the algorithm does not terminate), but it is not one of the orderings discussed above. Instead it is a "product ordering" in which one "distinguished" variable is handled differently than the others. This will be discussed in Section A.6.

There remain three issues concerning orderings that must be mentioned in passing. We have described our orderings by saying "x^m comes before x^μ" if such-and-such conditions hold. Parts of the literature describe this situation by saying x^m is *greater* than x^μ, while other parts say x^m is *smaller* than x^μ. This is purely a matter of language, and is not to be confused with the question of upwards and downwards orderings. We will continue to avoid the terms smaller and larger in this context, preferring such words as precedes, succeeds, comes before or after.

Related to the last issue is the question of well-orderings. A totally ordered set is defined to be *well-ordered* if every (finite or infinite) subset has a least element; it is *anti-well-ordered* if every subset has a greatest element. (Many orderings are neither.) Because authors disagree about the meaning of greater and less for monomial orderings, they disagree about which orderings are well-ordered or anti-well-ordered. The only important fact is that under upward orderings, every power series has a first term (or leading term), whereas under downward orderings, every power series has a last term.

Finally, there exist orderings that are not "locally finite." This means that between two monomials there may exist infinitely many others. For example, in $\mathbb{R}[[x, y]]$ under pure downward lexicographic order (defined as $x^i y^j$ precedes $k^k y^\ell$ if $i > k$ or $i = k$ and $j > \ell$) x comes before $1 = x^0 y^0$, but y^j comes between these for any j. This situation does not arise for any order that is basically an upward or downward order by degree, with lexicographic order used only to break ties. Since we use only orders of this kind, except in the preparation theorem, this issue will arise for us only in Section A.6.

$$
\begin{array}{r}
2x \;+\; 1 \\[2pt]
\hline
x^2 + 3x + 1 \,\big)\; 2x^3 + 7x^2 + 6x + 9 \\[2pt]
2x^3 + 6x^2 + 2x \\[2pt]
\hline
x^2 + 4x + 9 \\[2pt]
x^2 + 3x + 1 \\[2pt]
\hline
x + 8
\end{array}
$$

Figure A.1. Polynomial division in one variable.

Notes and References

Orderings of monomials are treated at length in any book dealing with Gröbner bases, such as Cox, Little, and O'Shea [32] and Adams and Loustaunau [1].

A.5 Division in Polynomial Rings; Gröbner Bases

In Section A.2, we saw that in questions concerning division by monomials, or more generally by monomial ideals with several generators, it made little difference what ring we operated with: The proofs were harder in the case of functions or germs ($\mathcal{F}^n$ or $\mathcal{E}^n$), but most of the results were similar, at least for ideals with finite codimension. Once we go beyond monomial ideals, this situation changes drastically, and we take up the study of division questions one ring at a time. This is seen most readily for $n = 1$, where polynomials mod $\langle 1 + x \rangle$ form a vector space of one dimension, but in the ring of power series $\langle 1 + x \rangle$ is everything (so the dimension of the quotient is zero). This section is devoted to the polynomial case $\mathcal{F}^n_*$.

The Case $n = 1$

Everyone is familiar with the usual method for division of polynomials, as illustrated by Figure A.1. The conclusion of this particular calculation is that

$$2x^3 + 7x^2 + 6x + 9 = (2x + 1)(x^2 + 3x + 1) + (x + 8),$$

which has the general form

$$g = \varphi(g) + \psi(g)$$

expected of a division process according to (A.1.2): Here

$$\varphi(g) = (2x + 1)(x^2 + 3x + 1) \in \mathfrak{I} = \langle x^2 + 3x + 1 \rangle$$

and

$$\psi(g) = x + 8.$$

Our interest is focused especially on the image of ψ, which gives a concrete realization of the quotient ring $\mathcal{R}/\mathcal{J}$. In the present example, as g varies over all polynomials in the single variable x, $\psi(g)$ ranges over all polynomials $ax + b$ of degree ≤ 1; $\psi(g)$ is the unique representative of the equivalence class of g modulo $\langle x^2 + 3x + 1\rangle$ having degree ≤ 1. To calculate in the quotient ring using these representatives, multiply two degree-1 polynomials and then apply ψ again (i.e., divide by $x^2 + 3x + 1$ and take the remainder).

It is proved in any beginning course in abstract algebra that all ideals in $\mathcal{F}^1_* = \mathbb{R}[x]$ are principal ideals $\mathcal{J} = \langle f\rangle$. Given polynomials g and f, with $f \neq 0$, there exist unique polynomials q and r (found by the division algorithm illustrated above) such that

$$g = qf + r \quad \text{with} \quad r = 0 \quad \text{or} \quad \deg r < \deg f. \qquad (\text{A.5.1})$$

Setting $\varphi(g) = qf$ and $\psi(g) = r$ gives a division process (in our sense) for the ideal. The image of ψ is the space of all polynomials of degree less than the degree of f, and this space forms a concrete representation of the ring $\mathcal{F}^1_*/\mathcal{J}$ (or $\mathbb{R}[x]/\langle f\rangle$).

For our purposes, it is important to point out certain features of this situation that are not usually mentioned in elementary abstract algebra. First, it is important to notice that the terms of the polynomials in Figure A.1 are ordered from the highest degree downward. This is in contrast to the ordering of terms used in most of this book, beginning with equation (1.1.1). It will become clear in the next section of this appendix that the correct ordering of monomials depends strongly on the ring. For polynomials, the downward ordering works, but for power series the upward ordering is needed. This has a strong effect on the correct division theory in each case.

Second, notice that $\operatorname{im}\psi$ in the case of $\mathcal{F}^1_*$ depends solely on the degree of f. In the example of $f(x) = x^2 + 3x + 1$, $\operatorname{im}\psi$ is all polynomials of degree 1, but the same would have been true if we had divided by $\tilde{f}(x) = x^2$. To be more precise, let $\psi(g)$ and $\tilde{\psi}(g)$ be the remainders upon dividing g by f and $\tilde{f}$, respectively; then ψ and $\tilde{\psi}$ are not the same map, but the images of these maps are the same:

$$\operatorname{im}\psi = \operatorname{im}\tilde{\psi} = \operatorname{span}\{1, x\},$$

the real vector subspace of $\mathcal{F}^1_*$ spanned by 1 and x. The equality of $\operatorname{im}\psi$ and $\operatorname{im}\tilde{\psi}$ means, in the first place, that they are equal *as sets*. They are also equal *as vector spaces*; that is, there is only one natural vector space structure (addition and multiplication by scalars) on the set $\operatorname{im}\psi = \operatorname{im}\tilde{\psi}$. But when $\operatorname{im}\psi$ and $\operatorname{im}\tilde{\psi}$ are viewed as rings (that is, as concrete representations of $\mathbb{R}[x]/\langle f\rangle$ and $\mathbb{R}[x]/\langle\tilde{f}\rangle$), they are no longer the same. The reason is that the multiplication is different: We multiply degree-1 polynomials (getting quadratic polynomials) and then apply either ψ or $\tilde{\psi}$, obtaining different results.

The observations in the last paragraph are crucial to all further considerations in this appendix. Notice that in passing from $\langle f \rangle$ to $\langle \widetilde{f} \rangle$, we have replaced an arbitrary ideal in $\mathcal{F}^1_*$ by a monomial ideal. In doing so we have not changed $\operatorname{im} \psi$, when it is regarded as a vector space, although we have changed its ring structure when regarded as a concrete representation of the quotient ring. This will be our general strategy in more complicated situations: We will attempt to replace an ideal $\mathfrak{I}$ in a ring $\mathcal{R}$ with an *associated monomial ideal* $\widetilde{\mathfrak{I}}$ such that the quotient rings $\mathcal{R}/\mathfrak{I}$ and $\mathcal{R}/\widetilde{\mathfrak{I}}$ are identical *as vector spaces*, although not as rings. In Section A.2 we saw that the computation of $\operatorname{im} \psi$ for monomial ideals in $\mathcal{F}^n_*$ is simple: It is the vector space spanned by the *standard monomials*, those monomials *not belonging to the ideal*. (See in particular Lemma A.2.12, Definition A.2.13, and the discussion following it.) In the present example, we replace $\mathfrak{I} = \langle f \rangle$ with $\widetilde{\mathfrak{I}} = \langle \widetilde{f} \rangle = \langle x^2 \rangle$. The standard monomials for $\widetilde{\mathfrak{I}}$, not belonging to $\langle x^2 \rangle$, are 1 and x. These span the image of the remainder map $\widetilde{\psi}$ for $\widetilde{\mathfrak{I}}$, and therefore they also span the image of the remainder map ψ for $\mathfrak{I}$. Once we have computed $\operatorname{im} \psi$ as a vector space, there is no difficulty determining the ring structure that it carries as a representation of $\mathcal{R}/\mathfrak{I}$.

Principal Ideals for $n > 1$

One obstacle to generalizing the previous ideas from polynomials in one variable ($\mathcal{F}^1_* = \mathbb{R}[x]$) to polynomials in n variables ($\mathcal{F}^n_* = \mathbb{R}[x_1, \ldots, x_n]$) is that not all ideals in $\mathcal{F}^n_*$ are generated by a single element. For a first step toward the general case, we can avoid this difficulty by confining ourselves to principal ideals $\mathfrak{I} = \langle f \rangle$ in $\mathcal{F}^n_*$. It turns out that after selecting a suitable ordering of the terms of f (the generator of $\mathfrak{I}$) and g (an arbitrary polynomial in $\mathcal{F}^n_*$), it is possible to divide g by f and obtain a unique remainder $\psi(g)$, with the property that $g \in \mathfrak{I}$ if and only if $\psi(g) = 0$. The vector space $\operatorname{im} \psi$ is unchanged if f is replaced by its leading term (in the chosen term ordering), and is spanned by an easily determined set of "standard monomials." Once $\operatorname{im} \psi$ has been found (as a vector space), it provides a concrete representation of $\mathcal{F}^n_*/\mathfrak{I}$ by calculating "modulo f" within this space.

There are a number of possible term orderings in $\mathcal{F}^n_*$ that can be used for this purpose, but we will concentrate on only one of them, the *graded lexicographic ordering* (or *degree lexicographic ordering*). In this ordering, a term $a_m x^m$ is written before the term $a_\mu x^\mu$ if $|m| > |\mu|$, so we say that the ordering is *downward* by degree. If $|m| = |\mu|$, the term $a_m x^m$ is written first if the first entry of m that differs from the corresponding entry of μ is greater than the entry in μ, that is, if $m_1 > \mu_1$, or if $m_1 = \mu_1$ but $m_2 > \mu_2$, or if $m_1 = \mu_1$ and $m_2 = \mu_2$ but $m_3 > \mu_3$, or if .... For the case $n = 2$, the lattice points (m_1, m_2) can be numbered as follows; the terms are then written in decreasing order by this numbering:

$$\begin{array}{llll} 7 & & & \\ 4 & 8 & & \\ 2 & 5 & 9 & \\ 1 & 3 & 6 & 10 \end{array} \qquad \text{(A.5.2)}$$

An example of a polynomial in $\mathcal{F}_*^2$ written in graded lexicographic ordering is

$$g(x) = x^3 + 3xy^2 + y^3 + 7x^2 - 3x + y + 2. \qquad \text{(A.5.3)}$$

Notice that the term $3xy^2$ comes before $7x^2$ because it is a cubic term; by pure lexicographic ordering of the exponents, $7x^2$ would come first, but the lexicographic ordering is used only to "break ties" in degree.

The graded lexicographic ordering has the following properties:

1. It is a total order; any two monomials occur in a definite order.

2. It is multiplicative. This means that if a polynomial written in this order is multiplied by a monomial, the terms automatically remain in the correct order.

3. Any *formal power series* written in this order has a *last* nonzero term, although it need not have a *first* nonzero term. (Of course, any *polynomial* written in this order has both a first and a last term. We mention power series here only as a convenient way of describing one of the well-ordering properties mentioned at the end of Section A.4.)

4. The ordering is locally finite: Given any monomial x^m, there are only finitely many monomials that can occur after it. Thus, if we know the leading term of a polynomial, we can write down all the terms that can occur in the polynomial, with unknown coefficients (some of which could be zero).

Property 4 in this list implies property 3, but not conversely. Only the first three of these properties are essential for the existence of a division process based on a term ordering, but the fourth property is rather convenient, as we will see.

The division of g, given by (A.5.3), by $f(x) = x + y + 1$ is illustrated in Figure A.2. The first step in this calculation is to write out the "missing" terms of $g(x)$ after the leading term, giving them zero coefficients. This is possible only because the graded lexicographic order satisfies property 4, and so is not possible when some other orders are used, but when it is possible, it is convenient because it identifies the columns into which all subsequent calculations are to be placed. Next we proceed as usual, attempting to divide the leading term of f into successive terms of g. Whenever this division is possible, the quotient (of a particular term of g by the leading term of f) is written above the term of g, and the product of this quotient

$$
\begin{array}{r}
x^2 - 1xy + 4y^2 \qquad\quad + 6x - 5y \qquad\qquad - 9 \\[2pt]
\hline
\end{array}
$$

$$x + y + 1 \,\big)\; x^3 + 0x^2y + 3xy^2 + 1y^3 + 7x^2 + 0xy + 0y^2 - 3x + 1y + 2$$

$$x^3 + 1x^2y \qquad\qquad\qquad + 1x^2$$

$$-1x^2y + 3xy^2 \qquad\qquad + 6x^2$$
$$-1x^2y - 1xy^2 \qquad\qquad\qquad\; - 1xy$$

$$+4xy^2 \qquad\qquad + 6x^2 + 1xy$$
$$+4xy^2 + 4y^3 \qquad\qquad\qquad + 4y^2$$

$$-3y^3 + 6x^2 + 1xy - 4y^2$$
$$+6x^2 + 6xy \qquad\quad + 6x$$

$$-3y^3 \qquad\qquad - 5xy - 4y^2 - 9x$$
$$-5xy - 5y^2 \qquad\quad - 5y$$

$$-3y^3 \qquad\qquad\qquad + 1y^2 - 9x + 6y$$
$$-9x - 9y - 9$$

$$-3y^3 \qquad\qquad\qquad\qquad + 1y^2 \qquad\quad + 15y + 11$$

Figure A.2. Polynomial division in two variables.

by f is subtracted. The multiplicative property (property 2) of the ordering guarantees that both of these steps (the division and the multiplication) produce terms that are automatically correctly ordered. One difference between this process and ordinary division is that it is sometimes necessary to "skip over" certain terms in g that are not divisible by the leading term of f. In ordinary polynomial division (in one variable), all such terms of g occur at the end, and the first time such a term is encountered, the division is finished. In several variables we may encounter a term of g that is not divisible by the leading term of f, even though some subsequent term is divisible. Such terms are simply skipped over, and eventually become part of the remainder (which is the bottom line of the calculation). The algorithm terminates because there are only finitely many columns in which a division could take place (namely, the columns that are set up at the first step using property 4).

A.5.1. Remark. If an ordering is used that does not satisfy property 4, then property 3 must be used instead in proving that the algorithm terminates. In this case the argument is that even though there can be infinitely many monomials that come after a given monomial in the term ordering, there cannot be any infinite *sequences* of such monomials. (For instance, in the pure lexicographic ordering on $\mathbb{R}[x, y]$, x comes before any y^i. But in any *sequence* containing x, some definite y^i must follow x, and there are only finitely many terms after that.) So we argue as follows: The leading terms of the lines that follow the subtractions in the division algorithm form a *sequence* of successive monomials and must therefore terminate.

Having established the division process, the next step is to identify $\operatorname{im}\psi$, the set of possible remainders. (Remember that ψ is defined with a fixed principal ideal $\mathfrak{I} = \langle f \rangle$ in mind, so we mean the set of remainders obtained in dividing all possible g by a single f.) In Figure A.2, it is clear that the terms that are "passed over" (and become part of the remainder) are those that are not divisible by x, and therefore have the form y^i for some i. It is equally clear that any y^i can appear in such a remainder; for instance, any polynomial in y alone could be taken as g, and then no division would take place, so that $\psi(g) = g$. Thus, in this example,

$$\operatorname{im}\psi = \mathbb{R}[y],$$

the set of polynomials in y. Notice that in contrast to the case $n = 1$, the space of possible remainders is infinite-dimensional.

It is easy to see that this procedure generalizes. Suppose the leading term of f were $x^2 y$; then the terms that are passed over and enter the remainder are the terms that are *not divisible* by $x^2 y$. These are just what we have called the *standard monomials* defined by the *associated monomial ideal* $\widetilde{\mathfrak{I}} = \langle x^2 y \rangle$ (Definition A.2.13). Monomials that are not divisible by $x^2 y$ are those marked by a dot on the following Newton diagram (continuing upwards and to the right in the same manner):

That is, they are the elements that do *not* lie in the infinite rectangle with lower-left corner $*$. These standard monomials can be divided into three classes: those that contain only a power of x (which may be the zeroth power); those that contain at least one factor of y and no x; and those that contain at least a factor of xy but not x^2. These three sets of monomials span the three vector spaces $\mathbb{R}[x]$, $\mathbb{R}[y]y$, and $\mathbb{R}[y]xy$. (Here $\mathbb{R}[y]$ is the space of polynomials $p(y)$ in y, and $\mathbb{R}[y]y$ denotes the space of polynomials $p(y)y$ containing at least one factor of y.) Thus, we can write

$$\operatorname{im}\psi = \mathbb{R}[x] \oplus \mathbb{R}[y]y \oplus \mathbb{R}[y]xy \qquad (A.5.4)$$

if the leading monomial of f is $x^2 y$. The reader can check that the following also works:

$$\operatorname{im}\psi = \mathbb{R}[y] \oplus \mathbb{R}[y]x \oplus \mathbb{R}[x]x^2. \qquad (A.5.5)$$

These expressions (A.5.4) and (A.5.5) are called *Stanley decompositions* of $\operatorname{im}\psi$. They are usually regarded as Stanley decompositions of the ring $\mathbb{R}[x,y]/\langle f \rangle$, but they do not actually contain any information about the ring structure. They are purely descriptions of $\operatorname{im}\psi$, which is the underlying vector space of a concrete realization of $\mathbb{R}[x,y]/\langle f \rangle$.

A.5.2. Remark. A formal definition of Stanley decomposition goes as follows. Let $\mathcal{R} \subset \mathbb{R}[x_1, \ldots, x_n]$ be a subring of the ring of polynomials in n variables. Let $\mathcal{R}_1, \ldots, \mathcal{R}_k$ be subrings of $\mathcal{R}$, each of which is a full polynomial ring in certain algebraically independent elements of $\mathcal{R}$. That is, for each $i = 1, \ldots, k$, there exist algebraically independent polynomials $g_{i1}, \ldots, g_{ij_i}$ such that $\mathcal{R}_i = \mathbb{R}[g_{i1}, \ldots, g_{ij_i}]$. Let $f_1, \ldots, f_k$ be elements of $\mathcal{R}$ such that the vector spaces $\mathcal{R}_i f_i$ (consisting of products of elements of $\mathcal{R}_i$ times f_i) are linearly independent, meaning that the sum $\mathcal{R}_1 f_1 + \cdots + \mathcal{R}_k f_k$ is a direct sum. Then if this sum equals $\mathcal{R}$, we say that

$$\mathcal{R} = \mathcal{R}_1 f_1 \oplus \cdots \oplus \mathcal{R}_k f_k$$

is a Stanley decomposition of $\mathcal{R}$. In the special case that all the $\mathcal{R}_i$ are equal, and equal to $\mathbb{R}[g_1, \ldots, g_j]$ (where it is no longer necessary to use double subscripts), the Stanley decomposition is called a *Hironaka decomposition*, with *primary generators* $g_1, \ldots, g_j$ and *secondary generators* $f_1, \ldots, f_k$.

The previous discussion was stated in terms of examples, but the ideas suffice to establish the following theorem:

A.5.3. Theorem. Let $\mathcal{I} = \langle f \rangle$ be a principal ideal in the polynomial ring $\mathcal{F}_*^n$. Choose an ordering of the monomials in $\mathcal{F}_*^n$ satisfying (at least) properties 1–3 in the list above, and define ψ to be the division process defined by this ordering, following the pattern of Figure A.2. Let $\tilde{f}$ be the leading monomial of f in the chosen ordering, and let $\tilde{\mathcal{I}} = \langle \tilde{f} \rangle$. Then im ψ is spanned (as a vector space over $\mathbb{R}$) by the monomials not divisible by $\tilde{t}$ (which are the standard monomials defined by $\tilde{\mathcal{I}}$). This vector space im ψ can be described by a Stanley decomposition, and when it is given a ring structure as in Lemma A.1.3, using the ideal $\mathcal{I}$ (not $\tilde{\mathcal{I}}$), it becomes a concrete representation of $\mathcal{F}_*^n/\mathcal{I}$.

Ideals That Are Not Principal

The ideal $\mathcal{I} = \langle xy - 1, y^2 - 1 \rangle \subset \mathcal{R} = \mathbb{R}[x, y] = \mathcal{F}_*^2$ is not a principal ideal; that is, it cannot be written using a single generator. As a first try at developing a division algorithm for this ideal, let us attempt to divide $x^2 y + xy^2 + y^2$ by the two divisors $xy - 1$ and $y^2 - 1$ simultaneously, recording the quotients on two different lines above the division sign. The procedure is to divide the successive terms by the leading term xy of the first divisor whenever possible, recording the quotient in the top line; if this is not possible, we divide by the leading term y^2 of the second divisor and record the quotient in the bottom line; if neither is possible, the term is skipped over and added to the remainder. After each step, the most recent quotient is multiplied by the divisor that was used, and the product is subtracted from what is left of the dividend. See Figure A.3.

$$
\begin{array}{r}
\left\{\begin{array}{c} x + y \\ \\ 1 \end{array}\right. \\
\hline
\end{array}
$$

Figure with long division:

$$xy - 1\Big)$$
$$y^2 - 1\Big)\; x^2y + xy^2 + y^2$$
$$\underline{x^2 y \qquad\qquad -\,x}$$
$$+\,xy^2 + y^2 + x$$
$$\underline{+\,xy^2 \qquad\qquad -\,y}$$
$$+\,y^2 + x + y$$
$$\underline{+\,y^2 \qquad\qquad -\,1}$$
$$+\,x + y + 1$$

Figure A.3. Division by an ideal with two generators.

$$
\left\{\begin{array}{c} x + 1 \\ \\ x \end{array}\right.
$$

$$y^2 - 1\Big)$$
$$xy - 1\Big)\; x^2y + xy^2 + y^2$$
$$\underline{x^2 y \qquad\qquad -\,x}$$
$$+\,x^2y + y^2 + x$$
$$\underline{+\,xy^2 \qquad\qquad -\,x}$$
$$y^2 \qquad\quad +\,2x$$
$$\underline{y^2 \qquad\qquad -\,1}$$
$$+\,2x + 1$$

Figure A.4. The same division with the generators reversed.

This calculation suffices to establish the result that

$$x^2y + xy^2 + y^2 = (x + y)(xy - 1) + (1)(y^2 - 1) + (x + y + 1).$$

Everything seems to work smoothly, until we try to do the same division with the two divisors listed in the opposite order; see Figure A.4. This calculation establishes that

$$x^2y + xy^2 + y^2 = (x + 1)(y^2 - 1) + (x)(xy - 1) + (2x + 1).$$

Therefore, this procedure does not result in a unique way of writing each element $g \in \mathcal{F}_*^n$ as the sum of an element $\varphi(g)$ of the ideal $\mathfrak{I} = \langle f_1, f_2 \rangle = \langle xy - 1, y^2 - 1 \rangle$ and a remainder $\psi(g)$. Or more precisely, there are two procedures, depending on the order in which f_1 and f_2 are used as divisors. Once the ordering of divisors is specified, then a unique remainder function

ψ is defined. (It is assumed throughout that a systematic way of ordering the terms within each polynomial has been specified, satisfying at least properties 1 through 3 for a term ordering, as listed above.)

However, fixing an ordering of f_1 and f_2 does not put an end to the difficulties. It is possible to find g such that $g \in \mathcal{I}$ but $\psi(g) \neq 0$. In fact, the polynomial $g(x, y) = x - y$ can be written

$$x - y = y(xy - 1) + x(y^2 - 1), \tag{A.5.6}$$

showing that $x - y \in \mathcal{I}$; but no term of $x - y$ is divisible by either xy or y^2, and $\psi(x - y) = x - y$ (whichever ordering of the divisors is used).

The fact is that ψ, as we have defined it (using a fixed ordering of f_1 and f_2), *does not provide a strong division process* for $\mathcal{R}$ with respect to $\mathcal{I}$ according to Definition A.1.2. However, properties 1 and 5' still hold, and such a ψ is an example of a *weak division process*. A weak division process cannot be used to provide a concrete representation for $\mathcal{R}/\mathcal{I}$.

There turns out to be a way to remedy this difficulty: If the polynomial $x - y$ is added to the set of generators of $\mathcal{I}$, all of the difficulties we have been discussing disappear. The ideal is not changed;

$$\mathcal{I} = \langle xy - 1, y^2 - 1, x - y \rangle \tag{A.5.7}$$

is the same ideal as before. But now, if any element $g \in \mathbb{R}[x, y]$ is divided simultaneously by the three polynomials $xy - 1$, $y^2 - 1$, and $x - y$, using exactly the same division scheme described above, it is found that

1. the remainder does not depend on how the three divisors are ordered, and

2. g belongs to $\mathcal{I}$ if and only if $\psi(g) = 0$.

The three generators given in (A.5.7) constitute what is called a *Gröbner basis* for the ideal $\mathcal{I}$.

All ideals $\mathcal{I} \subset \mathcal{F}_*^n$ are finitely generated, and may be written as $\mathcal{I} = \langle f_1, \ldots, f_s \rangle$. If a term ordering satisfying conditions 1 to 3 is specified, and an ordering of the divisors $f_1, \ldots, f_s$ is specified, there exists an algorithm (the obvious generalization of the examples above) that divides any g by all of $f_1, \ldots, f_s$ simultaneously, producing unique quotients $q_1, \ldots, q_s$ and a unique remainder $r = \psi(g)$ such that

$$g = q_1 f_1 + \cdots + q_s f_s + r. \tag{A.5.8}$$

However, in general ψ is only a weak division process. We say that $f_1, \ldots, f_s$ form a *Gröbner basis* for $\mathcal{I}$ if the monomial ideal $\langle \widetilde{f_1}, \ldots, \widetilde{f_s} \rangle$ generated by the leading terms of $f_1, \ldots, f_s$ (with respect to a chosen term ordering) is the same as the ideal $\widetilde{\mathcal{I}}$ generated by the leading terms of *all* the elements of $\mathcal{I}$. If $f_1, \ldots, f_s$ is a Gröbner basis for $\mathcal{I}$, then ψ is a strong division process, and in addition ψ does not depend on the order in which the generators are listed. Being a strong division process, $\operatorname{im} \psi$ gives a concrete representation of $\mathcal{R}/\mathcal{I}$. Also, im ψ is spanned by the standard monomials associated with the monomial ideal $\widetilde{\mathcal{I}}$, and these are easily computable from $\widetilde{\mathcal{I}} = \langle \widetilde{f}_1, \ldots, \widetilde{f}_s \rangle$. Furthermore, one can write down Stanley decompositions from this information.

The final fact that makes this approach work is that every ideal has a Gröbner basis. In fact, there exists an algorithm (Buchberger's algorithm) that, given a set of generators for an ideal, will create a Gröbner basis by adding additional generators to those that are given. We will not develop this algorithm in detail, but it is illustrated in (A.5.6) above. The idea is to create elements of $\mathcal{I}$ by using all possible cancellations of leading terms of the generators. When it is not possible to enlarge the set of generators any further by canceling leading terms, a Gröbner basis has been found.

How Gröbner Bases Are Used in This Book

In the main text of this book (outside of the appendices), Gröbner bases are mentioned only in side remarks. This is possible because the Gröbner basis calculations that arise are simple enough that they can be carried out in an ad hoc manner, without invoking the general theory. However, as soon as one goes on to harder examples, this theory becomes helpful, if not essential. The following discussion is intended to show the roles that Gröbner bases play in determining the Stanley decompositions of the rings of invariants that arise in Sections 4.6 through 4.9.

The starting point for this discussion is a collection of homogeneous polynomials $I_1, \ldots, I_r$ in $\mathcal{F}_*^n$. In the applications in Chapter 4, it is essential that these polynomials are invariant under a certain linear flow (for instance, $e^{A^* t}$ in the case of the inner product normal form) and that they are sufficient to generate the entire ring of invariants; these two facts play no role in the present discussion. All that matters is that each I_i is homogeneous (with degree depending on i). The goal is to obtain a Stanley decomposition of the ring of formal power series $\mathbb{R}[[I_1, \ldots, I_r]]$ in these polynomials. If $I_1, \ldots, I_r$ are algebraically independent, there is no difficulty: Each element of the ring is uniquely expressible as a formal power series in $I_1, \ldots, I_r$, and the Stanley decomposition is trivial (it is just $\mathbb{R}[[I_1, \ldots, I_r]]$). The problems arise when there are relations among $I_1, \ldots, I_r$, making the formal power series no longer unique.

There is one question that must be disposed of immediately before we proceed: Why do we pose this problem now, in the context of polynomial rings, when it clearly seems to be a problem in power series rings? Why do we not postpone the question until the next section of this appendix, which is devoted to power series rings? The answer is contained in the following lemma, which reduces the question from power series rings to polynomial rings.

A.5.4. Lemma. A power series belongs to $\mathbb{R}[[I_1, \ldots, I_r]]$ if and only if each truncation of the power series (with respect to the x variables) belongs to $\mathbb{R}[I_1, \ldots, I_r]$.

Proof. Each product $I^m = I_1^{m_1} \cdots I_r^{m_r}$ is a polynomial in $x = (x_1, \ldots, x_n)$ that is homogeneous of some degree, namely $m_1 d_1 + \cdots + m_r d_r$, where d_i is the degree of I_i. Therefore, when a power series in $I_1, \ldots, I_r$ is expanded as a power series in x and then truncated at a given degree in x, each I^m is either deleted or retained as a whole. Therefore, the truncated series is a polynomial in $I_1, \ldots, I_r$, that is, an element of $\mathbb{R}[I_1, \ldots, I_r]$. The converse is obvious. $\square$

Because of this lemma, it is sufficient to find a Stanley decomposition of $\mathbb{R}[I_1, \ldots, I_r]$. The first step in doing this is to find the ideal of relations satisfied by $I_1, \ldots, I_r$. As explained in Remark 4.5.15, this means finding the kernel of the ring homomorphism $\mathbb{R}[t_1, \ldots, t_r] \to \mathbb{R}[I_1, \ldots, I_r]$ defined by $t_i \mapsto I_i$. There exist algorithms, implemented in many computer algebra systems, that, given a set of polynomials $I_1, \ldots, I_r$, will return a Gröbner basis for the ideal of relations satisfied these polynomials. (In the examples in this book, this ideal is almost always principal, and therefore any single generator for the ideal automatically constitutes a Gröbner basis.)

Suppose that a Gröbner basis $\mathfrak{I} = \langle \sigma_1, \ldots, \sigma_s \rangle$ has been found for the ideal $\mathfrak{I}$ of relations among $I_1, \ldots, I_r$; this is an ideal in the ring $\mathbb{R}[t_1, \ldots, t_r]$. Then it follows that

$$\mathbb{R}[I_1, \ldots, I_r] \cong \mathbb{R}[t_1, \ldots, t_r]/\mathfrak{I}. \tag{A.5.9}$$

The next step is to find a concrete representation of this quotient ring. To do so, we pass from $\mathfrak{I}$ to $\widetilde{\mathfrak{I}} = \langle \widetilde{\sigma}_1, \ldots, \widetilde{\sigma}_r \rangle$, the ideal generated by the leading terms of the relations (with respect to a chosen term order on the variables $t_1, \ldots, t_r$). Then the standard monomials (in $t_1, \ldots, t_r$) with respect to $\widetilde{\mathfrak{I}}$ span a vector subspace of $\mathbb{R}[t_1, \ldots, t_r]$ that is the underlying space of a concrete representation of $\mathbb{R}[t_1, \ldots, t_r]/\mathfrak{I}$. (It is also a concrete representation of $\mathbb{R}[t_1, \ldots, t_r]/\widetilde{\mathfrak{I}}$, but the ring structures are different.) Once the standard monomials are known, a Stanley decomposition can be written down, since this is just a concise way of describing the standard monomials. Either in the standard monomials, or in the Stanley decomposition, the t_i can be replaced by the I_i. The representation of elements of $\mathbb{R}[I_1, \ldots, I_r]$ is now unique.

By Lemma A.5.4, the standard monomials for $\mathbb{R}[[I_1, \ldots, I_r]]$ are the same as those for $\mathbb{R}[I_1, \ldots, I_r]$. The only difference is that we allow infinite linear combinations of these standard monomials.

Examples of the procedure described here are found in (4.5.74) and (4.7.23).

$$
\begin{array}{r}
3x \quad -2 \\[2pt]
x^3 + x^2 \overline{\smash{\big)}\ 3x^4 + 1x^3 - 5x^2 + 2x + 1} \\
\underline{3x^4 + 3x^3\phantom{{}-5x^2 + 2x + 1}} \\
-2x^3 - 5x^2\phantom{{}+2x+1} \\
\underline{-2x^3 - 2x^2\phantom{{}+2x+1}} \\
-3x^2 + 2x + 1
\end{array}
$$

Figure A.5. Polynomial division in one variable.

Notes and References

The best beginning treatments of Gröbner bases are found in Cox, Little, and O'Shea [32] and Adams and Loustaunau [1]. There is a useful chapter on the subject in Eisenbud [40], much of which is readable independently of the rest of the book. An abstract treatment of Gröbner bases that includes many generalizations (including the standard bases of Section A.6 below and Gröbner bases for modules, Appendix B below) is given in Lunter [74].

A.6   Division in Power Series Rings; Standard Bases

When we pass from the ring $\mathcal{F}^n_*$ of polynomials to the ring $\mathcal{F}^n_{**}$ of formal power series, many things change; others remain somewhat as they were. This section is divided into subsections paralleling those in the last section: the special case $n = 1$, followed by principal ideals for arbitrary n, and then the general case (which is only touched on briefly).

The Case $n = 1$

To motivate the discussion, consider the example of ordinary polynomial division shown in Figure A.5. The conclusion of this calculation is that

$$
3x^4 + x^3 - 5x^2 + 2x + 1 = (3x - 2)\left(x^3 + x^2\right) - 3x^2 + 2x + 1.
$$

If we were to reverse the ordering of the terms and try to carry out the same division, we would obtain the result shown in Figure A.6. Notice that (as in some problems in the last section) it is necessary to "skip over" certain terms when it is not possible to divide. This division leads to the conclusion that

$$
1 + 2x - 5x^2 + x^3 + 3x
$$
$$
= 1 + 2x + \left(-5 + 6x = 3x^2 + 3x^3 + \cdots\right)\left(x^2 + x^3\right),
$$

which is a correct (and meaningful) conclusion as long as both sides are understood as elements of $\mathcal{F}^n_{**}$. It happens in this example that we began

$$\begin{array}{r} -\,5x^0 + 6x^1 - 3x^2 + 3x^3 + \cdots \\[2pt] \hline x^2 + x^3\,\big)\quad 1 + 2x - 5x^2 + 1x^3 + 3x^4 \\[2pt] -\,5x^2 - 5x^3 \\[2pt] \hline +\,6x^3 + 3x^4 \\[2pt] +\,6x^3 + 6x^4 \\[2pt] \hline -\,3x^4 \\[2pt] -\,3x^4 - 3x^5 \\[2pt] \hline +\,3x^5 \\[2pt] \vdots \end{array}$$

Figure A.6. Power series division in one variable.

with polynomials, but it is clear that the same procedure would work if the divisor and dividend were formal power series, and in this case it would be essential to order the terms by increasing powers, since if we attempted to divide with the terms arranged in decreasing order, there would be no leading term to divide by at each step.

Reflecting on this example suggests the following observations. First, the division process can never be completed in finite time, although (in the usual manner of mathematicians, see Remark 1.1.6) we are entitled to *imagine* it completed, and to regard the quotient as a well-defined formal power series. Second, the "remainder" $1 + 2x$ (which is skipped over at the beginning, rather than left over at the end) is finite, i.e., a polynomial, and consists of low-order terms. (One might have imagined that reversing the ordering would yield a remainder consisting of high-order terms, but that is not what happens.) More specifically, the remainder consists of terms having degree less than the *lowest-degree term* of the divisor, whereas the remainder in Figure A.5 consists of terms having degree less than the *highest-degree term* of the divisor. In spite of this difference, there is a way to state both results in the same words: In each case the degree of the remainder is less than the degree of the *leading term* of the divisor. Although the divisors are the same in both problems, their leading terms are different because of the different orderings.

There is another way to understand the division process in Figure A.6. The divisor $x^2 + x^3$ can be factored as $x^2(1 + x)$. Now, $1 + x$ is a unit in $\mathcal{F}^1_{**}$, that is, it is an invertible element, with reciprocal given by the geometric series

$$(1 + x)^{-1} = 1 - x + x^2 - x^3 + \cdots .$$

Therefore, to divide by $x^2 + x^3$, we can first multiply by this series, and then divide by x^2 (which can be done term by term, without long division). Every division in $\mathcal{F}^1_{**}$ can be viewed in this way, because any formal power series beginning with a nonzero constant term is a unit, and therefore any

$$\begin{array}{r} 1 + x \qquad\qquad\qquad + y \\ \overline{x^2 + y\,\big)\,1 + 2x + x^2 + x^3 + 2y + 2yx + yx^2 + y^2 + y^2x} \\ x^2 \qquad + 1y \end{array}$$

$$\begin{array}{r} 1 + 2x \qquad x^3 + 1y \\ x^3 \qquad + 1yx \\ \hline + 1y + 1yx + yx^2 \\ yx^2 + y^2 \\ \hline + y^2x \end{array}$$

Figure A.7. Weierstrass division (an artificial example).

formal power series at all is a monomial x^i times a unit, where i is the lowest degree appearing in the series.

Principal Ideals with $n > 1$

Just as there are a variety of possible "downward" term orders that can be used for the division of polynomials, there are (for $n > 1$) many possible "upward" orders that can be used for division of formal power series. Motivated by the needs of Section 6.1, we focus on one particular type of ordering that leads to a division process called *Weierstrass division*. The ordering will be described below. A somewhat artificial example (with $n = 2$), specially set up so that all of the formal power series are actually polynomials, is shown in Figure A.7. The remainder in this example is $1 + 2x + y + yx + y^2x$, these being the terms that are left unsubtracted in various lines of the calculation. At first sight, this division does not seem to be different from any of those carried out in the last section, except for the unusual ordering of the terms. The important thing to notice is that the remainder can be arranged as

$$(1 + y) + \left(2 + y + y^2\right)x, \qquad\qquad (A.6.1)$$

which is *a polynomial in x* whose coefficients are formal power series in y. (In the example, these formal power series happen to be polynomials as well. This will not usually be the case.) Notice that the divisor in Figure A.7 has x^2 for its leading term, and the remainder (A.6.1) is of lower degree in x, as should be expected.

The general statement of Weierstrass division in formal power series rings is as follows:

A.6.1. Theorem (Formal Weierstrass division). Let $g \in \mathcal{F}^n_{**}$ be any formal power series, and let $f \in \mathcal{F}^n_{**}$ be a formal power series satisfying

$$f(x_1, 0, \ldots, 0) = x_1^k h(x_1).$$

Then there exist $q, r \in \mathcal{F}_{**}^n$ such that

$$g = fq + r,$$

where r is a polynomial in x_1 of degree $\leq k - 1$ with coefficients that are formal power series in $x_2, \ldots, x_n$:

$$r(x) = r_0(x_2, \ldots, x_n) + x_1 r_1(x_2, \ldots, x_n) + \cdots + x_1^{k-1} r_{k-1}(x_2, \ldots, x_n).$$

Proof. Weierstrass division may be viewed simply as the division algorithm for formal power series that is dictated by a particular choice of term ordering, which requires the selection of a *distinguished variable* that is treated differently than the other variables. For $n = 2$ with variables (x, y), the distinguished variable will usually be x, and for arbitrary n it will be x_1. (Of course, in applications this may differ.) The term ordering is then as follows, for $n = 2$:

$$1, x, x^2, x^3, \ldots; y, yx, yx^2, yx^3, \ldots; y^2, y^2 x, y^2 x^2, y^2 x^3, \ldots; \ldots.$$

That is, all powers of x in ascending order; then y times these; then y^2 times these, etc. Now it should be clear why we picked an artificial polynomial example to present in Figure A.7: the general example contains an infinite series in x before the first term with y ever appears, so we could not write the problem down. (The resolution of this problem, for practical purposes, is to work in truncated power series rings. But we will come back to this later. As always, it is possible to imagine an infinite operation having been done.) Under this ordering, the leading term of the divisor will be assumed to be x^k for some k. (That is, we assume that the divisor is not divisible by y.) Then the remainder is guaranteed to consist of terms not divisible by x^k. Therefore, the remainder is a formal power series in x and y, each term of which has the form $x^i y^j$ with $i < k$. Such a series can always be rearranged as a polynomial in x of degree at most $k - 1$, having coefficients that are power series in y.

For $n > 2$, the term ordering is similar. The first step is to choose an upwards ordering for the monomials in the variables $x_2, \ldots, x_n$; for instance, we can order them upwards by total degree (in those variables), breaking ties lexicographically. Once this is done, these monomials can be enumerated as $M_1, M_2, \ldots$. Then all monomials in $x_1, \ldots, x_n$ can be ordered as follows: first $1, x_1, x_1^2, x_1^3, \ldots$; then this first string times M_1, that is, $M_1, x_1 M_1, x_1^2 M_1, x_1^3 M_1, \ldots$; then the first string times M_2, and so forth. □

As usual, formal results such as this are essentially trivial; it is just a matter of arranging things correctly. The much deeper fact (which is what Weierstrass actually proved, in the setting of complex variables) is that if f and g are convergent, so are q and r. Still deeper (and not discovered until the late twentieth century) is the fact that the same theorem is true for smooth functions. This will be stated (but not proved) in the next section

of this appendix. The following corollary of Theorem A.6.1 is especially
important.

A.6.2. Corollary (Formal Weierstrass preparation theorem). Let
$f \in \mathcal{F}^n_{**}$ be a formal power series satisfying

$$f(x_1, 0, \ldots, 0) = x_1^k h(x_1).$$

Then there exists a formal power series q with $q(0) \neq 0$ (q is a unit in the
ring $\mathcal{F}^n_{**}$) such that

$$q(x)f(x) = p(x)$$

is a polynomial of degree k in x_1 with coefficients that are power series in
$x_2, \ldots, x_n$.

Proof. Taking $g(x) = x_1^k$, Theorem A.6.1 yields

$$x_1^k = q(x)f(x) + r_0 + x_1 r_1 + \cdots + x_1^{k-1} r_{k-1},$$

where the r_i are power series in $x_2, \ldots, x_n$. It follows that

$$q(x)f(x) = p(x) = x_1^k - r_{k-1}x_1^{k-1} - \cdots - x_1 r_1 - r_0.$$

It remains to check that $q(0) \neq 0$. Set $x_2 = \cdots = \xi_n = 0$; then
$q(x_1, 0, \ldots, 0)x_1^k h(x_1) = p(x_1, 0, \ldots, 0) = x_1^k$. (The last equality follows
because the first expression cannot contain any terms of degree less than
k and the second cannot contain any terms of degree greater than k.) This
would be impossible if $q(0) = 0$.   □

Ideals That Are Not Principal

In the general case, generators $f_1, \ldots, f_s$ of an ideal $\mathfrak{I} \subset \mathcal{F}^n_{**}$ are given and
we would like a procedure to resolve any $g \in \mathcal{F}^n_{**}$ into

$$g = q_1 f_1 + \cdots + q_s f_s + r,$$

exactly as in (A.5.8). In general, the same two steps are necessary as in
the polynomial case: A term order (in this case, an upward order) must
be chosen, and additional generators must be added to the original set
(without enlarging the ideal). The second step is necessary in order that
the division procedure resulting from the term order be a strong division
process, and in particular, one that gives a unique remainder regardless of
the order of the divisors. In the setting of formal power series and upward
orders, the enlarged generating set is commonly known as a *standard basis*,
rather than a Gröbner basis, but it is characterized in the same way: A
generating set for an ideal $\mathfrak{I}$ is a standard basis if *the leading terms of the
generating set* generate the same ideal as that generated by *all the leading
terms of elements of* $\mathfrak{I}$. Once again this "initial ideal," denoted by $\widetilde{\mathfrak{I}}$, defines
a set of standard monomials (those that do not belong to $\widetilde{\mathfrak{I}}$), and these in
turn provide a Stanley decomposition for $\mathcal{R}/\mathfrak{I}$.

We will not develop this topic further here, except to say that this method gives an alternative approach to the problem of finding Stanley decompositions for rings of formal power series invariants, a problem that we solved in the last section by reducing it to the polynomial case (see Lemma A.5.4). It can be shown that with a suitable choice of term ordering, standard bases lead to the same solution of this problem (namely, the same standard monomials and Stanley decomposition) as do Gröbner bases. It is also possible to choose special term orders so that the remainders have special forms if the divisors $f_1, \ldots, f_s$ have special forms, generalizing the results obtained above for Weierstrass division in the case of a principal ideal.

Truncated Power Series Rings

It has already been mentioned that it is usually impossible to complete a division calculation in a power series ring, because it is necessary to work upwards through an infinite number of degrees. When using an ordering such as the one proposed for Weierstrass division, it is even necessary to work through infinitely many powers of x_1 before arriving at the first power of x_2. In terms of a Newton diagram (for $n = 2$), this means computing the entries across the bottom row before computing the first entry in the second row. Under these circumstances, such division does not seem to be practical.

The resolution of the difficulty lies in the fact that we are usually interested in results that are *finitely determined*; in the case $n = 2$ this means, roughly, that conclusions can be deduced from a knowledge of a small number of terms near the origin of Newton's diagram. Under these circumstances the problems addressed in this section can be reconsidered in one of the rings $\mathcal{F}^n_{[k]}$ for some k (see Section A.1). Computations in these rings are the same as in $\mathcal{F}^n_{**}$ except that all terms of total degree greater than k are ignored. (This is not the same k as in Theorem A.6.1 and Corollary A.6.2, but some higher number.) Thus, it becomes possible to compute the bottom row of Newton's diagram up to k, followed by the second row up to $k - 1$, and so forth; these values do not depend on the higher-order terms that are suppressed.

In addition to this, it should be mentioned that for most applications in Section 6.1 it is not actually necessary to carry out the computations at all. It is only the *possibility* of the division that matters (and for that matter it is the possibility of the division in $\mathcal{E}^n$ that matters, not in formal power series rings); see Remark 6.1.8. From this point of view, the principal value of our discussion of Weierstrass division in power series rings is to motivate the statement of the Malgrange preparation theorem in the next section. Nonetheless, there may be cases in which computation of a Weierstrass division modulo some power $\mathcal{M}^{k+1}$ of the maximal ideal might be advisable, and it is worth remarking that the computation is feasible.

Notes and References

A different proof of the Weierstrass division theorem for formal power series is given in Zariski and Samuel [115]. Standard bases in local rings are treated in Cox, Little, and O'Shea [33], but not in a form that is directly applicable to formal power series rings; for this case, see Becker [13], which includes an explicit algorithmic version of the Weierstrass preparation theorem.

A.7   Division in the Ring of Germs

Theorems that are easy or trivial for polynomials and formal power series can be quite difficult for smooth functions (if they remain true at all). An intimation of this fact appears already in Lemma A.2.5; although the proof of this lemma is simple, it is quite different from the proof of the corresponding fact for formal power series. On the other hand, the Borel–Ritt theorem has shown us that the only real difference between the classes of formal power series and smooth germs lies in the existence of germs of flat functions. Therefore, given any theorem about formal power series, it is not unreasonable to ask whether the same theorem (appropriately formulated) remains true for smooth germs. In the case of the Weierstrass division and preparation theorems (Theorem A.6.1 and Corollary A.6.2), the answer is yes. We will merely formulate the theorems. The proofs are difficult and will not be given. (See Notes and References.) These theorems are not algorithmic in character; they assert the existence of certain functions, but do not give a way to compute them. However, the infinity-jets (formal power series in $\mathcal{F}^n_{**}$) or k-jets (truncated formal power series in $\mathcal{F}^n_{[k]}$) of the functions whose existence is asserted are exactly those that would be calculated by first taking the jets of the given functions and then applying the theorems of Section A.6.

A.7.1. Theorem (Mather division theorem). Let g be any smooth function or germ (in $\mathcal{F}^n$ or $\mathcal{E}^n$), and let f be a smooth function or germ satisfying

$$f(x_1, 0, \ldots, 0) = x_1^k h(x_1)$$

for some smooth h. Then there exist $q, r \in \mathcal{F}^n$ or $\mathcal{E}^n$ such that

$$g = fq + r,$$

where r is a polynomial in x_1 of degree $\leq k - 1$ with coefficients that are smooth functions (or germs) in $x_2, \ldots, x_n$:

$$r(x) = r_0(x_2, \ldots, x_n) + x_1 r_1(x_2, \ldots, x_n) + \cdots + x_1^{k-1} r_{k-1}(x_2, \ldots, x_n).$$

A.7.2. Corollary (Malgrange preparation theorem). Let $f \in \mathcal{F}^n$ or $\mathcal{E}^n$ be a smooth function or germ satisfying

$$f(x_1, 0, \ldots, 0) = x_1^k h(x_1).$$

Then there exists a smooth function or germ q with $q(0) \neq 0$ such that

$$q(x)f(x) = p(x)$$

is a polynomial of degree k in x_1 with coefficients that are smooth functions or germs in $x_2, \ldots, x_n$. In the case of germs, q will be a unit in $\mathcal{E}^n$.

The simplest applications of the Malgrange preparation theorem (in Section 6.1) rely on the fact that since $q(0) \neq 0$, $q(x) \neq 0$ for all x near zero, and therefore the solution set of $f(x) = 0$ is locally the same as that of the simpler equation $q(x) = 0$. This application explains why q is a unit in $\mathcal{E}^n$, but not necessarily in $\mathcal{F}^n$: q may have zeros at a distance from the origin, so that $1/q$ is not defined as a (global) function, but $1/q$ is well defined as a germ at the origin.

An Argument Used in Section 6.1

In Section 6.1, the Malgrange preparation theorem is used together with various ad hoc arguments (for instance, arguments based on the quadratic formula) to show that there cannot exist more than a specified number of bifurcating rest points in certain situations. In Remark 6.1.8 it was pointed out that the ad hoc portion of these proofs can be replaced by a more abstract argument, which we now discuss. In the general case, $p(\eta, x)$ in (6.1.14) is a polynomial in x of degree k, where k is the smallest integer such that $a_{0k} \neq 0$. Then $p(\eta, x) = 0$ has exactly k complex solutions for each η, and those that are real satisfy $a(\eta, x)$ if x and η are small. So it would suffice to determine the Puiseux solutions of $p = 0$, and eliminate those that are nonreal. Nonreal solutions are easy to detect, because they contain nonreal terms. (Puiseux series, like Taylor series, are asymptotically valid even if they do not converge, so the existence of a first nonreal term is sufficient to show that the root being represented is nonreal.) On the other hand, a Puiseux series cannot be shown to be real by calculating a finite number of terms. So the general strategy is this: Prove the existence of j real branches by the implicit function theorem, and the nonexistence of $k - j$ branches by showing that their Puiseux series are nonreal. This strategy is greatly facilitated by showing that the Puiseux series obtained from $p = 0$ coincide exactly with those obtained from $a = 0$. This makes it unnecessary to compute p; it is enough to know that p exists. The remainder of this remark is devoted to the required proof. It should be noted that the statement in question is initially doubtful, or even meaningless: How can the Puiseux solutions of $p = 0$ coincide with those of $a = 0$, when a involves real variables only? (An immediate answer is that the series are constructed formally, without regard to the meaning of the variables.) We proceed to a more careful discussion.

The first step is to consider the case in which $a(\eta, x)$ is a polynomial in x and η. In this case, a does make sense for complex values of x and even of η. According to the Weierstrass preparation theorem (the older, analytic version of the Malgrange theorem), there exists a complex-analytic function $u(\eta, x)$, nonvanishing near $(0,0)$, such that $ua = p$, where p is a polynomial of degree k in x (with k defined as above) with coefficients that are analytic in η. This implies that a and p define the same level sets (complex varieties) near $(0,0)$ and hence have the same Puiseux series (since these are convergent and describe the branches of the variety).

The next (and final) step is to extend this result to the smooth case. The central observation is that the algorithm for passing from a to p is such that any particular term in the double power series of p depends on only finitely many terms in the double series of a; similarly, any particular term in a Puiseux solution of $p = 0$ depends on only finitely many terms in the double series of p. Therefore, we can truncate a at some high power of x and η without affecting any given term in the Puiseux solution of $p = 0$. But once a is truncated, the previous (analytic) argument applies, and the desired term in the Puiseux series can be computed directly from a. Restoring the deleted terms of a does not affect the calculation. Since we have proved that each individual term in the Puiseux solution of $p = 0$ can be calculated from $a = 0$, this is true of the entire series.

Notes and References

The Malgrange preparation theorem is proved in Golubitsky and Guillemin [49]. A generalized Malgrange theorem applicable to nonprincipal ideals is also stated there. Another form of the generalized version is given in Lojasiewicz [73] under the name Weierstrass preparation theorem (Thom–Martinet version), although this is stated for the complex-analytic rather than the real smooth case.

Appendix B
Modules

This appendix will introduce the notion of module. There are two kinds of modules that are relevant to this book: submodules of $\mathbb{Z}^n$, and submodules of various modules of vector fields.

B.1 Submodules of $\mathbb{Z}^n$

The set $\mathbb{Z}^n$ of integer vectors $(k_1, \ldots, k_n)$ with $k_i \in \mathbb{Z}$, is a module over the ring $\mathbb{Z}$ of integers. This means that it satisfies the same algebraic rules as a vector space over a field. The only difference concerns the fact that elements of the ring $\mathbb{Z}$ do not have multiplicative inverses (except for the elements ± 1), so it is not possible to "divide" an integer vector by an integer. The impact of this restriction is most visible when it comes to submodules of $\mathbb{Z}^n$.

Consider the subset M of $\mathbb{Z}^2$ consisting of integer linear combinations of the vectors $(1, 0)$ and $(0, 2)$, that is, the set of vectors of the form

$$m(1, 0) + n(0, 2) = (m, 2n).$$

An element of $\mathbb{Z}^2$ belongs to M if and only if its second component is even. First, observe that M is a module over $\mathbb{Z}$; it is closed under addition (of elements of M) and under multiplication (of elements of $\mathbb{Z}$ times elements of M). Now, if we were looking at the subspace of the vector space $\mathbb{R}^2$ generated by $(1, 0)$ and $(0, 2)$, it would be all of $\mathbb{R}^2$: we can multiply the real number $\frac{1}{2}$ times $(0, 2)$ to obtain $(0, 1)$, and of course $(1, 0)$ and $(0, 1)$ is the standard basis for $\mathbb{R}^2$. But we cannot do this with M, since $\frac{1}{2}$ is not an element of $\mathbb{Z}$. The submodule M of $\mathbb{Z}^2$ seems to have "dimension two" (because it has two linearly independent generators"), but it is not all of $\mathbb{Z}^2$.

But it gets worse. Consider the submodule M of $\mathbb{Z}^2$ generated by $(2,0)$, $(0,4)$, and $(1,1)$; these are the elements

$$\ell(2,0) + m(0,4) + n(1,1) = (2\ell + n, 4m + n).$$

The two entries in any element of M must have the same parity (they are both even or both odd, and the parity is determined by n). It follows that M is not all of $\mathbb{Z}^2$ even though it has more than two generators. Also the three generators are linearly dependent, since they satisfy

$$2(2,0) + 1(0,4) + (-4)(1,1) = (0,0), \tag{B.1.1}$$

but none of them can be eliminated; in particular, it is not possible to write any one of the generators as a linear combination of the other two. With vector spaces, any linear dependence relation holding among a set of generators implies that a generator can be eliminated; when the process of eliminating generators is complete, one is left with a basis for the vector space. The dimension is defined to be the number of elements in a basis, a definition that depends on a theorem stating that any two bases have the same number of elements. None of this holds up for modules:

1. A relation such as (B.1.1), called a *syzygy*, does not imply that one of the generators can be eliminated.

2. Two generating sets for the same module need not have the same number of elements.

3. The dimension of a module cannot (in general) be defined.

There is one special kind of submodule of $\mathbb{Z}^n$ for which things are not so bad. An element $k = (k_1, \ldots, k_n)$ of $\mathbb{Z}^n$ is called *divisible* by an integer r if each component k_i is divisible by r (with no remainder). A submodule M of $\mathbb{Z}^n$ is called *pure* if whenever an element of M is divisible by an integer r, the quotient belongs to M. For pure submodules, the notions of basis and rank (similar to dimension), and even complementary submodule, are well-defined and behave much as they do for vector spaces. This is relevant to the study of resonances (in any of the various senses of that word), since the set of integer vectors k such that $\langle k, \lambda \rangle = k_1\lambda_1 + \cdots + k_n\lambda_n = 0$ is a pure submodule of $\mathbb{Z}^n$, for any given set of real numbers $\lambda_1, \ldots, \lambda_n$.

Notes and References

The references given in Notes and References for Section A.1, except for the first volume [32] by Cox, Little, and O'Shea, also deal with modules. An application of pure submodules to resonances (in the context of multifrequency averaging) is given in Murdock [81], Section 2, and [84], Section 2.

The same technique used there should be applicable to determining the Stanley decompositions of the normal form module for multiple centers; it would replace the elementary number theory used to analyze the solutions of (4.5.63) in the case of the double center.

B.2   Modules of Vector Fields

If $\mathcal{R}$ is any ring (meaning commutative ring with identity), the set of n-tuples $\mathcal{R}^n$ of elements of $\mathcal{R}$ is a module over $\mathcal{R}$, just as $\mathbb{R}^n$ is a vector space over $\mathbb{R}$. A module of this type is called a *free module*. In this way, letting $\mathcal{R}$ be any of the rings $\mathcal{F}_*^n$, $\mathcal{F}^n$, $\mathcal{F}_{**}^n$ discussed in Appendix A (notice that the superscript n here does *not* denote a set of n-tuples), we obtain the modules $\mathcal{V}_*^n$, $\mathcal{V}^n$, and $\mathcal{V}_{**}^n$ of polynomial vector fields, smooth vector fields, and formal vector fields (or formal power series vector fields) defined at the beginning of Chapter 4. It is also possible to define the module of germs of vector fields, and the modules of truncated formal power series vector fields for various k, but these will not be discussed here.

The modules that are of greatest interest to us in this book are obtained from these three modules of vector fields in the following way. For convenience, consider the polynomial case, although there is no difference in the other cases. First, from the ring $\mathcal{F}_*^n$ of polynomial scalar fields we select a subring to be called $\mathcal{R}$, consisting of those polynomial scalar fields that are invariant under some linear flow e^{At} or e^{A^*t}, where A is the linear part of a system of differential equations that is to be put into normal form. Notice that $\mathcal{V}_*^n$ is still a module over the smaller ring $\mathcal{R}$, just as it was over $\mathcal{F}_*^n$. Next, from $\mathcal{V}_*^n$ we select the subset $\mathcal{M}$ consisting of vector fields that are equivariant under the same linear flow. Then, according to Lemma 4.5.7, $\mathcal{M}$ is a module over $\mathcal{R}$. (In this lemma, the argument was presented for the case that A is semisimple and the flow is e^{At}. It is remarked in Section 4.6 that the same proof works for general A, but in that case we are interested instead in the flow e^{A^*t}.) Notice that $\mathcal{M}$ will *not* be a module over $\mathcal{F}_*^n$: the product of a function that is not invariant and a vector field that is equivariant will not be equivariant.

The ring $\mathcal{R}$ in the last paragraph can be viewed in two ways. On the one hand it is a subring of $\mathcal{F}_*^n$; on the other hand, it is a quotient ring of some other ring $\mathbb{R}[t_1, \ldots, t_r]$ by an ideal of relations, as expressed in (A.5.9). It is in this latter context that one uses the theory of Gröbner bases for ideals to define standard monomials and obtain a Stanley decomposition for $\mathcal{R}$. There exists a theory of Gröbner bases for modules, which closely resembles the theory for ideals. We mention this only to emphasize that the Gröbner basis theory for modules *is not relevant* to obtaining Stanley decompositions for the modules that we study. The usual theory of Gröbner bases for modules over a ring $\mathcal{R}$ concerns submodules of $\mathcal{R}^n$; our modules are not of this form. (The components of an equivariant vector are not invariant functions.)

The notion of Stanley decomposition for a module of equivariants, on the other hand, is very relevant to the study of normal forms. The approach taken in this book is to develop these Stanley decompositions in a concrete way that requires no additional theory of abstract modules. A complete theory is presented in Section 4.7 for the case of equivariants of a nilpotent matrix. The semisimple case is handled on a case-by-case basis in Section 4.5.

Notes and References

The Gröbner basis theory for modules (which we do not need) is presented in Adams and Loustaunau [1], Section 3.5. Further information is found in Cox, Little, and O'Shea [33] and Lunter [74].

There does exist a general method by which the problem of finding Stanley decompositions of modules of equivariants can be reduced to finding Stanley decompositions of rings of invariants. The idea is to associate with each module element $(f_1, \ldots, f_n)$ an element $z_1 f_1 + \cdots + z_n f_n$ of a ring of polynomials in the original variables plus new "slack variables" $z_1, \ldots, z_n$, and define an appropriate group action on this larger ring; the set of invariants of the new action that are linear in the slack variables then corresponds to the set of equivariants. Details are given in Gatermann [47].

Besides Gröbner bases, there is another method for efficiently computing with rings of invariants and modules of equivariants. Instead of slack variables, it computes with the original variables only. Analogous to the division algorithm for Gröbner bases there is a *subduction algorithm*, which computes a representation of an invariant in terms of a basis of the ring of invariants if this basis is of the kind called a *sagbi basis*. For details, see Chapter 11 of Sturmfels [105].

Appendix C

Format 2b: Generated Recursive (Hori)

This appendix contains the detailed treatment of format 2b in both linear and nonlinear settings. Section C.1 can be read with Chapter 3, after Section 3.6. Section C.2 can be read with Chapter 4, after Section 4.4; it depends on Sections 4.4 and C.1.

C.1 Format 2b, Linear Case (for Chapter 3)

Format 2b is our name for the set of computational procedures ("bookkeeping devices") needed to handle transformations generated according to the formula

$$T(\varepsilon) = e^{\varepsilon V(\varepsilon)}, \tag{C.1.1}$$

where

$$V(\varepsilon) = V_1 + \varepsilon V_2 + \varepsilon^2 V_3 + \cdots. \tag{C.1.2}$$

Throughout this section, we will regard all functions of ε as formal power series, and use $=$ to mean equality of formal power series. (Alternatively, the functions could be regarded as smooth, with $=$ replaced by $\sim$, or we could operate "modulo ε^{k+1}" for some fixed k, with $=$ replaced by $\equiv$.) For instance, if we say $V(\varepsilon) \in \mathfrak{g}$ for some Lie algebra $\mathfrak{g}$, it cannot mean that $V(\varepsilon)$ takes values in $\mathfrak{g}$, since formal power series do not "have values." Instead, $V(\varepsilon) \in \mathfrak{g}$ means that $V_j \in \mathfrak{g}$ for each j. The reader is expected to be able to make such interpretations as needed, and they will be mentioned only when it is crucial to do so.

Format 2b maintains all of the advantages of format 2a. In particular, $T(\varepsilon)$ will lie in a specified Lie subgroup $\mathfrak{G}$ of $GL(n)$ if $V(\varepsilon)$ lies in the associated Lie algebra $\mathfrak{g}$, so it is possible to normalize matrix series belonging to specific classes (representation spaces of $\mathfrak{G}$ under similarity) without leaving the class.

Algorithms I and II

It is convenient to introduce the following notation for the expansions of the powers of $\varepsilon V(\varepsilon)$, beginning with the zeroth power:

$$
\begin{aligned}
\varepsilon^0 V(\varepsilon)^0 &= I + 0 && + 0 && + 0 && + \cdots, \\
\varepsilon V(\varepsilon) &= && \varepsilon V_0^{(1)} + && \varepsilon^2 V_1^{(1)} + && \varepsilon^3 V_2^{(1)} + \cdots, \\
\varepsilon^2 V(\varepsilon)^2 &= && && \varepsilon^2 V_0^{(2)} + && \varepsilon^3 V_1^{(2)} + \cdots, \\
\varepsilon^3 V(\varepsilon)^3 &= && && && \varepsilon^3 V_0^{(3)} + \cdots.
\end{aligned}
$$

$$\text{(C.1.3)}$$

Notice that the sum $i + j$ of the indices on $V_j^{(i)}$ equals the power of ε in the coefficient, and that according to (C.1.2),

$$V_j^{(1)} = V_{j+1}. \tag{C.1.4}$$

Another way to write the definition of $V_j^{(i)}$ is

$$\varepsilon^i V(\varepsilon)^i = \varepsilon^i \left(V_0^{(i)} + \varepsilon V_1^{(i)} + \varepsilon^2 V_2^{(i)} + \cdots \right). \tag{C.1.5}$$

(Thus, the top line of (C.1.3) is $V_0^{(0)} + \varepsilon V_1^{(0)} + \cdots$, with $V_0^{(0)} = I$ and $V_j^{(0)} = 0$ for $j > 0$.) A recursion for the $V_j^{(i)}$ can be developed by writing

$$
\begin{aligned}
V(\varepsilon)^{i+1} &= V(\varepsilon)V(\varepsilon)^i \\
&= (V_1 + \varepsilon V_2 + \cdots)\left(V_0^{(i)} + \varepsilon V_1^{(i)} + \cdots \right) \\
&= \left(V_0^{(i+1)} + \varepsilon V_1^{(i+1)} + \varepsilon^2 V_2^{(i+1)} + \cdots \right).
\end{aligned}
$$

Comparison of the last two lines reveals that

$$V_j^{(i+1)} = \sum_{k=0}^{j} V_{k+1} V_{j-k}^{(i)}. \tag{C.1.6}$$

Equations (C.1.4) and (C.1.6) make it possible to develop a table of $V_j^{(i)}$ recursively. If it is laid out as

$$
\begin{array}{cccc}
I & 0 & 0 & 0 & \cdots \\
& V_0^{(1)} & V_1^{(1)} & V_2^{(1)} & \cdots \\
& & V_0^{(2)} & V_1^{(2)} & \cdots \\
& & & V_0^{(3)} & \cdots
\end{array}
\tag{C.1.7}
$$

to match the arrangement in (C.1.3), then the top row comes from (C.1.2), and each succeeding entry is derived from those in the row above it that are also strictly to its left, using (C.1.6).

From (C.1.1),

$$T(\varepsilon) = I + \varepsilon T_1 + \varepsilon^2 T_2 + \cdots$$

$$= I + \varepsilon V(\varepsilon) + \frac{\varepsilon^2}{2} V(\varepsilon)^2 + \frac{\varepsilon^3}{6} V(\varepsilon)^3 + \cdots .$$

Thus, if the fractions $1, 1, \frac{1}{2}, \frac{1}{6}, \ldots$ are multiplied by the rows of (C.1.3) to obtain

$$\varepsilon^0 V(\varepsilon)^0 = I + \varepsilon 0 + \varepsilon^2 0 + \varepsilon^3 0 + \cdots,$$
$$\varepsilon^1 V(\varepsilon)^1 = \varepsilon V_0^{(1)} + \varepsilon^2 V_1^{(1)} + \varepsilon^3 V_2^{(1)} + \cdots,$$
$$\tfrac{1}{2}\varepsilon^2 V(\varepsilon)^2 = \tfrac{1}{2}\varepsilon^2 V_0^{(2)} + \tfrac{1}{2}\varepsilon^3 V_1^{(2)} + \cdots,$$
$$\tfrac{1}{6}\varepsilon^3 V(\varepsilon)^3 = \tfrac{1}{6}\varepsilon^3 V_0^{(3)} + \cdots,$$

and the columns are added, the result will be $I + \varepsilon T_1 + \varepsilon^2 T_2 + \cdots$, automatically collected by powers of ε. In other words, T_r is the weighted sum of the rth column of the table (C.1.7):

$$T_r = \sum_{s=0}^{r} \frac{1}{s!} V_{r-s}^{(s)}. \qquad (C.1.8)$$

(For $r > 0$, the summation could start at $s = 1$, but the given form better matches similar formulas appearing later, such as (C.1.17) and others in Section C.2.) Thus, we have the following:

C.1.1. Algorithm (Algorithm I for format 2b). Given the generator $V(\varepsilon)$, the transformation $T(\varepsilon)$ that it generates can be obtained by using (C.1.4) and (C.1.6) to develop table (C.1.7), and then using (C.1.8) to obtain the coefficients T_r.

C.1.2. Algorithm (Algorithm II for format 2b). Given the generator $V(\varepsilon)$, the coefficients of $T(\varepsilon)^{-1}$ are obtained by changing the signs of the odd-numbered rows in table (C.1.7) and then forming the sums in (C.1.8).

Proof. In format 2b, as in format 2a, the generator of the inverse is the negative of the generator:

$$T(\varepsilon)^{-1} = e^{-\varepsilon V(\varepsilon)}.$$

Therefore, the inverse can be obtained by applying Algorithm I to the negative of the generator. This means building a table like (C.1.7) beginning with a top row that is the negative of the top row in the original table. The successive rows will be built using (C.1.6) with V_j replaced by $-V_j$. It follows that the second row will equal the second row of the original table, the third row will be the negative of the original third row, and so forth. □

Algorithms III and IV

Algorithm III, for computing the transform $B(\varepsilon) = \mathbb{S}_{T(\varepsilon)}A(\varepsilon)$ of a series $A(\varepsilon)$ given the generator $V(\varepsilon)$, is similar to Algorithm I. We begin with

$$A(\varepsilon) = A_0 + \varepsilon A_1 + \varepsilon^2 A_2 + \cdots. \tag{C.1.9}$$

The notations

$$\mathbb{L}_j = \mathbb{L}_{V_j} \tag{C.1.10}$$

and

$$\mathbb{L}(\varepsilon) = \varepsilon \mathbb{L}_1 + \varepsilon^2 \mathbb{L}_2 + \cdots \tag{C.1.11}$$

have already been introduced in the introductory discussion of format 2b in Section 3.2, and will be used again here. Notice that

$$\mathbb{L}(\varepsilon) = \mathbb{L}_{\varepsilon V(\varepsilon)} = \varepsilon \mathbb{L}_{V(\varepsilon)},$$

so the indices in (C.1.11) match the powers of ε. Introduce the following notation for the iterates of $\mathbb{L}(\varepsilon)$ on $A(\varepsilon)$:

$$
\begin{aligned}
A(\varepsilon) &= A_0^{(0)} &+&\ \varepsilon A_1^{(0)} &+&\ \varepsilon^2 A_2^{(0)} &+&\ \cdots, \\
\mathbb{L}(\varepsilon)A(\varepsilon) &= & &\ \varepsilon A_0^{(1)} &+&\ \varepsilon^2 A_1^{(1)} &+&\ \cdots, \\
\mathbb{L}(\varepsilon)^2 A(\varepsilon) &= & & & &\ \varepsilon^2 A_0^{(2)} &+&\ \cdots.
\end{aligned}
\tag{C.1.12}
$$

The first equation is the same as (C.1.9), so

$$A_j^{(0)} = A_j. \tag{C.1.13}$$

A recursion for the $A_j^{(i)}$ is developed as follows:

$$
\begin{aligned}
\mathbb{L}(\varepsilon)^{i+1}A(\varepsilon) &= \mathbb{L}(\varepsilon)(\mathbb{L}(\varepsilon)^i A(\varepsilon)) \\
&= (\varepsilon \mathbb{L}_1 + \varepsilon^2 \mathbb{L}_2 + \cdots)\left(\varepsilon^i A_0^{(i)} + \varepsilon^{i+1} A_1^{(i)} + \cdots\right) \\
&= \varepsilon^{i+1}(\mathbb{L}_1 + \varepsilon \mathbb{L}_2 + \cdots)\left(A_0^{(i)} + \varepsilon A_1^{(i)} + \cdots\right) \\
&= \varepsilon^{i+1}\left(A_0^{(i+1)} + \varepsilon A_1^{(i+1)} + \varepsilon^2 A_2^{(i+1)} + \cdots\right).
\end{aligned}
$$

Comparing the last two lines gives

$$A_j^{(i+1)} = \sum_{k=0}^{j} \mathbb{L}_{k+1}A_{j-k}^{(i)}. \tag{C.1.14}$$

Thus, from (C.1.13) and (C.1.14) it is possible to develop the following table:

$$
\begin{matrix}
A_0^{(0)} & A_1^{(0)} & A_2^{(0)} & \cdots \\
 & A_0^{(1)} & A_1^{(1)} & \cdots \\
 & & A_0^{(2)} & \cdots
\end{matrix}
\tag{C.1.15}
$$

By the fundamental theorem of Lie series, the transform of $A(\varepsilon)$ is

$$B(\varepsilon) = B_0 + \varepsilon B_1 + \varepsilon^2 B_2 + \cdots \tag{C.1.16}$$

$$= \mathbb{S}_{T(\varepsilon)} A(\varepsilon) = e^{\mathbb{L}(\varepsilon)} A(\varepsilon)$$

$$= A(\varepsilon) + \mathbb{L}(\varepsilon) A(\varepsilon) + \frac{1}{2} \mathbb{L}(\varepsilon)^2 A(\varepsilon) + \cdots .$$

It follows from (C.1.12) that the coefficients B_r are the following weighted sums of the entries in table (C.1.15):

$$B_r = \sum_{s=0}^{r} \frac{1}{s!} A_{r-s}^{(s)}. \tag{C.1.17}$$

C.1.3. Algorithm (Algorithm III for format 2b). Given the matrix series $A(\varepsilon)$ and the generator $V(\varepsilon)$, the transformed matrix series $B(\varepsilon)$ can be obtained by using (C.1.13) and (C.1.14) to develop table (C.1.15), and then using (C.1.17) to obtain the coefficients B_r.

C.1.4. Algorithm (Algorithm IV for format 2b). Assume that we are given "transformed" matrix series $B(\varepsilon)$ and the generator $V(\varepsilon)$ by which it was transformed. Then the original series $A(\varepsilon)$ can be obtained by creating a table like (C.1.15) with top row

$$B_j^{(0)} = B_j$$

and successive rows

$$B_j^{(i+1)} = \sum_{k=0}^{j} -\mathbb{L}_{k+1} B_{j-k}^{(i)}.$$

Then the A_r will be the weighted sums of the columns:

$$A_r = \sum_{s=0}^{r} \frac{1}{s!} B_{r-s}^{(s)}.$$

Proof. Since the generator of the inverse is $-V(\varepsilon)$, it is only necessary to replace each $\mathbb{L}_j$ by $-\mathbb{L}_j$ in the recursion that generates the table.   □

Algorithm V

Finally, we come to Algorithm V. This, of course, will be the same as Algorithm III, except that the V_j are not known in advance and must be determined by solving certain homological equations. It should be clear (see Section 3.2) that B_1 and V_1 are determined by

$$\mathcal{L} V_1 = A_1 - B_1. \tag{C.1.18}$$

Suppose, inductively, that V_i is known for $i < j$. It is clear that the first j columns (or better, columns 0 through $j - 1$) of table (C.1.15) must be computable from this much knowledge of $V(\varepsilon)$, since B_{j-1}, which is the weighted sum of the jth (or "$(j-1)$st", starting with zero) column, must be computable from $V_1, \ldots, V_{j-1}$ and $A_0, \ldots, A_{j-1}$. (This heuristic argument can be verified from equation (C.1.14).) To see what happens in the next column (the $(j+1)$st, or the "jth"), consider the case $j = 3$. According to (C.1.14), the entries of the fourth ("third") column are

$$A_3^{(0)} = A_3 = \text{given},$$
$$A_2^{(1)} = \mathbb{L}_1 A_2^{(0)} + \mathbb{L}_2 A_1^{(0)} + \mathbb{L}_3 A_0^{(0)},$$
$$A_1^{(2)} = \mathbb{L}_1 A_1^{(1)} + \mathbb{L}_2 A_0^{(1)},$$
$$A_0^{(3)} = \mathbb{L}_1 A_0^{(2)}.$$

If V_1 and V_2 are known, then $\mathbb{L}_1$ and $\mathbb{L}_2$ are known, but $\mathbb{L}_3$ is not. Therefore, all entries in this column are known except the entry in the second row, $A_2^{(1)}$, and in fact, all of the terms of that entry except the last term, $\mathbb{L}_3 A_0^{(0)}$, are known. By (C.1.17), B_3 is the weighted sum of the entries in this column:

$$B_3 = A_3^{(0)} + A_2^{(1)} + \frac{1}{2} A_1^{(2)} + \frac{1}{6} A_0^{(3)}.$$

Letting K_3 be the weighted sum of the known terms (that is, all except the last term of $A_2^{(1)}$), and writing the unknown term explicitly, this becomes

$$B_3 = K_3 + \mathbb{L}_3 A_0^{(0)}.$$

But

$$\mathbb{L}_3 A_0^{(0)} = [A_0, V_3] = -[V_3, A_0] = -\pounds V_3.$$

Therefore,

$$\pounds V_3 = K_3 - B_3.$$

This pattern continues for all j.

C.1.5. Algorithm (Algorithm V for format 2b). Let us assume that $V_1, \ldots, V_{j-1}$ have been determined. Then the column $A_j^{(0)}, \ldots, A_0^{(j)}$ of table (C.1.15) is computable from (C.1.14) except for the term $\mathbb{L}_j A_0^{(0)}$ in $A_{j-1}^{(1)}$. Let K_j be the weighted sum (according to (C.1.17)) of the known terms. Then B_j and V_j are determined by the homological equation

$$\pounds V_j = K_j - B_j.$$

An Alternative Formulation of Algorithm I

It is possible to formulate Algorithm I for format 2b in a way that makes it almost exactly the same as Algorithm III. The notations needed may seem unmotivated at first, but they will appear natural after one has read Section C.2 below. Introduce the *right multiplication* operator $\mathbb{D}$ by

$$\mathbb{D}_Q P = PQ.$$

(See (D.1.4) and Remark D.1.1 in the next section for further discussion of this operator.) Then let

$$\mathbb{D}_j = \mathbb{D}_{V_j} \tag{C.1.19}$$

and

$$\mathbb{D}(\varepsilon) = \varepsilon \mathbb{D}_1 + \varepsilon^2 \mathbb{D}_2 + \cdots, \tag{C.1.20}$$

by analogy with (C.1.10) and (C.1.11). Consider an arbitrary matrix series

$$C(\varepsilon) = C_0 + \varepsilon C_1 + \varepsilon^2 C_2 + \cdots,$$

and introduce the following notation for the iterates of $\mathcal{D}(\varepsilon)$ on $C(\varepsilon)$, by analogy with (C.1.12):

$$
\begin{aligned}
C(\varepsilon) &= C_0^{(0)} + \varepsilon C_1^{(0)} + \varepsilon^2 C_2^{(0)} + \cdots, \\
\mathbb{D}(\varepsilon)C(\varepsilon) &= \varepsilon C_0^{(1)} + \varepsilon^2 C_1^{(1)} + \cdots, \\
\mathbb{D}(\varepsilon)^2 C(\varepsilon) &= \varepsilon^2 C_0^{(2)} + \cdots.
\end{aligned}
\tag{C.1.21}
$$

Taking $C(\varepsilon) = I$, (C.1.12) becomes equivalent to (C.1.3), and the subsequent development from (C.1.3) to C.1.8) coincides exactly with the development of Algorithm III from (C.1.12) to (C.1.17). Just as

$$B(\varepsilon) = e^{\mathbb{L}(\varepsilon)} A(\varepsilon),$$

we have

$$T(\varepsilon) = e^{\mathbb{D}(\varepsilon)} I, \tag{C.1.22}$$

and when it is written this way (rather than as (C.1.1)), the algorithms to produce $B(\varepsilon)$ from $A(\varepsilon)$, and $T(\varepsilon)$ from I, are formally identical except that $\mathbb{L}(\varepsilon)$ becomes $\mathbb{D}(\varepsilon)$.

C.2 Format 2b, Nonlinear Case (for Chapter 4)

Format 2b for matrices (Section C.1) is characterized by the following relationship between the generator and the transformation:

$$T(\varepsilon) = e^{\varepsilon V(\varepsilon)}, \qquad V(\varepsilon) = V_1 + \varepsilon V_2 + \cdots + \varepsilon^k V_k, \tag{C.2.1}$$

where k is the predetermined order to which normalization is to be accomplished. In order to arrive at a version of format 2b for vector fields, the exponential must be replaced by a flow, as discussed in Section 4.4. We begin by reexpressing the exponential relationship (C.2.1) for the linear case by a system of differential equations. Removing one factor of ε from the exponent, consider the matrix differential equation

$$\frac{dT}{ds} = V(\varepsilon)T(s),$$

having solution

$$T(s, \varepsilon) = e^{sV(\varepsilon)}.$$

This expression reduces to the desired $T(\varepsilon)$ upon setting $s = \varepsilon$. The flow of the vector differential equation

$$\frac{dx}{ds} = v(x, \varepsilon) = V(\varepsilon)x$$

can be expressed as

$$\varphi_\varepsilon^s(y) = T(s, \varepsilon)y;$$

this is the solution passing through the point y at "time" zero. Here s is the "time" along the solution, and ε is just a parameter. Setting $s = \varepsilon$ gives the coordinate transformation defined by the generator $V(\varepsilon)$:

$$x = \varphi_\varepsilon^\varepsilon(y) = T(\varepsilon, \varepsilon)y.$$

Once format 2b for matrices has been formulated in this manner, it passes over naturally to the vector field context by allowing $v(x, \varepsilon)$ to become a nonlinear vector field. The version that we develop is suited for the "universal" setting of (4.2.1).

We begin with a vector field of the form

$$v_\varepsilon(x) = v(x, \varepsilon) = v_1(x) + \varepsilon v_2(x) + \cdots + \varepsilon^{k-1}v_k(x), \qquad \text{(C.2.2)}$$

which is to be the generator of a near-identity transformation depending on ε. (The notation v_ε will be used when we need an accurate name for the vector field $x \mapsto v(x, \varepsilon)$ without specifying the variable.) Let the solution of the differential equation

$$\frac{dx}{ds} = v(x, \varepsilon) \qquad \text{(C.2.3)}$$

with initial condition $x(0) = y$ be denoted by $\varphi_\varepsilon^s(y)$; in other words, let φ_ε^s be the flow of v_ε. Then the transformation generated by $v(x, \varepsilon)$ is defined to be

$$x = \psi_\varepsilon(y) = \varphi_\varepsilon^\varepsilon(y). \qquad \text{(C.2.4)}$$

Although φ_ε^s is a (local) flow in the sense of Section 4.4 for each ε, ψ_ε is not a flow. The letter y (rather than ξ as in Section 4.4) is used for the new variable because it is intended that with the proper choice of v, a vector field $a(x, \varepsilon)$ will be normalized to order k in one step by the transformation (C.2.4) (rather than after a sequence of steps), and we frequently use y as the variable in the normalized equations. (See the discussion surrounding equation (4.3.7).)

C.2.1. Remark. In Section C.1, everything was expressed in terms of formal power series. This was possible because it is clear how to construct the exponential of a matrix formal power series. Here we have instead used the finite form of (C.2.2), ending at some k, so that $v(x, \varepsilon)$ will be an actual vector field and (C.2.3) will have actual (smooth) solutions. When Algorithm I has been developed, there will be a procedure to construct the formal power series of the transformation ψ_ε from $v(x, \varepsilon)$. This procedure works if $v(x, \varepsilon)$ is a formal power series in ε.

Let $a(x, \varepsilon)$ be a vector field in universal form (4.2.1), and let $b(y, \varepsilon)$ be the transformed vector field under (C.2.4), so that

$$b(y, \varepsilon) = (S_{\psi_\varepsilon} a)(y, \varepsilon) = \psi'_\varepsilon(y)^{-1} a(\psi_\varepsilon(y), \varepsilon).$$

For $j = 1, 2, \ldots$, define

$$L_j = L_{v_j} \tag{C.2.5}$$

and

$$L(\varepsilon) = \varepsilon L_1 + \cdots + \varepsilon^k L_k. \tag{C.2.6}$$

(Compare (C.1.10) and (C.1.11).)

C.2.2. Lemma. The effect of the transformation (C.2.4) defined by the generator (C.2.2) is to transform the vector field $a(x, \varepsilon)$ into

$$b(y, \varepsilon) = (e^{L(\varepsilon)} a)(y, \varepsilon),$$

where $\mathcal{L}(\varepsilon)$ is defined by (C.2.6) and (C.2.5).

Before proving this lemma, it may be helpful to point out that it is stated in the "operators on functions" notation discussed in Remark 4.3.2 (contrasted with "operators on expressions"). That is, $L(\varepsilon)$ is understood as an operator on vector fields, without any reference to a specific variable; after $e^{L(\varepsilon)}$ is applied to a, the resulting vector field is evaluated at y. In practical terms it is more useful to work with "operators on expressions," as follows: Let

$$L_j(y) = L_{v_j(y)} \tag{C.2.7}$$

and

$$L(y, \varepsilon) = \varepsilon L_1(y) + \varepsilon^2 L_2(y) + \cdots$$

be operators on expressions in the vector variable y. Then Lemma C.2.2 becomes

$$b(y, \varepsilon) = e^{L(y,\varepsilon)} a(y, \varepsilon). \tag{C.2.8}$$

Proof. The fundamental theorem of Lie series, Theorem 4.4.8, implies that for any fixed ε,

$$(\varphi_\varepsilon^s)'(y, \varepsilon)^{-1} a(\varphi_\varepsilon^s(y), \varepsilon) \sim (e^{sL_{v_\varepsilon}} a)(y, \varepsilon). \tag{C.2.9}$$

If it is permissible to substitute $s = \varepsilon$ into this equation, the result will be

$$\psi'_\varepsilon(y, \varepsilon) a(\psi_\varepsilon(y), \varepsilon) \sim (e^{\mathsf{L}(\varepsilon)} a)(y, \varepsilon), \tag{C.2.10}$$

because $\varepsilon \mathsf{L}_{v_\varepsilon} = \mathsf{L}_{\varepsilon v_\varepsilon} = \mathsf{L}_{\varepsilon v_1 + \cdots + \varepsilon^k v_k} = \mathsf{L}(\varepsilon)$. Equation (C.2.10) is just what was to be proved. So it remains only to investigate the validity of substituting $s = \varepsilon$ into (C.2.9). Now, the left-hand side of (C.2.9) is a smooth function of s and ε (and, of course, y, but we may treat y as fixed for present purposes). The right-hand side is a formal power series in s, with coefficients which are smooth functions of ε (and y). The rules for operating with Taylor series of smooth functions permit expanding the right-hand side in ε to obtain a double series (in s and ε) and then substituting $s = \varepsilon$ and collapsing the result into a single power series in ε, since only finitely many terms must be added to obtain the coefficient of any given power of ε. (In fact, the triangle algorithm given later in this section, based on (C.2.10), may be viewed as the procedure for collecting the appropriate terms in (C.2.9) when s is set equal to ε. The complexity of the algorithm reflects the fact that ε enters in three ways: through v_ε, $a(y, \varepsilon)$, and by setting $s = \varepsilon$.)   □

Lemma C.2.2 is formally equivalent to (C.1.16), and therefore Algorithm III will be formally equivalent to Algorithm C.1.3, with A, V, and B_r replaced by a, v, and b_r. A precise statement will be given in the list of algorithms at the end of this section.

Algorithm I will be developed in a manner parallel to the "alternative formulation" of Algorithm I in Section C.1 (equations (C.1.19) through (C.1.22)). Let $f(x)$ be any (scalar- or vector-valued) function that "transforms as a function" (rather than as a vector field). (This just means that in case f is vector-valued, we are interested in finding $f(\psi_\varepsilon(y))$ rather than $\psi'_\varepsilon(y)^{-1} f(\psi_\varepsilon(y))$.) Then, by Lemma 4.4.9,

$$f(\varphi_\varepsilon^s(y)) \sim (e^{s\mathcal{D}_{v_\varepsilon}} f)(y),$$

where the right-hand side is a formal power series in s. This is formally the same as (C.2.9), with $\psi'_\varepsilon(y)^{-1}$ deleted from the left-hand side and L replaced by $\mathcal{D}$. (We could have allowed f to depend on ε as well as x, just as a depends on ε in (C.2.9).) Thus, the same argument as in Lemma C.2.2 shows that

$$f(\psi_\varepsilon(y)) \sim (e^{s\mathcal{D}(\varepsilon)} f)(y), \tag{C.2.11}$$

where

$$\mathcal{D}(\varepsilon) = \varepsilon \mathcal{D}_1 + \varepsilon^2 \mathcal{D}_2 + \cdots + \varepsilon^k \mathcal{D}_k$$

with $\mathcal{D}_j = \mathcal{D}_{v_j}$. Taking f to be the identity function, the left-hand side of (C.2.11) becomes $\psi_\varepsilon(y)$. This may be expressed in the manner of (C.2.8), using operators on expressions in the vector variable y, as follows. Let

$$\mathcal{D}_j(y) = \mathcal{D}_{v_j(y)}$$

and

$$\mathcal{D}(y,\varepsilon) = \varepsilon\mathcal{D}_1(y) + \varepsilon^2\mathcal{D}_2(y) + \cdots.$$

The identity transformation $f(y) = y$ may be written simply as y. Then the transformation generated by $v(x,\varepsilon)$ is

$$x = \psi_\varepsilon(y) \sim e^{\mathcal{D}(y,\varepsilon)}y. \qquad (C.2.12)$$

Equation (C.2.12) is formally equivalent to (C.1.22) and leads to a formally equivalent algorithm, stated below as Algorithm I.

The Five Algorithms

I. Given a generator as in (C.2.2), define $\mathcal{D}_j(y) = \mathcal{D}_{v_j(y)}$ as above. Create a table

$$\begin{array}{cccc} f_0^{(0)} & f_1^{(0)} & f_2^{(0)} & \cdots \\ & f_0^{(1)} & f_1^{(1)} & \cdots \\ & & f_0^{(2)} & \cdots \end{array} \qquad (C.2.13)$$

in which the first row is the (trivial) expansion in ε of the identity function $f(y,\varepsilon) = y = y + 0\varepsilon + 0\varepsilon^2 + \cdots$; that is, $f_0^{(0)}(y) = y$ and $f_j^{(0)}(y) = 0$ for $j > 0$. The remaining rows of (C.2.13) are generated according to the rule

$$f_j^{(i+1)}(y) = \sum_{k=0}^{j} \mathcal{D}_{k+1}(y) f_{j-k}^{(i)}(y). \qquad (C.2.14)$$

Then the coefficients ψ_r of the formal power series of the transformation

$$x = \psi_\varepsilon(y) = y + \varepsilon\psi_1(y) + \varepsilon^2\psi_2(y) + \cdots \qquad (C.2.15)$$

are given by

$$\psi_r(y) = \sum_{s=0}^{r} \frac{1}{s!} f_{r-s}^{(s)}(y). \qquad (C.2.16)$$

(These formulas are taken over exactly from equations (C.1.14) through (C.1.17), changing A to f, $\mathbb{L}$ to $\mathcal{D}$, and B to ψ.)

II. The inverse transformation is computed by the same algorithm with v replaced by $-v$.

III. Given a generator as in (C.2.2) and a vector field $a(x,\varepsilon)$ as in (4.2.1), define $\mathsf{L}_j(y) = \mathsf{L}_{v_j(y)}$ as in (C.2.7) and create a table

$$\begin{array}{cccc} a_0^{(0)} & a_1^{(0)} & a_2^{(0)} & \cdots \\ & a_0^{(1)} & a_1^{(1)} & \cdots \\ & & a_0^{(2)} & \cdots \end{array} \qquad (C.2.17)$$

in which the first row is the expansion of $a(y, \varepsilon)$, so that $a_j^{(0)}(y) = a_j(y)$. The remaining rows are generated according to the rule

$$a_j^{(i+1)}(y) = \sum_{k=0}^{j} \mathsf{L}_{k+1}(y) a_{j-k}^{(i)}(y). \qquad (\text{C.2.18})$$

Then the coefficients of the formal power series expansion of $b(y, \varepsilon)$ are given by

$$b_r(y) = \sum_{s=0}^{r} \frac{1}{s!} a_{r-s}^{(s)}(y). \qquad (\text{C.2.19})$$

As in Algorithm I, these formulas are taken over from (C.1.14) through (C.1.17) with appropriate changes.

IV. The inverse transform, from b back to a, is computed as in Algorithm III but with v replaced by $-v$.

V. Algorithm V is the same as Algorithm III, except that the v_j must be determined along the way by solving homological equations. The required discussion is exactly the same as in Section C.1 beginning with equation (C.1.18), provided that A, B, V, and $\mathbb{L}$ are replaced with a, b, v, and L.

Format 2b in Setting 1 Without Dilation

It was shown in Section 4.4, beginning with equation (4.4.34), that format 2a in setting 1 without dilation amounts to applying the time-one map of the flow produced by a given generator $u_j \in \mathcal{V}_j^n$. In the same way, format 2b (in setting 1 without dilation) is equivalent to applying the time-one map of the flow generated by a vector field $v \in \mathcal{V}_{**}^n$ having no constant term. The simplest way to see this is to set the "time" ε equal to 1 in (C.2.2) and in Lemma C.2.2; if the time-one map of $v = v_1 + v_2 + \cdots$ is ψ, then ψ carries a into

$$b = e^{\mathsf{L}v} a.$$

This can be expanded and arranged in algorithmic form exactly as before; the only difference is that we do not have the powers of ε to handle the bookkeeping, and must collect the terms of a given grade ourselves. This version of format 2b is used in many of the advanced papers on normal forms, such as those of Alberto Baider (see Section 4.10), and may be introduced with words such as the following: Regard $\mathcal{V}_{**}^n$ as a formal graded Lie algebra; then the subalgebra spanned by terms of grade ≥ 1 acts on the full algebra by $(v, a) \mapsto e^{\mathsf{L}v} a$. See also Remark 3.2.3.

Appendix D
Format 2c: Generated Recursive (Deprit)

This appendix presents format 2c for the linear and nonlinear settings. Section D.1 can be read with Chapter 3 after Section 3.6; there are some references to Section C.1, but Appendix D is mostly independent of Appendix C. Similarly, Section D.2 can be read with Chapter 4 after Sections 4.4 and D.1; there are some references to Section C.2.

D.1   Format 2c, Linear Case (for Chapter 3)

Format 2c is our name for the set of computational procedures ("bookkeeping devices") needed to handle transformations generated according to the matrix differential equation

$$\frac{dT(\varepsilon)}{d\varepsilon} = W(\varepsilon)T(\varepsilon) \tag{D.1.1}$$

with initial condition

$$T(0) = I, \tag{D.1.2}$$

where

$$W(\varepsilon) = W_1 + \varepsilon W_2 + \varepsilon^2 W_3 + \cdots . \tag{D.1.3}$$

(The power of ε lags behind the index because, in effect, an extra ε is added in the process of solving equation (D.1.1), as explained in Section 3.2.) As in Section C.1, all functions in this section are to be understood as formal power series in ε. (See the first paragraph of Section C.1 for further explanation of this point.)

Format 2c retains the advantage (common to the generated formats) that if the generator $W(\varepsilon)$ belongs to a Lie algebra $\mathfrak{g}$, then the transformation $T(\varepsilon)$ that it generates belongs to the associated Lie group $\mathfrak{G}$. This is the converse of Remark 3.6.5, and we will simply sketch the proof as follows, for those familiar with a little manifold theory. For $\mathfrak{g} = \mathrm{gl}(n)$ there is nothing to prove except that $T(\varepsilon)$ is invertible, which follows from Wronski's formula (3.6.8). Each other Lie group $\mathfrak{G}$ is a submanifold of $\mathrm{gl}(n, \mathbb{R}) \cong \mathbb{R}^{n^2}$ (or, in the case of some Lie groups we have not discussed, of $\mathrm{gl}(n, \mathbb{C}) \cong \mathbb{C}^{n^2} \cong \mathbb{R}^{n^4}$, regarded as a real vector space). The tangent space of $\mathfrak{G}$ at I is $\mathfrak{g}$, and the tangent space at any other point $T \in \mathfrak{G}$ is the set of matrices WT for $W \in \mathfrak{g}$. Therefore, the vector field defined by (D.1.1) is tangent to $\mathfrak{G}$ at each point, and can be regarded as a vector field on the manifold. Therefore, its integral curves (the solutions of the differential equation) lie in the manifold.

Algorithms I to V for format 2c are more difficult to derive than those of format 2b, for two reasons: The fundamental theorem of Lie series does not apply directly, and must be replaced by a similar theorem proved by similar methods; and the generator of the inverse is not the negative of the generator. Rather than attempt to construct the generator of the inverse, it is easier to obtain an algorithm for the inverse by running the algorithm for the direct transformation "in reverse" using the generator of the direct transformation. In spite of these additional difficulties, format 2c seems to be extremely popular (in the setting of Chapter 4, under the name of Deprit's method), while format 2b (under the name of Hori's method) is much less well known. For this reason we give a complete treatment of format 2c, both in the linear (Chapter 3) and nonlinear (Chapter 4) situations.

The traditional notation used in "Deprit's method" calls for the inclusion of factorials in the series expansions for all functions (so that our ε^j is always replaced by $\varepsilon^j/j!$). We will instead express format 2c in the same notation as the other formats. The resulting equations actually turn out to be simpler than those of Deprit's method. At the end of this section the main results will be restated in the traditional notation.

Algorithms I and II

Algorithms I and II are based on the following lemma. Define the *right multiplication* operator $\mathbb{D}_W : \mathrm{gl}(n) \to \mathrm{gl}(n)$ by

$$\mathbb{D}_W C = CW. \tag{D.1.4}$$

D.1.1. Remark. The reason right multiplication is denoted by $\mathbb{D}$ is as follows. In Chapter 4 the directional derivative of a scalar function $f : \mathbb{R}^n \to \mathbb{R}$ along a vector field v is written $(\mathcal{D}_v f)(x) = f'(x)v(x)$; here $f'(x)$ is a row vector. The same operator $\mathcal{D}_v$ can be applied to a vector-valued function $f : \mathbb{R}^n \to \mathbb{R}^n$ (provided that f is understood as a function, not as a vector field; $\mathcal{D}_v f$ does not transform correctly for a vector field under change of coordinates), with $f'(x)$ now being an $n \times n$ matrix. Now suppose v and f are linear, so that $v(x) = Wx$ and $f(x) = Cx$. Then $(\mathcal{D}_v f)(x) = (Cx)'(Wx) = (CW)x = (\mathbb{D}_W C)x$.

D.1.2. Lemma. Let $C(\varepsilon)$ be any (formal power series) matrix function of ε. Then for each $j = 1, 2, \ldots,$

$$\left(\frac{d}{d\varepsilon}\right)^j C(\varepsilon)T(\varepsilon) = \left[\left(\frac{d}{d\varepsilon} + \mathbb{D}_{W(\varepsilon)}\right)^j C(\varepsilon)\right] T(\varepsilon).$$

Proof. For $j = 1$ the lemma takes the form

$$\left(\frac{d}{d\varepsilon}\right) C(\varepsilon)T(\varepsilon) = \left[\left(\frac{d}{d\varepsilon} + \mathbb{D}_{W(\varepsilon)}\right) C(\varepsilon)\right] T(\varepsilon). \tag{D.1.5}$$

This is proved as follows, using (D.1.1), with $' = d/d\varepsilon$: $(CT)' = C'T + CT' = C'T + CWT = (C' + CW)T$. Equation (D.1.5) says that to differentiate any product of a matrix times $T(\varepsilon)$, the operator $d/d\varepsilon + \mathbb{D}_{W(\varepsilon)}$ should be applied to the matrix. But the right-hand side of (D.1.5) again has the form of a matrix times $T(\varepsilon)$, so the process can be repeated, leading to the general case of the lemma.   □

Introduce the notation

$$C^{[j]}(\varepsilon) = \left(\frac{d}{d\varepsilon} + \mathbb{D}_{W(\varepsilon)}\right)^j C(\varepsilon) = C_0^{[j]} + \varepsilon C_1^{[j]} + \varepsilon^2 C_2^{[j]} + \cdots . \tag{D.1.6}$$

Then Lemma D.1.2 says that

$$\left(\frac{d}{d\varepsilon}\right)^j C(\varepsilon)T(\varepsilon) = C^{[j]}(\varepsilon)T(\varepsilon). \tag{D.1.7}$$

Up to this point, $C(\varepsilon)$ has been arbitrary. To obtain Algorithm I, take

$$C(\varepsilon) = I; \tag{D.1.8}$$

this fixes the meaning of $C^{[j]}(\varepsilon)$, and (D.1.7) becomes

$$T^{(j)}(\varepsilon) = C^{[j]}(\varepsilon)T(\varepsilon), \tag{D.1.9}$$

where $T^{(j)}(\varepsilon)$ is the jth derivative of $T(\varepsilon)$. To develop a recursion for the $C_i^{[j]}$, notice that on the one hand,

$$T^{(j+1)}(\varepsilon) = C^{[j+1]}(\varepsilon)T(\varepsilon),$$

while on the other, by another application of Lemma D.1.2,

$$T^{(j+1)}(\varepsilon) = \left[\left(\frac{d}{d\varepsilon} + \mathbb{D}_{W(\varepsilon)}\right) C^{[j]}(\varepsilon)\right] T(\varepsilon).$$

It follows that

$$C^{[j+1]} = \left(\frac{d}{d\varepsilon} + \mathbb{D}_{W(\varepsilon)}\right) C^{[j]}(\varepsilon). \qquad (D.1.10)$$

Writing

$$\mathbb{D}_i = \mathbb{D}_{W_i}, \qquad (D.1.11)$$

the right-hand side may be expanded as

$$\left(\frac{d}{d\varepsilon} + \mathbb{D}_1 + \varepsilon\mathbb{D}_2 + \cdots\right)\left(C_0^{[j]} + \varepsilon C_1^{[j]} + \varepsilon^2 C_2^{[j]} + \cdots\right) \qquad (D.1.12)$$

$$= \left(C_1^{[j]} + \mathbb{D}_1 C_0^{[j]}\right) + \varepsilon\left(2C_2^{[j]} + \mathbb{D}_1 C_1^{[j]} + \mathbb{D}_2 C_0^{[j]}\right)$$

$$+ \varepsilon^2(eC_3^{[j]} + \mathbb{D}_1 C_2^{[j]} + \mathbb{D}_2 C_1^{[j]} + \mathbb{D}_3 C_0^{[j]}) + \cdots.$$

Comparing this with the expansion of the left-hand side of (D.1.10) gives

$$C_{i-1}^{[j+1]} = iC_i^{[j]} + \sum_{k=1}^{i} \mathbb{D}_k C_{i-k}^{[j]}. \qquad (D.1.13)$$

It is customary to arrange the $C_i^{[j]}$ in a table in such a way that the coefficients of $C^{[j]}(\varepsilon)$ appear in a column rather than a row (as we did in similar situations in format 2b). This results in

$$
\begin{array}{llll}
C_0^{[0]} & & & \\
C_1^{[0]} & C_0^{[1]} & & \\
C_2^{[0]} & C_1^{[1]} & C_0^{[2]} & \\
C_3^{[0]} & C_2^{[1]} & C_1^{[2]} & C_0^{[3]}.
\end{array}
\qquad (D.1.14)
$$

According to (D.1.8) the first column contains the entries $I, 0, 0, 0, \ldots$. The recursion (D.1.13) allows the rest of the table to be built from this starting point; each entry $C_{i-1}^{[j+1]}$ depends on the entry $C_i^{[j]}$ to its left and the entries above $C_i^{[j]}$, making an L-shaped dependence pattern. The diagonal entries $C_0^{[j]}$ are the constant terms of $C^{[j]}(\varepsilon)$, so that by (D.1.9) and (D.1.2),

$$C_0^{[j]} = C^{[j]}(0) = T^{(j)}(0). \qquad (D.1.15)$$

But by Taylor's theorem

$$T(\varepsilon) = I + \varepsilon T_1 + \varepsilon^2 T_2 + \cdots = I + \varepsilon T^{(1)}(0) + \varepsilon^2 \frac{1}{2} T^{(2)}(0) + \cdots,$$

so

$$T_j = \frac{1}{j!} C_0^{[j]}. \qquad (D.1.16)$$

D.1.3. Algorithm (Algorithm I for format 2c). Given the generator $W(\varepsilon)$, build the table (D.1.14) by entering $I, 0, 0, 0, \ldots$ as the first column and generating the successive columns according to (D.1.13). Then the coefficients of the transformation generated by $W(\varepsilon)$ are given by (D.1.16).

The same arguments show that if $C(\varepsilon)$ is arbitrary, the right-hand side of (D.1.16) gives the expansion of $C(\varepsilon)T(\varepsilon)$ rather than of $T(\varepsilon)$. Therefore, if we take

$$C(\varepsilon) = T^{-1}(\varepsilon) \qquad (D.1.17)$$

instead of $C(\varepsilon) = I$, the diagonal entries of (D.1.14) will be $I, 0, 0, 0, \ldots$. Notice that the entries in the first column will be the exact coefficients of (D.1.17), and do not require division by factorials.

D.1.4. Algorithm (Algorithm II for format 2c). Given the generator $W(\varepsilon)$, build the table (D.1.14) by entering $I, 0, 0, 0, \ldots$ as the diagonal elements $C_0^{[j]}$ and constructing the rest of the table recursively by

$$C_i^{[j]} = \frac{1}{i}\left(C_{i-1}^{[j+1]} - \sum_{k=1}^{i} \mathbb{D}_k C_{i-k}^{[j]}\right).$$

(This constructs the *corner* element of each L-shaped portion from those above it and to its right.) Then the entries in the first column give the coefficients of $T(\varepsilon)^{-1}$, the inverse of the transformation generated by $W(\varepsilon)$.

Algorithms III and IV

Turning to Algorithms III and IV, the following lemma is the replacement for the fundamental theorem of Lie series that has been mentioned several times.

D.1.5. Lemma. Let $A(\varepsilon)$ be any (formal power series) matrix function of ε. Then for each $j = 1, 2, \ldots$,

$$\left(\frac{d}{d\varepsilon}\right)^j T(\varepsilon)^{-1} A(\varepsilon) T(\varepsilon) = T(\varepsilon)^{-1}\left[\left(\frac{d}{d\varepsilon} + \mathbb{L}_{W(\varepsilon)}\right)^j A(\varepsilon)\right] T(\varepsilon).$$

Equivalently,

$$\left(\frac{d}{d\varepsilon}\right)^j \mathbb{S}_{T(\varepsilon)} A(\varepsilon) = \mathbb{S}_{T(\varepsilon)}\left(\frac{d}{d\varepsilon} + \mathbb{L}_{W(\varepsilon)}\right)^j A(\varepsilon).$$

Proof. From (D.1.1), we have

$$0 = \frac{d}{d\varepsilon}(TT^{-1}) = (WT)T^{-1} + T\frac{dT^{-1}}{d\varepsilon} = W + T\frac{dT^{-1}}{d\varepsilon},$$

so that

$$\frac{dT^{-1}}{d\varepsilon} = -T^{-1}W.$$

Now

$$\frac{d}{d\varepsilon}T^{-1}AT = (-T^{-1}W)AT + T^{-1}A'T + T^{-1}A(WT)$$
$$= T^{-1}(A' + AW - WA)T.$$

Therefore,

$$\left(\frac{d}{d\varepsilon}\right)T(\varepsilon)^{-1}A(\varepsilon)T(\varepsilon) = T(\varepsilon)^{-1}\left[\left(\frac{d}{d\varepsilon} + \mathbb{L}_{W(\varepsilon)}\right)A(\varepsilon)\right]T(\varepsilon).$$

This proves the lemma for $j = 1$. Just as with (D.1.5), this formula is "recursible." That is, it states that to differentiate a matrix nested between T^{-1} and T, apply the operator $d/d\varepsilon + \mathbb{L}_W$ to the matrix. Since the result has the same form (a matrix nested between T^{-1} and T), the process can be repeated, proving the lemma.   □

Since Lemma D.1.5 has exactly the same form as Lemma D.1.2, with C replaced by A and $\mathbb{D}_W$ by $\mathbb{L}_W$, it is not necessary to repeat the arguments leading to Algorithms III and IV, but only to note the slight differences from those for Algorithms I and II. The chief difference is that for Algorithm I we took $C(\varepsilon) = I$, while for Algorithm III we take $A(\varepsilon)$ to be the matrix series that is to be transformed. As in (D.1.6), we define

$$A^{[j]}(\varepsilon) = \left(\frac{d}{d\varepsilon} + \mathbb{L}_{W(\varepsilon)}\right)^j A(\varepsilon) = A_0^{[j]} + \varepsilon A_1^{[j]} + \varepsilon^2 A_2^{[j]} + \cdots. \quad \text{(D.1.18)}$$

As in (D.1.13), it follows that

$$A_{i-1}^{[j+1]} = iA_i^{[j]} + \sum_{k=1}^{i}\mathbb{L}_k A_{i-k}^{[j]}. \quad \text{(D.1.19)}$$

The matrices $A_i^{[j]}$ can be arranged as in (D.1.14):

$$\begin{array}{llll} A_0^{[0]} & & & \\ A_1^{[0]} & A_0^{[1]} & & \\ A_2^{[0]} & A_1^{[1]} & A_0^{[2]} & \\ A_3^{[0]} & A_2^{[1]} & A_1^{[2]} & A_0^{[3]}. \end{array} \quad \text{(D.1.20)}$$

The rule of construction of this table, and the L-shaped dependence pattern, are the same as before, the only difference being the use of the operators $\mathbb{L}_k$ in place of $\mathbb{D}_k$. Lemma D.1.5 says that the transformed series

$$B(\varepsilon) = \mathbb{T}(\varepsilon)^{-1}A(\varepsilon)T(\varepsilon) = B_0 + \varepsilon B_1 + \varepsilon^2 B_2 + \cdots \quad \text{(D.1.21)}$$

satisfies

$$B^{(j)}(\varepsilon) = T(\varepsilon)^{-1}A^{[j]}(\varepsilon)T(\varepsilon), \quad \text{(D.1.22)}$$

so that by Taylor's theorem and (D.1.2),

$$B_j = \frac{1}{j!}B^{(j)}(0) = \frac{1}{j!}A_0^{[j]}.$$ (D.1.23)

D.1.6. Algorithm (Algorithm III for format 2c). Given a series $A(\varepsilon)$ and a generator $W(\varepsilon)$ as in (D.1.3), build the table (D.1.20) according to the rule (D.1.19), beginning with the coefficients of $A(\varepsilon)$ in the first column. Then the diagonal entries give the coefficients of the transformed series $B(\varepsilon)$ after division by factorials, as in (D.1.23).

D.1.7. Algorithm (Algorithm IV for format 2c). If we are given a "transformed" series $B(\varepsilon)$ and a generator $W(\varepsilon)$, the "original" series $A(\varepsilon)$ that transforms into $B(\varepsilon)$ is obtained by entering the matrices $j!B_j$ into the diagonal of table (D.1.20), and constructing the remainder of the table by the rule

$$A_i^{[j]} = \frac{1}{i}\left(A_{i-1}^{[j+1]} - \sum_{k=1}^{i}\mathbb{L}_k A_{i-k}^{[j]}\right).$$

Then the first column gives the coefficients of $A(\varepsilon)$ (with no division by factorials).

Algorithm V

As in format 2b, Algorithm V is the same as Algorithm III except that the W_j (and hence the $\mathbb{L}_j$) are not all known at the beginning of the operation. The matrices W_1 and B_1 are determined from the homological equation

$$\pounds W_1 = A_1 - B_1,$$ (D.1.24)

as explained in Section 3.2. Suppose, inductively, that $W_1, \ldots, W_{j-1}$ are known; then the first j rows (rows 0 through $j-1$) of table (D.1.20) are known. Let us calculate the row beginning with the known entry $A_j = A_j^{[0]}$. The entry next to this in the second row is

$$A_{j-1}^{[1]} = jA_j^{[0]} + \mathbb{L}_1 A_{j-1}^{[0]} + \mathbb{L}_2 A_{j-2}^{[0]} + \cdots + \mathbb{L}_j A_0^{[0]};$$ (D.1.25)

all of this is computable except the last term, which depends upon W_j. The next entry to the right, $A_{j-2}^{[2]}$, equals $(j-1)$ times expression (D.1.25) plus some terms that are computable from the entries above (D.1.25) using only the known operators $\mathbb{L}_1, \ldots, \mathbb{L}_{j-1}$. That is,

$$A_{j-2}^{[2]} = j(j-1)\mathbb{L}_j A_0^{[0]} + \text{computable terms}.$$

Continuing in this way across the row, we see that the last entry will be

$$A_0^{[j]} = j!\mathbb{L}_j A_0^{[0]} + \text{computable terms}.$$

Dividing by $j!$ and using (D.1.23), we obtain

$$B_j = \mathbb{L}_j A_0^{[0]} + K_j,$$

where K_j is computable. By the usual argument,

$$\mathbb{L}_j A_0^{[0]} = [A_0, W_j] = -[W_j, A_0] = -\pounds W_j,$$

so

$$\pounds W_j = K_j - B_j. \tag{D.1.26}$$

D.1.8. Algorithm (Algorithm V for format 2c). Given the matrix function $A(\varepsilon)$, build table (D.1.20) in the following way. Enter the coefficients of $A(\varepsilon)$ in the first column. Choose W_1 and B_1 from the homological equation (D.1.24), together with a specific choice of normal form style; enter B_1 in the position of $A_0^{[1]}$ (this will agree with the entry in that position calculated using W_1 and (D.1.19)). When $W_1, \ldots, W_{j-1}$ have been determined, calculate the entries of the next row (beginning with $A_j = A_j^{[0]}$) using (D.1.19), carrying the unknown term (containing $\mathbb{L}_j$) in symbolic form. Let K_j be $(1/j!)$ times the known part of the final entry $A_0^{[j]}$. Determine W_j and B_j from (D.1.26) and the chosen normal form style. Then go back over the same row of the table, computing the unknown term in each entry using W_j. The final entry $A_0^{[j]}$ will then automatically equal $j! B_j$, and the inductive step is completed.

Format 2c in the Traditional Notation

The notation in which we have developed format 2c contains an obvious asymmetry: The entries in the first column of the triangular tables for algorithms I and III are the actual coefficients of certain power series, whereas the entries in the diagonal must be divided by factorials to obtain the new coefficients. The traditional (Deprit) notation avoids this asymmetry, at the expense of a large number of extra multiplication operations that must be performed in the course of constructing the table. We will give the formulas for Algorithm III; those for Algorithm I are entirely similar.

First, it is necessary to change the notation used for representing all power series. (Warning: The equations below must not be mixed with the equations developed earlier in this section. To avoid confusion, none of the these equations will be numbered.) For Algorithm III we write

$$A(\varepsilon) = A_0 + \varepsilon A_1 + \frac{\varepsilon^2}{2!} A_2 + \frac{\varepsilon^3}{3!} A_3 + \cdots$$

and similarly for $B(\varepsilon)$; also

$$W(\varepsilon) = W_1 + \varepsilon W_2 + \frac{\varepsilon^2}{2!} W_3 + \cdots,$$

with the familiar shift of index. (The factorials match the power of ε, not the index.) The first column of the triangular table is defined by

$$A_j^{[0]} = A_j$$

(which is, of course, *not the same* as before, because the A_j are different). The recursion to build the rest of the table is

$$A_\ell^{[j+1]} = A_{\ell+1}^{[j]} + \sum_{k=0}^{\ell} \binom{\ell}{k} \mathbb{L}_{k+1} A_{\ell-k}^{[j]},$$

where

$$\mathbb{L}_i = \mathbb{L}_{W_i}.$$

(Remember that these are also not the same as our $\mathbb{L}_i$, because the W_i are different.) Finally, the diagonal entries give the B_j:

$$B_j = A_0^{[j]}.$$

D.2   Format 2c, Nonlinear Case (for Chapter 4)

Given an ε-dependent vector field

$$w(x, \varepsilon) = w_1(x) + \varepsilon w_2(x) + \varepsilon^2 w_3(x) + \cdots, \qquad (D.2.1)$$

either truncated at some ε^k or else formal, the near-identity transformation generated by w according to format 2c is the solution operator

$$x = \psi_\varepsilon(y) \qquad (D.2.2)$$

of the (actual or formal) differential equation

$$\frac{dx}{d\varepsilon} = w(x, \varepsilon). \qquad (D.2.3)$$

That is, $\psi_\varepsilon(y)$ is the solution of (D.2.3) passing through y at "time" $\varepsilon = 0$. Notice that, unlike the differential equation (C.2.3) in format 2b, (D.2.3) is a nonautonomous differential equation, because the parameter ε appearing in the vector field also serves as the independent variable. Therefore, ψ_ε will not be a flow.

The algorithms developed below are expressed in the notation used throughout most of Section D.1. These results can be reexpressed in the traditional (but more complicated) notation containing many factorials and binomial coefficients, in the manner described at the end of Section D.1.

Algorithms I and II

Suppose that $f(x, \varepsilon)$ is a scalar- or vector-valued function, and write f_1 and f_2 for the partial derivatives of f with respect to x and ε. (If f is a scalar, f_1 is a row vector; if f is a vector, f_1 is a matrix.) Then

$$\frac{d}{d\varepsilon} f(\psi_\varepsilon(y), \varepsilon) = f_1(\psi_\varepsilon(y)) w(\psi_\varepsilon(y), \varepsilon) + f_2(\psi_\varepsilon(y), \varepsilon). \qquad (D.2.4)$$

We are going to rewrite this equation in a form suitable for recursion, and also suitable for programming in a symbolic processing language. The latter requires that we use "operators on expressions" (as discussed in Remark 4.3.2 and in Section C.2). The operator $\mathcal{D}_{w(x,\varepsilon)}$ has already been introduced, and is defined by

$$\mathcal{D}_{w(x,\varepsilon)} f(x, \varepsilon) = f_1(x, \varepsilon) w(x, \varepsilon).$$

The operator $\partial/\partial\varepsilon$ satisfies

$$\frac{\partial}{\partial\varepsilon} f(x, \varepsilon) = f_2(x, \varepsilon).$$

(It is important that x be treated simply as a symbol, not equal to $\psi_\varepsilon(y)$, so that $\partial/\partial\varepsilon$ does not act on x.) Now (D.2.4) may be written as

$$\left(\frac{d}{d\varepsilon}\right)\left(f(x,\varepsilon)|_{x=\psi_\varepsilon(y)}\right) = \left[\left(\frac{\partial}{\partial\varepsilon} + \mathcal{D}_{w(x,\varepsilon)}\right) f(x,\varepsilon)\right]\Bigg|_{x=\psi_\varepsilon(y)}. \qquad (D.2.5)$$

Notice that the replacement of x by $\psi_\varepsilon(y)$ is done before the differentiation on the left-hand side of this equation, and after it on the right. Therefore, the "output" of the operation $d/d\varepsilon$ (that is, the right-hand side) has the same form as the "input," namely, a function of (x, ε) evaluated at $x = \psi_\varepsilon(y)$. Therefore, the equation is recursible, resulting in

$$\left(\frac{d}{d\varepsilon}\right)^j\left(f(x,\varepsilon)|_{x=\psi_\varepsilon(y)}\right) = \left[\left(\frac{\partial}{\partial\varepsilon} + \mathcal{D}_{w(x,\varepsilon)}\right)^j f(x,\varepsilon)\right]\Bigg|_{x=\psi_\varepsilon(y)}. \qquad (D.2.6)$$

Introduce the notation

$$f^{[j]}(x,\varepsilon) = \left(\frac{\partial}{\partial\varepsilon} + \mathcal{D}_{w(x,\varepsilon)}\right)^j f(x,\varepsilon) = f_0^{[j]}(x) + \varepsilon f_1^{[j]}(x) + \varepsilon^2 f_2^{[j]}(x) + \cdots.$$

$$(D.2.7)$$

Then equation (D.2.6) says that

$$\left(\frac{d}{d\varepsilon}\right)^j f(\psi_\varepsilon(y), \varepsilon) = f^{[j]}(x, \varepsilon)\Big|_{x=\psi_\varepsilon(y)}. \qquad (D.2.8)$$

Now take f to be the identity function

$$f(x, \varepsilon) = x. \qquad (D.2.9)$$

This fixes the meaning of the functions $f^{[j]}(x, \varepsilon)$. (As a technical remark, note that these depend on ε even though f does not, so it was essential to allow f to depend on ε in developing the recursive argument above.)

Equation (D.2.8) becomes

$$\left(\frac{d}{d\varepsilon}\right)^j \psi_\varepsilon(y) = f^{[j]}(x,\varepsilon)\Big|_{x=\psi_\varepsilon(y)},$$

which implies that

$$\left(\frac{d}{d\varepsilon}\right)^j \psi_\varepsilon(y)\Big|_{\varepsilon=0} = f^{[j]}(y,0). \tag{D.2.10}$$

This in turn implies that the Taylor series of the transformation

$$x = \psi_\varepsilon(y) = y + \varepsilon\psi_1(y) + \varepsilon^2\psi_2(y) + \cdots \tag{D.2.11}$$

is given by

$$\psi_j(y) = \frac{1}{j!}f^{[j]}(y,0) = \frac{1}{j!}f_0^{[j]}(y). \tag{D.2.12}$$

The discussion so far is parallel to the development of Algorithm I in Section D.1 beginning with equation (D.1.4). To be precise, if $w(x,\varepsilon)$ is a linear vector field $W(\varepsilon)x$, $\psi_\varepsilon(x) = T(\varepsilon)x$, and $f(x,\varepsilon) = C(\varepsilon)x$, then the equations of this section specialize to those of Section 6.8. It remains only to point out that with

$$\mathcal{D}_i = \mathcal{D}_{w_i},$$

the recursion

$$f_{i-1}^{[j+1]} = if_i^{[j]} + \sum_{k=1}^{i} \mathcal{D}_k f_{i-k}^{[j]} \tag{D.2.13}$$

follows exactly as does (D.1.14), and the results can be arranged in the triangle

$$\begin{array}{llll} f_0^{[0]} & & & \\ f_1^{[0]} & f_0^{[1]} & & \\ f_2^{[0]} & f_1^{[1]} & f_0^{[2]} & \\ f_3^{[0]} & f_2^{[1]} & f_1^{[2]} & f_0^{[3]}, \end{array} \tag{D.2.14}$$

having the same L-shaped dependence pattern as before. The calculations for the triangle can be done using either the variable x (as would be natural from (D.2.7) or y, but in the former case, x must be replaced by y (and not by $\psi_\varepsilon(y)$) in order to obtain (D.2.12). The first column will be x (or y) with zeros under it.

Algorithm II is related to Algorithm I exactly as in Section D.1. That is, x (or y) is entered in the $f_0^{[0]}$ position and there are zeros down the diagonal. The remaining entries are constructed by the formula in Algorithm D.1.4, with C replaced by f and $\mathbb{D}$ by $\mathcal{D}$. The details are left to the reader.

Algorithms III, IV, and V

It should be clear by now how the remaining algorithms will go; there is only one lemma that requires proof. Given a vector field $a(x, \varepsilon)$, its transform under the coordinate change (D.2.2) will be

$$b(y, \varepsilon) = \psi_\varepsilon'(y)^{-1} a(\psi_\varepsilon(y), \varepsilon). \tag{D.2.15}$$

Then the crucial fact is that

$$\left(\frac{\partial}{\partial \varepsilon}\right)^j b(y, \varepsilon) \bigg|_{\varepsilon=0} = \left(\frac{\partial}{\partial \varepsilon} + \mathsf{L}_{w(x, \varepsilon)}\right)^j a(x, \varepsilon) \bigg|_{x=y, \varepsilon=0}. \tag{D.2.16}$$

Once this is proved, Algorithm III follows as expected: The given coefficients $a_j(x)$ are entered into the first column of a table

$$
\begin{array}{llll}
a_0^{[0]} & & & \\
a_1^{[0]} & a_0^{[1]} & & \\
a_2^{[0]} & a_1^{[1]} & a_0^{[2]} & \\
a_3^{[0]} & a_2^{[1]} & a_1^{[2]} & a_0^{[3]},
\end{array}
$$

and the remaining entries generated according to

$$a_{i-1}^{[j+1]} = i a_i^{[j]} + \sum_{k=1}^{i} \mathsf{L}_k a_{i-k}^{[j]},$$

where

$$\mathsf{L}_k = \mathsf{L}_{w_k(x)}.$$

The coefficients of $b(y, \varepsilon)$ are then

$$b_j(y) = \frac{1}{j!} a_0^{[j]}(y).$$

Algorithm IV will be the exact analogue of Algorithm D.1.7, and Algorithm V will be like Algorithm D.1.8. That is, Algorithm V is the same as III except that the calculations are interrupted periodically by the need to solve a homological equation in a manner consistent with the normal form style that is adopted.

It remains, then, to prove (D.2.16). This could be done by proving a version of Lemma D.1.5 in the present setting, but we prefer to illustrate a different technique. In Chapter 3, Lemma D.1.5 was needed because the fundamental theorem of Lie series did not apply directly to format 2c, because the differential equation (D.1.1) is not autonomous. The same problem confronts us with (D.2.3). One way of dealing with nonautonomous equations is to create an equivalent autonomous equation in one additional dimension. This trick allows the fundamental theorem of Lie series to be used to prove (D.2.16).

Along with the differential equations

$$\dot{x} = a(x, \varepsilon)$$

and

$$\frac{dx}{de} = w(x, \varepsilon),$$

consider the following systems in one higher dimension (with $\tau \in \mathbb{R}$):

$$\begin{bmatrix} \dot{x} \\ \dot{\tau} \end{bmatrix} = \widehat{a}(x, \tau) = \begin{bmatrix} a(x, \tau) \\ 0 \end{bmatrix} \tag{D.2.17}$$

and

$$\frac{d}{d\varepsilon} \begin{bmatrix} x \\ \tau \end{bmatrix} = \widehat{w}(x, \tau) = \begin{bmatrix} w(x, \tau) \\ 1 \end{bmatrix}. \tag{D.2.18}$$

Equation (D.2.18) is autonomous, and its flow may be denoted by Ψ_ε; the solution of (D.2.18) with initial conditions $x(0) = y$, $\tau(0) = \sigma$ will then be

$$(x, \tau) = \Psi_\varepsilon(y, \sigma).$$

Notice that $\tau = \sigma + \varepsilon$, regardless of y; it follows that

$$\Psi_\varepsilon'(y, \sigma) = \begin{bmatrix} \partial x/\partial y & \partial x/\partial \sigma \\ 0 & 1 \end{bmatrix}.$$

Notice also that if we take $\sigma = 0$, then

$$\Psi_\varepsilon(y, 0) = (\psi_\varepsilon(y), \varepsilon),$$

where ψ_ε is as in (D.2.2), and the entry $\partial x/\partial y$ in $\Psi_\varepsilon'(y, 0)$ will be $\psi_\varepsilon'(y)$. Now the transform $\dot{y} = \widehat{b}(y, \sigma, \varepsilon)$ of (D.2.17) under Ψ_ε will be

$$\widehat{b}(y, \sigma, \varepsilon) = \Psi_\varepsilon'(y, \sigma)^{-1}\widehat{a}(\Psi_\varepsilon(y, \sigma)).$$

When $\sigma = 0$ this can be calculated as

$$\widehat{b}(y, 0, \varepsilon) = \Psi_\varepsilon'(y, 0)^{-1}\widehat{a}(\Psi_\varepsilon(y, 0)) \tag{D.2.19}$$

$$= \begin{bmatrix} \psi_\varepsilon'(y) & * \\ 0 & 1 \end{bmatrix}^{-1} \begin{bmatrix} a(\psi_\varepsilon(y), \varepsilon)0 \end{bmatrix}$$

$$= \begin{bmatrix} \psi_\varepsilon'(y)^{-1}a(\psi_\varepsilon(y), \varepsilon) \\ 0 \end{bmatrix}$$

$$= \begin{bmatrix} b(y, \varepsilon) \\ 0 \end{bmatrix},$$

where $b(y, \varepsilon)$ is as in (D.2.15). On the other hand, (D.2.18) is autonomous, and the (formal power series of the) transform of (D.2.17) by Ψ_ε can be calculated by the fundamental theorem of Lie series:

$$\widehat{b}(y, \sigma, \varepsilon) = e^{\varepsilon L_{\widehat{w}(x, \tau)}}\widehat{a}(x, \tau)\big|_{(x, \tau)=(y, \sigma)}. \tag{D.2.20}$$

The operator $\mathsf{L}_{\widehat{w}}$ appearing in this expression can be calculated as follows:

$$
\begin{aligned}
\mathsf{L}_{\widehat{w}}\widehat{a} &= \begin{bmatrix} a \\ 0 \end{bmatrix}' \begin{bmatrix} w \\ 1 \end{bmatrix} - \begin{bmatrix} w \\ 1 \end{bmatrix}' \begin{bmatrix} a \\ 0 \end{bmatrix} \qquad\qquad (D.2.21) \\[2mm]
&= \begin{bmatrix} a_x & a_\tau \\ 0 & 0 \end{bmatrix} \begin{bmatrix} w \\ 1 \end{bmatrix} - \begin{bmatrix} w_x & w_\tau \\ 0 & 0 \end{bmatrix} \begin{bmatrix} a \\ 0 \end{bmatrix} \\[2mm]
&= \begin{bmatrix} a_x w - w_x a + a_\tau \\ 0 \end{bmatrix} \\[2mm]
&= \begin{bmatrix} \mathsf{L}_w a + \partial a / \partial \tau \\ 0 \end{bmatrix}.
\end{aligned}
$$

Comparing the last lines of (D.2.19), (D.2.20), and (D.2.21) shows that the derivatives of $b(y, \varepsilon)$ with respect to ε are obtained by iterating the operator

$$
\frac{\partial}{\partial \varepsilon} + \mathsf{L}_{w(x,\varepsilon)}
$$

on $a(x, \varepsilon)$ and replacing y by x. This fact is equivalent to (D.2.16), which was the equation to be proved. (We have eliminated τ entirely here, replacing it by ε and changing $\partial/\partial\tau$ to $\partial/\partial\varepsilon$, because the result is the same.)

Appendix E
On Some Algorithms in Linear Algebra

There are circumstances in which one may wish to split a matrix A into its semisimple and nilpotent parts $A = A_s + A_n$ without changing coordinates. There are at least three algorithms that achieve this with less work than finding the Jordan form. The one that we present here is the simplest, but not the fastest; see Notes and References for further information. The following theorem provides the theoretical foundation for Algorithm E.1.2, given below.

E.1.1. Theorem. Let A be a matrix having eigenvalues $\lambda_{(1)}, \ldots, \lambda_{(r)}$ with multiplicities $m_1, \ldots, m_r$, and let σ be the polynomial (in one variable x) defined by

$$\sigma(x) = (x - \lambda_{(1)}) \cdots (x - \lambda_{(r)}),$$

which we call the *square-free polynomial* of A. (Its roots are the same as those of the characteristic and minimal polynomials of A, but it is square-free because it contains no repeated linear factors.) Suppose that $f(x)$ is a polynomial j is a nonnegative integer satisfying the following conditions:

1. $\sigma(f(x)) \equiv 0 \bmod \sigma(x)^j$,

2. $\sigma(A)^j = 0$, and

3. $f(x) \equiv x \bmod \sigma(x)$.

Then

$$S = f(A)$$

is the semisimple part of A.

Proof. The first of the listed properties implies that there exists a polynomial $a(x)$ such that $\sigma(f(x)) = a(x)\sigma(x)^j$; then $\sigma(f(A)) = 0$ by the second property. It follows from this that each generalized eigenspace $E_{\lambda_{(i)}}$ of A is actually annihilated by $(f(A) - \lambda_{(i)})I$, so each generalized eigenspace of A is a (true) eigenspace of $f(A)$; that is, $f(A)$ is semisimple. The third listed property implies that there is a polynomial $b(x)$ such that $f(x) = x + b(x)\sigma(x)$, which in turn implies $f(\lambda_{(i)}) = \lambda_{(i)}$ for $i = 1, \ldots, r$. Let $g(x) = x - f(x)$; then $g(\lambda_{(i)}) = 0$. Since the eigenvalues of $g(A)$ are the images under g of the eigenvalues of A (this statement, which is easy to check, is the polynomial case of the spectral mapping theorem), it follows that $g(A)$ has only the eigenvalue zero; hence it is nilpotent. Since $f(x) + g(x) = x$, $f(A) + g(A) = A$. Finally, $f(A)$ and $g(A)$ clearly commute. Therefore $S = f(A)$ and $N = g(A)$ gives the semisimple/nilpotent splitting of A. □

Theorem E.1.1 reduces the problem of finding a semisimple/nilpotent splitting to that of finding f and j satisfying the three listed properties. The following algorithm solves this problem by recursively generating functions f_j, for $j = 1, 2, \ldots$, satisfying the first and third of these conditions. This algorithm can be iterated until a value of j is reached such that $\sigma(A)^j = 0$; this is guaranteed to occur for $j \le n$, since $\sigma(x)^n = (x - \lambda_{(1)})^n \cdots (x - \lambda_{(r)})^n$ is divisible by the characteristic polynomial of A, defined as $\chi(A) = (x - \lambda_{(1)})^{m_1} \cdots (x - \lambda_{(r)})^{m_r}$. (The Cayley–Hamilton theorem, which says that $\chi(A) = 0$, follows easily from the Jordan normal form theorem.)

E.1.2. Algorithm. A sequence of polynomials $f_j(x)$ satisfying $\sigma(f(x)) \equiv 0 \bmod \sigma(x)^j$ and $f(x) \equiv x \bmod \sigma(x)$ can be constructed by taking $f_1(x) = x$ and, recursively,

$$f_{j+1}(x) = f_j(x) + u_j(x)\sigma(x)^j,$$

where $u_j(x)$ is obtained as described in the proof below.

Proof. That $f_1(x) = x$ satisfies the requirements for $j = 1$ is trivial. Suppose that f_j is known, and define f_{j+1} as above, with u_j to be determined. Since $u_j\sigma^j \equiv 0 \bmod \sigma$ regardless of u_j, all that is needed is to choose u_j such that

$$\sigma(f_j + u_j\sigma^j) \equiv 0 \mod \sigma^{j+1}.$$

This is equivalent to

$$\sigma(f_j(x)) + u_j(x)\sigma'(f_j(x))\sigma(x)^j \equiv 0 \mod \sigma(x)^{j+1}, \qquad \text{(E.1.1)}$$

since the (finitely many) omitted terms in the Taylor expansion are already $\equiv 0$. By the induction hypotheses on f_j, there exist polynomials $a(x)$ and $b(x)$ such that $\sigma(f_j(x)) = a(x)\sigma(x)^j$ and $f_j(x) = x + b(x)\sigma(x)$. The first of these allows (E.1.1) to be reduced to

$$a(x) + u_j(x)\sigma'(f_{j-1}(x)) \equiv 0 \mod \sigma(x), \qquad \text{(E.1.2)}$$

while the second provides a method (explained below) to solve (E.1.2) for u_j.

Since $f_j(x) = x + b(x)\sigma(x)$, Taylor's theorem (noting that there are only finitely many terms in the expansion) implies

$$
\begin{aligned}
\sigma'(f_j(x)) &= \sigma'(x + b(x)\sigma(x)) \\
&= \sigma'(x) + \sigma''(x)b(x)\sigma(x) + \cdots \\
&= \sigma'(x) + c(x)\sigma(x)
\end{aligned}
$$

for some polynomial $c(x)$ (which we do not need to compute). It follows that any nonconstant common factor of $\sigma(x)$ and $\sigma'(f_j(x))$ would divide $\sigma'(x)$ as well; but this is impossible, since $\sigma(x)$ has no repeated roots and therefore $\sigma(x)$ and $\sigma'(x)$ have no nonconstant common factors. We conclude that $\sigma(x)$ and $\sigma'(f_j(x))$ are relatively prime, so there exist polynomials $\widetilde{r}(x)$ and $\widetilde{s}(x)$, which may be computed by the Euclidean algorithm, such that

$$
\widetilde{r}(x)\sigma(x) + \widetilde{s}(x)\sigma'(f_j(x)) = 1.
$$

Setting $r = a\widetilde{r}$ and $s = a\widetilde{s}$ gives

$$
r(x)\sigma(x) + s(x)\sigma'(f_j(x)) = a(x),
$$

or

$$
a(x) - s(x)\sigma'(f_j(x)) \equiv 0 \mod \sigma(x).
$$

In other words, $u_j(x) = -s(x)$ is a solution of (E.1.2). $\qquad\square$

Notes and References

Algorithm E.1.2 is found in Levelt [71]. A variant of this approach, said (by some) to be faster, is found in Burgoyne and Cushman [23]. The polynomial $f(x)$ such that $f(A) = S$ is sought in the form

$$
f(x) = x + u_1\sigma(x) + u_2\sigma(x)^2 + \cdots + u_{n-1}(x)\sigma(x)^{n-1},
$$

which is the form given by Algorithm E.1.2 for $j = n$. The coefficients $u_1, \ldots, u_n$ are determined recursively, using the Euclidean algorithm only once, to find u_1; the remaining u_n are determined by rather complicated formulas. A still faster algorithm, using Newton's method, is given in Schmidt [101]. All of these algorithms use Theorem E.1.1 as their starting point.

References

Numbers in parentheses indicate the sections of this book in which a reference is cited (under the subheading Notes and References, at the end of the section).

[1] William W. Adams and Philippe Loustaunau. *An Introduction to Gröbner Bases*. American Mathematical Society, Providence, 1994 (**A.1, A.4, A.5, B.2**).

[2] A. Algaba, E. Freire, and E. Gamero. Hypernormal forms for equilibria of vector fields. codimension one linear degeneracies. *Rocky Mountain Journal of Mathematics*, 29:13–45, 1999 (**4.10**).

[3] A. Algaba, E. Freire, E. Gamero, and C. García. Quasi-homogeneous normal forms. 2001. preprint (**4.10**).

[4] V.I. Arnold. *Mathematical Methods of Classical Mechanics*. Springer, New York, 1978 (**4.9**).

[5] V.I. Arnold. Spectral sequences for reducing functions to normal forms. *Selecta Mathematica Sovietica*, 1:3–17, 1981 (**4.10**).

[6] V.I. Arnold. *Geometrical Methods in the Theory of Ordinary Differential Equations*. Springer, New York, second edition, 1988 (**3.4, 4.3**).

[7] D. Arrowsmith and C. Place. *An Introduction to Dynamical Systems*. Cambridge University Press, Cambridge, 1990 (**1.1**).

[8] Sheldon Axler. *Linear Algebra Done Right*. Springer, New York, 1995 (**2.1**).

[9] Alberto Baider. Unique normal forms for vector fields and Hamiltonians. *Journal of Differential Equations*, 78:33–52, 1989 (**4.10**).

[10] Alberto Baider and Richard Churchill. Unique normal forms for planar vector fields. *Mathematische Zeitschrift*, 199:303–310, 1988 (**4.10**).

[11] Alberto Baider and Jan Sanders. Unique normal forms: the nilpotent Hamiltonian case. *Journal of Differential Equations*, 92:282–304, 1991 (**4.10**).

[12] Alberto Baider and Jan Sanders. Further reduction of the Takens–Bogdanov normal form. *Journal of Differential Equations*, 99:205–244, 1992 (**4.10**).

[13] Thomas Becker. Standard bases and some computations in rings of power series. *Journal of Symbolic Computation*, 10:165–178, 1990 (**A.6**).

[14] G.R. Belitskii. C^∞-normal forms of local vector fields. *Acta Applicandae Mathematicae*, 70:23–41, 2002 (**5.1**).

[15] G.R. Belitskii. Invariant normal forms of formal series. *Functional Analysis and Applications*, 13:46–67, 1979 (**3.4, 4.10**).

[16] George D. Birkhoff. *Dynamical Systems*. American Mathematical Society, Providence, 1927 (**4.9**).

[17] V.N. Bogaevski and A. Povzner. *Algebraic Methods in Nonlinear Perturbation Theory*. Springer, New York, 1991 (**3.3, 3.7**).

[18] I.U. Bronstein and A.Ya. Kopanskii. *Smooth Invariant Manifolds and Normal Forms*. World Scientific, Singapore, 1994 (**5.2**).

[19] J.W. Bruce and P.G. Giblin. *Curves and Singularities*. Cambridge University Press, Cambridge, England, second edition, 1992 (**6.3**).

[20] Alexander D. Bruno. Analytical form of differential equations. *Transactions of the Moscow Mathematical Society*, 25:131–288, 1971. Translation published by the American Mathematical Society, Providence, Rhode Island, 1973. The author's name is spelled Brjuno in this source (**4.5, 5.1**).

[21] Alexander D. Bruno. *Local Methods in Nonlinear Differential Equations*. Springer, New York, 1989 (**5.1, 5.4**).

[22] Alexander D. Bruno. *Power Geometry in Algebraic and Differential Equations*. Elsevier, Amsterdam, 2000 (**5.1, 5.4**).

[23] N. Burgoyne and R. Cushman. The decomposition of a linear mapping. *Linear Algebra and its Applications*, 8:515–519, 1974 (**E.1**).

[24] Jack Carr. *Applications of Centre Manifold Theory*. Springer, New York, 1981 (**5.2**).

[25] Guoting Chen and Jean Della Dora. An algorithm for computing a new normal form for dynamical systems. *Journal of Symbolic Computation*, 29:393–418, 2000 (**4.10**).

[26] Guoting Chen and Jean Della Dora. Further reductions of normal forms for dynamical systems. *Journal of Differential Equations*, 166:79–106, 2000 (**4.10**).

[27] C. Chicone and W. Liu. On the continuation of an invariant torus in a family with rapid oscillation. *SIAM Journal of Mathematical Analysis*, 31:386–415, 1999/2000 (**5.2**).

[28] Carmen Chicone. *Ordinary Differential Equations with Applications*. Springer, New York, 1999 (**5.2**).

[29] Shui-Nee Chow, Chengzhi Li, and Duo Wang. *Normal Forms and Bifurcation of Planar Vector Fields*. Cambridge University Press, Cambridge, 1994 (**1.1, 6.6**).

[30] Neil Chriss and Victor Ginzburg. *Representation Theory and Complex Geometry*. Birkhäuser, Boston, 1997 (**2.7**).

[31] Earl A. Coddington and Norman Levinson. *Theory of Ordinary Differential Equations*. McGraw-Hill, New York, 1955 (**5.2**).

[32] David Cox, John Little, and Donal O'Shea. *Ideals, Varieties, and Algorithms*. Springer, New York, 1997 (**A.1, A.2, A.4, A.5**).

[33] David Cox, John Little, and Donal O'Shea. *Using Algebraic Geometry*. Springer, New York, 1998 (**A.1, A.6, B.2**).

[34] R. Cushman and J.A. Sanders. Nilpotent normal forms and representation theory of sl(2, $\mathbb{R}$). In M. Golubitsky and J Guckenheimer, editors, *Multiparameter Bifurcation Theory*, volume 56 of *Contemporary Mathematics*, pages 31–51. American Mathematical Society, Providence, 1986 (**4.8**).

[35] R. Cushman and J.A. Sanders. Nilpotent normal form in dimension 4. In S.-N. Chow and J.K. Hale, editors, *Dynamics of Infinite Dimensional Systems*, volume F37 of *NATO ASI series*, pages 61–66. Springer, Berlin, 1987 (**4.8**).

[36] R. Cushman and J.A. Sanders. A survey of invariant theory applied to normal forms of vectorfields with nilpotent linear part. In Dennis Stanton, editor, *Invariant Theory and Tableaux*, pages 82–106. Springer, New York, 1990 (**4.7, 4.8**).

[37] Richard Cushman. *Global Aspects of Classical Integrable Systems*. Birkhäuser, Basel, 1997 (**4.9**).

[38] Richard Cushman, André Deprit, and Richard Mosak. Normal form and representation theory. *Journal of Mathematical Physics*, 24:2102–2117, 1983 (**4.9**).

[39] Freddy Dumortier. *Singularities of Vector Fields*. Instituto de Matemática Pura e Aplicada, Rio de Janeiro, 1978 (**5.4, 6.6**).

[40] David Eisenbud. *Commutative Algebra with a View toward Algebraic Geometry*. Springer, New York, 1995 (**A.1**).

[41] C. Elphick et al. A simple global characterization for normal forms of singular vector fields. *Physica D*, 29:95–127, 1987 (**1.1, 3.4, 4.6, 5.1**).

[42] A. Erdélyi. *Asymptotic Expansions*. Dover, New York, 1956 (**A.3**).

[43] Neil Fenichel. Persistence and smoothness of invariant manifolds for flows. *Indiana University Mathematics Journal*, 21:193–226, 1971 (**5.2**).

[44] Neil Fenichel. Asymptotic stability with rate conditions. *Indiana University Mathematics Journal*, 23:1109–1137, 1974 (**5.2**).

[45] J.M. Finn. Lie transforms: a perspective. In A.W. Sáenz, W.W. Zachary, and R. Cawley, editors, *Local and Global Methods of Nonlinear Dynamics*, pages 63–86. Springer, New york, 1986. Proceedings of a workshop held at the Naval Surface Weapons Center, Silver Spring, MD, July 23–26, 1984 (**3.2, 4.4, 4.9**).

[46] William Fulton and Joe Harris. *Representation Theory: A First Course.* Springer, New York, 1991 (**2.5**).

[47] Karin Gaternann. *Computer Algebra Methods for Equivariant Dynamical Systems.* Number 1728 in Lecture Notes in Mathematics. Springer, New York, 2000 (**B.2**).

[48] Herbert Goldstein. *Classical Mechanics.* Addison-Wesley, Reading, Massachusetts, 1950 (**4.9**).

[49] Martin Golubitsky and Victor Guillemin. *Stable Mappings and Their Singularities.* Springer, New York, 1973 (**A.7**).

[50] Martin Golubitsky and David G. Schaeffer. *Singularities and Groups in Bifurcation Theory,* volume 1. Springer, New York, 1985 (**6.1, 6.2, 6.3, 6.5, A.2**).

[51] Willy J.F. Govaerts. *Numerical Methods for Bifurcations of Dynamic Equilibria.* Society for Industrial and Applied Mathematics, Philadelphia, 2000 (**6.2**).

[52] John Guckenheimer and Philip Holmes. *Nonlinear Oscillations, Dynamical Systems, and Bifurcations of Vector Fields.* Springer, New york, 1983. Second printing, revised and corrected (**1.1, 6.5, 6.6, 6.7**).

[53] Jack K. Hale. *Ordinary Differential Equations.* Wiley, New York, 1969 (**5.2, 6.5**).

[54] G. Haller. *Chaos Near Resonance.* Springer, New York, 1999 (**5.2**).

[55] M.M. Hapaev. *Averaging in Stability Theory.* Kluwer, Dordrecht, 1993 (**4.2**).

[56] Melvin Hausner and Jacob Schwartz. *Lie Groups; Lie Alebras.* Gordon and Breach, London, 1968 (**3.2, 3.6**).

[57] Sigurdur Helgason. *Differential Geometry, Lie Groups, and Symmetric Spaces.* Academic Press, New York, 1978 (**2.7**).

[58] Einar Hille. *Analytic Function Theory.* Chelsea, New York, second edition, 1973 (**6.1**).

[59] Morris W. Hirsch, Charles C. Pugh, and Michael Shub. *Invariant Manifolds.* Springer, New York, 1977. Lecture notes in mathematics 583 (**5.2**).

[60] Kenneth Hoffman and Ray Kunze. *Linear Algebra.* Prentice-Hall, Englewood Cliffs, N.J., 1961 (**2.1**).

[61] Roger Howe. Very basic Lie theory. *American Mathematical Monthly,* 90:600–623, 1983. See volume 91 p. 247 for an erratum to the proof of a topological lemma (**3.6**).

[62] Peter B. Kahn and Yair Zarmi. *Nonlinear Dynamics: Exploration through Normal Forms.* Wiley, New York, 1998 (**4.5**).

[63] Tosio Kato. *Perturbation Theory for Linear Operators.* Springer, New York, corrected printing of second edition, 1980 (**3.1**).

[64] Urs Kirchgraber and K.J. Palmer. *Geometry in the Neighborhood of Invariant Manifolds of Maps and Flows and Linearization.* Longman Scientific and Technical, Harlow, England, 1990 (**5.2**).

[65] Urs Kirchgraber and Eduard Stiefel. *Methoden der analytischen Störungsrechnung und ihre Anwendungen.* Teubner, Stuttgart, 1978 (**3.2**).

[66] Hiroshi Kokubu, Hiroe Oka, and Duo Wang. Linear grading function and further reduction of normal forms. *Journal of Differential Equations*, 132:293–318, 1996 (**4.10**).

[67] Yuri A. Kuznetsov. *Elements of Applied Bifurcation Theory*. Springer, New York, second edition, 1998 (**1.1, 6.5, 6.6, 6.7**).

[68] W.F. Langford and K. Zhan. Hopf bifurcation near 0:1 resonance. In Chen, Chow, and Li, editors, *Bifurcation Theory and Its Numerical Analysis*, pages 79–96. Springer, New York, 1999 (**6.7**).

[69] W.F. Langford and K. Zhan. Interactions of Andronov–Hopf and Bogdanov–Takens bifurcations. In B. Bierstone, B. Khesin, A. Khovanskii, and J.E. Marsden, editors, *The Arnoldfest: Proceedings of a Conference in Honour of V.I. Arnold for his Sixtieth Birthday*, pages 365–383. Fields Institute, 1999. Fields Institute Communications, volume 24 (**6.7**).

[70] Solomon Lefschetz. *Differential Equations: Geometric Theory*. Wiley, New York, second edition, 1963 (**5.3**).

[71] A.H.M. Levelt. The semi-simple part of a matrix. In A.H.M. Levelt, editor, *Algoritmen In De Algebra: A Seminar on Algebraic Algorithms*. Department of Mathematics, University of Nijmegen, Nijmegen, The Netherlands, 1993 (**E.1**).

[72] P. Lochak and C. Meunier. *Multiphase Averaging for Classical Systems*. Springer, New York, 1980 (**4.2**).

[73] Stanislaw Lojasiewicz. *Introduction to Complex Analytic Geometry*. Birkhäuser, Basel, 1991 (**A.7**).

[74] Gerard Lunter. *Bifurcations in Hamiltonian Systems: Computing singularities by Gröbner bases*. PhD thesis, Rijksuniversiteit Groningen, 1999. Available on the Web at www.ub.rug.nl/eldoc/dis/science/g.a.lunter/ (**A.1, B.2**).

[75] R.S. MacKay and J.D. Meiss. *Hamiltonian Dynamical Systems*. Adam Hilger, Bristol, 1987. A collection of reprinted articles by many authors, including the main authors listed above, compiled and introduced by these authors (**4.9**).

[76] J.E. Marsden and M. McCracken. *The Hopf Bifurcation and Its Applications*. Springer, New York, 1976 (**6.5**).

[77] Ian Melbourne. The recognition problem for equivariant singularities. *Nonlinearity*, 1:215–240, 1987 (**6.2**).

[78] William Mersman. A new algorithm for the Lie transformation. *Celestial Mechanics*, 3:81–89, 1970 (**3.2**).

[79] Kenneth R. Meyer and Glen R. Hall. *Introduction to Hamiltonian Dynamical Systems and the N-Body Problem*. Springer, New York, 1992 (**4.9**).

[80] Jürgen Moser. *Stable and Random Motions in Dynamical Systems*. Princeton University Press, Princeton, 1973 (**4.9**).

[81] James Murdock. Nearly Hamiltonian systems in nonlinear mechanics: averaging and energy methods. *Indiana University Mathematics Journal*, 25:499–523, 1976 (**4.5, B.1**).

[82] James Murdock. Some mathematical aspects of spin-orbit resonance. *Celestial Mechanics*, 18:237–253, 1978 (**3.7**).

[83] James Murdock. On the length of validity of averaging and the uniform approximation of elbow orbits, with an application to delayed passage through resonance. *Journal of Applied Mathematics and Physics (ZAMP)*, 39:586–596, 1988 (**5.3**).

[84] James Murdock. Qualitative theory of nonlinear resonance by averaging and dynamical systems methods. In U. Kirchgraber and H.O. Walther, editors, *Dynamics Reported, Volume I*, pages 91–172. Wiley, New York, 1988 (**3.7, 4.2, 4.5, B.1**).

[85] James Murdock. Shadowing multiple elbow orbits: an application of dynamcial systems theory to perturbation theory. *Journal of Differential Equations*, 119:224–247, 1995 (**5.3**).

[86] James Murdock. Shadowing in perturbation theory. *Applicable Analysis*, 62:161–179, 1996 (**5.3**).

[87] James Murdock. Asymptotic unfoldings of dynamical systems by normalizing beyond the normal form. *Journal of Differential Equations*, 143:151–190, 1998 (**3.4, 6.3, 6.4**).

[88] James Murdock. *Perturbations: Theory and Methods*. SIAM, Philadelphia, 1999 (**3.1, 4.2**).

[89] James Murdock. On the structure of nilpotent normal form modules. *Journal of Differential Equations*, 180:198–237, 2002 (**3.4, 4.7, 4.8**).

[90] James Murdock and Clark Robinson. A note on the asymptotic expansion of eigenvalues. *SIAM Journdal of Mathematical Analysis*, 11:458–459, 1980 (**3.7**).

[91] James Murdock and Clark Robinson. Qualitative dynamics from asymptotic expansions: local theory. *Journal of Differential Equations*, 36:425–441, 1980 (**3.7**).

[92] Raghavan Narasimhan. *Analysis on Real and Complex Manifolds*. North-Holland, Amsterdam, 1968 (**A.3**).

[93] Ali Nayfeh. *Perturbation Methods*. Wiley, New York, 1973 (**3.2, 4.3**).

[94] Ali Nayfeh. *Method of Normal Forms*. Wiley, New York, 1993 (**4.5**).

[95] Jacob Palis and Welington de Melo. *Geometric Theory of Dynamical Systems*. Springer, New York, 1982 (**5.2**).

[96] Lawrence M. Perko. Higher order averaging and related methods for perturbed periodic and quasi-periodic systems. *SIAM Journal of Applied Mathematics*, 17:698–724, 1968 (**4.2, 5.3**).

[97] Charles C. Pugh and Michael Shub. Linearization of normally hyperbolic diffeomorphisms and flows. *Inventiones Mathematicae*, 10:187–198, 1970 (**5.2**).

[98] Jan Sanders. Normal form theory and spectral sequences. In preparation (**4.10**).

[99] Jan Sanders. Versal normal form computations and representation theory. In E. Tournier, editor, *Computer Algebra and Differential Equations*, pages

185–210. Cambridge University Press, Cambridge, 1994 (**2.6**, **3.2**, **3.4**, **3.5**).

[100] Jan Sanders and Ferdinand Verhulst. *Averaging Methods in Nonlinear Dynamical Systems.* Springer, New York, 1985 (**4.2**).

[101] Dieter Schmidt. Construction of the Jordan decomposition by means of Newton's method. *Linear Algebra and Its Applications*, 314:75–89, 2000 (**E.1**).

[102] Michael Shub. *Global Stability of Dynamical Systems.* Springer, New York, 1987 (**5.2**).

[103] C.L. Siegel and J.K. Moser. *Lectures on Celestial Mechanics.* Springer, New York, 1971 (**4.9**).

[104] Bernd Sturmfels. *Algorithms in Invariant Theory.* Springer, New York, 1993 (**4.7**).

[105] Bernd Sturmfels. *Gröbner Bases and Convex Polytopes.* Number 8 in University Lecture Series. American Mathematical Society, Providence, 1996 (**B.2**).

[106] Shigehiro Ushiki. Normal forms for singularities of vector fields. *Japan Journal of Applied Mathematics*, 1:1–37, 1984 (**4.10**).

[107] M.M. Vainberg and V.A. Trenogin. *Theory of Branching of Solutions of Non-linear Equations.* Noordhoff, Leyden, 1974 (**6.1**).

[108] Jan Cornelis van der Meer. *The Hamiltonian Hopf Bifurcation.* Number 1110 in Lecture Notes in Mathematics. Springer, New York, 1985 (**4.9**).

[109] Duo Wang, Jing Li, Huang Minhai, and Young Jiang. Unique normal form of Bogdanov–Takens singularities. *Journal of Differential Equations*, 163:223–238, 2000 (**4.10**).

[110] Frank Warner. *Foundations of Differentiable Manifolds and Lie Groups.* Springer, New York, 1983 (**3.6**).

[111] Stephen Wiggins. *Introduction to Applied Nonlinear Dynamical Systems and Chaos.* Springer, New York, 1990 (**1.1**, **3.4**, **6.3**, **6.5**, **6.6**, **6.7**).

[112] Stephen Wiggins. *Normally Hyperbolic Invariant Manifolds in Dynamical Systems.* Springer, New York, 1994 (**5.2**).

[113] Pei Yu. Simplest normal forms of Hopf and generalized Hopf bifurcations. *International Journal of Bifurcation and Chaos*, 9:1917–1939, 1999 (**4.10**).

[114] Yuan Yuan and Pei Yu. Computation of simplest normal forms of differential equations associated with a double zero eigenvalue. *International Journal of Bifurcation and Chaos*, 11:1307–1330, 2001 (**4.10**).

[115] Oscar Zariski and Pierre Samuel. *Commutative Algebra.* Van Nostrand, Princeton, N.J., 1958–1960 (**A.1**, **A.6**).

Index

(**Boldface indicates a major
definition**)

adiabatic invariant, 322
adjoint
    with respect to an inner product,
        32, 100
    in a Lie algebra, 67, 136
algebra, 410
algebraically independent, **200**, 214,
    255
antihomomorphism, 67, 189
asymptotic expansion, 162, 164
asymptotic phase, 310
averaging
    method of, 162, 164, 170
    operator, 171

basis
    chain, 35, 38, 41
    chain-weight, 48, **49**
    formal, 159, 198
    Gröbner, 202, 215, 255, 435, 436
    Hamel, 422
    for an ideal, 406
    standard, 192, 442

Stanley, 256
    weight, **54**
biased averaging (of stripes), 118
bias factor, 118, 122, **123**
bifurcation, 16
    degenerate Hopf, 386
    fold, 342, 347
    Hamiltonian Hopf, 271
    Hopf, 16, 20, 386
    imperfect, 366
    isola, 353
    pitchfork, 341, 342, 350
        imperfect, 367
    saddle-node, *see* bifurcation, fold
    Takens–Bogdanov, 26, 389
    transcritical, 342, 353
bifurcation theorem for a Newton
        diagram, **344**
block
    Jordan, **42**
    large, **42**, 44
    small, **42**, 44
blowing up, xii, 338
bootstrappping, 326
Borel–Ritt theorem, 12, 158, **422,
    423**
bottom (of a chain), 34, 49

bracket
  commutator, 47, 77, 174
  Lie, 175, 182, 274, 275
  Poisson, **275**
Buchberger's algorithm, 436

Caesari–Hale method, 387
Campbell–Baker–Hausdorff, 84
canonical form
  Jordan, 35
    lower, **42**
    upper, **42**
  modified Jordan, **48**, 244
  real, **44**
  real semisimple, **30**
canonical transformation, **272**, 273,
    274
center
  compared with center manifold,
    299
  double, **203**
    Hamiltonian, 278
    nonresonant, 217, 311, 314
    nonsemisimple 1 : −1 resonance,
      280
    nonsemisimple 1 : 1 resonance,
      262, 264
    1 : 1 resonance, 217
    1 : 2 resonance, 207
    $p : q$ resonance, 213
  multiple, 203
  single, 2, 203, 382
    Hamiltonian, 277, 308
center manifold reduction, 295, 320
codimension, 401
  of an ideal, 417
  of a matrix, **111**, 377, 380
  of an unfolding, 17, 369, 381
  of a vector space, 419
commuting vector fields, **189**, 196
completely split, **58**
computation problem, ix, 6, 94, 124,
    167, 218, 229, 277
congruent, 407
conjugation, **134**, 140, 175, 176, **179**,
    273
conserved quantity, 302
contact, **317**
continuation, 341, 346

coset, 407
cross-section method, 250

degeneracy condition, 344
degrees of freedom, **271**
Deprit's method, 76, 463
depth, **51**, **250**
description problem, ix, 6, 87, 90,
    167, 197, 227, 230, 277
determinacy
  finite, 13
  of hyperbolicity, 72, 154
  of a jet, 8, 9, 11, 13, 21, 345
determining function, 388
diffeomorphism, **139**, 175, **177**
dilation, **161**, 164
direct product, 158
divisible, 405, 406
division process
  strong, **407**, 418
  weak, 407, 435
  Weierstrass, 440
division theorem
  formal Weierstrass, 440, 444
  Mather, 444
  monomial, **420**

entry point, 35
equivariant, 15, 23, **189**, 196, **197**
  basic, 15, 212, 216
  complex, 205
  module of equivariants, 15
  monomial, **198**
  real, 203
  restricted complex, 210, 216
error estimates, 323–335

fibration, 300
  preserved, 301
  stable, 301, 321
  unstable, 301, 321
flat function, 11, 13, 70, 158, 413,
    414, **421**
flow, 175, 176, **178**, 272
flow property, 178
foliation, 301
  invariant, 304
  $M-$, 336
  preserved, 302

multiple, 313
  by tori, 312, 315
$S-$, 302
formal
  power series, 158
  scalar field, 158
  vector field, 158, 169
format, ix
  1a, 3, 76, 165
  1b, 78, 172
  2a, 79, 184, 187
  2b, 82, 451
  2c, 84, 463
  in general, 74, 157
  iterative, 76
  recursive, 76
fractional power series, 71, 349, 445
Fredholm alternative theorem, 33
full system, 296
functional independence, 200
fundamental theorem of Lie series,
    175, 176, 182
  for matrices, 80, 131
  for scalar fields, 186, 275
  for vector fields, 183, 275

general linear group, 130, 132
generalized eigenspace, 40
generalized eigenvector, 40
generalized kernel, 34
generator
  algebraic, 214
  of a flow, 178
  of an ideal, 406
  of a module, 201, 214, 215
  of a one-parameter group, 130
generic, 345
germ, 170, 187, 409
GL(n), 130
gl(n), 78
grade, 158
  for Hamiltonians, 275
Gröbner basis, see basis, Gröbner
group, 130

Hamiltonian (function), 272
Hamiltonian system, 271, 272
Hartman's theorem, 319
height, 34

relative, 37
height-one sorting algorithm, 58
high term, 343
Hironaka decomposition, 433
homological
  algebra, 4, 293
  equation, 4, 19, 22, 78, 185
    general, 285
    of type (j, k), 284
  operator, 3, 4, 19, 78, 80, 166, 169,
    171, 172, 191
homomorphism
  of an algebra, 411
  of a ring, 407
Hori's method, 76, 86, 451
hypernormal form, 25, 73, 97, 99, 139,
    283

ideal, 360, 406
  associated monomial, 429
  of finite codimension, 417
  finitely generated, 406
  maximal, 416
  monomial, 414, 417, 418
  principal, 406
  product of ideals, 416
  of relations, 214
index of nilpotence, 34
inner product, 32
  for $\mathcal{F}_*^n$, 222
  Frobenius, 100
  standard, 100
  for $\mathcal{V}_*^n$, 224
integral, 302
intermediate term, 343
invariant, 15, 23, 195, 197, 232, 249,
    302
  basic, 15, 200, 211, 213, 215, 249,
    254
  complex, 205
  monomial, 197
  real, 203, 211
  ring of invariants, 15
invariant embedding, 357
invariant foliation, 304
invariant local manifold, 317
invariant subset, 297
invariant subspace, 297, 298
invariant torus, 304

Jacobi identity, **132**, 182
Jacobson–Morosov lemma, 67
jet, 8, 158, 187, 410
jet sufficiency, *see* determinacy, of a
    jet
Jordan chain
  lower, 34
  upper, 35
Jordan form, *see* canonical form

large block, **42**, **44**
leaf, 301
Lie algebra, 47, **131**, 174, 175, 177,
    271, 273, 411
  reductive, 65
Lie bracket, *see* bracket, Lie
Lie differentiation, 176
Lie group, **131**, 132, 174, 175, 177,
    271, 273
Lie operator, 175
  for matrices, 66, **80**, 134, 135
Lie series, 76, *see also* fundamental
    theorem of Lie series, **80**
Lie transform, 76
local, 4, 21
low term, 343

main row, **233**, 236, 243, 374
manifold
  invariant, **317**
  local, **317**
  strong stable, 198
matrix
  Hamiltonian, **132**
  Hermitian, 139
  normal, 33, 105
  orthogonal, 101, 138
  seminormal, 105, 110, 113, 228, 266
  simple striped, **107**
  skew-Hermitian, 139
  skew-symmetric, 132, 138
  striped, **111**
  subordinate, 312
  symmetric, 137
  symplectic, **132**, 272
  unitary, 101, 139
  weakly unitary, 101
matrix series, 75
  Hermitian, 99

symmetric, 99
Melnikov function, 395, 396
metanormal form, 73, 144
M-foliation, 336
modding out, 407
mode interaction, 396
  double Hopf, 399
  0/Hopf, 397
  00/Hopf, 402
module, 14, 159
  of equivariants, 196, 226
  free, 201, 449
  generator of, 201
  of syzygies, 201
  of vector fields, 449
modulo, 407
monomial, 412
  nonstandard, 419
  ordering, 424
    anti-well-ordered, 426
    graded lexicographic, 429
    lexicographic, 426
    locally finite, 426
    well-ordered, **426**
  standard, 215, 256, 258, 419
  vector, 192
monomial division theorem, 9, 348,
    350, 354, **420**
multi-index, 192, 410
multiplication principle, 254
multiplicity (of an eigenvalue)
  algebraic, **41**
  geometric, **41**

Nakayama's lemma, 340, 361
Newton diagram, 72, 343, 344, 352,
    413, 414
  secondary, 354, 355
nilpotent, **34**
nilpotent part, 45, **46**, 227, 477
nondegeneracy condition, 344
normal form
  to all orders, 12, 13
  extended semisimple, **89**, 104, 114,
    116, 194, 241, 267
    not a style, 89
  module, 15
  nilpotent
    N_3, 231, 233

N_4, 246, 247, 260
$N_{2,2}$, 262
$N_{2,3}$, 234
space, 87, 191, 196
style, ix, 7, 22, 87, 157, 191, 277
    inner product, 23, **101**, 221, **226**
    inner product, 277
    semisimple, 7, 89, **194**, 277
    simplified, 23, **115**, 221, 240, 372
    sl(2), 24, **120**, 244, 247, 265,
        267, 277

one-parameter group, **130**, 175, 176,
    178
operator, **28**
    on expressions, 167
    on functions, 167
    homological, *see* homological,
        operator
    Lie, 176, 179, **181**, 182, 273
organizing center, xii, 143, 340, 393
orthogonal group, 132

partial inverse, 88, 95
periodic solution, 304
pounds operator, *see* homological,
    operator
power transformation, xiii
prefix, 259
preparation theorem
    formal, **442**
    Malgrange, 339, 349, 351, 354, 371,
        426, **444**
    Weierstrass, 339
preserved
    fibration, 301
    foliation, 302
pressure, **49**, 50, 251
principal row, 374
projection
    into inner product normal form,
        113, 239
    into simplified normal form, 116,
        241
    into striped matrices, 112
pseudotranspose, **119**, 265
Puiseux series, *see* fractional power
    series
pushdown algorithm, 60, **61**, 115

quadratic convergence, 87
quotient ring, 407

reality
    condition, 91, 92, 205
        for scalar fields, 205, 210
    subspace, 92, 204
recognition problem, 366
reduction (of a flow to a manifold),
    318, 320
relation, 214, 215, 254
remainder, **407**
removable, 357, 362
representation, 48
    irreducible, 50
representation space, 48, 137, 452
representative (of a germ), 409
resonance, 198, 207, 213, 217, 262
    number of resonances, 313
    strong, 217
resonant monomial, 198
restricted tangent space, 360
right multiplication, 457, 464
ring, 14, 159, 360, 405, 409
    of invariants, 196, 226
    local, 416
    prefix, 259
    principal ideal, 406

scalar field
    formal, 158
    smooth, **158**
semisimple, 7, 28, 29, **30**, 31, 88, 193
semisimple part, 45, **46**, 227, 477
setting, 157, 160
    universal, **160**
S-foliation, 302
shadowing, 323
shearing, xii, 392
shearing matrix, 142
shift
    primary, 24, 372, **373**, 374
    secondary, 25, 372, **373**, 374
Shoshitaishvili's theorem, 323
significant degeneration, 393
similarity, **130**, 134, 135, 140, 175,
    176, **179**, 273
sl(2) representation, 27, 48
small block, **42**, **44**

smooth, **158**
special linear group, 132
special orthogonal group, 134
spectral theorem, 33, 105, 138
square-free polynomial, 477
Stanley decomposition, 202, 214, 217,
        254, 256, 258–260, 432, 449
string of weights, *see* weight
stripe, **112**, 235
striped vector field, 235
stripe structure, **111**
stripe subspace, **112**, 121
stripe sum, 108, **112**, 124, 242
style, *see* normal form, style
subcritical, 347
submodule
    pure, 448
    of $\mathbb{Z}^n$, 447
subordinate matrix, **312**
subring, 406
subspace
    center, **299**
    center-stable, **299**
    center-unstable, **299**
    stable, **299**
    unstable, **299**
subtended, **90**
sufficiency, *see* determinacy
suffix, 259
supercritical, 347
symplectic group, 132
syzygy, 14, **201**, 214, 215, 448

table function, 255, 256
Taylor's theorem, 420
top (of a chain), 34, 45, 49
top weight list, *see* weight, top weight
        list

topologically conjugate, **319**
    locally, 319
transcritical, 347
transplanting, xii, 143, 392
transvectant, 250
triad, 27, 47, 48, 244
truncated system, 8, 296

unfolding, x, 16, 17, 24, 340
    asymptotic, 340, 369, 370, 380, 381
    first-order, 372
    higher-order, 379
    parameter, 17, 19, 377
    with reality conditions, 383
    universal, 366
unit (in a ring), 411
unperturbed problem, *see* organizing
        center

vector field, 139, 176
    flat, 158
    formal, **158**, 169
    polynomial, 158
    smooth, 158

Weierstrass division, 440
weight, 48, 247
    basis, *see* basis, weight
    matrix, 121
    polynomial, 248
    space, 48, 121
    string of weights, 49
    table, **54**, 56, 121, 247, 248
    top, **49**
    top weight list, **54**, 56, 247, 248
    vector, 48, 247
winding line, 312